Functional Fluorescent Materials

Functional Fluorescent Materials: Applications in Sensing, Bioimaging, and Optoelectronics explains functional molecular probes (organic/inorganic materials, polymers, nanomaterials), with a focus on those that represent spectroscopic properties with detection of different analytes and specific roles in molecular recognition and their applications. It broadly covers molecular recognition to applications of fluorescence reporters, starting from optoelectronic properties of materials, detection of heavy metals, through biological macromolecules, and further to a living cell, tissue imaging, and theranostics.

Features:

- Covers different aspects of fluorescence spectroscopy ranging from chemical, physical, and biological aspects along with optoelectronic properties, mechanisms, and applications.
- Describes all types of chemical and functionalized fluorescent nanomaterials.
- Provides additional information on different kinds of fluorescence reporters.
- Explains the concept of fluorescence spectroscopy and its role in human health care.
- Discusses changes in static and dynamic properties of fluorescent probes and molecular recognitions.

This book is aimed at graduate students and researchers in materials, chemical engineering, and engineering physics.

Advances in Bionanotechnology

Series Editors: Ravindra Pratap Singh, Department of Biotechnology, Indira Gandhi National Tribal University, Anuppur, Madhya Pradesh, India; Jay Singh, Department of Chemistry, Institute of Science, Banaras Hindu University, Varanasi, Uttar Pradesh, India; and Charles Oluwaseun Adetunji, Department of Microbiology, Edo State University Uzairue, Iyamho, Edo State, Nigeria

Bionanotechnology is a multi-disciplinary field that shows immense applicability in different domains, namely chemistry, physics, material sciences, biomedical sciences, agriculture, environment, robotics, aeronautics, energy, electronics, and so forth. This book series will explore the enormous utility of bionanotechnology for biomedical, agricultural, environmental, food technology, space industry, and many other fields. It aims to highlight all the spheres of bionanotechnological applications and its safety and regulations for using biogenic nanomaterials that are a key focus of researchers globally.

Bionanotechnology Towards Sustainable Management of Environmental Pollution
Edited by Naveen Dwivedi and Shubha Dwivedi

Natural Products and Nano-formulations in Cancer Chemoprevention
Edited by Shiv Kumar Dubey

Bionanotechnology Towards Green Energy: Innovative and Sustainable Approach
Edited by Shubha Dwivedi and Naveen Dwivedi

Biotic Stress Management of Crop Plants using Nanomaterials
Edited by Krishna Kant Mishra and Santosh Kumar

Bionanotechnology for Advanced Applications
Edited by Ajaya Kumar Singh and Bhawana Jain

Nanoarchitectonics for Brain Drug Delivery
Edited by Anurag Kumar Singh, Vivek K. Chaturvedi and Jay Singh

Functional Fluorescent Materials: Applications in Sensing, Bioimaging, and Optoelectronics
Edited by Vivek Mishra, Syed Sibtay Razi and Ajit Kumar

For more information about this series, please visit: www.routledge.com/Advances-in-Bionanotechnology/book-series/CRCBIONAN

Functional Fluorescent Materials
Applications in Sensing, Bioimaging, and Optoelectronics

Edited by
Vivek Mishra, Syed Sibtay Razi, and
Ajit Kumar

CRC Press
Taylor & Francis Group
Boca Raton London New York

CRC Press is an imprint of the
Taylor & Francis Group, an **informa** business

Designed cover image: © Vivek Mishra

First edition published 2024
by CRC Press
2385 NW Executive Center Drive, Suite 320, Boca Raton FL 33431

and by CRC Press
4 Park Square, Milton Park, Abingdon, Oxon, OX14 4RN

CRC Press is an imprint of Taylor & Francis Group, LLC

Library of Congress Cataloging-in-Publication Data
Names: Mishra, Vivek (Chemist) editor. | Razi, Syed Sibtay, editor. |
Kumar, Ajit (Chemist), editor.
Title: Functional fluorescent materials : applications in sensing,
bioimaging, and optoelectronics / edited by Vivek Mishra, Syed Sibtay Razi and Ajit Kumar.
Description: First edition. | Boca Raton : CRC Press, 2024. |
Series: Advances in bionanotechnology | Includes bibliographical references and index. |
Identifiers: LCCN 2023056513 (print) | LCCN 2023056514 (ebook) |
ISBN 9781032402970 (hardback) | ISBN 9781032402987 (paperback) |
ISBN 9781003352372 (ebook)
Subjects: LCSH: Fluorescence spectroscopy. | Fluorescent probes. |
Biotechnology.
Classification: LCC QP519.9.F56 F86 2024 (print) | LCC QP519.9.F56 (ebook) |
DDC 616.07/58–dc23/eng/20240314
LC record available at https://lccn.loc.gov/2023056513
LC ebook record available at https://lccn.loc.gov/2023056514

ISBN: 9781032402970 (hbk)
ISBN: 9781032402987 (pbk)
ISBN: 9781003352372 (ebk)

DOI: 10.1201/9781003352372

Typeset in Times
by codeMantra

Contents

Preface

Fluorescence spectroscopy is the most important tool of any fluorescence applications like sensing, recognition, detection, and imaging systems. It's a unique approach towards molecular or nanoscale materials, devices, probes, scaffolds, and modern techniques that shows photophysical changes (absorption, emission, lifetimes, rate of reactions, electron transfer mechanisms, etc.), molecular interactions and dynamics approach towards an experimental fluorescence emission signal. Some of the most alluring features of fluorescence technologies are their extreme sensitivity (limit of detection) for the detection of biological analytes (proteins and amino acids), alkali, alkaline earth metals, transition elements, lanthanides, and actinides, as well as their ultrahigh resolution in time for biological images, potential for real sample analysis applications, and remote accessibility with potential applications in medical theranostics. The main objective of this volume with the title, "Functional Fluorescent Materials: Applications in Sensing, Bioimaging, and Optoelectronics," is to provide a concept of fluorescence reporters, with a focus on those that represent organic fluorophores or have similar spectroscopic properties with detection of different analytes and specific roles in molecular recognition and their applications. Organic dyes, as the first in history, continue to hold the lead in both basic research and technological advancements. This volume is devoted to the fundamentals and design of organic fluorophores, as well as the optimization of their useful molecular recognition and applications in biotechnology. The content of this book mainly focuses on the overview of fluorescence spectroscopy and its role in molecular recognitions of alkali, alkaline earth metals, transition metals detections, lanthanides/actinides role, and triplet annihilation and also on the detection of biological analytes (amino acids, proteins, and plasma), structures based on organic fluorophores, and systems that behave similarly to these reporters, highlighting the applications of fluorescence in sensing and imaging, from the molecular to nanoscale materials. Any fluorescent molecular and nanosystems have heterogeneity in structural and dynamic features. Such specific designing of systems and their binding properties are demanding to investigate the molecular recognition with specific changes of molecular signals (generally emission properties). The basic essential topic of fluorescence is addressed in the first chapter, which mainly focuses on the Jablonski photophysical concept and theory behind the detection of analytes and molecular designing, and critically examines several techniques based on physical theory, empirical correlations, and specific role of molecular recognition. A concerted effort is undertaken to reduce uncertainty in experimental data interpretation. Fluorescence techniques must evolve in order to improve existing materials and design new ones. The special chapter focuses on the changes in static and dynamic properties of fluorescent probes and molecular recognitions as well as how they change optoelectronic behaviours and related applications. Readers interested in the unique features of fluorescence spectroscopy and its role in molecular recognition in different approaches as a potential application will find useful information gathered through molecular systems. The most widespread and effective applications of fluorescence techniques are in the study of biological images in various aspects – for example, for sensing other molecules as targets and collecting distributions of sensed molecules within living cells and tissues. The design of biosynthetically manufactured molecular probes, protein, and peptide tags can help in operating with organic fluorophores within living cells, which is a difficult operation. Fluorescence probes have a wide range of applications at the tissue and whole-body levels. They were made possible by the development of strong fluorescent dyes that absorbed and emitted light in the near-IR band. Successful applications of organic dyes as fluorescence probes include in vivo imaging of vascular targets and detection of analytes and cancer cells. The book's last chapters demonstrate current development in these practical areas. This volume illustrates progress in a fast-growing interdisciplinary field of research and development. As a result, it is aimed at a multidisciplinary readership ranging from photophysicists, chemical engineers, organic chemists to material scientists, postgraduate students, and researchers working with living objects at the molecular recognitions and cellular levels. This understanding will serve as a further impetus for continued growth in the industrial and biotechnological fields.

Dr. Vivek Mishra
*Amity Institute of Click Chemistry Research and Studies, Amity University,
Noida, India*
Dr. Syed Sibtay Razi
Gaya College Gaya, Magadh University, Bodhgaya, Bihar, India
Dr. Ajit Kumar
Department of Chemistry, Banaras Hindu University, Varanasi, India

About the Editors

Vivek Mishra is currently working as an Assistant Professor (grade III) in the Amity Institute of Click Chemistry Research and Studies (AICCRS) under the umbrella of Amity University Noida Campus. Before this, he joined the Department of Chemistry as a DST SERB-National Post-Doctoral Fellow in July 2017 from the Science and Engineering Research Board, New Delhi, Government of India. Dr. Mishra has completed his Ph.D. in Chemistry at the Institute of Science-Banaras Hindu University (BHU) in 2012. After that, he was offered three postdoctoral fellowships: one from the Indian Institute of Technology (IIT) Indore, MP, India and the other two are from South Korea. He was a postdoctoral research associate under the Brain Korea-21 programme at the University of Ulsan, Ulsan, South Korea and a Specialist (PDF) at the Korea Institute of Industrial Technology, Cheonan-si, South Korea for 2 years. Dr. Mishra research is mainly focused on synthesis and characterization of polymers with selected functionality, composition, and molecular architecture. His research focuses on drug delivery, hydrogel and nanogel synthesis, stimuli-responsive polymers, magnetic nanoparticles for water remediation, catalysis, dye removal and degradation, and biowaste/plastic waste utilization for the fruitful by-products. He has published 52 papers in international journals of high repute with an H index of 20 and an i-10 index of 25, and also presented a dozens of papers in national and international symposia/conferences.

Syed Sibtay Razi has been an Assistant Professor since 2017 in the Department of Chemistry, Gaya College Gaya, Magadh University Bodhgaya, Bihar, India. His main research interests are the design and synthesis of organic triplet sensitizers, focused on the intersystem crossing (ISC) and triplet excited state, photophysics and photochemistry of organic compounds, and application as triplet–triplet annihilation. The research in the group is based on synthetic chemistry, steady-state and time-resolved transient optical spectroscopy, theoretical computations, fluorescent materials, and biological applications. Electron transfer, energy transfer, photodynamic therapy, and photogenetics of organic compounds are the main research in the group. Dr. Syed Sibtay Razi earned his Bachelor's, Master's, and Ph.D. degrees in Chemistry at Banaras Hindu University, Varanasi. He has been awarded Ph.D. degree from Banaras Hindu University in the year 2015. The topic of his research work was "Synthesis, Photophysical Behaviour and Application of Some Organic Scaffolds." During his research work, hc has synthesized various molecular organic scaffolds based on anthracene, naphthalene, benzothiazole, benzimidazole, azo dye, coumarin, and pyrene systems. He studied meticulously the photophysical behaviour of these molecules in different medium and utilized some of these molecules as chemosensors in the recognition of hazardous and biologically important ions as well as biomolecules. He has published 28 papers in international journals of high repute with an H index of 16 and an i-10 index of 19, and also presented a dozen papers in national and international symposia/conferences. He already got the best oral presentation award in the International Symposium on Functional Materials (ISFM-2018): Energy and Biomedical Applications, at Chandigarh, India, organized by the University of Illinois and IIT Kanpur. He recently got Prof. Anisuddin Malik Memorial Award for the best oral presentation in the 2nd International Conference on Chemistry, Industry and Environment-2019 at Aligarh Muslim University, India. He has established Photochemistry Research Lab at Gaya College, Gaya and got a UGC-startup grant for starting his own independent research. Currently, Syed's group has close collaboration with international groups in the UK, the USA, Japan, Korea, and China.

Ajit Kumar obtained his B.Sc. and M.Sc. degrees from Ranchi University, Ranchi and received a Ph.D. degree in Chemistry from Banaras Hindu University, Varanasi. He has worked at the National Institute of Foundry & Forge Technology, Ranchi as a DST Young Scientist Fellow. He has published 50 papers in international journals of high repute with an H index of 18 and an i-10 index of 28, and also presented a dozen papers in national and international symposia/conferences. His research interests include exploration of some new moieties for the development of efficient synthetic optical receptors for ionic analytes.

Contributors

Mohammad Asif Ali
Graduate School of Advanced Science
  and Technology, Energy, and
  Environment Area
Japan Advanced Institute of Science and
  Technology
Nomi, Japan
and
School of Chemical and Material
  Engineering
Jiangnan University
Wuxi, China

Thomas D. Anthopoulos
KAUST Solar Centre
King Abdullah University of Science
  and Technology (KAUST)
Thuwal, Saudi Arabia

Kavya Bhakuni
Department of Chemistry, St. Stephen's
  College
University of Delhi
Delhi, India

Xiaoman Bi
School of Materials Science and
  Engineering, Henan Institute of
  Advanced Technology
Zhengzhou University
Zhengzhou, P.R. China

Amit Chauhan
Department of Chemistry
L. N. Mithila University
Darbhanga, India

Hendrik Faber
KAUST Solar Centre
King Abdullah University of Science
  and Technology (KAUST)
Thuwal, Saudi Arabia

Anas D. Fazal
Tarsadia Institute of Chemical Science
Uka Tarsadia University
Surat, India

Meenakshi Gupta
Amity Institute of Pharmacy
Amity University
Noida, India

Neeraj Mohan Gupta
Department of Chemistry
Govt. P. G. College
Guna, India

Sruthi Guru
Women Scientist -C, KIRAN-IPR
  Scheme,
Department of Science and Technology
Govt. of India
India

Preeti Kasana
Department of Chemistry
J. C. Bose University of Science and
  Technology, YMCA
Faridabad, India

Deepak Kumar
Department of Physics
Panjab University
Chandigarh, India
and
Department of Physics
Guru Jambheshwar University of
  Science and Technology
Hisar, India

Vinod Kumar
Department of Chemistry
J. C. Bose University of Science and
  Technology, YMCA
Faridabad, India

Pragati Kushwaha
Department of Chemistry
University of Lucknow
Lucknow, India

Xuying Liu
School of Materials Science and
    Engineering, Henan Institute of
    Advanced Technology
Zhengzhou University
Zhengzhou, P. R. China

Rita Mahapatra
Krishna School of Science
KPGU, Varnama
Vadodara, India

Poonam Mishra
Krishna School of Science
Drs. Kiran and Pallavi Patel Global
    University
Vadodara, India

Vivek Mishra
Amity Institute of Click Chemistry
    Research and Studies
Amity University
Noida, India

Aniruddha Nag
Graduate School of Science and
    Technology
Nara Institute of Science and
    Technology
Ikoma, Japan

Rishi Pal
Department of Physics
Guru Jambheshwar University of
    Science and Technology
Hisar, India

Sumit Kumar Panja
Tarsadia Institute of Chemical Science
Uka Tarsadia University
Surat, India

Darshankumar Prajapati
Krishna School of Science
Drs. Kiran and Pallavi Patel Global
    University
Vadodara, India

Abhishek Rai
Department of Chemistry
L. N. Mithila University
Darbhanga, India

Ashish Kumar Rajayan
University Institute of Engineering and
    Technology
Panjab University
Chandigarh, India

Ishpal Rawal
Department of Physics, Kirori Mal
    College
University of Delhi
Delhi, India

Syed Sibtay Razi
Department of Chemistry, Gaya College
Magadh University
Bodh Gaya, India
and
University of Murcia
Murcia, Spain

Pawan Kumar Sada
Department of Chemistry
L. N. Mithila University
Darbhanga, India

Maryam Sarwat
Amity Institute of Pharmacy
Amity University
Noida, India

Abhinav Sharma
KAUST Solar Centre
King Abdullah University of Science
    and Technology (KAUST)
Thuwal, Saudi Arabia

Ashima Sharma
Department of Life Sciences
J.C. Bose University of Science and
    Technology, YMCA
Faridabad, Haryana

Ruchi Sharma
Department of Life Sciences
J.C. Bose University of Science and
    Technology, YMCA,
Faridabad, India

Sonkeshwar Sharma
Chemistry and Bioprospecting Division
Rain Forest Research Institute
Jorhat, India
and
Indian Council of Forestry Research
    and Education
Dehradun, India

Parul Shrivastava
Krishna School of Science
Drs. Kiran and Pallavi Patel Global
    University
Vadodara, India

Indu Tucker Sidhwani
Department of Chemistry, Gargi
    College
University of Delhi
Delhi, India

Alok Kumar Singh
Department of Chemistry
Deen Dayal Upadhyaya Gorakhpur
    University
Gorakhpur, India

Maninder Singh
Graduate School of Advanced Science
    and Technology, Energy, and
    Environment Area
Japan Advanced Institute of Science and
    Technology
Nomi, Japan

Rishi Singh
Amity Institute of Biotechnology
Amity University Uttar Pradesh
Noida, India

Shilpi Thakur
Krishna School of Science
Drs. Kiran and Pallavi Patel Global
    University
Vadodara, India

Qingyong Tian
School of Materials Science and
    Engineering, Henan Institute of
    Advanced Technology
Zhengzhou University
Zhengzhou, P. R. China

Deepshikha Verma
Florida Department of Agriculture and
    Consumer Services
Kissimmee, Florida

Youfusheng Wu
Laboratory of Printable Functional
    Nanomaterials and Printed
    Electronics, School of Printing and
    Packaging
Wuhan University
Wuhan, P. R. China

Weijing Yao
School of Materials Science and
    Engineering, Henan Institute of
    Advanced Technology
Zhengzhou University
Zhengzhou, P. R. China

1 Fluorescence Spectroscopy
An Overview

Sumit Kumar Panja and Anas D. Fazal

1.1 INTRODUCTION

Luminescence is the emission of light from any substance and occurs from electronically excited states. Luminescence is formally divided into two categories – fluorescence and phosphorescence – depending on the nature of the excited state (Figure 1.1).[1] In excited singlet states, the electron in the excited orbital is paired (by the opposite spin) to the second electron in the ground-state orbital. Consequently, return to the ground state is spin-allowed and occurs rapidly by emission of a photon. At room temperature, most molecules occupy the lowest vibrational level of the ground electronic state, and on absorption of light, they are elevated to produce excited states. The simplified diagram below shows absorption by molecules to produce either the first, S_1, or second S_2, excited state.

Excitation can result in the molecule reaching any of the vibrational sublevels associated with each electronic state. Since the energy is absorbed as discrete quanta, this should result in a series of distinct absorption bands. However, the simple diagram

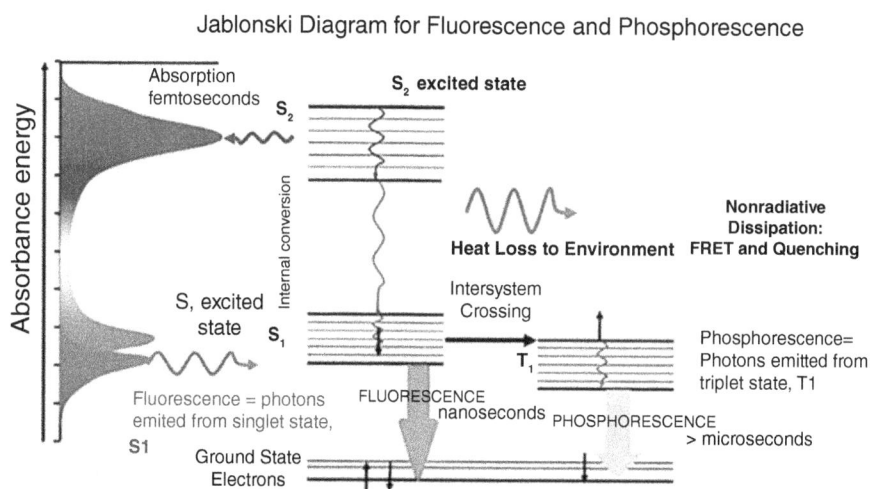

Jablonski Diagram for Fluorescence and Phosphorescence

FIGURE 1.1   Electronic transition and different processes.[2]

DOI: 10.1201/9781003352372-1

1

above neglects the rotational levels associated with each vibrational level which normally increase the number of possible absorption bands to such an extent that it becomes impossible to resolve individual transitions. Having absorbed energy and reached one of the higher vibrational levels of an excited state, the molecule rapidly loses its excess of vibrational energy by collision and falls to the lowest vibrational level of the excited state. In addition, almost all molecules occupying an electronic state higher than the second undergo internal conversion and pass from the lowest vibrational level of the upper state to a higher vibrational level of a lower excited state which has the same energy. From there, the molecules again lose energy until the lowest vibrational level of the first excited state is reached. From this level, the molecule can return to any of the vibrational levels of the ground state, emitting its energy in the form of fluorescence.[3]

The emission rates of fluorescence are typically $10^8 s^{-1}$, so that a typical fluorescence lifetime is near 10 ns (10×10^{-9} s). The lifetime (τ) of a fluorophore is the average time between its excitation and return to the ground state. It is valuable to consider a 1-ns lifetime within the context of the speed of many fluorophores displaying subnanosecond lifetimes. Because of the short timescale of fluorescence, measurement of the time-resolved emission requires sophisticated optics and electronics. In spite of the added complexity, time-resolved fluorescence is widely used because of the increased information available from the data, as compared with stationary or steady-state measurements. Additionally, advances in technology have made time-resolved measurements easier, even when using microscopes. Fluorescence typically occurs from aromatic molecules.[4]

1.1.1 The Perrin–Jablonski Diagram

For describing the processes subsequent to light absorption by a molecule, it was found convenient to use an energy diagram in which the electronic states of the molecule are represented together with arrows indicating the possible transitions between them. A modern and more detailed diagram is shown in Figure 1.2. Since the 1970s, this diagram is most often called the Jablonski diagram (from the name of the Polish physicist Aleksander Jablonski).[5] However, it should be called the Perrin–Jablonski diagram in order to give appropriate credit to the contributions of the French physicists Jean and Francis Perrin.[6,7]

It is important to note that a Jablonski diagram shows what sorts of transitions can possibly happen in a particular molecule. Each of these possibilities is dependent on the time scales of each transition. The faster the transition, the more likely it is to happen as determined by selection rules. Therefore, understanding the time scales each process can happen is imperative to understanding if the process may happen.[8] The time scale for basic radiative and non-radiative processes is tabulated in Table 1.1.

1.1.2 On the Distinction between Fluorescence and Phosphorescence: Decay Time Measurements

Fluorescence and phosphorescence are types of photoluminescence. Photoluminescence refers to radiative emissions where the absorbance of a photon is followed by the emission of a lower energy photon. The main empirical difference between fluorescence

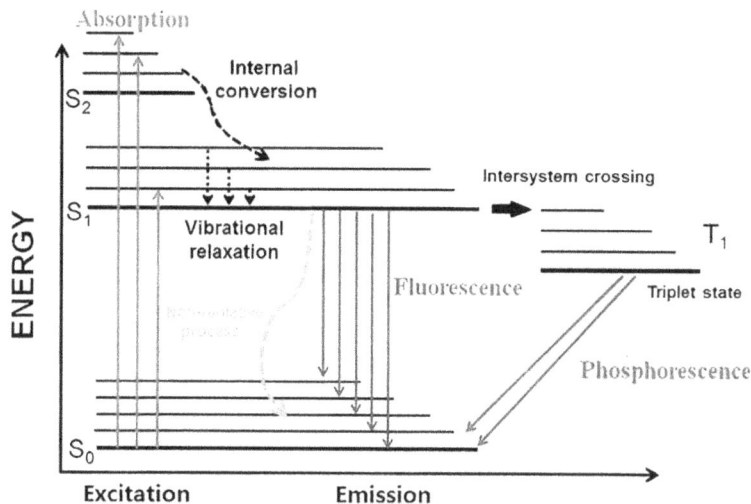

FIGURE 1.2 Perrin–Jablonski diagram.

TABLE 1.1
Electronic Transition and Process with Time Scale

Transition	Time scale	Radiative process
Absorption	10^{-15}s	Yes
Internal conversion	10^{-14} to 10^{-11}s	No
Vibrational relaxation	10^{-14} to 10^{-11}s	No
Fluorescence	10^{-9} to 10^{-7}s	Yes
Intersystem crossing	10^{-8}s-10^{-3}s	No
Phosphorescence	10^{-4} to 10^{-1}s	Yes

and phosphorescence is the time in between absorbance and the emission of photons.[9] Fluorescence is where a material absorbs a photon and almost immediately emits a lower energy photon. Phosphorescence occurs over a longer period as it requires a forbidden transition.

On an atomic level, the distinction between phosphorescence and fluorescence depends on the electron spin state during energy level transitions. In fluorescence transitions, this electron spin state is maintained throughout. Phosphorescence, however, requires the spin state to change when a photon is absorbed. This is reversed when the subsequent emission occurs. One key difference is that fluorescence only occurs when light is incident on the material, while phosphorescent materials can continue to glow sometime after the light source has been removed. Just like for fluorescence emissions, in phosphorescence, an electron is initially excited through absorbance of a photon.

The electron can then non-radiatively relax into the triplet state through intersystem crossing (ISC). This transition is not spin-allowed but can occur due to spin–orbit coupling. Once the electron is in the triplet state, it will eventually radiatively relax into the ground state (S_0) or one of the vibrational states above. Although singlet–triplet transitions are forbidden by spin selection rules, they can still occur over a longer time period. This type of emission is termed phosphorescence and can have lifetimes of seconds or even as long as several hours.

However, such a distinction only based on the duration of emission is not sound. In fact, we now know that there is long-lived fluorescence whose decay times are comparable to those of short-lived phosphorescence (ca. 0.1–1 μs).

1.1.3 EMISSION SPECTRA ARE TYPICALLY INDEPENDENT OF THE EXCITATION WAVELENGTH

Another general property of fluorescence is that the same fluorescence emission spectrum is generally observed irrespective of the excitation wavelength. This is known as Kasha's rule,[10] although Vavilov reported in 1926 that quantum yields were generally independent of excitation wavelength.[11] Upon excitation into higher electronic and vibrational levels, the excess energy is quickly dissipated, leaving the fluorophore in the lowest vibrational level of S_1. This relaxation occurs in about 10^{-12} s (1 ps = 10^{-12} s) and is presumably a result of a strong overlap among numerous states of nearly equal energy. Because of this rapid relaxation, emission spectra are usually independent of the excitation wavelength.

Exceptions exist, such as fluorophores that exist in two ionization states, each of which displays distinct absorption and emission spectra. Also, some molecules are known to emit from the S_2 level, but such emission is rare and generally not observed in biological molecules.

According to the Franck–Condon principle, all electronic transitions are vertical, that is, they occur without a change in the position of the nuclei. As a result, if a particular transition probability (Franck–Condon factor) between the 0th and 1st vibrational levels is largest in absorption, the reciprocal transition is also most probable in emission.

1.1.4 STOKES SHIFT IN FLUORESCENCE SPECTROSCOPY

In fluorescence spectroscopy, the Stokes shift is the difference between the spectral position of the maximum of the first absorption band and the maximum of the fluorescence emission and can be expressed in either wavelength or wavenumber units as shown in Figure 1.3.

Examination of the Jablonski diagram (Figure 1.3) reveals that the energy of the emission is typically less than that of absorption. Fluorescence typically occurs at lower energies or longer wavelengths. This phenomenon was first observed by Sir. G. G. Stokes in 1852 at the University of Cambridge.[12]

Stokes Shift (Wavelength): $\Delta\lambda = \left(\lambda_{max}^{Fl} - \lambda_{max}^{Abs}\right)$

Stokes Shift (Wavenumber: cm^{-1}): $\Delta\overline{v} = \overline{v}_{max}^{Abs} - \overline{v}_{max}^{Fl} = \dfrac{1}{\lambda_{max}^{Abs}} - \dfrac{1}{\lambda_{max}^{Fl}}$

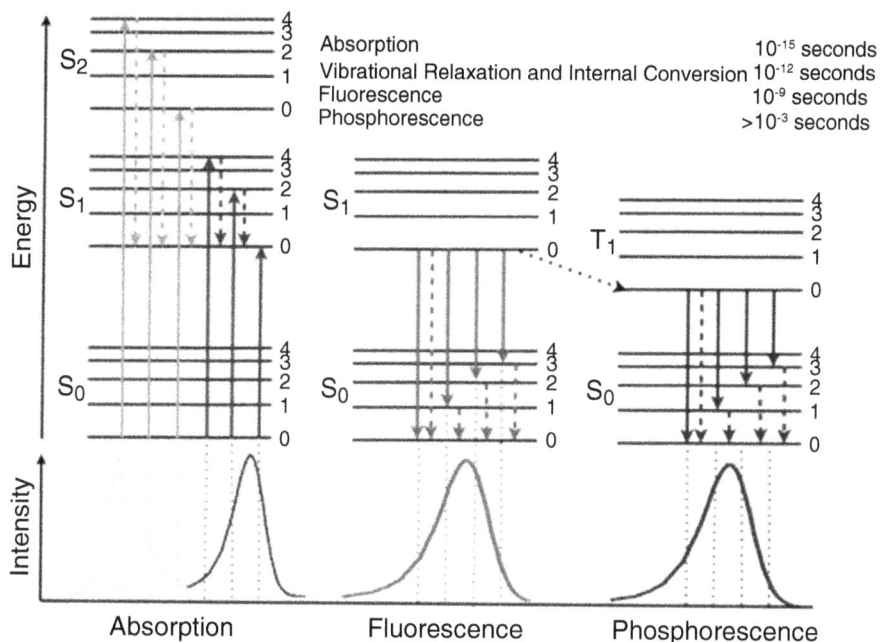

Absorption	10^{-15} seconds
Vibrational Relaxation and Internal Conversion	10^{-12} seconds
Fluorescence	10^{-9} seconds
Phosphorescence	$>10^{-3}$ seconds

FIGURE 1.3 Pictorial presentation of electronic processes.

It should be noted that the wavenumber Stokes shift expression written above is only an approximation since it assumes that the wavenumber maxima are at the same position as the wavelength maxima which is not strictly true. When fluorescence spectra are converted from a wavelength scale to a wavenumber scale, the position of the maxima slightly shifts since the spectral bandpass of the measurement is constant in wavelength but not in wavenumber. For very accurate wavenumber Stokes shift calculations, one should therefore first convert the spectra to a wavenumber scale and locate the maxima from the wavenumber spectra.

The extent of the Stokes shift depends on the particular fluorophore and its solvation environment, with more polar solvents typically giving larger Stokes shifts. The emission and absorption spectra of two fluorophores with a small and large Stokes shift are shown in Figure 1.3.

Energy losses between excitation and emission are observed universally for fluorescent molecules in solution. The origin of the Stokes shift is commonly represented in a Perrin–Jablonski diagram as an initial excitation to a higher vibrational level of the S_1 followed by a rapid non-radiative decay to the vibrational ground state of the S_1 (Figure 1.4), which means that the fluorescence will have a lower energy than the absorbed photon and therefore a longer wavelength.

One common cause of the Stokes shift is the rapid decay to the lowest vibrational level of S_1. Furthermore, fluorophores generally decay to higher vibrational levels of S_0 (Figure 1.5), resulting in further loss of excitation energy by thermalization of

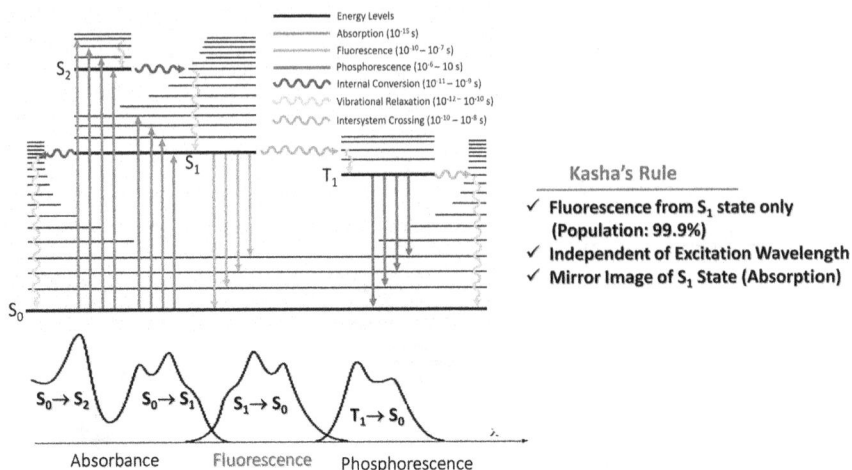

FIGURE 1.4 Pictorial presentation of Kasha's rule related to the emission process.

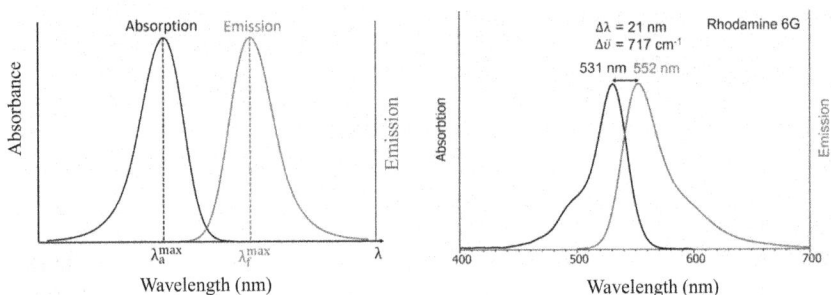

FIGURE 1.5 Representation of Stokes shift for fluorescence.

the excess vibrational energy. In addition to these effects, fluorophores can display further Stokes shifts due to solvent effects, excited-state reactions, complex formation, and/or energy transfer.[13]

1.1.5 Fluorescence Quantum Yield from Fluorophore (Φ)

Fluorescence quantum yield (Φ_F) is a characteristic property of a fluorescent species and is denoted as the ratio of the number of photons emitted through fluorescence to the number of photons absorbed by the fluorophore. Φ_F ultimately relates to the efficiency of the pathways leading to emission of fluorescence (Figure 1.1), providing the probability of the excited state being deactivated by fluorescence rather than by other competing relaxation processes. The magnitude of Φ_F is directly related to the intensity of the observed fluorescence.[14]

$$\Phi_F = \frac{\text{Number of photons emitted through fluorescence}}{\text{Number of photons absorbed}}$$

Fluorescence quantum yield can be measured using two methods: the absolute method and the relative method. Relative Φ_F measurements are achieved using the comparative method. Here, the Φ_F of a sample is calculated by comparing its fluorescence intensity to another sample of known Φ_F (the reference). Unlike absolute quantum yield measurements, which require an integrating sphere, the relative method uses conventional fluorescence spectrometers with a standard single cell holder.[15] The relative method does, however, require knowledge of the absorbance of both the reference and the sample.

The Φ_{sample} of a fluorophore is determined relative to a reference compound of known Φ_{ref}. If the same excitation wavelength, gain, and slit bandwidths are applied for the two samples, then the Φ_{sample} is calculated as

$$\Phi_{sample} = \Phi_{ref} \frac{\eta_{sample}^2}{\eta_{ref}^2} \frac{I_{ref}}{I_{sample}} \frac{A_{ref}}{A_{sample}}$$

where Φ_{ref} is the quantum yield of the reference compound, h is the refractive index of the solvent, I is the integrated fluorescence intensity, and A is the absorbance at the excitation wavelength.

The fluorescence lifetime and quantum yield are perhaps the most important characteristics of a fluorophore. Quantum yield is the number of emitted photons relative to the number of absorbed photons. Substances with the largest quantum yields, approaching unity, such as rhodamines, display the brightest emissions.

1.1.6 FLUORESCENCE QUENCHING: STATIC AND DYNAMICS

The intensity of fluorescence can be decreased by a wide variety of processes. Such decreases in intensity are called quenching. Quenching can occur by different mechanisms as shown in Figure 1.6.

Collisional quenching occurs when the excited-state fluorophore is deactivated upon contact with some other molecules in solution, which is called the quencher.

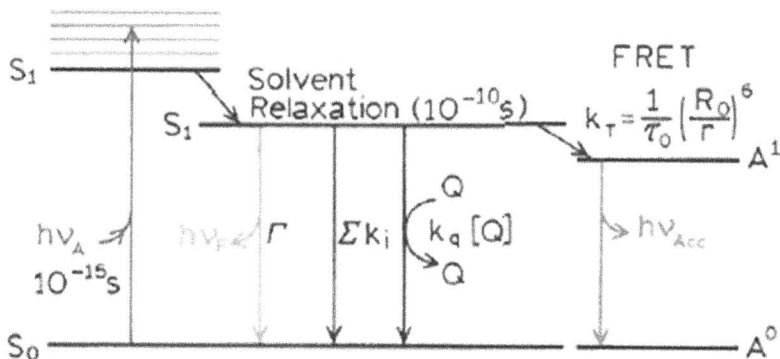

FIGURE 1.6 Jablonski diagram with collisional quenching and fluorescence resonance energy transfer (FRET). The term $\sum k_i$ is used to represent non-radiative paths to the ground state aside from quenching and FRET.

In this case, the fluorophore is returned to the ground state during a diffusive encounter with the quencher.

The molecules are not chemically altered in the process. For collisional quenching, the decrease in intensity is described by the well-known Stern–Volmer equation:

$$\frac{F_0}{F} = 1 + k_q \tau_0 |Q| = 1 + K_D |Q|$$

In this expression, K is the Stern–Volmer quenching constant, k_q is the bimolecular quenching constant, τ_0 is the unquenched lifetime, and $[Q]$ is the quencher concentration.

The Stern–Volmer quenching constant K_D indicates the sensitivity of the fluorophore to a quencher. A fluorophore buried in a macromolecule is usually inaccessible to water-soluble quenchers, so that the value of K_D is low. Larger values of K_D are found if the fluorophore is free in solution or on the surface of a biomolecule.

A wide variety of molecules can act as collisional quenchers. Examples include oxygen, halogens, amines, and electron-deficient molecules like acrylamide. The mechanism of quenching varies with the fluorophore–quencher pair. For instance, quenching of indole by acrylamide is probably due to electron transfer from indole to acrylamide, which does not occur in the ground state.

Quenching by halogen and heavy atoms occurs due to spin–orbit coupling and ISC to the triplet state. Aside from collisional quenching, fluorescence quenching can occur by a variety of other processes. Fluorophores can form non-fluorescent complexes with quenchers. This process is referred to as static quenching since it occurs in the ground state and does not rely on diffusion or molecular collisions.

Quenching can also occur by a variety of trivial, i.e., non-molecular mechanisms, such as attenuation of the incident light by the fluorophore itself or other absorbing species.

It is important to recognize that observation of a linear Stern–Volmer plot does not prove that collisional quenching of fluorescence has occurred. Static quenching also results in linear Stern–Volmer plots.

Static and dynamic quenching can be distinguished by their differing dependence on temperature and viscosity, or preferably by lifetime measurements. Higher temperatures result in faster diffusion and hence larger amounts of collisional quenching (Figure 1.7). Higher temperatures will typically result in the dissociation of weakly bound complexes and hence smaller amounts of static quenching.

1.1.7 FLUORESCENCE POLARIZATION OR ANISOTROPY

The polarization state of fluorescence is an important aspect that was investigated almost from the beginning of fluorescence studies.

Fluorescence polarization measurements provide information on molecular orientation and mobility and processes that modulate them, including receptor–ligand interactions, protein–DNA interactions, proteolysis, membrane fluidity, and muscle contraction. Because polarization is a general property of fluorescent molecules (with certain exceptions such as lanthanide chelates), polarization-based readouts are somewhat less dye dependent and less susceptible to environmental interferences such as pH changes than assays based on fluorescence intensity measurements.

Collisional   Quenching   Static   Quenching

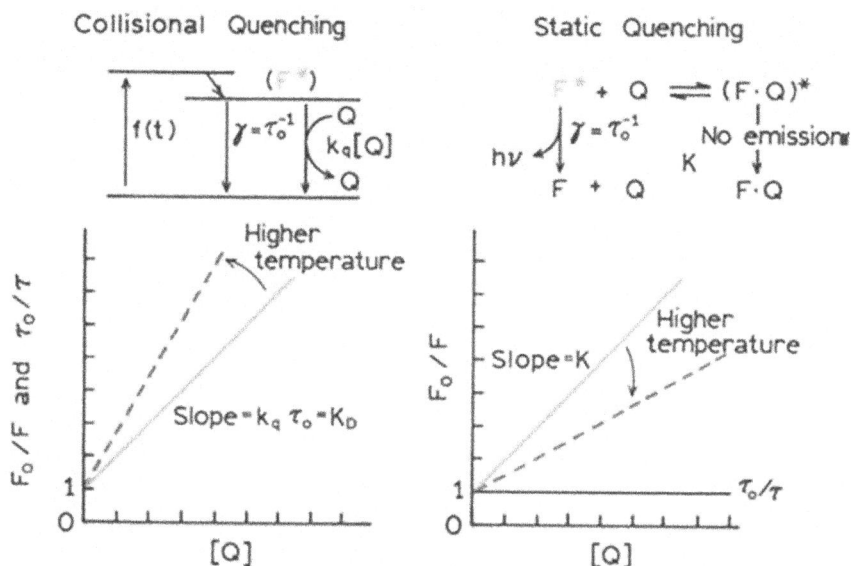

FIGURE 1.7   Comparison of dynamic and static quenching.

Experimentally, the degree of polarization is determined from measurements of fluorescence intensities parallel and perpendicular with respect to the plane of linearly polarized excitation light (Figure 1.8), and is expressed in terms of fluorescence polarization (P) or anisotropy (r).

$$P = \frac{\left(F_\| - F_\perp\right)}{\left(F_\| + F_\perp\right)} \quad r = \frac{\left(F_\| - F_\perp\right)}{\left(F_\| + 2F_\perp\right)}$$

$F_\|$ = Fluorescence intensity parallel to the excitation plane
$F_\perp$ = Fluorescence intensity perpendicular to the excitation plane

Here, it should be noted that both P and r are ratio quantities with no nominal dependence on dye concentration. Because of the ratio formulation, fluorescence intensity variations are due to the presence of colored sample additives tend to cancel and produce relatively minor interferences. P has physically possible values ranging from -0.33 to 0.5.

1.1.8   FÖRSTER RESONANCE ENERGY TRANSFER OR RESONANCE ENERGY TRANSFER (RET)

RET, the transport of electronic energy from one atom or molecule to another, has significant importance to a number of diverse areas of science. Since the pioneering experiments on RET by Cario and Franck in 1922, the theoretical understanding of the process has been continually refined.[16] A quantum theory of RET via dipole–dipole interaction in the gas phase was developed by H. Kallman and F. London in 1928. The concept of critical radius (distance at which transfer and spontaneous

Photoselected
Fluorophores

Randomized
Fluorophores

Rotational
Diffusion

$I_{\perp}$

$I_{\parallel}$

Polarized
Excitation

Polarized
Emission

Unpolarized
Emission

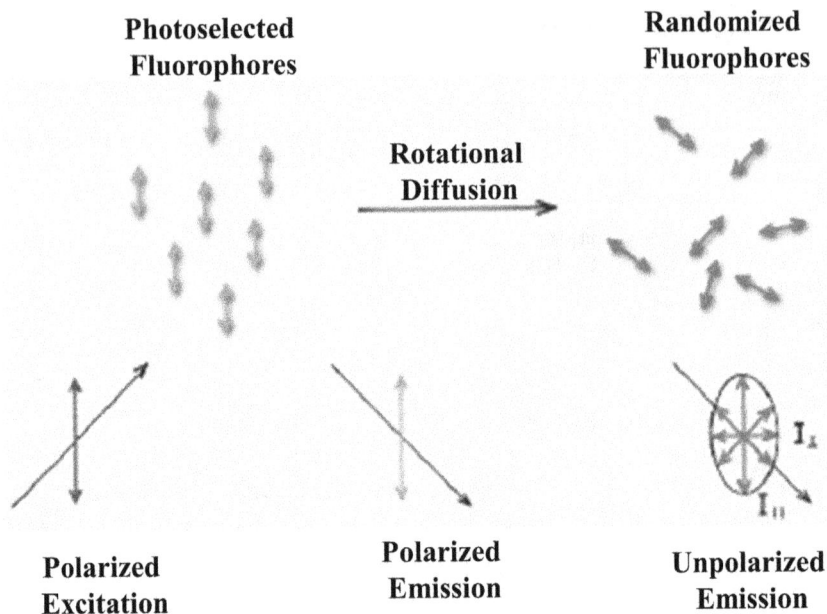

FIGURE 1.8   Effects of polarized excitation and rotational diffusion on the polarization or anisotropy of the emission.

decay of the excited donor are equally probable) was introduced for the first time. It is through this quantum framework that the short-range, R^{-6} distance dependence of the Förster theory was unified with the long-range, radiative transfer governed by the inverse-square law.[17] Crucial to the theoretical knowledge of RET is the electric dipole–electric dipole coupling tensor. The higher order interactions that involve magnetic dipoles and electric quadrupoles are also important for the efficient energy transfer process. However, RET is not limited to the Förster-type transfer, that is, via dipole–dipole interaction. Since the end of the 1970s, (F)RET has been used as a "spectroscopic ruler": in fact, it allows one to measure the distance between a donor chromophore and an acceptor chromophore in the 1–10 nm range.[18]

It has provided the latest research, which includes transfer between nanomaterials, enhancement due to surface plasmons, possibilities outside the usual ultraviolet or visible range, and RET within a cavity. It also permits monitoring of the approach or separation of two species. (F)RET has found numerous applications in photophysics, photochemistry, and photobiology.[19]

1.1.9   FLUORESCENCE LIFETIMES AND VARIOUS DE-EXCITATION PROCESSES OF EXCITED MOLECULES

The lifetime is also important, as it determines the time available for the fluorophore. Once a molecule is excited by absorption of a photon, it can return to the ground state with emission of fluorescence, or phosphorescence after ISC, but it can also undergo intramolecular charge transfer and conformational change. Interactions in the excited state with other molecules may also compete with de-excitation: elec-tron transfer, proton transfer, energy transfer, and excimer or exciplex formation (Figure 1.9). These de-excitation pathways may compete with fluorescence emission if they take place on a time scale comparable with the average time (lifetime) dur-ing which the molecules stay in the excited state. This average time represents the experimental time window for observation of dynamic processes. The characteristics of fluorescence (spectrum, quantum yield, lifetime),

which are affected by any excited-state process involving interactions of the excited molecule with its close environment, can then provide information on such a microenvironment. It should be noted that some excited-state processes (conforma-tional change, electron transfer, proton transfer, energy transfer, excimer or exciplex formation) may lead to a fluorescent species whose emission can superimpose that of the initially excited molecule (Figure 1.10).

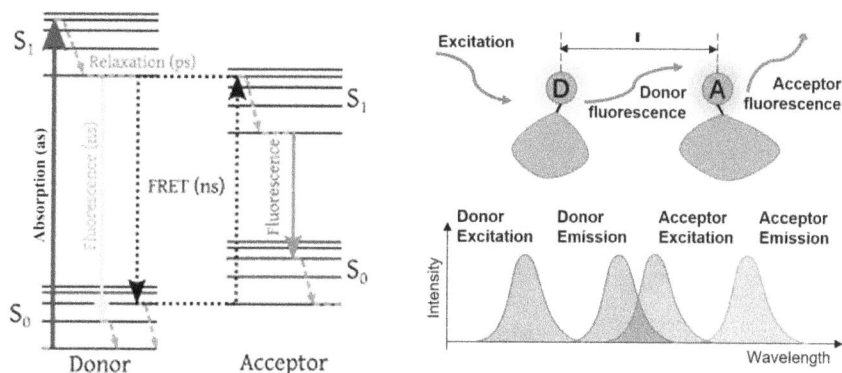

FIGURE 1.9   Förster resonance energy transfer (FRET).

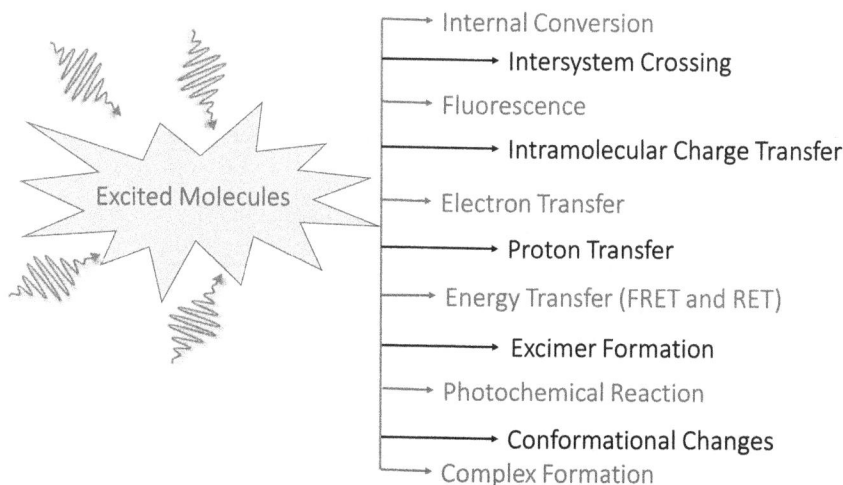

FIGURE 1.10   Possible de-excitation pathways of excited molecules.

Such an emission should be distinguished from the "primary" fluorescence arising from the excited molecule. The success of fluorescence as an investigative tool in studying the structure and dynamics of matter or living systems arises from the high sensitivity of fluorometric techniques, the specificity of fluorescence characteristics due to the microenvironment of the emitting molecule, and the ability of the latter to provide spatial and temporal information (Figures 1.11 and 1.12). It shows the physical and chemical parameters that characterize the microenvironment and can thus affect the fluorescence characteristics of a molecule.[20]

1.1.10 FLUORESCENT PROBES, INDICATORS, LABELS, AND TRACERS

Due to the environmental impact of fluorescence emissions, fluorescent substances, often called fluorophores, are used to obtain information about physical, structural, or chemical locations (Figure 1.13). The term fluorescent test is commonly utilized, but within the specific case of a chemical parameter like pH or the concentration of a species, the term fluorescent marker may be favored (e.g., fluorescent pH indicator).[21]

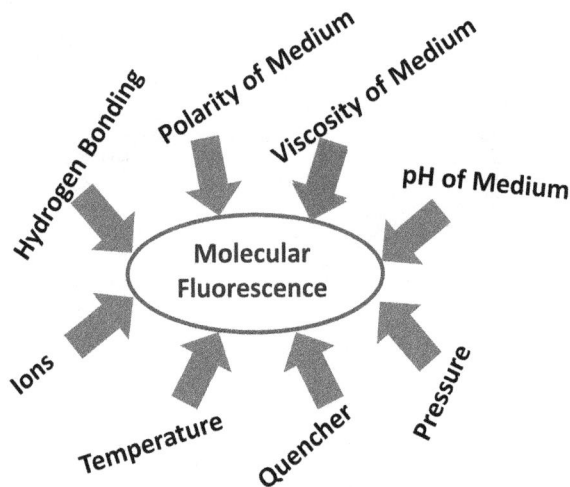

FIGURE 1.11 Various parameters influencing the emission of fluorescence.

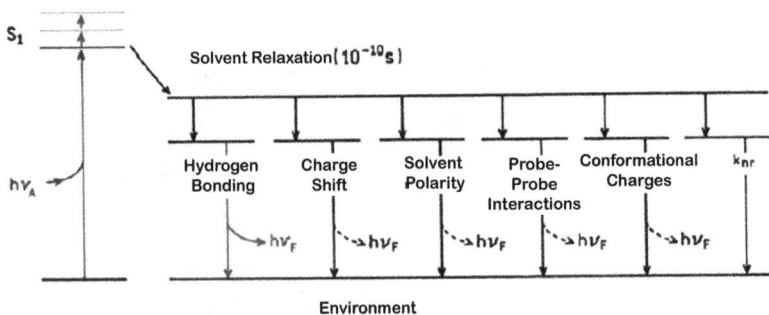

FIGURE 1.12 Effect of environment on the energy of excited state.

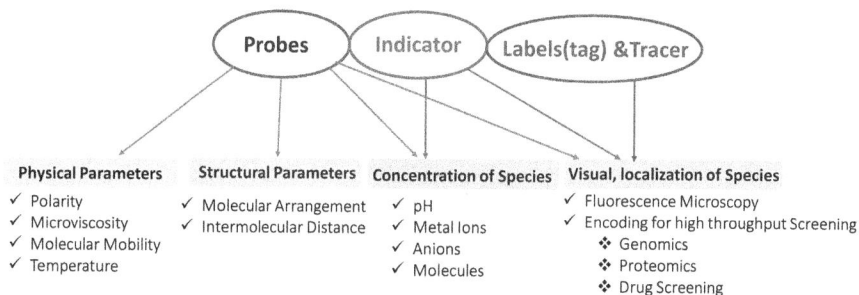

FIGURE 1.13 Information provided by fluorescent probes, indicators, labels, and tracers.

On the other hand, when a fluorescent atom is utilized to imagine or localize a species, for illustration, by utilizing microscopy, the terms fluorescent names (or labels) and tracers are regularly utilized. This suggests that a fluorescent particle is covalently bound to the species of intrigued: surfactants, polymer chains, phospholipids, proteins, oligonucleotides, and so on. For occasion, protein labeling can be effortlessly accomplished by implies of labeling reagents having appropriate utilitarian bunches; for occasion, covalent authoritative is conceivable on amino bunches.[22]

The hydrophilic, hydrophobic, or amphiphilic character of a fluorophore is basic. In microscopy, a specific interaction of the fluorophore with particular parts of the framework beneath consider (cell, tissue, etc.), permitting their visualization, is regularly called recoloring, a term customarily utilized for colors.

Intrinsic fluorophores are ideal as tests and tracers but there are as it were a number of cases found in science (e.g., tryptophan, NADH, flavins). Owing to the trouble of union of atoms or macromolecules with covalently bound fluorophores, numerous investigations are carried out with non-covalently associating fluorophores. The locales of solubilization of such outward tests are represented by their chemical nature and the coming about particular intuition that can be built up inside the locale of the framework to be tested. The hydrophilic, hydrophobic, or amphiphilic character of a fluorophore is fundamental. Feedback frequently pointed at the utilization of outward fluorescent tests is the conceivable nearby irritation actuated by the test itself on the microenvironment to be tested. There are undoubtedly a few cases of frameworks irritated by fluorescent tests. Be that as it may, it ought to be emphasized that numerous illustrations of what comes about steady with those obtained by other strategies can be found within the literature (transition temperature in lipid bilayer, adaptability of polymer chains, etc.). To play down the irritation, consideration must be paid to the estimate and shape of the test with regard to the examined locale. In the event that is conceivable, more than one test ought to be utilized for a consistency check.

1.1.11 ULTIMATE TEMPORAL AND SPATIAL RESOLUTION: FEMTOSECONDS, FEMTOLITERS, FEMTOMOLES, AND SINGLE-MOLECULE DETECTION

The ability of fluorescence to provide transient data is of major significance. Awesome advances have been made since the primary assurance of an excited-state lifetime by Gaviola in 1926 employing a stage fluorometer. A time resolution of a few tens of picoseconds can effortlessly be accomplished in both beat and phase fluorometries by using high repetition rate picosecond lasers and microchannel plate photomultipliers.[23] Such a time resolution is limited by the response of the photo-multiplier but not by the width of the laser pulse, which can be as short as 50–100 fs (1 femtosecond $= 10^{-15}$ s) (e.g., with a titanium:sapphire laser). The time resolution can be reduced to a few picoseconds with a streak camera.[24] To get an even better time resolution (100–200 fs), a more recent technique based on fluorescence up-conversion has been developed.

Regarding spatial resolution, fluorescence microscopy in confocal configuration or with two-photon excitation allows the diffraction limit to be approached, which is approximately half the wavelength of the excitation light (0.2–0.3 μm for visible radiation) with the advantage of three-dimensional resolution.

The excitation volume can be as small as 0.1 fL (femtoliter). Compared to con-ventional fluorometers, this represents a reduction by a factor of 10 of the excitation volume. At high dilution (~10^{-9} M or less), fluorophores entering and leaving such a small volume cause changes in fluorescence intensity.[25] Analysis of these fluctuations (which is the object of fluorescence correlation spectroscopy) in terms of autocor-relation function can provide information on translational diffusion, flow rates, and molecular aggregation.[26] Fluctuations can also be caused by chemical reactions or rotational diffusion. The typical lower limit concentration is ~1 fM (femtomol L^{-1}). The progress of these techniques allows studying molecular interactions at the unsur-passed sensitivity of single-molecule detection.[27]

The diffraction limit can be overcome by using a subwavelength light source and by placing the sample very close to this source (i.e., in the near field). The rele-vant domain is near-field optics (as opposed to far-field conventional optics), which has been applied in particular to fluorescence microscopy.[28] This technique, called scanning near-field optical microscopy (SNOM), is an outstanding tool in physical, chemical, and life sciences for probing the structure of matter or living systems. The resolution is higher than in confocal microscopy, with the additional capability of force mapping of the surface topography and the advantage of reduced photo-bleach-ing. Single-molecule detection is of course possible by this technique. Recent far-field techniques like stimulated emission depletion and stochastic optical reconstruction microscopy also allow breaking the diffraction limit.

Detection and spectroscopy of individual fluorescent molecules thus provide new tools not only in basic research but also in biotechnology and pharmaceutical indus-tries (e.g., drug screening).[29]

1.2 CHALLENGES AND FUTURE DIRECTIONS

Fluorescence spectroscopy is a useful and highly efficient technique in present research fields like biological sciences, material science, chemistry, and biophysics. Design and synthesis of fluorophores with higher fluorescence efficiency or quan-tum yield are still challenging. Further, self-quenching is a big problem for organic chromophore. Solid-state fluorescence property is one of the biggest challenges in material chemistry. Application of organic chromophores in biological systems for cell imaging is an important research field. Aggregation-induced fluorescence enhancement and its application is an attractive research area in material chemistry. Design, synthesis, and applications of suitable chromophores for different research areas like biological sciences, material science, biochemistry, and biophysics are still attractive in the present and future.

ACKNOWLEDGMENT

SKP acknowledges Tarsadia Institute of Chemical Science, Uka Tarsadia University, Maliba Campus, Gopal Vidyanagar, Bardoli, Mahuva Road, Surat-394350, Gujrat, India.

CONFLICT OF INTEREST

The author confirms that he has no conflict of interest to declare for this publication.

REFERENCES

1. Braslavsky, S. et al. Glossary of terms used in photochemistry. *Pure Appl. Chem.* (2007), 79, 293–465.
2. Lakowicz, J. R. (1983, 1999, 2006) *Principles of Fluorescence Spectroscopy*, Springer, Berlin.
3. Strickler, S. J. and Berg, R. A. Relationship between absorption intensity and fluorescence lifetime of molecules. *J. Chem. Phys.* (1962), 37(4), 814–822.
4. Berlman, I. B. (1971) *Handbook of Fluorescence Spectra of Aromatic Molecules*, 2nd ed., Academic Press, New York.
5. Jabłoński, A. (1935). Über den mechanismus der photolumineszenz von farbstoffphosphoren. *Zeitschrift für Physik*, 94(1), 38–46.
6. Nickel, B. From the Perrin diagram to the Jablonski diagram. *EPA Newslett.* (1996), 58 (Part 1), 9–38.
7. Guliani, E., Taneja, A., Ranjan, K. R., and Mishra, V. Luminous insights: Exploring organic fluorescent "Turn-On" chemosensors for metal-ion (Cu^{+2}, Al^{+3}, Zn^{+2}, Fe^{+3}) detection. *J. Fluoresc.* (2023), 1–37. doi: 10.1007/s10895-023-03419-5.
8. Birks, J. B. (1973) *Organic Molecular Photophysics*, John Wiley & Sons, New York.
9. Valeur, B. and Berberan-Santos, M. N. A brief history of fluorescence and phosphorescence before the emergence of quantum theory. *J. Chem. Educ.* (2011), 88, 731–738.
10. Kasha, M. Characterization of electronic transitions in complex molecules. *Disc. Faraday Soc.* (1950), 9, 14–19.
11. Berlman, I. B. (1971) *Handbook of Fluorescence Spectra of Aromatic Molecules*, 2nd ed. Academic Press, New York.
12. Stokes, G. G. On the change of refrangibility of light. *Phil. Trans. R Soc. (London)* (1852), 142, 463–562.
13. Klessinger, M. and Michl, J. (1995) *Excited States and Photochemistry of Organic Molecules*, John Wiley & Sons, Inc., New York.
14. Atkins, P., De Paula, J., and Keeler, J. (2018) *Atkins' Physical Chemistry*, Oxford University Press, Oxford.
15. Würth, C., Grabolle, M., Pauli, J., Spieles, M., and Resch-Genger, U. Relative and absolute determination of fluorescence quantum yields of transparent samples. *Nature Protocol* (2013), 8, 1535–1550.
16. Kramer, H. E., and Fischer, P. (2011). The scientific work of Theodor Förster: a brief sketch of his life and personality. *ChemPhysChem*, 12(3), 555–558.

17. Summaries of Förster's biography and scientific achievements can be found in Porter, G. Naturwiss, (1976) 63, 207.
18. Förster, Th. Intermolecular energy migration and fluorescence (Transl RS Knox). *Ann. Phys. (Leipzig)* (1948), 2, 55–75.
19. Stryer L. Fluorescence energy transfer as a spectroscopic ruler. *Ann. Rev. Biochem.* (1978), 47, 819–846.
20. Turro, N. J., Ramamurthy, V., and Scaiano, J. C. (2009) *Principles of Molecular Photochemistry*, University Science Books, Sausalito, CA.
21. Lakowicz, J. R. (1995) Fluorescence spectroscopy of biomolecules. In *Encyclopedia of Molecular Biology and Molecular Medicine*, pp. 294–306. Ed R. A. Meyers. VCH Publishers, New York.
22. Hof, M., Hutterer, R., and Fidler, V. (eds.) (2005) *Fluorescence Spectroscopy in Biology*, Springer Series on Fluorescence, vol. 3, Springer, Berlin.
23. O'Connor, D. V. and Phillips D. (1984) *Time-Correlated Single-Photon Counting*, Academic Press, New York.
24. Birch, D. J. S., and Imhof, R. E. (1991) Time-domain fluorescence spectroscopy using time-correlated single-photon counting. In *Topics in Fluorescence Spectroscopy*, vol. 1: Techniques, pp. 1–95. Boston, MA: Springer US.
25. Rigler, R., Orrit, M., and Basché, T. (2001) *Single Molecule Spectroscopy*, Springer, Berlin.
26. Yadav, N., Gaikwad, R. P., Mishra, V., and Gawande, M. B. (2022) Synthesis and photo-catalytic applications of functionalized carbon quantum dots. *Bulletin of the Chemical Society of Japan*, 95(11), 1638–1679.
27. Zander, Ch., Enderlein, J., and Keller, R. A., eds. (2002) *Single Molecule Detection in Solution, Methods and Applications*, Wiley-VCH, Darmstadt, Germany.
28. Sauer, M., Hofkens, J., and Enderlein, J. (2011) *Handbook of Fluorescence Spectroscopy and Imaging*, Wiley-VCH Verlag GmbH, Weinheim.
29. Birch, D. J. S., and Imhof, R. E. (1991) Time-domain fluorescence spectroscopy using time-correlated single-photon counting. In *Topics in Fluorescence Spectroscopy*, vol. 1: Techniques, pp. 1–95. Ed J. R. Lakowicz. Plenum Press, New York.

2 Fluorescence Probes to Detect Transition Metal ions

Pawan Kumar Sada, Amit Chauhan,
Alok Kumar Singh, and Abhishek Rai

2.1 INTRODUCTION

Recent human achievements, such as industrial growth and its consequent impact on water streams, have a negative impact on the appearance of transition metal ions in soil and water [1]. Because transition metal ions are not biodegradable, they can accumulate in edible animals and plants and eventually find their way into the food chain for humans [2]. Effects of these ions that are harmful to both people and the environment [3] prompted researchers to create low-cost, fluorescent chemosensors that take advantage of their binding sites to provide analyte detection with a change in fluorescence response [4,5] and an effective separation capacity [6]. A significant amount of attention has been paid over the past 10 years to the outline and production of selective and sensitive fluorescence probes for trace amounts of transition metallic ions, like Fe^{3+}, Zn^{2+}, Hg^{2+}, Pb^{2+}, Cu^{2+}, Cd^{2+}, and Cr^{3+} [7–10]. There are numerous instrument-intensive techniques available for the identification of analytes, including voltametric techniques [11], ion-selective electrodes [12], atomic absorption/emission spectrometry, and inductively coupled plasma mass spectrometry [13,14]. However, the majority of these conventional analytical methods frequently have significant drawbacks such as costly apparatus setup and labour-intensive sample pretreatment procedures. Because of their ease of use, low detection threshold, capacity for specialized identification, and outstanding spectroscopic properties, like emission profiles and long wavelength excitation coefficients and high fluorescence quantum yields, fluorescent detection holds more promise than other analytical techniques [15,16]. Numerous fluorescence sensors have been created and characterized during the past 20 years using a variety of fluorophores, including quinolone, rhodamine, pyrene, coumarin, fluorescein, cyanine, and azo dye, among others [17–30]. In environmental and biological chemistry, it is highly interesting to construct fluorescent molecular sensors to find these metal cations, because it has been shown that diseases in humans or animals are closely connected with either a lack or excess of trace levels of transition metal ions [31–36].

Usually, a fluorescence sensor is split into a minimum of two working parts (i.e., the signalling unit and the receptor unit). The interaction between the receptor and the target metal ion determines the spectroscopic signal generated by the signal

DOI: 10.1201/9781003352372-2

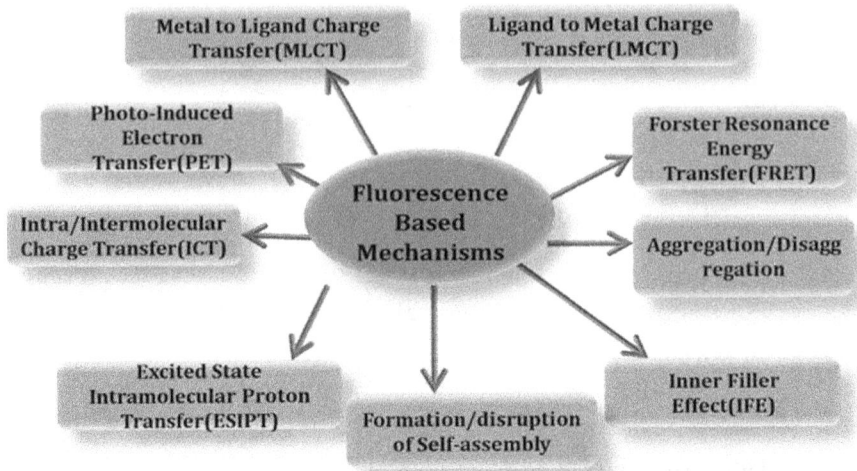

FIGURE 2.1 Mechanisms of fluorescence sensing and the interaction between a receptor and a substrate are shown schematically.

moiety, whereas the reaction location where metal ions can be attached is the receptor moiety. For the exposure of a wide variety of analytes employing a variety of emission processes, there are many different types of fluorescence probes available, concentrating on specific analyte classes [37–42] architecture of a receptor or transducer [43–45], and other elements of fluorescence sensing [46–48]. Chemosensing is a simple, quick, and sensitive method because fluorophores have variable optical characteristics in their free and analyte-bound states due to various photophysical mechanisms. The recognized photophysical mechanisms include excimer production, charge transfer, photo-induced electron transfer (PET), Förster resonance energy transfer (FRET), and the more recently identified aggregation-induced emission (Figure 2.1) [49–51].

The highest occupied molecular orbital (HOMO) of the receptor transfers a single electron from a photo-excited fluorophore to its energetically closed HOMO during the PET procedure. The excited electron from the lowest unoccupied molecular orbital (LUMO) can no longer fall down to the HOMO after the transfer since the HOMO of the fluorophore is fully occupied; instead, it is transferred back to the HOMO of the receptor, quenching the fluorescence ("turn off"). The energy of the receptor HOMO is significantly lower than that of the fluorophore HOMO upon cation binding because the receptor transfers its electron to the cation, raising the reduction potential of the receptor. The PET process is now constrained, and the excited electron of the fluorophore returns to its ground state, causing fluorescence emission ("turn on") to be visible [52–54].

Sometimes, the fluorescence emission of chemosensors is caused by internal (intramolecular) charge transfer (ICT) [55,56], where the conjugated system that joins the fluorophore (transducer) and receptor is the chemosensor, which combines groups that accept and donate electrons. On activation, an electron redistribution to the electron-acceptor from the electron-donor section occurs, producing a dipole moment in the molecule. Depending on the analyte's nature and the electrical configuration of the fluorophore and receptor, the dipole moment may be increased or decreased when an analyte binds to the molecule.

Other charge transfer techniques include MLCT (metal to ligand charge transfer) and TICT (twisted internal charge transfer) [57]. Due to the ability of the solvent arrangement to stabilize the dipoles, CT routes are extremely solvent polarity dependent.

Through a long-range dipole–dipole interaction, an excited fluorophore (donor) transfers non-radiative energy to a suitable energy acceptor in the FRET process [58,59]. It is a fluorescence quenching mechanism in which the excited electron cannot return to its ground state but the acceptor is excited instead, causing any chosen fluorophore to emit at a wavelength that is significantly red shifted from the excitation wavelength. To create a significant spectrum overlapping between the acceptor's absorption and the donor's emission profiles, the separation between the donor and acceptor units must be in the range of 10–100 Å for an efficient FRET process. Thus, the FRET process makes ratiometric analyte detection possible while controlling the "on" or "off" state of chemosensor.

Even though a lot of experimental research has been carried out to identify using multiple sensors to study transition metal ions and to gain a mechanistic knowledge of ON–OFF fluorescence emission is still in its infancy, there are still many unanswered concerns that cannot be fully addressed by experimental studies. High-level electronic structure computations must be used to better investigate and improve the intricate ON–OFF mechanism of fluorescence emission. It should be noted that because several complicated fluorescence quenching involve excitation patterns, it is still difficult to calculate the excited state for transition metal complexes accurately.

The development of fluorescence probes is essentially founded on the most adaptable fluorophores, namely, coumarin, fluorescein, rhodamine, cyanine, and azo dye, and to figure out how they recognize different transition metal ions (Figure 2.2).

2.2 RHODAMINE-BASED FLUORESCENCE PROBES

Rhodamine's use in fluorescent probes is receiving more and more attention as a result of its favourable spectroscopic characteristics, which include a high fluorescence quantum yield, a large absorption coefficient, and a long emission wavelength [60–62]. Additionally, it is intriguing that some transition metal ions can be recognized by rhodamine derivatives in an "off-on" mode. Because, in the vicinity of the appropriate metal ions, the revised spirolactam structure that enables colourless and nonfluorescent can be converted into the colourful and extremely fluorescent amide form with the ring-opened [63]. Rhodamine derivatives have emerged as the best option for developing innovative chemosensors to sense certain ions of transition metals because of the distinct "off-on" action mechanism.

Xu et al. synthesized rhodamine 6G appended hydrazone **1** using 7-diethylamino -3-(1-hydroxy-3-oxobut-1-enyl)-2H-chromen-2-one (Figure 2.3) [64]. The **1** displays ratiometric fluorescence in neutral aqueous environments with remarkable selectivity towards Cu^{2+}. At pH 10, the probe detected Hg^{2+} with a fluorescence increase. The FRET process was triggered by the hydrazone **1**'s binding mechanism with Cu^{2+}.

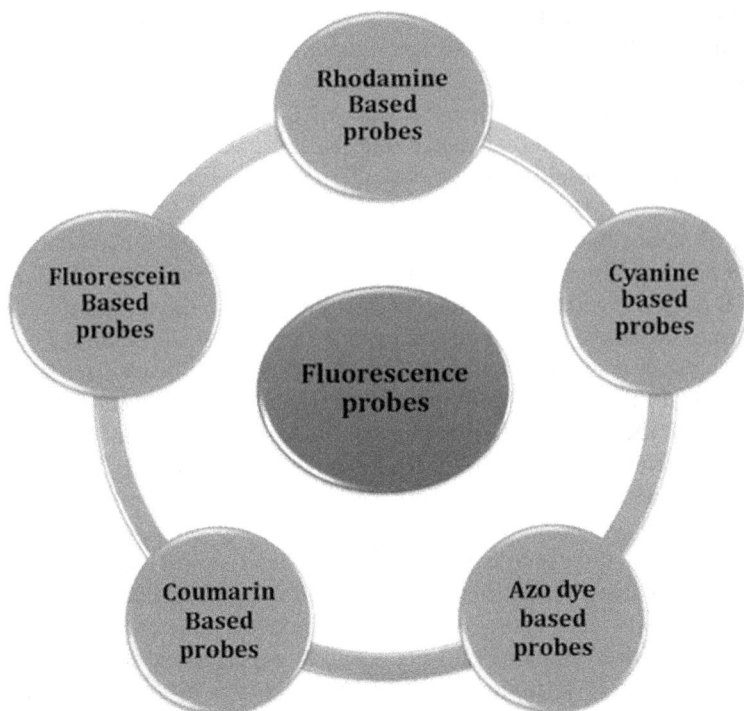

FIGURE 2.2 Schematic diagram for different fluorophores like rhodamine, fluorescein, coumarin, cyanine, azo dye, etc. based fluorescent probes.

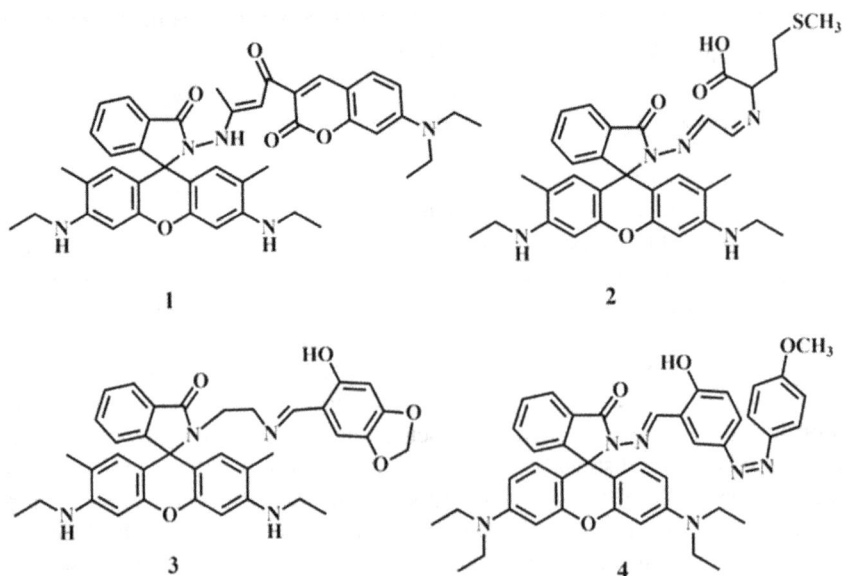

FIGURE 2.3 Structure of probes **1–4**.

The free probe **1** showed one fluorescence emission peak at 498 nm when excited at 445 nm, which decreases significantly by the addition of 5 equiv. Cu^{2+}. The results indicated that the Cu^{2+} detection limit is 6.88 mM and the Hg^{2+} detection limit is 2.96 mM. The Hg^{2+} binding constant for **1** was 3.34×10^4 M^{-1} ($R = 0.99768$). According to Job's plot experiment, the stoichiometry for the binding of **1** to Hg^{2+} is 1:1. Using HeLa cells as a test subject, the capability of **1** to detect Cu^{2+} was examined. The results indicated that it may be utilized to photograph intracellular Cu^{2+} in living cells.

Sikdar et al. synthesized rhodamine 6G appended probe **2** with methionine conjugate (Figure 2.3) [65]. In a Tris-HCl buffer, the probe **2** was discovered to be very sensitive and selective to Hg^{2+}. **2** exhibits relatively weak wide peaks at 545 nm upon excitation at 500 nm, but in the presence of Hg^{2+}, **2** demonstrated an astounding 27-fold amplification in the emission band at 545 nm. 2.63×10^{-8} M was discovered to be the hydrazone **2**'s detection threshold for Hg^{2+}. The probe **2** was effectively used to find Hg^{2+} in biological cells, water samples, and strip tests.

Based on rhodamine 6G derivatives' spirolactam ring-opening reaction, Diao et al. created a unique "turn-on" fluorescent probe **3** (Figure 2.3) [66]. The probe had a number of advantages, including excellent water solubility, and high sensitivity and selectivity for Fe^{3+}. These findings suggest that probe **3** can be employed to detect Fe^{3+} without being affected by competing metal ions. **3**'s intensity of emission at 556 nm was greatly increased by Fe^{3+} ($\Phi = 0.74$), despite the fact that **3** had a very small emission peak and a low quantum yield ($\Phi = 0.02$) ($\lambda_{ex} = 525$ nm). The probe's limit of detection (LOD) and tested concentration range were 0–30.00 and 0.030 μmol/L, respectively. The current probe **3** was used to assess the presence of Fe^{3+} in drinking water samples. The fluorescent probe **3** can be used for a bioimaging device for the exposure of Fe^{3+} in BEL-7402 cells.

A unique fluorescent probe **4** (Figure 2.3) created by Yang et al. showed a good "off-on-type" fluorescence change with a strong selectivity towards Cu^{2+} [67]. Upon excitation at $\lambda_{ex} = 535$ nm, Cu^{2+} caused a new emission peak to arise at 588 nm, which was attributed to rhodamine's lactam ring opening. Curiously, the probe–Cu^{2+} complex was able to identify cysteine because of how strongly this amino acid can coordinate Cu^{2+}, and no overt interference from other amino acids or anions was seen. Further demonstrated the probe **4**'s capacity for bioimaging, both in cells and on living mice.

A novel solvent-dependent chemosensor **5** created by Zhao et al. was constructed from a diarylethene containing a rhodamine B unit (Figure 2.4) [68]. By observing the variations in different solvents in the fluorescence and UV-vis spectra, it may be employed as a binary-functional chemosensor used for the discriminatory detection of Hg^{2+} and Cu^{2+}. Chemosensor **5** by itself did not demonstrate a detectable excitation of the emission signal at 520 nm, but when 10.0 equivalents of Hg^{2+} were added, the intensity of the fluorescence at 617 nm increased approximately 34-fold. Job's plot supported a 1:1 binding stoichiometry between Cu^{2+} and the Hg^{2+} ions present in solution. Hg^{2+} and Cu^{2+} were found to have limits of detection of 0.14 and 0.51 mM, respectively. The **5**–Hg^{2+} complex's binding constant was 0.42×10^4 M^{-1}. **5**-Hg^{2+} has an absolute fluorescent quantum yield that has been calculated to be 0.35. Based on the fluorescence behaviour generated by lights and chemical stimuli, a molecular logic circuit built with four inputs and one output.

Based on a selenide that contains rhodamine, Paulino et al. created the novel reversible fluorescent probe **6** (Figure 2.4) [69]. Probe **6** has a detection limit of 32 nM for Pd^{2+} ions, demonstrating excellent selectivity and sensitivity. Additionally, there was a peak emission at 588 nm, on λ_{ex} (nm) = 515 nm. The system also displayed a strong bathochromic shift of 137 nm together with a Stokes shift of 23 nm and a fluorescence quantum yield of $\Phi = 0.17$. The classic continuous variation method or Job's plot, which presents a 1:1 relationship (**6**: Pd^{2+}), was used to determine the stoichiometric relationship between **6** and Pd^{2+}. According to DFT simulations, the metallic complex **6**/Pd^{2+} is most stable in the square planar geometry that has been proposed.

A unique "off-on" colorimetric and fluorescent chemosensor **7** (Figure 2.4) [70] was created and synthesized by Wang et al. With a roughly 75-fold growth in intensity of the fluorescence emission at 585 nm over a variety of naturally significant metal cations in PBS reaction media, it specifically detects the presence of Fe^{3+}. Fluorescence emission at 585 nm was not noticed in the absence of Fe^{3+}. The fluorescent emission was steadily enhanced at 585 nm when Fe^{3+} was introduced. In the fluorescence measurement, the Fe^{3+} ultrasensitive detection threshold limit is as low as 2.0×10^{-8} mol/L. Job's plot experiment demonstrated the binding ratio among Fe^{3+} and **7** was 1:1. Exogenous assays were used to examine **7**'s capacity to identify Fe^{3+} ions in human hepatocarcinoma cell culture. This study provides a reliable analytical technique for detecting Fe^{3+} in real time in biological systems.

A quinoline-conjugated fluorescence chemosensor based on rhodamine has been published by Murugan et al. (Figure 2.4) for the recognition of the two (Fe^{3+}/Cu^{2+}) paramagnetic ions [71]. A significant fluorescence increase was also elicited by Fe^{3+}. This is an excellent example of a dual chemosensor that may be used to selectively detect transition metal ions that are important to biology and the environment. Fe^{3+} ions in an aqueous solution excited the fluorescence of **8** at λ_{ex} 510 nm, which was then

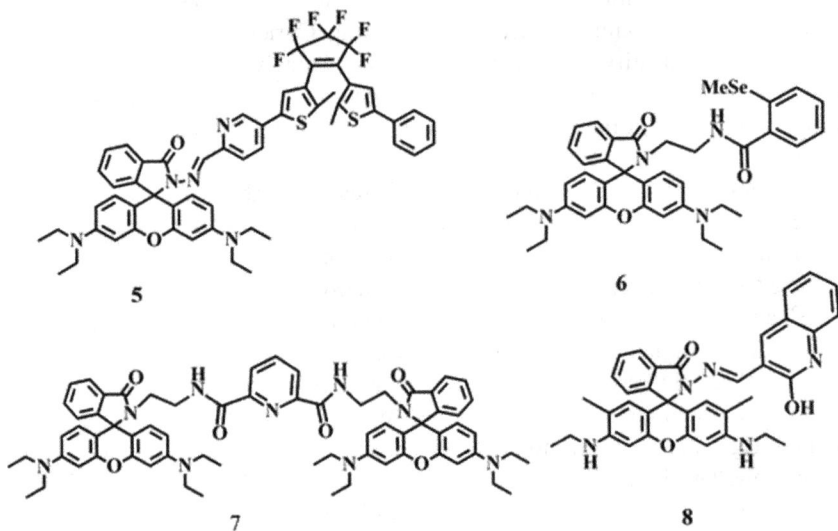

FIGURE 2.4   Structure of probes **5–8**.

increased at 572 nm. The discovered linear connection with Fe^{2+}/Cu^{3+} ions proved its exceptional sensitivity, the computed LOD values was 1.8×10^{-8} M for Cu^{2+} and 3.3×10^{-8} M for Fe^{3+}. Job's plot was used to calculate the 1:1 stoichiometry among the **8** and Fe^{3+} ions, and the B-H equation was used to calculate the binding constant, which yielded the result of 1.3×10^{-4}. It was evident from the probe's increased density of electron on the xanthenyl moiety in the HOMO state and its complete transfer to the LUMO state of the Schiff base in that probe was engaged in ICT. What's more intriguing is that this chemosensor, **8**, was used to identify intracellular Fe^{3+} ions in zebrafish embryos.

Two new rhodamine B chemosensors of the pyridine type, **9** and **10**, have been developed and produced by Song et al. (Figure 2.5) [72]. In the solution of EtOH/ H_2O (3:1, v/v, 0.5 mM, HEPES, pH = 7.33), both **9** and **10** exhibit strong spectrum responses to Fe^{3+} with high binding constants and low detection limits. **9** and **10** displayed no colour and no distinct at 582 nm ($\lambda_{ex} = 560$ nm) fluorescence intensity in the absence of Fe^{3+}. The measuring solution's colour shifted from colourless to pale pink as the concentration of Fe^{3+} increased. The formation of the open-ring amide of **9** and **10** with Fe^{3+} binding can be used to explain how **9** and **10** changed towards Fe^{3+}. Under saturation circumstances, a fluorescence development of nearly 100-fold was seen. As a result, it is possible to compute the association constant K_a as 2.70×10^4 M^{-1} for the **9**/Fe^{3+} complex and 1.97×10^4 M^{-1} for the **10**/Fe^{3+} complex. It was also possible to discover that the detection threshold for Fe^{3+} for both **9** and **10** was 0.067 and 0.345 μM. While **10** displayed spectral behaviour with Fe^{3+} following a binding stoichiometry of 1:2, **9** demonstrated a sensing mechanism towards Fe^{3+} with a 1:1 binding stoichiometry. The utility of **9** and **10** as fluorescent chemosensors to detect Fe^{3+} in active human breast cancer (MCF-7) cells was established by intracellular imaging applications.

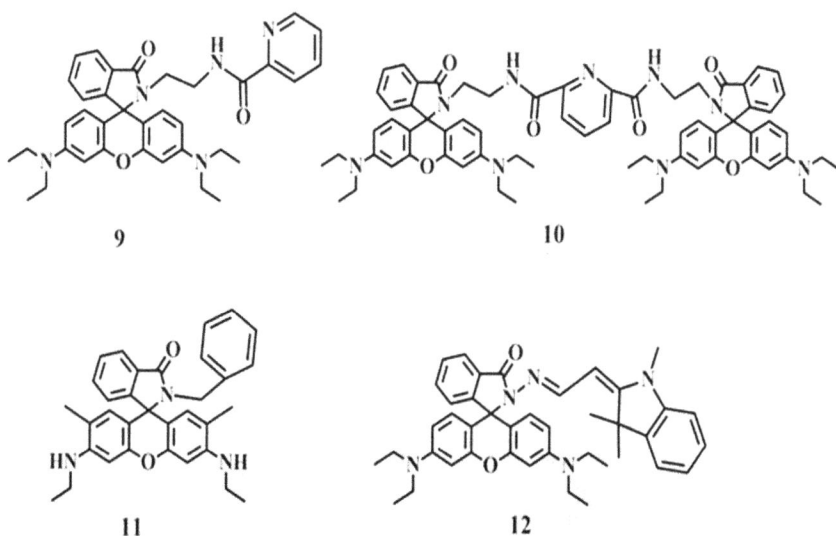

FIGURE 2.5 Structure of probes **9–12**.

A novel rhodamine 6G-benzylamine-based sensor **11** with just hydrocarbon frames in the stretched part was created by Das et al. (Figure 2.5) [73]. It demonstrated good Fe^{3+} and Cr^{3+} identification that was sensitive and selective. Fluorescence investigation revealed that **11** has good reversible fluorescence ON–OFF properties. Fe^{3+} (0–3 equivalent) and Cr^{3+} (0–3 equivalent) were gradually added to the nonfluorescent solution, and a 41-fold and a 26-fold increase in fluorescence intensity was observed, respectively, at 558 nm following excitation at $\lambda_{ex} = 502$ nm. This also implies that coordination to Fe^{3+} and Cr^{3+} will cause the spirolactam ring in **11** to open. When 3.0 equivalents of these metal ions were added to probe **11** in H_2O/CH_3CN (v/v, 4:1, pH 7.2), it was possible to see a significant growth in fluorescence intensity for Fe^{3+} (41-fold) and Cr^{3+} (26-fold) with open eye detection. The equivalent K_f values for Fe^{3+} and Cr^{3+} were found to be 9.4×10^3 and 8.7×10^3 M^{-1}, respectively. LODs for both Fe^{3+} and Cr^{3+} were calculated using the 3σ method, and they were discovered to be 1.28 and 2.28 μM, respectively. Probe **11**'s fluorescence intensity increased significantly upon complexation with Fe^{3+} and Cr^{3+}, indicating that the probe can be employed for use in living cells for bioimaging applications.

A brand-new rhodamine-based probe **12** was created by Zhang et al. (Figure 2.5). It exhibits remarkable Cu^{2+} selectivity and sensitivity [74]. It is based on a phenomenon called FRET. A significant emission at 503 nm in the fluorescence spectra of compound **12** is due to the trimethylindolin moiety when stimulated at 345 nm. It suggests that the rhodamine moiety in compound **12** still exists as a ring-closed compound devoid of metal ions. Due to the inhibition of FRET, only the donor moiety's emission can be seen at 503 nm. For Cu^{2+}-based fluorescence spectra, the detection limit of **12** is calculated to be 1.168×10^{-8} mol/L. The analysis result from Job's plot shows that the **12**–Cu^{2+} complex has a 1:1 stoichiometric ratio. Actual water samples can be effectively identified by the **12** for Cu^{2+}. Test paper strips can also be used to implement some of the probe's possible applications.

2.3   FLUORESCEIN DYE-BASED FLUORESCENCE PROBES

In 1871, von Bayer created fluorescein, with a yellowish green fluorescence form of a xanthene dye, by utilizing phthalic anhydride and resorcinol in a Friedel–Craft cyclodegradation/acylation synthesis [75]. Two aryl groups are fused together to form a pyran ring, giving it a solid coplanar tricyclic structure. Two different structures make up the fluorescein family of dyes: a fluorescent ring-opened carboxylic acid form. The significance of the fluorescein structure is found in the spirolactam structure, which is different from the colourless and nonfluorescent spirocyclic structure. This is a result of its distinctive "open and close" reaction to particular situations. Specific analytes generating strong signals with great molar absorptivity and high quantum yield can also open up the ring, causing colour shifts and fluorescence intensification, thanks to the spirolactam structure of fluorescein [76,77]. Fluorescein is sensitive to the pH of the medium because of its open–close equilibrium structure [78]. By keeping track of the equilibrium between the opening and shutting of its rings, as well as other spectroscopic effects, fluorescein's pH sensitivity can be used to identify industrial and commercial samples' metal ions [79].

By combining fluorescein hydrazide with 2-(Pyridin-2-ylmethoxy)-napthlene-1-carbaldehyde [80], Hasan et al. were able to create a fluorescein hydrazone-based conjugate **13** (Figure 2.6). In semi-aqueous media (10 mM HEPES, pH 7.2), probe **13** showed high sensitivity and great selectivity to hazardous Hg^{2+} from another metallic ion. The spirolactam moiety in the fluorescein structure was subject to Hg^{2+}-induced ring opening, which was the cause of the substantial enrichment in fluorescence emission centred at 520 nm. It may be said that the interaction between Hg^{2+} and the chelating probe results in a rigid complex called $[Hg(13)]^+$ that tends to promote chelation of the fluorescence (CHEF). With detection limits of 1.24 μM, Job's plot and ESI-MS+(m/z) experiments proved the 1:1 binding of **13** to Hg^{2+}, and the binding constant was determined as $(0.43 \pm 0.04) \times 10^4$ M^{-1}. The **13** is suited for fluorescent cell imaging of Hg^{2+} ions in live HepG2 cells and demonstrates biocompatibility with low cytotoxicity.

Phenolphthalein-fluorescein dye derivative was synthesized and studied by Erdemir et al. (Figure 2.6). Over other cations, Zn^{2+} and Hg^{2+} ions were efficiently recognized by **14** as a dual channel probe [81] in EtOH-H$_2$O (pH 7.0, v/v = 8/2, HEPES, 5 mM). The cations, Zn^{2+} and Hg^{2+}, resulted in a notable enrichment in fluorescence emission centred at 500 and 520 nm, respectively. Studies on fluorescence titration were conducted by varying the amounts of Zn^{2+} ion (0–5.0 eq.) and Hg^{2+} ion (0–20.0 eq.) in a solution of **14**. Moving the emission band and emission intensity to 500 nm augmentation when **14** was titrated with Zn^{2+}. While for Hg^{2+}, the emission band is at 520 nm, which is associated with the emission of ring-opened xanthenes. These phenomena provided additional evidence that the Hg^{2+} addition enhanced the fluorescein spirolactam's ring-opened reaction. The FRET to the fluorescein acceptor from the conjugated phenolphthalein donor also explains this selectivity. Job's plot revealed that probe **14** binds to Zn^{2+} as well as Hg^{2+} in a 1:2 stoichiometry, and the computed binding constants are 6.45×10^{10} and 2.11×10^{10} M^{-2}, respectively.

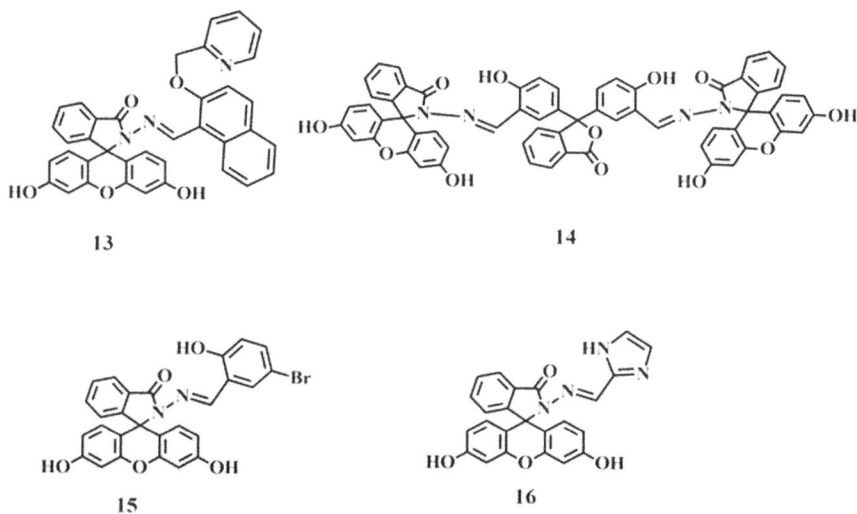

FIGURE 2.6 Structure of probes **13–16**.

The detection limits for Zn^{2+} and Hg^{2+} were 0.54 and 1.16 μM, respectively. In practical applications, probe **14** can be accurately used to sense Zn^{2+} or Hg^{2+} in samples of water, indicating its future potential and implication in analysis and detection.

To detect Zn^{2+} in alcoholic aqueous solution within a physiological pH variety, Das et al. developed and created a new Schiff base **15** (Figure 2.6), which is made up of fluorescein hydrazine and a phenol functionalized moiety [82]. The fluorescein fluorophore was excited at 410 nm to produce the fluorescence spectra. Zn^{2+} selective fluorescence signalling behaviour is indicated by a consistent increase in the intensity of emission at 520 nm with increasing concentrations of Zn^{2+} (0–20 μM) to **15** (20 μM). It is discovered that the fluorescence of **15** is enhanced 23-fold when Zn^{2+} binds to the receptor **15**. For the strong fluorescent response to Zn^{2+}, C=N isomerization has been suppressed and with the introduction of the PET (photo-induced electron transfer) mechanism, additional photophysical research proposes the introduction of CHEF (chelation-enhanced fluorescence). Job's plot established the metal complex's 1:1 binding mode. LOD and complexation binding constants for Zn^{2+} are determined to be 2.86×10^4 M^{-1} and 1.59 μM, respectively. The **15** is utilized in the cell imaging investigation employing kidney cells from an African green monkey (Vero cells)

Fluorescein dye containing imidazole **16** (Figure 2.6) was created by Vidya et al. as a reliable colorimetric and "turn-on" fluorescent sensor for Zn^{2+} [83]. Probe **16** exhibits a more sensitive fluorescence "turn-on" response for Zn^{2+} as well as a dramatic and selective change from colourless to yellow colour shift. When compound **16**'s fluorescence spectra were stimulated at 365 nm in an aqueous buffer, there was no detectable emission. When Zn^{2+} ions were added to receptor number **16**, a 532 nm fluorescence emission that is growing with a 567 nm shoulder was the result. An outstanding "turn-on" ratio for Zn^{2+} sensing is provided by the fluorescence's 200-fold rise in intensity. Job's plot supported the 1:1 binding stoichiometry between the probe and Zn^{2+}. It was discovered that **16**'s detection threshold for Zn^{2+} ions was 5.49×10^{-8} M, while **16**'s binding constant to Zn^{2+} was predicted to be 9.28×10^6 M^{-1}. The live-cell imaging in HeLa cells demonstrated how the probe **16** might be used for Zn^{2+} detection. Additionally, probe **16** was used to image the Zn^{2+} ions released during apoptosis.

2.4   COUMARIN-BASED FLUORESCENCE PROBES

The use of coumarins in biology, perfumes, medicine, fluorescent dyes, and cosmetics is widespread [84]. For the detection of hydrogen peroxide, nitroxide, and nitric oxide, coumarin derivatives are the most widely employed fluorescent probes. Additionally, coumarin derivatives have shown to be effective chemosensors for a variety of metal ions such as Cu^{2+}, Zn^{2+}, Hg^{2+}, Ni^{2+}, Fe^{3+}, Cr^{3+}, Pb^{2+}, and Ag^+ [85] and anions such as cyanide, benzoate, pyrophosphate, fluoride, dihydrogen phosphate, and acetate. An intriguing organic fluorophore for the identification and measurement of trace quantities of ecologically significant metallic ions and disease indicators in physiological samples is the class of phytochemicals known as coumarins, which include benzopyrone [86]. Along with a wide range of anti-microbial and anti-cancer effects, pi-extended coumarin analogues made by chemical synthesis and natural processes have also been shown to have superior and controllable photophysical characteristics that could make them useful as chemical sensors.

Bhasin et al. expressed potentially novel Schiff base **17** (Figure 2.7) derived from coumarins was chemically synthesized and physico-chemically characterized as a possible sensor for the medically important Zn^{2+} as an analyte in a semi-aqueous media [87]. By adding an aqueous solution of Zn^{2+} ions (700 nM) to an aqueous solution of **17** (100 ppm), a remarkable 2.5-fold increase in the fluorescence emission of **17** ($\lambda_{ex.} = 470$ nm) was observed from the fluorescence spectrum ($\lambda_{ex.} = 380$ nm), introducing ligand **17** as a potential "turn-on" fluorescent ligand in the background of a large library of the traditional fluorescence "turn-off" metal. This particular switch-on response may be ascribed to a suppression of a likely PET-induced fluorescence quenching inside the native ligand **17** via a likely Zn^{2+}-mediated complexation with a perhaps zinc-binding N,N'-ditopic motif. According to our research, Zn^{2+} causes a **17**-fold increase in fluorescence on its own, against the background of an aqueous pool containing several competing cations, with the chemically related Cd^{2+}. The stoichiometry of **17**: Zn^{2+} coordination was estimated to be 1:1 (B-H studies), with a remarkably low detection threshold of 0.272 nM (Stern–Volmer plot). The Benesi–Hildebrand (B–H) plot was used to determine the stoichiometry of the **17**: Zn^{2+} binding system, which was determined to be 1:1 and the binding constant being 0.0037 nM^{-1}. In addition, recovery investigations show that spiked natural water samples of **17** have a substantial sensing efficacy towards Zn^{2+}.

Chen et al. for the purpose of detecting Hg^{2+}, a reaction-based fluorescent probe **18** (Figure 2.7) with a pyrimidine moiety was devised and created [88]. At physiological pH, the probe can preferentially identify Hg^{2+} from the other metal ions with a 42-fold fluorescence increase. Hg^{2+} induced fluorescence change of **18** (10 μM) in a CH_3CN/H_2O (v/v = 3/7) solution. The emission gradually increased at 475 nm, with the excitation wavelength being 396 nm. The outcome demonstrates that the maximum intensity was attained after the addition of 6 Hg^{2+} equivalents. The detection threshold for Hg^{2+} using this probe is 1.08 μM. It was discovered that the fluorescence quantum yield of **18** with Hg^{2+} was 0.042, which was around 42 times greater than that of **18** ($\Phi = 0.001$). Sensor **18** can be used to detect Hg^{2+} in live cells that emit blue light. Probe **18** exhibits little cytotoxicity and has potential use in zebrafish and HeLa cell imaging of Hg^{2+}.

Mani and coworkers reported that N,N-diethylamino-3-acetyl coumarin and 2-hydrazinobenzothiazole were combined to develop a novel coumarin-based fluorescent probe **19** (Figure 2.7) with ICT character [89]. The chemosensor can uniquely distinguish Cu^{2+} ions from other distinct metal ions, according to the absorbance and fluorescence spectral features of **19**, and the smallest detectable limit was discovered in the nanomolar range. Emission spectra of probe **19** show an emission maxima at 536 nm ($\lambda_{ex} = 420$ nm) prior to adding Cu^{2+} ions. When Cu^{2+} ions were added to the **19** solution, the emission intensity (572 nm) steadily dropped and red shifted from 536 to 572 nm, which also caused the colour change from yellow to wine red, clearly showing that Cu^{2+} was coordinated with the **19** receptor. Chelating enhanced emission quenching as well as ICT are the causes of the observed emission quenching. The lowermost recognition limit was evaluated as 40 nM. The K_S value was estimated to

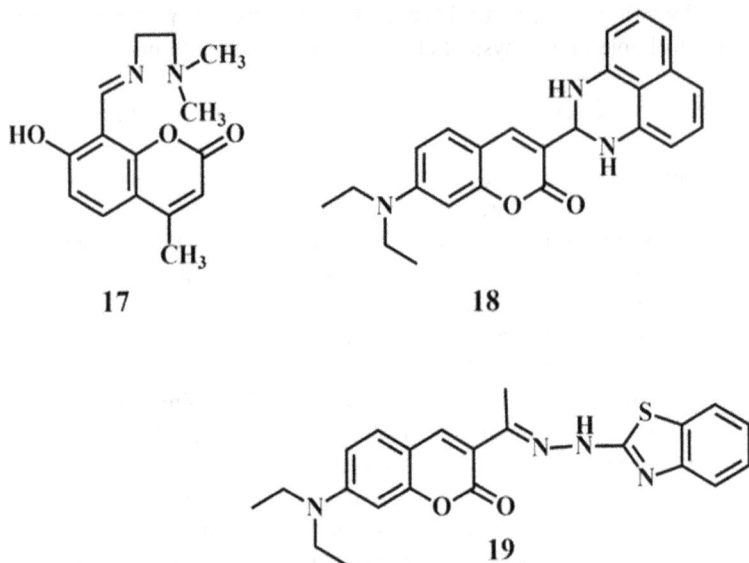

FIGURE 2.7 Structure of probes **17–19**.

be 4.09×10^4 L/mol according to the Stern–Volmer linear plot's slope. These attained results indicated that during their complexation, **19** and Cu^{2+} formed a 1:1 stoichiometry. Experiments using fluorescent microscopy showed that probe **19** might be used like fluorophore to find Cu^{2+} in cellular life.

2.5 CYANINE-BASED FLUORESCENCE PROBES

Near-infrared (NIR) dyes have attracted a lot of attention in spectroscopic, biological, and analytical research. NIR dyes that emit between 650 and 900 nm are a lot better for in vivo imaging than ultraviolet-visible (UV-vis) light because the light in this region has a deeper tissue penetration depth and causes less cellular auto-fluorescence [90–93]. Tricarbocyanines were selected as the NIR dye platform because they are frequently used as fluorescent labels and sensors in biomolecular imaging [94–96]. Better photostability and higher fluorescence quantum yield are specifically provided by tricarbocyanines with a stiff chlorocyclohexenyl ring in the methine chain.

The NIR fluorescent probe **20** was created by Li et al. and utilized as a NIR fluorescence indicator to measure mercury ions in water and live cells [97]. The probe **20** showed a modest emission because the PET process might be involved. When Hg^{2+} was added to the probe's solution, a stronger emission peak at 790 nm was visible. After interacting with Hg^{2+}, the emission at 790 nm in particular experienced a dramatic increase of more than 25 times. The binding test revealed that the indicator and Hg^{2+} formed a 1:1 complex. According to calculations, probe **20**'s detection threshold is 7.3 nM. The binding constant among host **20** and guest Hg^{2+} was 1.0×10^6 M^{-1}. This indicator may also be utilized to track Hg^{2+} in live cells and assess the Hg^{2+} levels in a sample of tap water (Figure 2.8).

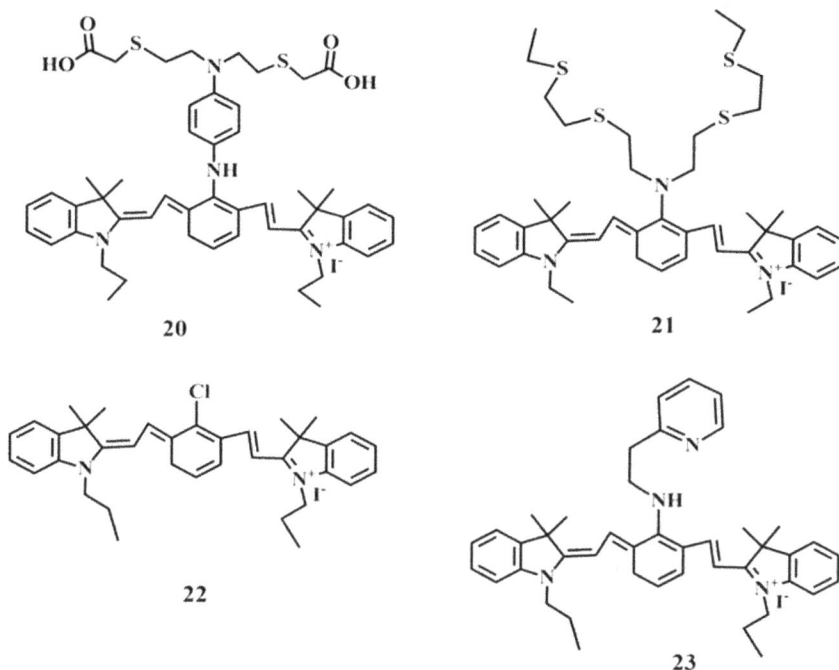

FIGURE 2.8 Structure of probes **20–23**.

In order to identify the occurrence of Cu$^+$ [98], Cao and coworkers introduced a NIR region fluorescence probe **21** (Figure 2.8), the fluorescent group being an indole heptamethine cyanine dye, and the monitoring group being a thioether ring. The recognition group supplied electrons and reduced the cyanine dyes' fluorescence signal prior to binding to Cu$^+$. The fluorescence signal increased and the absorption spectra moved towards the red following the binding to Cu$^+$.

Han et al. [99] produced the fluorescent probe **23** (Figure 2.8) for the highly selective recognition of Cu^{2+}. The probe's reaction was according to the fluorescence quenching after attaching to Cu^{2+}, and the probe **23** was made up of 2-(2-Amino-ethyl) pyridine and IR-780 iodide (probe **22**, Figure 2.8). The sensing capabilities of the suggested Cu^{2+}-sensitive probe were next examined. The probe has a linear concentration range of 4.8×10^{-7} to 1.6×10^{-4} mol/L and a detection limit of 9.3×10^{-8} mol/L, making it applicable to the quantitative detection of Cu^{2+}. According to the experimental findings, the probe responded to Cu^{2+} in medium conditions regardless of pH and with outstanding selectivity for Cu^{2+} than other cations of common metals.

A new NIR fluorescence probe **24** for the uncovering of Zn^{2+} was reported by Tang Bo's team [100] (Figure 2.9). The indole heptamethine cyanine dyes served as the fluorophore and the recognition group for the probe, which was 2,2-dimethyl-1-pyridine. When the fluorescent probe and Zn^{2+} are combined, and the probe's PET is blocked, fluorescence quenching occurs.

A brand-new NIR fluorescence probe **25** (Figure 2.9) was published by Zhu et al. [101] to measure Hg^{2+} concentrations. The dioxane crown ether recognition group

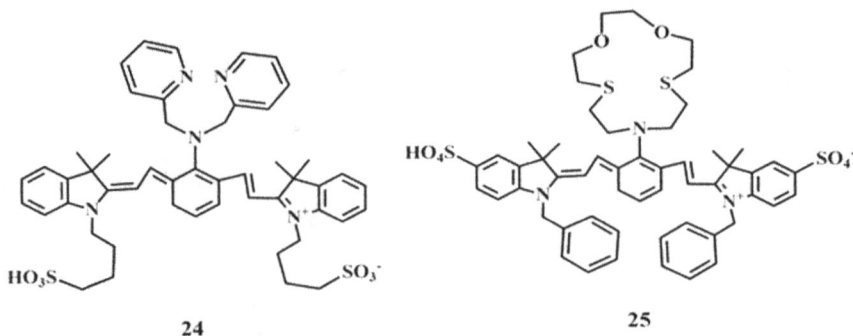

FIGURE 2.9 Structure of probes **24–25**.

was added to the fluorescence group to produce the Hg^{2+} concentrations. The macromolecule's entire conjugate system was affected by the probe's recognition group and Hg^{2+}, which also altered the fluorescence signal. The findings demonstrated that the probe had excellent Hg^{2+} selectivity and could therefore quickly determine its presence.

2.6   AZO DYE-BASED FLUORESCENCE PROBES

Due to the presence of heteroatoms (S, N, and O) in their structures, rhodamine-based azo compounds can interact with metal ions fairly easily. These substances can form stable six-membered rings after complexing with a metal ion, allowing them to chelate with a variety of metal ions.

Akram et al. designed and synthesized novel derivatives for naphthalene-ringed rhodamine azo compounds **26–28** (Figure 2.10) [102]. Using UV-vis and fluorescence spectroscopy methods, the sensitivity and selectivity of the newly created rhodamine azo compounds and transition metals Cu^{2+}, Co^{2+}, Zn^{2+}, Fe^{3+}, and Ni^{2+} were studied. When stimulated at 395 nm, azo **28**'s emission spectra demonstrated two distinct bands at 374 and 423 nm. With the addition of Fe^{3+}, azo **28**'s fluorescence intensity was reduced without changing its peak location. Through photo-induced electronic energy transfer or metal-to-fluorophore electron transfer pathways, paramagnetic ions like Fe^{3+} have the capacity to reduce the fluorescence intensity of a fluorophore. The probe **28** exhibited Fe^{3+} ion affinity, having a 4.63×10^8 M^{-1} association constant. Furthermore, with detection limits of 5.14 µM, this azo **28** demonstrated high sensitivity to Fe^{3+} ions. The production of complexes between azo **28** and Fe^{3+} with a 1:2 binding stoichiometry was established by the molar ratio and Benesi–Hildebrand methods. As a result, azo **28** demonstrated outstanding potential for producing effective Fe^{3+} chemosensors. The biological effects of the azo rhodamine compounds **26–28** on various harmful microorganisms were assessed. According to the findings, compounds **26–28** exhibit fair to excellent vital action against the pathogenic germs.

Bozkurt and others developed a novel rhodamine ring-based chemosensor **29** (Figure 2.10) [103]. Among other metal ions, chemosensor **29** is highly sensitive and specific for Fe^{3+} and Cu^{2+} ions. **29** displayed a fresh and high peak fluorescence at

589 and 595 nm with regard to Fe^{3+} and Cu^{2+} ions, respectively, while exhibiting very weak fluorescence at $\lambda_{ex} = 540$ nm with the presence of additional metal ions. Fe^{3+} and Cu^{2+} had LOD values of 0.91 and 1.04 µM, respectively. Fe^{3+} and Cu^{2+} ions had response times of **29** that were as quick as 0.5 min. With Fe^{3+} and Cu^{2+} ions present, the fluorescence quantum yield values of the chemosensor **29** dramatically enhanced ($\Phi = 0.40$ for Fe^{3+}, $\Phi = 0.57$ for Cu^{2+}). The Fe^{3+} and Cu^{2+} ions improved the fluorescence property of the chemosensor **29**, as evidenced by the considerable increases in fluorescence quantum yield values of **29** in the presence of these ions.

Two azo dyes **30** and **31** (Figure 2.10) were created and synthesized by Bartwal et al. [104]. The selective "turn-on" behaviour of **30** and **31** towards Al^{3+} and Co^{2+} ions was validated by the absorption and emission spectrum analyses. The emission spectrum of free probe **31** showed a weak emission maximum at 486 nm ($\Phi = 0.015$) when stimulated at $\lambda_{ex} = 400$ nm in buffered aqueous solution of $MeOH:H_2O$ (v/v = 1:1, HEPES buffer, pH = 7.2), which confirmed the absence of any strongly conjugated systems. Upon the addition of one equivalent of Co^{2+} ions, a bathochromic shift in the peak emission wavelength from 486 to 500 nm was observed. When Co^{2+} ions are present, **31**'s emission behaviour uses CHEF and ICT mechanisms, which causes a 26-fold rise in emission intensity. The 1:1 stoichiometric ratios for the complexes **30**-Al^{3+} and **31**-Co^{2+} were revealed by Job's plot analyses. Complexes **30**-Al^{3+} and **31**-Co^{2+} were discovered to have association constants of 1.2×10^6 and 4×10^6 M^{-1}, respectively. It was established that the thresholds for detection of **30** was 1×10^{-8} M and **31** was 9.1×10^{-8} M. Both probes **30** and **31** can be utilized as secondary sensors for the EDTA^{2-} anion, according to reversible studies.

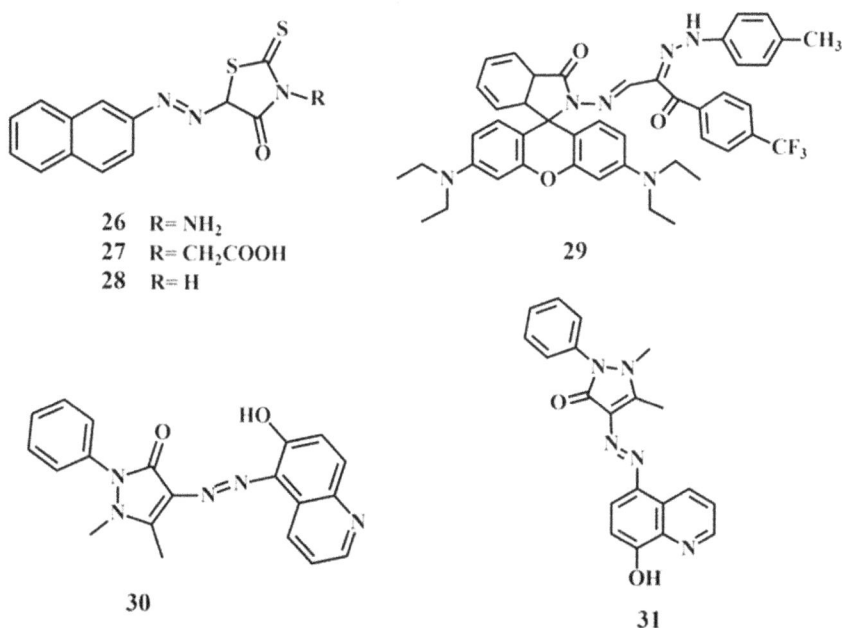

26 R= NH$_2$
27 R= CH$_2$COOH
28 R= H

29

30

31

FIGURE 2.10 Structure of probes **26–31**.

2.7 CONCLUSION

The development of smart, multi-stimuli and highly sensitive probes has encouraged researchers to focus on fluorescent probes. The physical and chemical characteristics of numerous fluorescence probes are now being identified by chemists, physicists, and material scientists [105,106]. Various examples of smart, multi-stimuli and highly sensitive probes have been documented. In this chapter, the recent work on fluorescence probes to detect transition metal ions has been reviewed. Efforts have been devoted to reviewing fluorescence probes of rhodamine, fluorescein, coumarin, cyanine, and azo dye. These probes are renowned for having rich optical characteristics that produce strong bands in the visible region. The fluorescence probes to detect transition metal ions (Zn^{2+}, Cu^{2+}, Fe^{3+}, Pb^{2+}, Cd^{2+}, Cr^{3+}, Hg^{2+}, etc.) are discussed in detail with focus on recent results. The probes based on rhodamine, fluorescein, coumarin, cyanine, and azo dye have been demonstrated as an ideal choice for the detection of transition metal ions.

REFERENCES

1. E. Guliani, A. Taneja, K. R. Ranjan, and V. Mishra. 2023. Luminous Insights: Exploring Organic Fluorescent "Turn-On" Chemosensors for Metal-Ion (Cu^{+2}, Al^{+3}, Zn^{+2}, Fe^{+3}) Detection. . *J. Fluoresc.* 1–37..
2. Z. Wang, M. Wang, G. Wu, D. Wu and A. Wu. 2014. Colorimetric detection of copper and efficient removal of heavy metal ions from water by diamine-functionalized SBA-15. *Dalton Trans.* 43: 8461–8468.
3. P. B. Tchounwou, C. G. Yedjou, A. K. Patlolla and D. J. Sutton. 2012. Heavy metal toxicity and the environment. *Exp. Suppl.* 101: 133–164.
4. V. Dujols, F. Ford and A. W. Czarnik. 1997. A long-wavelength fluorescent chemodosimeter selective for Cu(II) Ion in water. *J. Am. Chem. Soc.* 119: 7386–7387.
5. N. Yadav, D. Mudgal, and V. Mishra. 2023. In-situ synthesis of ionic liquid-based-carbon quantum dots as fluorescence probe for hemoglobin detection. *Anal. Chim. Acta* 1272 : 341502.
6. N. Yadav, R. P. Gaikwad, V. Mishra and M. B. Gawande. 2023. Synthesis and photocatalytic Applications of functionalized carbon quantum dots. *Bull. Chem. Soc. Jpn.* 95(11): 1638–1679.
7. (a) A. P. de Silva, H. Q. N. Gunaratne, T. Gunnlaugsson, A. J. M. Huxley, C. P. McCoy, J. T. Rademacher and T. E. Ric. 1997. Signaling recognition events with fluorescent sensors and switches. *Chem. Rev.* 97(5): 1515–1566. (b) E. M. Nolan and S. J. Lippard. 2008. Tools and tactics for the optical detection of mercuric ion. *Chem. Rev.* 108 (9): 3443–3480. (c) N. Kaur and S. Kumar. 2011. Colorimetric metal ion sensors. *Tetrahedron* 67: 9233–9264.
8. (a) V. K. Gupta, M. L. Yola and N. Atar. 2014. A novel molecular imprinted nanosensor based quartz crystal microbalance for determination of kaempferol. *Sens. Actuators B Chem.* 194: 79–85. (b) V. K. Gupta, A. K. Jain, S. Agarwal and G. Maheshwari. 2007. An iron(III) ion-selective sensor based on a μ-bis(tridentate) ligand. *Talanta* 71: 1964–1968. (c) R. Prasad, V. K. Gupta and A. Kumar. 2004. Metallo-tetraazaporphyrin based anion sensors: Regulation of sensor characteristics through central metal ion coordination. *Anal. Chim. Acta* 508: 61–70. (d) V. K. Gupta, L. P. Singh, R. Sμingh, N. Upadhyay, S. P. Kaur and B. Sethi. 2012. A novel copper (II) selective sensor based on dimethyl 4, 4′(o-phenylene) bis (3-thioallophanate) in PVC matrix. *J. Mol. Liq.* 174: 11–16. (e) R. Jain, V. K. Gupta, N. Jadon and K. Radhapyari. Voltammetric determination of

cefixime in pharmaceuticals and biological fluids. *Anal. Biochem.* 407: 79–88. (f) V. K. Gupta, B. Sethi, R. A. Sharma, S. Agarwal and A. Bharti. 2013. Mercury selective potentiometric sensor based on low rim functionalized thiacalix [4]-arene as a cationic receptor. *J. Mol. Liq.* 177: 114–118.

9. H. N. Kim, W. X. Ren, J. S. Kim and J. Yoon. 2012. Fluorescent and colorimetric sensors for detection of lead, cadmium, and mercury ions. *Chem. Soc. Rev.* 2012, 41: 3210–3244.

10. K. J. Wallace. 2009. Molecular dyes used for the detection of biological and environmental heavy metals: Highlights from 2004 to 2008. *Supramol. Chem.* 21: 89–102.

11. L. Mart, H. W. Nürnberg and P. Valenta. 1980. Prevention of contamination and other accuracy risks in voltammetric trace metal analysis of natural waters. *Fresenius' Z. Anal. Chem.* 300: 350–362.

12. S. T. Mensah, Y. Gonzalez, P. Calvo-Marzal and K. Y. Chumbimuni-Torres. 2014. Nanomolar detection limits of Cd^{2+}, Ag^+, and K^+ using paper-strip ion-selective electrodes. *Anal. Chem.* 86: 7269–7273.

13. (a) L. Dostál, W. M. Kohler, J. E. Penner-Hahn, R. A. Miller and C. A. Fierke. 2015. Fibroblasts from long-lived rodent species exclude cadmium. *J. Gerontol. A Biol. Sci. Med. Sci.* 70: 10–19. (b) N. Zhang and B. Hu. 2012. Cadmium (II) imprinted 3-mercaptopropyltrimethoxysilane coated stir bar for selective extraction of trace cadmium from environmental water samples followed by inductively coupled plasma mass spectrometry detection. *Anal. Chim. Acta* 723: 54–60.

14. (a) G. Kaya and M. Yaman. 2008. Online preconcentration for the determination of lead, cadmium and copper by slotted tube atom trap (STAT)-flame atomic absorption spectrometry. *Talanta* 75: 1127–1133. (b) A. C. Davis, C. P. Calloway and B. T. Jones. 2007. Direct determination of cadmium in urine by tungsten-coil inductively coupled plasma atomic emission spectrometry using palladium as a permanent modifier. *Talanta* 71: 1144–1149.

15. Q. Zou, J. Jin, B. Xu, L. Ding and H. Tian. 2011. New photochromic chemosensors for Hg^{2+} and F^-. *Tetrahedron* 67: 915–921.

16. R. W. Ramette and E. B. Sandell. 1956. Rhodamine B equilibriums. *J. Am. Chem. Soc.* 78: 4872–4878.

17. S. J. Ranee, G. Sivaraman, A. M. Pushpalatha and S. Muthusubramanian. 2018. Quinoline based sensors for bivalent copper ions in living cells. *Sens. Actuators B Chem.* 255: 630–637.

18. C. Zhou, N. Xiao and Y. Li. 2014. Simple quinoline-based turn-on fluorescent sensor for imaging copper (II) in living cells. *Can. J. Chem.* 92: 1092–1097.

19. A. Sikdar, S. S. Panja, P. Biswas and S. Roy. 2012. A rhodamine-based dual chemosensor for Cu(II) and Fe(III). *J. Fluoresc.* 22: 443–450.

20. Y. J. Gong, X. B. Zhang, C. C. Zhang, A.-L. Luo, T. Fu, W. Tan, G.-L. Shen and R.-Q. Yu. 2012. Through bond energy transfer: A convenient and universal strategy toward efficient ratiometric fluorescent probe for bioimaging applications. *Anal. Chem.* 84: 10777–10784.

21. Y. Ge, X. Zheng, R. Ji, S. Shen and X. Cao. 2017. A new pyrido[1,2-a]benzimidazole-rhodamine FRET system as an efficient ratiometric fluorescent probe for Cu^{2+} in living cells. *Anal. Chim. Acta*, 965: 103–110.

22. L. Xu, S. Wei, Q. Diao, P. Ma, X. Liu, Y. Sun, D. Song and X. Wang. 2017. Sensitive and selective rhodamine-derived probes for fluorometric sensing of pH and colorimetric sensing of Cu^{2+}. *Sens. Actuators B Chem.* 246: 395–401.

23. W. N. Wu, P. D. Mao, Y. Wang, X. J. Mao, Z. Q. Xu, Z. H Xu, X. L. Zhao, Y. C. Fan and X. F. Hou. 2018. AEE active Schiff base-bearing pyrene unit and further Cu^{2+}-induced self-assembly process. *Sens. Actuators B Chem.* 258: 393–401.

24. S. Sun, W. Hu, H. Gao, H. Qi and L. Ding. 2017. Luminescence of ferrocene-modified pyrene derivatives for turn-on sensing of Cu^{2+} and anions. *Spectrochim Acta A* 184: 30–37.
25. N. Mergu, M. Kim and Y. A. Son. 2018. A coumarin-derived Cu^{2+}-fluorescent chemosensor and its direct application in aqueous media. *Spectrochim Acta A*, 188: 571–580.
26. H. S. Jung, M. Park, J. H. Han, J. H. Lee, C. Kang, J. H. Jung and J. S. Kim. 2012. Selective removal and quantification of Cu(II) using fluorescent iminocoumarin-functionalized magnetic nanosilica. *Chem. Commun.* 48: 5082–5084.
27. K. Mariappan, M. Alaparthi, G. Caple, V. Balasubramanian, M. M. Hoffman, M. Hudspeth and A. G. Sykes. 2014. Selective fluorescence sensing of copper(II) and water via competing imine hydrolysis and alcohol oxidationpathways sensitive to water content in aqueous acetonitrile mixtures. *Inorg. Chem.* 53: 2953–2962.
28. Y. H. Lee, N. Park, Y. B. Park, Y. J. Hwang, C. Kang and J. S. Kim. 2014. Organelle-selective fluorescent Cu2+ ion probes: Revealing the endoplasmic reticulum as a reservoir for Cu-overloading, *Chem. Commun.* 50: 3197–3200.
29. B. Gu, L. Huang, W. Su, X. Duan, H. Li and S. Yao. 2017. A benzothiazole-based fluorescent probe for distinguishing and bioimaging of Hg^{2+} and Cu^{2+}. *Anal. Chim. Acta* 954: 97–104.
30. H. Ryu, M. G. Choi, E. J. Cho and S. K. Chang. 2018. Cu^{2+}-selective fluorescent probe basedon the hydrolysis of semicarbazide derivative of 2-(2-aminophenyl) benzothiazole. *Dyes Pigm.* 149: 620–625.
31. Y. Xiang, A. Tong, P. Jin and Y. Ju. 2006. New fluorescent rhodamine hydrazone chemosensor for Cu(II) with high selectivity and sensitivity. *Org. Lett.* 8 (13): 2863–2866.
32. L. Guo, T. Tang, L. Hu, M. Yang and X. Chen. 2017. Fluorescence assay of Fe (III) in human serum samples based on pH dependent silver nanoclusters. *Sens. Actuators B Chem.* 241: 773–778.
33. Y. M. Zhou, J. L. Zhang, H. Zhou, Q. Y. Zhang, T. S. Ma and J. Y. Niu, J. Lumin. 2012. A new rhodamine B based "off-on" fluorescent chemosensor for Cu^{2+} in aqueous media. *J. Lumin.* 132: 1837–1841.
34. F. Ge, H. Ye, J. Z. Luo, S. Wang, Y. J. Sun, B. X. Zhao, et al. 2013. A new fluorescent and colorimetric chemosensor for Cu(II) based on rhodamine hydrazone and ferrocene unit. *Sens. Actuators B Chem.* 181: 215–220.
35. H. Sasaki, K. Hanaoka, Y. Urano, T. Terai and T. Nagano. 2011. Design and synthesis of a novel fluorescence probe for Zn^{2+} based on the spirolactam ring-opening process of rhodamine derivatives. *Bioorg. Med. Chem.* 19: 1072–1078.
36. H. Liu, Y. Dong, B. Zhang, F. Liu, C. Tan, Y. Tan and Y. Jiang. 2016. An efficient quinoline-based fluorescence sensor for zinc(II) and its application in live-cell imaging. *Sens. Actuators B Chem.* 234: 616–624.
37. P. D. Beer and P. A. Gale. 2001. Anion recognition and sensing: The state of the art and future perspectives. *Angew. Chem. Int. Ed.* 40: 486–516.
38. M. Cametti and K. Rissanen. 2009. Recognition and sensing of fluoride anion. *Chem Commun.* 20: 2809–2829.
39. R. M. Duke, E. B. Veale, F. M. Pfeffer, P. E. Kruger and T. Gunnlaugsson. 2010. Colorimetric and fluorescent anion sensors: An overview of recent developments in the use of 1,8-naphthalimide-based chemosensors. *Chem. Soc. Rev.* 39: 3936–3953.
40. Z. Xu, J. Yoon and D. R. Spring. 2010. Fluorescent chemosensors for Zn2+. *Chem. Soc. Rev.* 39: 1996–2006.
41. N. H. Evans and P. D. Beer. 2014. Advances in anion supramolecular chemistry: From recognition to chemical applications. *Angew. Chem. Int. Ed.* 53: 11716–11754.
42. A. Kaur, J. L. Kolanowski and E. J. New. 2016. Reversible fluorescent probes for biological redox states. *Angew. Chem. Int. Ed.* 55: 1602–1613.

43. Y. H. Lau, P. J. Rutledge, M. Watkinson and M. H. Todd. 2011. Chemical sensors that incorporate click-derived triazoles. *Chem. Soc. Rev.* 40: 2848–2866.
44. X. Li, X. Gao, W. Shi and H. Ma. 2014. Design strategies for water soluble small molecular chromogenic and fluorogenic probes. *Chem. Rev.* 114: 590–659.
45. P. Alreja and N. Kaur. 2016. Recent advances in 1,10-phenanthroline ligands for chemosensing of cations and anions. *RSC. Adv.* 6: 23169–23217.
46. Y. Ding, Y. Tang, W. Zhu and Y. Xie. 2015. Fluorescent and colorimetric ion probes based on conjugated oligopyrroles. *Chem. Soc. Rev.* 44: 1101–1112.
47. J. F. Callan, A. P. de Silva and D. C. Magri. 2005. Luminescent sensors and switches in the early 21st century. *Tetrahedron* 61: 8551–8588.
48. M. E. Jun, B. Roy and K. H. Ahn. 2011. "Turn-on" fluorescent sensing with "reactive" probes. *Chem. Commun.* 47: 7583–7601.
49. J. Wu, W. Liu, J. Ge, H. Zhang and P. Wang. 2011. New sensing mechanisms for design of fluorescent chemosensors emerging in recent years. *Chem. Soc. Rev.* 40: 3483–3495.
50. R. A. Bissell, A. P. de Silva, H. Q. N. Gunaratne, P. L. M. Lynch, G. E. M. Maguire and K. R. A. S. Sandanayake. 1992. Molecular fluorescent signaling with 'fluor-spacer-receptor' systems: Approaches to sensing and switching devices via supramolecular photophysics. *Chem. Soc. Rev.* 21: 187–195.
51. A. P. de Silva, H. Q. N. Gunaratne, T. Gunnlaugsson, A. J. M. Huxley, C. P. McCoy, J. T. Rademacher and T. E. Rice. 1997. Signaling recognition events with fluorescent sensors and switches. *Chem. Rev.* 97: 1515–1566.
52. M. Formica, V. Fusi, L. Giorgi and M. Micheloni. 2012. New fluorescent chemosensors for metal ions in solution. *Coord. Chem. Rev.* 256: 170–192.
53. S. Ding, A. Xu, A. Sun, Y. Xia and Y. Liu. 2020. An excited state intramolecular proton transfer-based fluorescent probe with a large stokes shift for the turn-on detection of cysteine: A detailed theoretical exploration. *ACS Omega* 31: 19695–19701.
54. B. Valeur and I. Leray. 2000. Design principles of fluorescent molecular sensors for cation recognition. *Coord. Chem. Rev.* 205: 3–40.
55. S. Li, N. Shi, M. Zhang, Z. Chen, D. Xia, Q. Zheng, G. Feng and Z. Song. 2022. A novel benzotriazole derivate with twisted intramolecular charge transfer and Aggregation Induced emission features for proton determination. *Spectrochim. Acta Part A* 269: 120780.
56. S. Lee, M. Jen, T. Jang, G. Lee and Y. Pang. 2022. Twisted intramolecular charge transfer of nitroaromatic push-pull chromophores. *Sci. Rep.* 12: 6557.
57. A. Karak, S. K. Manna and A. K. Mahapatra. 2022. Triphenylamine-based small-molecule fluorescent probes. *Anal. Methods* 14: 972–1005.
58. L. Yuan, W. Lin, K. Zheng and S. Zhu. 2013. FRET-based small-molecule fluorescent probes: Rational design and bioimaging applications. *Acc. Chem. Res.* 46: 1462–1473.
59. K. E. Sapsford, L. Berti and I. L. Medintz. 2006. Materials for fluorescence resonance energy transfer analysis: Beyond traditional donor-acceptor combinations. *Angew. Chem. Int. Ed. Engl.* 45: 4562–4589.
60. G. Li, F. Tao, H. Wang, L. Wang, J. Zhang, P. Ge, L. Liu, Y. Tong and S. Sun. 2015. A novel reversible colorimetric chemosensor for the detection of Cu2+ based on water-soluble polymer containing rhodamine receptor pendants. *RSC Adv.* 5: 18983–18989.
61. G. Li, L. Bai, F. Tao, A. Deng and L. Wang. 2018. A dual chemosensor for Cu2+ and Hg2+ based on a rhodamine-terminated water-soluble polymer in 100% aqueous solution. *Analyst* 143: 5395–5403.
62. S. Pu, H. Ding, G. Liu, C. Zheng and H. Xu. 2014. Multiaddressing fluorescence switch based on a new photochromic diarylethene with a triazole-linked rhodamine B unit. *J. Phys. Chem. C* 118: 7010–7017.
63. H. Kang, C. Fan, H. Xu, G. Liu and S. Pu. 2018. A highly selective fluorescence switch for Cu2+ and Fe3+ based on a new diarylethene with a triazole-linked rhodamine 6G unit. *Tetrahedron* 74: 4390–4399.

64. Z. Q. Xu, X. J. Mao, Y. Wang, W. N. Wu, P. D. Mao, X. L. Zhao, Y. C. Fan and H. J. Li. 2017. Rhodamine 6G hydrazone with coumarin unit: A novel single-molecule multianalyte (Cu2+ and Hg2+) sensor at different pH value. *RSC Adv.* 7: 42312–42319.

65. A. Sikdar, S. Roy, S. Dasgupta, S. Mukherjee and S. S. Panja. 2018. Logic gate-based Rhodamine-methionine conjugate highly sensitive fluorescent probe for Hg2+ ion and its application: An experimental and theoretical. *Sens. Actuators B Chem.* 263: 298–311.

66. Q. Diao, H. Guo, Z. Yang, W. Luo, T. Li and D. Hou. 2019. A Rhodamine-6G-based "turn-on" fluorescent probe for selective detection Fe3+ in living cells. *Anal. Methods* 11: 794–799.

67. M. Yang, L. Ma, J. Li and L. Kang. 2019. Fluorescent probe for Cu2+ and the secondary application of the resultant complex to detect cysteine. *RSC Adv.* 9: 16812–16818.

68. H. Zhao, H. Ding, H. Kang, C. Fan, G. Liu and S. Pu. 2019. A solvent-dependent chemosensor for fluorimetric detection of Hg2+ and colorimetric detection of Cu2+ based on a new diarylethene with a rhodamine B unit. *RSC Adv.* 9: 42155–42162.

69. A. A. S. Paulino, L. Giroldo, N. A. Pradie, J. S. Reis, D. F. Back, A. A. C. Braga, H. A. Stefani, C. Lodeiro and A. A. D. Santos. 2020. Nanomolar Detection of Palladium (II) through a Novel Seleno-Rhodamine-based fluorescent and colorimetric chemosensor. *Dyes Pigm.* 179: 108355.

70. X. Wang and T. Li. 2020. A novel "off-on" rhodamine-based colorimetric and fluorescent chemosensor based on hydrolysis driven by aqueous medium for the detection of Fe3+. *Spectrochim. Acta Part A* 229: 117951.

71. A. S. Murugan, N. Vidhyalakshmi, U. Ramesh and J. Annaraj. 2018. *In vivo* bioimaging studies of highly selective, sensitive rhodamine based fluorescent chemosensor for the detection of Cu2+/Fe3+ ions. *Sens. Actuators B Chem.* 274: 22–29.

72. F. Song, C. Yang, H. Liu, Z. Gao, J. Zhu, X. Bao and C. Kan. 2019. Dual-binding pyridine and rhodamine B conjugate derivatives as fluorescent chemosensors for ferric ions in aqueous media and living cells. *Analyst* 144: 3094–3102.

73. D. Das, R. Alam, A. Katarkar and M. Ali. 2019. A differentially selective probe for trivalent chemosensor upon single excitation with cell imaging application: Potential applications in combinatorial logic circuit and memory devices, *Photochem. Photobiol. Sci.* 18: 242–252.

74. J. Zhang, M. Zhu, D. Jiang, H. zhang, L. Li, G. Zhang, Y. Wang, C. Feng and H. Zhao. 2019. FRET-based colorimetric and ratiometric fluorescent probe for Cu2+ with a new trimethylindolin fluorophore. *New J. Chem.* 43: 10176–10182.

75. J. B. Grim and L. D. Lavis. 2011. Synthesis of rhodamine from fluoresceins using Pd catalysed C-N Cross Coupling. *Org. Lett.* 13: 6354–6357.

76. M. Adamczyk and J. Grote. 2001. Efficient fluorescein spirolactam and bisspirolactam synthesis. *Synth. Commun.* 31: 2681–2690.

77. E. Oliveira, E. Bertolo, C. Nunez, V. Pilla, H. M. Santos, J. Fernandez-Lodeiro, A. Fernandez Lodeiro, J. Djafari, J. L. Capelo and C. Lodeiro. 2017. Green and Red fluorescent dyes for translational applications in imaging and sensing analytes: A dual-color flag. *Chem. Open* 7: 9–52.

78. M. Rajasekar. 2021. Recent development in fluorescein derivatives. *J. Mol. Struct.* 1224: 129085.

79. W. A. Banks, A. J. Kastin and D. A. Durham. 1989. Bidirectional transport of interleukin alpha across the blood-brain barrier. *Brain Res. Bull.* 23: 433–437.

80. M. Hasan, A. S. M. Islam, C. Prodhan and M. Ali. 2019. A fluorescein-based chemosensor for "turn-on" detection of Hg2+ and the resultant complex as a fluorescent sensor for S2– in semi-aqueous medium with cell-imaging application: Experimental and computational studies. *New J. Chem.* 43, 5297–5307.

81. S. Erdemir and O. Kocyigit. 2017. A novel dye based on phenolphthalein-fluorescein as a fluorescent probe for the dual-channel detection of Hg2+ and Zn2+. *Dyes Pigm.* 145: 72–79.

82. B. Das, A. Jana, A. D. Mahapatra, D. Chattopadhyay, A. Dhara, S. Mabhai and S. Dey. 2019. Fluorescein derived Schiff base as fluorimetric zinc (II) sensor via 'turn on' response and its application in live cell imaging. *Spectrochim. Acta Part A* 212: 222–231.

83. B. Vidya, G. Sivaraman, R. V. Sumesh and D Chellappa. 2016. Fluorescein-based "Turn On" fluorescence detection of Zn2+ and its applications in imaging of Zn2+ in apoptotic cells. *Chem. Sel.* 1: 4024–4029.

84. (a) K. Ajay Kumar, N. Renuka, G. Pavithra and G. Vasanth Kumar. 2015. Comprehensive review on coumarins: Molecules of potential chemical and pharmacological interest. *J. Chem. Pharm. Res.* 7(9): 67–81. (b) Y. Al-Majedy, A. Al-Amiery, A. A. Kadhum and A. BakarMohamad. 2017. Antioxidant activity of coumarins. *Sys. Rev. Pharm.* 8(1): 24–30.

85. Y. Song, Z. Chen and H. Li. 2012. Advances in coumarin-derived fluorescent chemosensors for metal ions. *Curr. Org.Chem.* 16: 2690–2707.

86. A. K. K. Bhasin, P. Raj, P. Chauhan, S. K. Mandal, S. Chaudhary, N. Singh and N. Kaur. 2020. Design and synthesis of a novel coumarin-based framework as a potential chemomarker of a neurotoxic insecticide, azamethiphos. *New J. Chem.* 44: 3341–3349.

87. A. K. K. Bhasin, P. Chauhan and S. Chaudhary. 2021. A novel coumarin-tagged ditopic scaffold as a selectively sensitive fluorogenic receptor of zinc (II) ion. *Sens. Actuators B Chem.* 330: 129328.

88. C.-G. Chen, N. Vijay, N. Thirumalaivasan, S. Velmathi and S. P. Wu. 2019. Coumarinbased Hg2+ fluorescent probe: Fluorescence turn-on detection for Hg2+ bioimaging in living cells and zebrafish. *Spectrochim. Acta Part A: Mol. Biomol. Spectrosc.*, 219: 135–140.

89. K. S. Mani, R. Rajamanikandan, B. Murugesapandian, R. Shankar, G. Sivaraman, M. Ilanchelian and S. P. Rajendran. 2019. Coumarin based hydrazone as an ICT-based fluorescence chemosensor for the detection of Cu2+ ions and the application in HeLa cells. *Spectrochim. Acta Part A: Mol. Biomol. Spectrosc.* 214: 170–176.

90. S. A. Hilderbrand and R. Weissleder. 2010. Near-infrared fluorescence: Application to in vivo molecular imaging. *Curr. Opin. Chem. Biol.* 14: 71–79.

91. J. V. Frangioni. 2003. In vivo near-infrared fluorescence imaging. *Curr. Opin. Chem. Biol.* 7: 626–634.

92. L. Yuan, W. Lin, K. Zheng, L. He and W. Huang. 2013. Far-red to near infrared analyte-responsive fluorescent probes based on organic fluorophore platforms for fluorescence imaging. *Chem. Soc. Rev.* 42: 622–661.

93. Z. Guo, S. Park, J. Yoon and I. Shin. 2014. Recent progress in the development of near-infrared fluorescent probes for bioimaging applications. *Chem. Soc. Rev.* 43: 16–29.

94. W. Sun, S. Guo, C. Hu, J. Fan and X. Peng. 2016. Recent development of chemosensors based on cyanine platforms. *Chem. Rev.* 116: 7768–7817.

95. J. Yin, Y. Kwon, D. Kim, D. Lee, G. Kim, Y. Hu, J. H. Ryu and J. Yoon. 2014. Cyanine-based fluorescent probe for highly selective detection of glutathione in cell cultures and live mouse tissues. *J. Am. Chem. Soc.* 136: 5351–5358.

96. X. Zhou, Y. Kwon, G. Kim, J. H. Ryu and J. Yoon. 2015. A ratiometric fluorescent probe based on a coumarin-hemicyanine scaffold for sensitive and selective detection of endogenous peroxynitrite. *Biosens. Bioelectron.* 64: 285–291.

97. G. Li, Y. Guan, F. Ye, S. H. Liu and J. Yin. 2020. Cyanine-based fluorescent indicator for mercury ion and bioimaging application in living cells. *Spectrochim. Acta Part A: Mol. Biomol. Spectrosc.* 239: 118465.

98. X. Cao, W. Lin and W Wan. 2012. Development of a near-infrared fluorescent probe for imaging of endogenous Cu+ in live cells. *Chem. Commun.* 48: 6247–6249.

99. Z. Han, Q. Yang, L. Liang and X. Zhang. 2013. A new heptamethine cyaninebased near-infrared fluorescent probe for divalent copper ions with high selectivity. *Adv. Mater. Phys. Chem.* 3: 314–319.

100. B. Tang, H. Huang, K. Xu, L. Tang, G. Yang, X. Liu and L. An. 2006. Highly sensitive and selective nearinfrared fluorescent probe for zinc and its application to macrophage cells. *Chem. Commun.* 34: 3609–3611.

101. M. Zhu, M. Yuan, X. Liu, J. Xu, J. Lv, C. Huang, H. Liu, Y. Li, S. Wang and D. Zhu. 2008. Visible near-infrared chemosensor for mercury ion. *Org. Lett.* 10: 1481–1484.

102. D. Akram, I. A. Elhaty and S. S. AlNeyadi. 2020. Synthesis and antibacterial activity of rhodanine based azo dyes and their use as spectrophotometric chemosensor for Fe3+ ions. *Chemosensors* 8(1): 16.

103. A. Sagırlı and E. Bozkurt. 2020. Rhodamine-based arylpropenone azo dyes as dual chemosensor for Cu2+/Fe3+ detection. *J. Photochem. Photobiol. A* 403: 112836.

104. G. Bartwal, K. Aggarwal and J. M. Khurana. 2020. Quinoline-ampyrone functionalized azo dyes as colorimetric and fluorescent enhancement probes for selective aluminium and cobalt ion detection in semi-aqueous media. *J. Photochem. Photobiol. A* 394: 112492.

105. N. Yadav, R. P. Gaikwad, V. Mishra and M. B. Gawande. 2022. Synthesis and photocatalytic applications of functionalized carbon quantum dots. *Bull. Chem. Soc. Jpn.* 95(11): 1638–1679.

106. N. Yadav, D. Mudgal and V. Mishra. 2023. In-situ synthesis of ionic liquid-based-carbon quantum dots as fluorescence probe for hemoglobin detection. *Anal. Chim. Acta* 7: 341502.

3 Alkali and Alkaline Earth Metal-Based Fluorescent Probes

Pragati Kushwaha and Neeraj Mohan Gupta

3.1 INTRODUCTION

All the alkali and alkaline earth metals are of great importance to both animal life and plant life, and these elements play a very important role in the existence and maintenance of the life cycle. Alkali metals include lithium (Li$^+$), sodium (Na$^+$), potassium (K$^+$), rubidium (Rb$^+$), cesium (Cs$^+$), and francium (Fr$^+$). Among these metals, rubidium can ameliorate metabolism and supersede potassium in biosystems having potassium deficiencies. Cesium is mildly toxic; it competes with potassium resulting in potassium deficiency. Francium having radioactivity displays no significant biological role. Moreover, the remaining metal ions lithium (Li$^+$), sodium (Na$^+$), and potassium (K$^+$) play a very important role in biological systems. Lithium is essential to humans though its high concentration has a toxic effect. In soil, it naturally occurs and hence it is included in the food chain. Speaking of its biological significance, lithium compounds are found in many medications used to treat depression and anxiety disorders and have an antioxidant effect. Likewise, sodium is crucial for maintaining the proper balance of water in the human body. The sodium ions are also involved in both muscular contraction and relaxation in addition to the transmission of nerve impulses. Potassium is also very crucial as it balances osmotic pressure between cells and interstitial fluid; it is used to treat many diseases like hypokalaemia, weakness of muscles, cardiac disorders, etc.

Alkaline earth metals include beryllium (Be), magnesium (Mg), calcium (Ca), strontium (Sr), barium (Ba), and radium (Ra). Beryllium is used to investigate blood samples detecting HIV and other diseases. Magnesium supports cell growth and the protein synthesis. It is also essential for blood pressure and muscular action. Calcium is crucial for bone strengthening; additionally, it creates cell walls and aids in blood clotting in case of injury. Strontium encourages calcium absorption and aids in the repair of cracked or damaged bones; it is mostly utilized in bone restoration. Because of the toxicity of barium and radioactivity of radium, these two are not biologically important as compared to other metals. The importance of alkali and alkali earth metal cations in biological systems is because variations in their concentration have a direct impact on the healthy body and physiological processes. Given the physiological relevance of these metal cations, they are among the most sought-after targets for

DOI: 10.1201/9781003352372-3

the development of new sensitive technologies and methods to keep track of changes in these metal ion concentrations. Significant progress has been made in the last 10 years in the development of sensors, switches, and molecular devices for the detection of these analytes.

Among various kinds of metal detection techniques, one of the most adaptable techniques is fluorescence imaging. This technique has distinctive advantages such as simplicity, high sensitivity, great selectivity, low cost, and non-invasive visualization features.[1-3] Having these properties, the fluorescence sensors are extremely significant in biological systems, since metal ions that play important roles are mostly present in μM quantities.[4] Fluorescence probes enable to convenient relay of the transmission of binding of an analyte to the receptor to the macroscopic world through the emission and quenching of fluorescence.[5] During the previous decade, various studies have been performed to search these fluorescent probes for biologically important alkali and alkali earth metal cations. In the current chapter, we will discuss current developments in fluorescence probe design and bioimaging applications for alkali metal and alkaline earth metal cations. In general, the fluorescent sensors for metal ions contain two essential features: a fluorescent carrier capable of absorbing and emitting light and a metal chelating or binding moiety. These two components may be either separate species connected through a spacer or connected by covalent bonds (Figure 3.1).

There are various types of fluorescent probes such as rhodamine, BODIPY (difluoro-boron-dipyrromethene), RI (rylenecarboximide), and Cy (cyanine). When a metal binds to a probe, it changes the electrical and/or molecular structure which correspondingly changes the fluorescence characteristics of the fluorophore. Changes in the molecular structure can vary the distance or orientation between a pair of fluorophores that act as a donor–acceptor pair, whereas changes in the electronic structure can change the intensity or wavelength of light that is absorbed or emitted. A fluorescence microscope enables the spatially detailed visualization of changes in fluorescence and, consequently, the target of a particular metal ion of interest. The change in the above characteristics indicates the presence of a particular metal binding.

There are two main categories in which fluorescent probes can be divided, i.e., reaction-based and recognition-based probes. Recognition-based probes react to the reversible coordination of a metal to the receptor, while reaction-based probes attached to the metal start a chemical reaction that leads to a change in the fluorescence. These indicators can be categorized according to various criteria or traits:

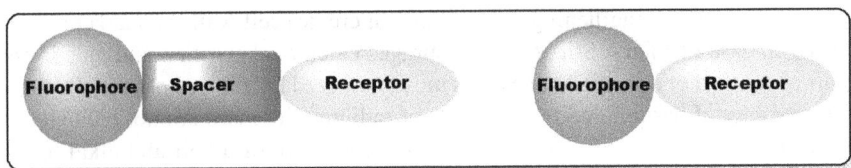

FIGURE 3.1 Components of fluorescent probes.

1. **Based on the responding mechanism:**
   - Photo-induced electron transfer (PET) probes.
   - Fluorescence resonance energy transfer probes.
   - Aggregation-induced emission (AIE) probes.
   - Intramolecular charge transfer (ICT) probes.
   - Twisted intramolecular charge transfer probes.
   - Bond energy transfer probes.
2. **Based on the optical performance:**
   - "Off–on" probes.
   - "On–off" probes.
   - Ratiometric probes.

The majority of fluorescent probes reported are fluorescent "off–on" probes and ratiometric probes. Overall, in this chapter, we present an overview of fluorescent sensors for imaging the distributions of alkali and alkaline earth metals in biological systems. Further, this chapter does not give an exhaustive list of sensors but covers recent developments in the detection and imaging of alkali and alkaline earth metal ions, such as Na^+, K^+, Mg^{2+}, and Ca^{2+}, in biological systems and offers an opinion on remaining challenges.

3.2 SENSORS FOR Li+

Li^+ is one of the most prominent alkali cations having biological significance. Medications containing lithium are frequently utilized in medical and therapeutic applications for the treatment of bipolar disorder and manic-depressive psychosis.[6] Patients are often required to take the medication for several months or even years. After taking the medication, the quantity of lithium ions in blood serum varies significantly from person to person and needs to be frequently checked in each patient. Thus, for effective and secure therapeutic uses, the accurate assessment of the lithium-ion concentration levels in blood samples is crucial since extremely low amounts have no impact at all and a lithium overdose can have harmful effects that are potentially fatal.

Obare et al. reported that increasing the aromatic nature of 1,10-phenanthroline through dipyrido[3,2-a:2'3'-c]-phenazine to benzodipyrido[3,2-a:2'3'-c]phenazine coupled with butyl substituents **1** yields a fluorescent sensor molecule that visibly changes the colour upon binding with Li^+.[7] The emission properties of **1** were investigated for the effect of Li^+ with excitation at 410 nm. The increase in Li^+ concentration resulted in a quenching of the fluorescence intensity that was accompanied by a red shift. The emission lifetime of **1** in ethanol was found to be 26.3 ns. A minor decrease was observed in the lifetime of **1** which became 23.0 ns. Similar experiments were performed in THF and the increase in Li^+ concentration had the opposite effect of the ethanol result: the fluorescence increased in intensity and was further accompanied by spectral broadening and a slight red shift. The emission lifetime of **1** was found to be 5.5 ns in THF. After complexation with Li^+, the lifetime of **1** increased to 26 ns.

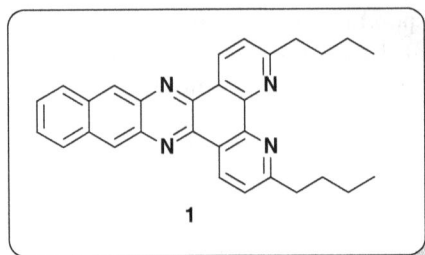

1

STRUCTURE 3.1

Citterio et al. reported an optical sensor approach for the determination of Li$^+$, a novel lithium fluoroionophore **2** and its polymer immobilizable derivative **3**.[8] The fluoroionophores displayed high selectivity for Li$^+$ with binding-induced blue shift in the fluorescence spectra. Further, they exhibited no response to major biological interfering cations such as K$^+$, Ca^{2+}, and Mg^{2+} and no response to pH between ranges of 3–10. A hydrophilic optode membrane with **2** immobilized also exhibited good selectivity for Li$^+$ and pH independence in the physiological range of 6–8.

Probes **2** and **3** are very good Li$^+$ fluorescent chemosensors that can be applied to quantitative measurements of lithium in clinical samples. However, at the lower therapeutic level of Li$^+$, it is necessary to consider the possible interference of Na$^+$.

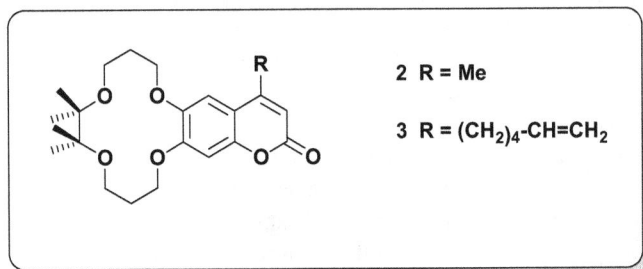

2 R = Me

3 R = $(CH_2)_4$-CH=CH$_2$

STRUCTURE 3.2

Rochat et al. developed very simple but powerful chemosensors **4–6** by taking account of self-assembly processes.[9] The group have reported a turn-on fluorescent sensor for lithium ion in the low millimolar concentration range. The sensor is suitable for use in purely aqueous solutions, and it exhibited a very high selectivity for lithium even in the presence of other alkali and alkaline earth metal ions. The fluorescent probe is based on 12-metallacrown-3 complex receptor unit, which is able to self-assemble in situ from simple building units. These macrocycles were synthesized by the reaction of organometallic half-sandwich complexes of Ru(II), Rh(III), or Ir(III) with organic ligands containing the 2,3-dihydroxypyridine motif.

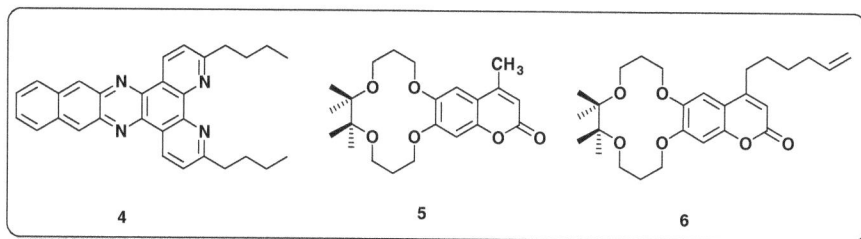

STRUCTURE 3.3

3.3 SENSORS FOR Na⁺

The most prevalent cation in extracellular fluid is sodium. It is essential for maintaining fluid and electrolyte balance as well as controlling blood pressure, pH levels, blood volume, and osmotic equilibrium. Absorption of sodium in the small intestine plays a critical role in the absorption of chloride, amino acids, glucose, and water. Therefore, the ability to identify Na⁺ ions is very important for both clinical biochemical applications and diagnosis. Numerous fluorescent probes were discovered in order to achieve Na⁺ detection and imaging in living cells. Tsien et al. created the first fluorescent probe for cytosolic sodium Na⁺ in 1989.[10] These probes had heteroaromatic fluorophores like quinolone and acridine attached to aza-crown ethers. Moreover, all of these probes had poor cell permeabilities. Gee et al. created two sodium-targeted aza-crowns with 1,7-diaza-4,10,13-trioxa-15-crown-5 skeletons by leaving out one fluorophore group. Through PET or ICT quenching processes, a group of probes containing aza-crowns demonstrated Na⁺ specificity and increased cell permeabilities with acetoxymethyl esters.[11]

Further, diaza-15-crown-5 undergoes drastic conformational change in the presence of Na⁺ ion which has been nicely used in designing the fluoroionophore **7**.[12] Binding of Na⁺ ion to the receptor in MeOH:H₂O (1:1, v/v) mixed solvent leads to a drastic conformational change to adopt a pseudo cryptand conformation. Consequently, N lone-pairs de-conjugate by the methoxy group from the rest of the system. Additionally, the PET process is suppressed by the binding of N lone-pairs to the metal ion. Fluorescence recovers as a result of both of these actions. However, this impact is not specific to Na⁺ and a comparable outcome is seen when Li⁺ ion is present.

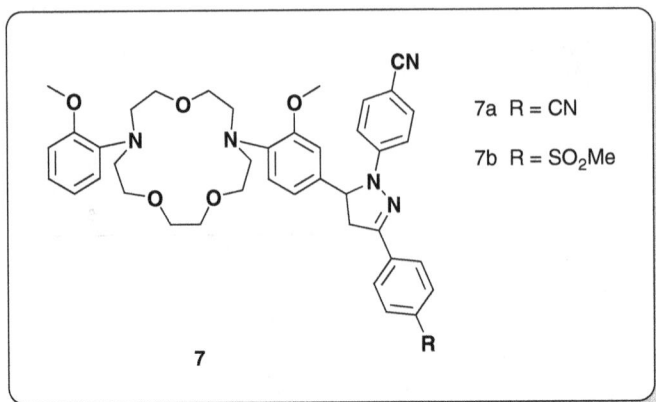

STRUCTURE 3.4

Poronik et al. developed a sensor **8** based on two-photon excited fluorescence (TPEF).[13] Some compounds can show significant two-photon fluorescence having a large two-photon absorption (TPA) cross-section. The ability to map an analyte in a biological system without interference from the natural fluorophores is great with such systems. The fluorescent probe **8** displayed high sensitivity and good selectivity for the Na^+ ion in acetonitrile. The TPA value in the biological window of 750–800 nm is significantly increased by complexation with Na^+ in the receptor. Additionally, this complex provides a strong TPEF signal.

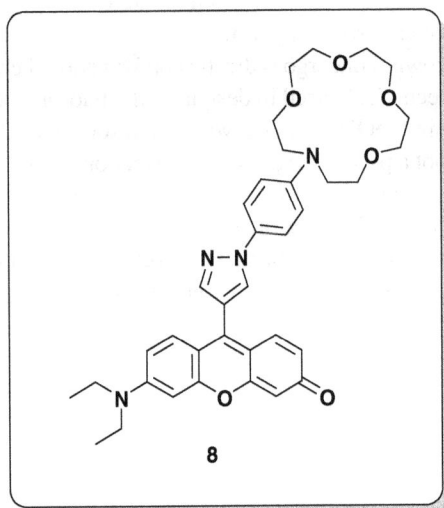

STRUCTURE 3.5

The well-known fluorophore biaryl borondipyrromethene (BODIPY) has a number of advantageous properties, including good photo- and chemical stability, high fluorescence quantum yields, the ability to excite with visible light, and narrow absorption and emission bands with strong peak intensities. Additionally, it is easily modifiable structurally, which enables spectral shifts in the absorption and emission bands to longer wavelengths.

Yamada et al. constructed a fluoroionophore by combining a BODIPY with an oligoethyleneglycol bridge acting as a binding site for metal cations.[14] Metal-free **9** has sharp absorption and emission bands with a maximum of 551 and 589 nm, respectively. Although there is not much selectivity for Na^+ over K^+, only the Na^+ complex exhibits a notable shift in absorbance to longer wavelengths. The two oxygen atoms at the benzyl locations are probably close to one another when they are complexed with the Na^+ ion, which results in a reduction in the dihedral angles between the aryl plane and the dipyrromethene plane. The orbital interaction is enhanced as a result, causing the red shift. However, unlike a PET sensor, the fluorescence quantum yield does not change considerably and the emission maxima shifts only slightly when Na^+ ions are added. However, a significant increase in emission was observed upon exciting at 575 nm, the highest absorption of the Na^+ complex.

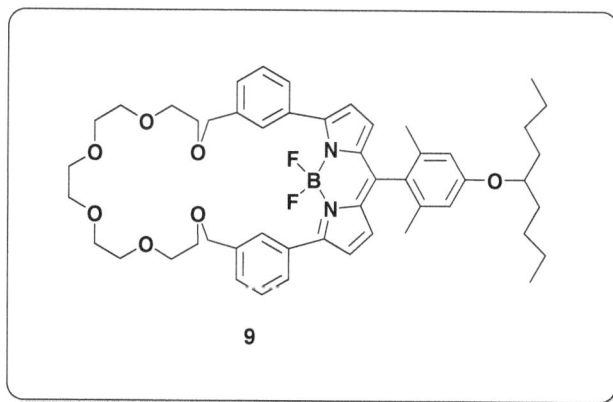

9

STRUCTURE 3.6

3.4 SENSORS FOR K^+

The potassium ion has a number of crucial functions in biological systems. Biomedical diagnosis requires the sensitive and selective detection of K^+ since variations in this ion's concentration in bodily fluids have been associated with hypertension, stroke, and seizures. K^+ ion has a variety of biological functions, hence it is problematic when it accumulates inside cells or extracellular fluid. However, due to its competition with Na^+, which is generally present in much larger concentrations, it is very challenging to detect K^+ ions. Therefore, a practical probe needs to prefer K^+ over Na^+ efficiently. Although, this kind of fluorescent probe with high selectivity for potassium over sodium is yet to be developed.

Lu et al. reported an AIE effect-based fluorescent probe for cellular K⁺ analysis and imaging.[15] Benefitting from the K⁺-induced AIE phenomenon, **the** designed TPE derivative modified guanine (G)-rich oligonucleotide fluorescent probe displayed high sensitivity with extended photo stability, which facilitated the prolonged fluorescence observations of K⁺ in living cells. With these advantages, the TPE-oligo nucleotide probe served as a promising candidate for the functional study and analysis of K⁺.

Further, phenylaza-18-crown-6 has been used to obtain probes with higher affinity to K⁺. Schwarze et al. discovered two molecular fluorescent probes **10** and **11** for the selective determination of physiologically relevant K⁺ levels in water on the basis of highly K⁺/Na⁺ building block, the *o*-(2-methoxyethoxy)phenylaza-18-crown-6 lariat ether unit.[16] In both fluorescent probes, 10 displayed K⁺-induced fluorescence enhancement (FE) by a factor of 7.7 of the anthracenic emission and a dissociation constant (K_d) value of 38 mM in water. Further, with probe **11**, a dual emission behaviour at 405 and 505 nm was observed. K⁺ increases the fluorescence intensity of **11** at 405 nm by a factor of ~4.6 and K⁺ decreases the fluorescence intensity at 505 nm by a factor of ~4.8. Thus, these probes are a promising fluorescent tool to measure ratio metrically and selectively physiologically relevant K⁺ levels.

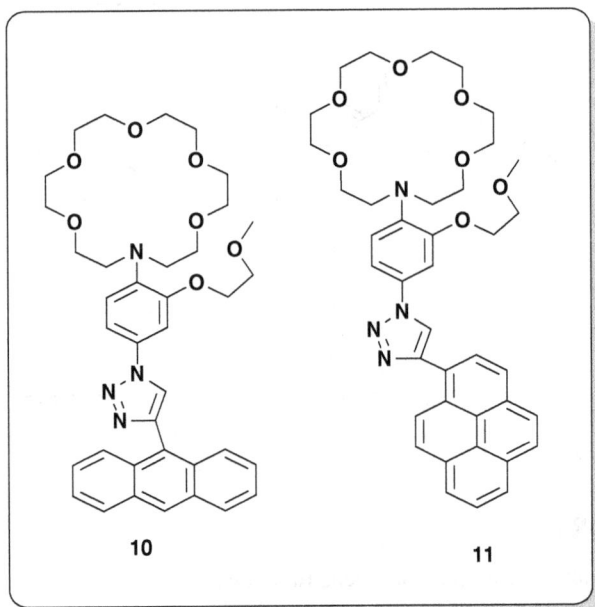

STRUCTURE 3.7

Xia et al. constructed a fluorescence signalling system by using a distyryl benzene derivative joined to a benzo-15-crown-5 at each end,[17] the metal-free fluorescent probe when excited in MeCN at 325 nm. It is easily converted from Z to E isomer, which is the major non-radiative deactivation pathway. After the addition of K+ ion, this probe forms intermolecular sandwich complexes. As a result, non-radiative deactivation is prevented, resulting in an emission increase of around 20 times.

On the basis of this principle, the same group reported the tris-crown chemosensor **12** as self-assembling induced fluorescence enhancement (*SAFE*) after coordination with potassium.[18]

This dendritic sensor has modest luminescence in acetonitrile solution, but displayed significant intensity when K+ ion is added. When the concentration of K+ ions reaches saturation, the sensor exhibits a linear fluorescence amplification with a sensor-to-metal ratio of 2:3. In the presence of a high excess of Na+ ions, it can detect minute amounts of K+; in addition, its performance is largely unaffected in the presence of a huge excess (about 60 times) of Na+ ions. However, its application is constrained by its insolubility in aqueous media.

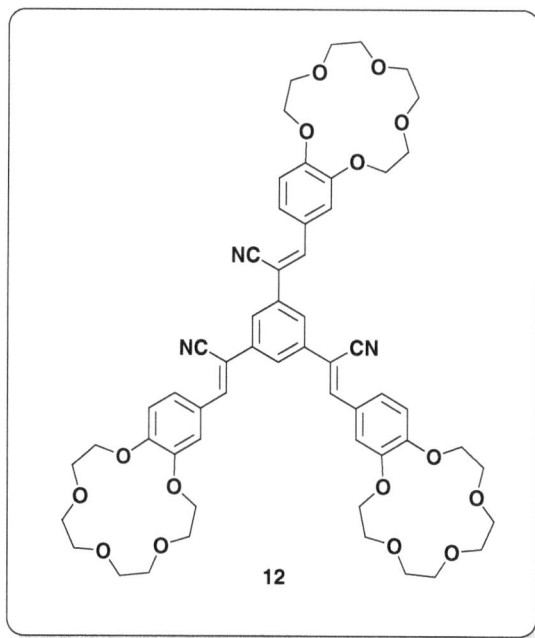

12

STRUCTURE 3.8

He et al. created a K+ ion optode **13**, utilizing a sensor that takes advantage of the inner filter effect (IFE) of fluorescence.[19] Fluorescence fluctuates due to the variable absorption of dye, which is controlled by the K+ concentration through the IFE. The sensor material responds to K+ reversibly over the concentration range of 1–10 µM with a suitable dynamic range from 1 µM to 1 mM.

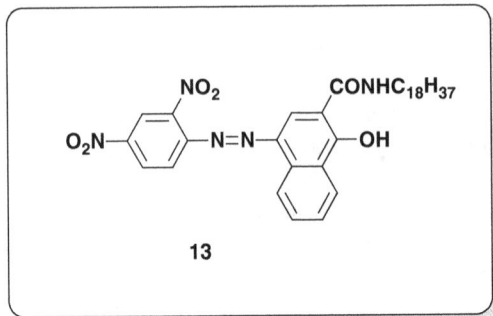

13

STRUCTURE 3.9

3.5 SENSORS FOR Mg²⁺

The fourth-most prevalent metal ion in cells is magnesium (per mole). More than 300 enzyme systems that control various metabolic activities use it as a cofactor. Magnesium is necessary for the oxidative phosphorylation, glycolysis, and generation of energy. It is necessary for the production of DNA, RNA, and the antioxidant glutathione and aids in the structural growth of bone. Magnesium also participates in the active transport of calcium and potassium ions across cell membranes, a procedure necessary for the conduction of nerve impulses, the contraction of muscles, and a regular heartbeat. However, understanding of the cellular homeostasis of Mg^{2+} molecular mechanisms remains difficult and requires molecular tools with high ion sensitivity. Since most fluorescent Mg^{2+} indicators have poor selectivity towards other divalent cations, particularly Ca^{2+}, they are unreliable markers of cellular Mg^{2+} concentrations in activities involving these metals. Lin et al. described a novel set of highly selective fluorescent indicators, using alkoxy styryl-functionalized BODIPY fluorophores embellished with a 4-oxo-4H-quinolizine-3-carboxylic acid metal binding moiety.[20] These novel sensors, **14** and **15**, exhibit absorption and emission maxima above 600 nm, with a 29-fold fluorescence increase and high quantum yields after coordination of Mg^{2+} in an aqueous buffer.

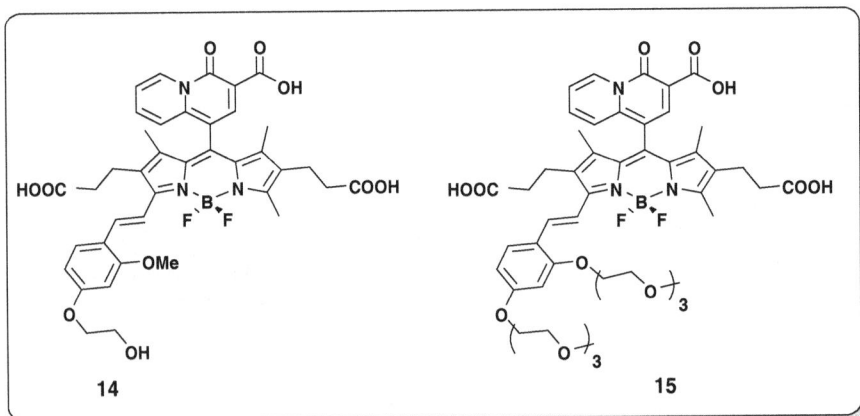

14   **15**

STRUCTURE 3.10

Further, Dong et al. reported two coumarin salen-based sensors **16** and **17**, in the presence of Na^+ as a synergic trigger. (A tetradentate C2-symmetric ligand known as salen is created by combining salicylaldehyde and ethylenediamine.[21] It may also refer to a group of substances, notably bis-Schiff bases, that are structurally linked to the traditional salen ligand.) These fluorescent probes demonstrated a potent fluorescence amplification response to Mg^{2+} up to 36-fold and 111-fold, respectively. More notably, the presence of Mg^{2+} and Na^+ together caused **16** to change from a dim yellow to a brilliant green fluorescent colour, which was visible to the human eye.

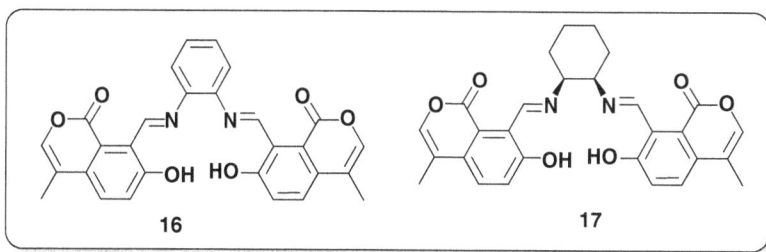

STRUCTURE 3.11

Kang et al. designed and synthesised sensor **18** for Mg^{2+} ion by covalently bonding bispicolylamine to coumarin for ratiometric FE technique.[22] When Mg^{2+} is added to MeCN solution of sensor **18**, the chemosensor's emission redshifts from 367 to 413 nm with increased intensity. Job's plot supports the establishment of a bis-complex. In addition, metal ions like Zn^{2+} and Ca^{2+} emit light at a somewhat longer wavelength while quenching the emission are paramagnetic transition metal ions. Alkali metal ions or heavy metal ions like Hg^{2+}, Ag^+, or Ag^+ do not exhibit any change in the emission band.

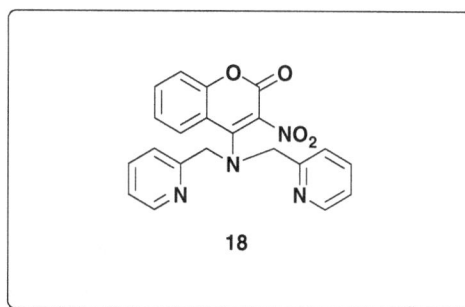

STRUCTURE 3.12

Liu et al. successfully reported a fluorescent indicator **19** with great selectivity to Mg^{2+} in an efficient way.[23] It is a great illustration of a fluorescence sensor that can differentiate Mg^{2+} from Ca^{2+}. The C=N isomerization was eliminated when the sensor connected with Mg^{2+}, which led to an increase in fluorescence intensity. It provides a 110-fold increase in fluorescence upon excitation at 395 nm in acetonitrile. Addition of Mg^{2+} ion only is favoured by the size of cavity and shape. Any other ion except Mg^{2+} did not bind at the site effectively and was unable to stop C=N isomerization.

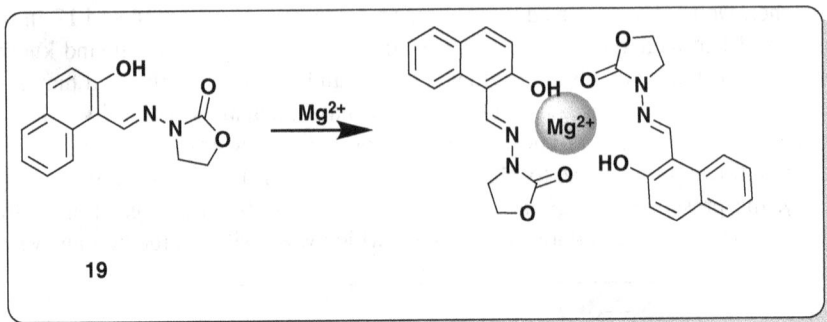

19

STRUCTURE 3.13

3.6   SENSORS FOR Ca²⁺

In biological systems, calcium ions serve a crucial role and exhibit the widest range of concentrations among all intracellular divalent cations. Fluctuations in calcium levels can be used to track a variety of biological activities, including excitability, neurotransmitter release, gene transcription, cell proliferation, synaptic plasticity, and hormone production. The concentration of calcium in living systems ranges from 1 to 7 mM, but 99.9% of the calcium in cells is found in organelles like the mitochondria and endoplasmic reticulum. As a result, there are only about 100 nM of free calcium. But different cellular processes can encourage a 10- to 100-fold release of free calcium from the intracellular organelles that store calcium. Understanding calcium transport, signal transduction, and illnesses associated with calcium imbalance might therefore benefit from monitoring dynamic variations in intracellular calcium concentration.

The very first calcium indicator for intracellular calcium was reported by Tsien in 1980, a variety of fluorophores have been combined with the 1,2-bis(2-aminophenoxy) ethane-N,N,N',N'-tetraacetic acid (BAPTA) binding unit to create fluorescent calcium indicators.[24] The concentration of extracellular Ca²⁺ is in the millimolar range. Because of their high ion binding, the fluorophores designed using BAPTA as the receptor are not suited for detecting this Ca²⁺. One of the aromatic rings has had electron-withdrawing groups like nitro or halogens added to it in order to reduce the binding force.[25,26] While they reduce the binding, fluorescence quenching and other issues start to show. Additionally, only half of the BAPTA binding unit, o-anisidine-N,N-diacetic acid, has been employed to weaken the binding. On the other hand, the ionophore disintegrates in aqueous media at pH 7.4 after storage. To solve these issues, He et al. created calcium fluoroionophore, **20** by covalently immobilizing a fresh fluoroionophore on an aminofunctionalized cellulose matrix enclosed in a hydrophilic polymer matrix.[27]

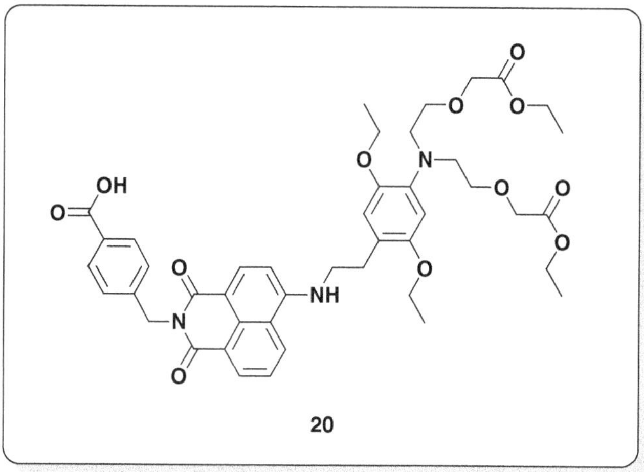

20

STRUCTURE 3.14

As calcium concentrations rise, the fluorescence intensity increases noticeably. As predicted, the cation's attachment to the ionophore prevents the anisidine donor from quenching fluorescence. The cation and the fluorophore do not directly interact sterically or electrically. The excitation and emission maxima are hence almost invariant with the fluctuating calcium concentrations typical of a PET indication.

Cho and coworkers developed the series of (two-photon) TP turn-on fluorescence probes **21–27** using BAPTA based on the excellent selectivity of the BAPTA ligand. In this group, **21** shows a 44-fold increase in emission in response to calcium ions, and when excited with light at 780 nm, the TP action spectrum of its complex with calcium in MOPS buffer solution (pH 7.2) is 110 GM.[28–30] The addition of the acetoxymethyl group in **22** improves cell permeability while having no impact on emission characteristics. The responses of probes **21–27** are moderate for the ions of zinc and manganese and weaker for the ions of magnesium, iron(II), and cobalt. However, all of the probes may be used to specifically monitor intracellular free calcium levels in the areas with lower levels of zinc and manganese.

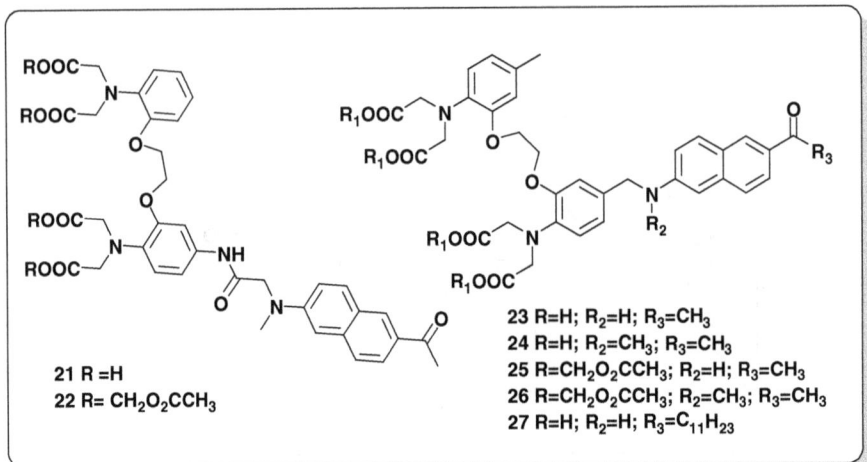

STRUCTURE 3.15

BAPTA-based probes typically use the PET mechanism to function. Two novel BAPTA-based calcium probes with TP emission that use the ICT principle were described by Liu et al. in 2009. Probes **28** and **29** have a push–pull electron system and exhibit strong affinities for calcium.[31] It is noteworthy that probe **28**, maybe as a result of the D-p-A-p-D scaffolds, exhibits a significant TP action cross-section '12324421' of 917 GM under 800 nm excitation. Additionally, this research showed that **28** could be used to capture green fluorescence pictures of HeLa cells in the presence of calcium.

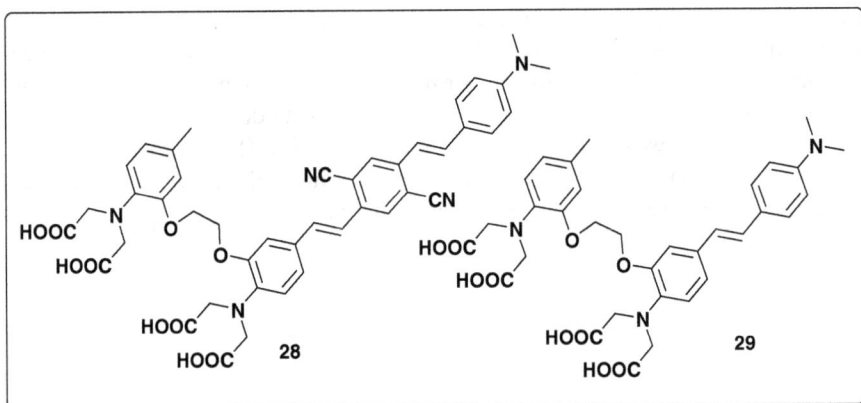

STRUCTURE 3.16

Despite the fact that BAPTA is a highly selective ligand for calcium, it is still necessary to increase the sensitivity of probes that include this binding moiety. Johnsson designed the probe **30**, which displays a 250-fold increase in fluorescence intensity following calcium binding in MOPS buffer solution, by linking the fluorophore BODIPY to BAPTA (pH 7.2). The moiety of probe **31**, which may be covalently and precisely coupled to SNAP-tag fusion proteins in live cells, is benzylguanine. The conjugate of this moiety increases its fluorescence intensity 180 times when calcium is introduced. Further research revealed that **31** can be used to track changes in calcium content in the CHO-K1 cells' nucleus and cytoplasm. Suzuki and Oka then described a BODIPY calcium probe 50 that was furan-fused.[32] In the presence of calcium, this probe experiences a 120-fold increase in its near-infrared fluorescence at 670 nm. Probe **32** is further compatible with real-time multiplex imaging when used in conjunction with commercial visible-light emitting probes like Cy5 and Fluo-4. This means that it may be used to scan intracellular calcium in HeLa cells as well.

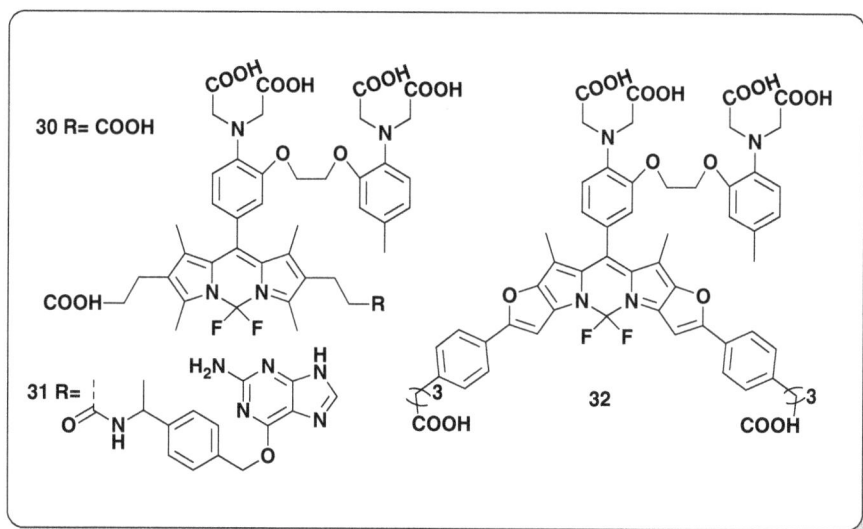

STRUCTURE 3.17

3.7   SENSORS FOR Ba^{2+}

The most common alkaline earth metal ion in nuclear reactors is barium, which is also the heaviest. There aren't many sensors on the market that can detect this ion. Anthracene, pyrene, and other fluorophores are the foundation of the sensor **33**.[33] In order to encourage intramolecular interactions between the chromophores, the sensor **33** is built with an electron donor (naphthalene) and an electron acceptor (p-substituted benzene). The linear ethereal moiety bonds to the metal in the presence of Ba^{2+} and, to a lesser degree, Ca^{2+} ions, bringing the two chromophores closer together and revealing an exciplex band.

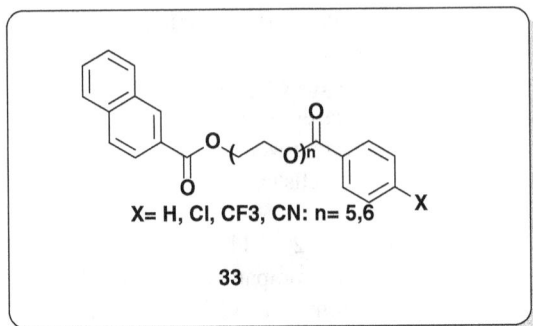

X= H, Cl, CF3, CN: n= 5,6

33

STRUCTURE 3.18

The usage of cryptands as alkali and alkaline earth metal ion sensors is extremely common. Methylene spacer **34** has been used to covalently attach a monoaza cryptand with carbon bridgeheads to a pyrene chromophore.[34] The cryptand is easier to dissolve in water thanks to its aliphatic backbone. As a result, the stiff cavity absorbs a Ba^{2+} ion inside it in aqueous solution at pH 10.2.

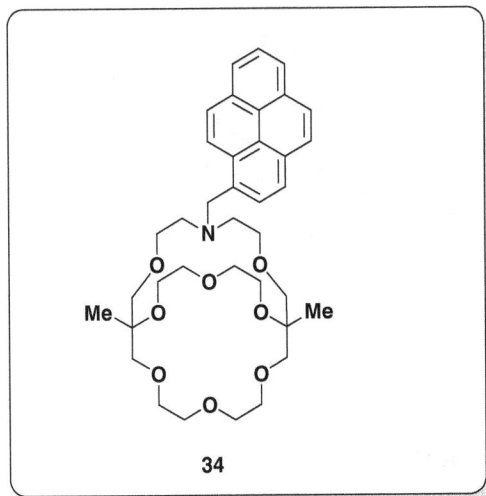

34

STRUCTURE 3.19

The metal ion inside the cavity activates the PET that suppresses N lone-pairs. As a result, fluorescence is seen to increase. The complexing capacity of the cryptand towards Ba^{2+} is enhanced when the sensor is introduced to Triton X-100 surfactant micelles, making it a sensitive detection device for Ba^{2+} ion in aqueous medium.

The same group has also substituted aza-crown ether macrocycles for the cryptand mentioned above in order to create sensors **35** and **36** for the Ba^{2+} ion.[35] Both of these may be used to detect the Ba^{2+} ion in an aqueous media when Triton X-100 micelles are present, just as the sensor **34**.

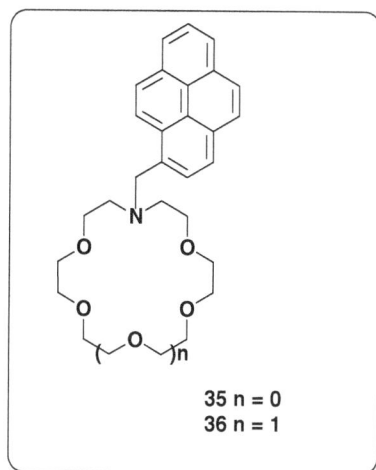

35 n = 0
36 n = 1

STRUCTURE 3.20

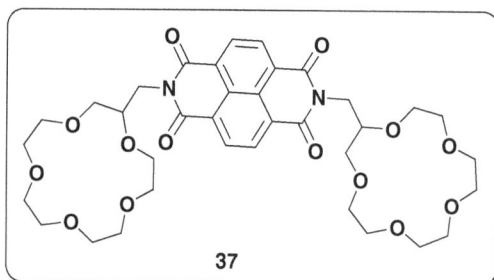

37

STRUCTURE 3.21

Utilizing the sensor **37** in acetonitrile medium, the Ba^{2+} ion was detected using the SAFE approach.[36] Metal-free **37** has a low emissivity, but the addition of Ba^{2+} increases its fluorescence by a factor of 35, producing a visible blue-green glow.

3.8 CONCLUSION

Alkali and alkali metal ions have a variety of fluorescence signalling systems that are described in the literature. More work should be put towards developing new receptor designs with strong selectivity, though. Although there are some general guidelines, additional work should be done to find receptors that are more selective for a certain ion and at a specific concentration range. This is necessary for the great spatial and temporal fidelity of ion concentration measurement that is necessary to understand the physiological state of a biosystem. It is important to keep in mind while designing sensors that these systems must operate in aqueous medium devoid of interference from transition metal ions for research enhancement as well as for confocal microscopic analysis. For practical applications, fluorescence amplification over quenching should be the preferable goal.

REFERENCES

1. Yue, D., Wang, M., Deng, F., Yin, W., Zhao, H., Zhao, X., & Xu, Z. 2018. Biomarker-targeted fluorescent probes for breast cancer imaging. *Chin. Chem. Lett.* 29(5): 648–656.
2. Han, X., Liu, Y., Liu, G., Luo, J., Liu, S. H., Zhao, W., & Yin, J. 2019. A versatile naphthalimide-sulfonamide coated tetraphenylethene: Aggregation induced emission behavior, mechanochromism, and tracking glutathione in Living Cells. *Chem. Asian. J.* 14(6): 890–895.
3. Wang, M., Yue, D., Qiao, Q., Miao, L., Zhao, H., & Xu, Z. 2018. Aptamer based fluorescent probe for serum HER2-ECD detection: The clinical utility in breast cancer. *Chin. Chem. Lett.* 29(5): 703–706.
4. Frausto da Silva, J. J. R., & Williams, R. J. P. 1993. *The Biological Chemistry of the Elements. The Inorganic Chemistry of Life.* Clarendon Press, Oxford, XXI, 561. ISBN 0-19-855598-9.
5. Xu, Z., & Xu, L. 2016. Fluorescent probes for the selective detection of chemical species inside mitochondria. *Chem. Commun.* 52(6): 1094–1119.
6. Birch, N. J. 1999. Biomedical uses of Lithium. *Uses of Inorganic Chemistry in Medicine,* pp. 11–25. doi: 10.1039/9781847552242-00011.
7. Murphy, C. J., & Obare, S. O. 2001. A two-color fluorescent lithium ion sensor. *Inorg. Chem. Commun.* 40(23): 6080–6082.
8. Citterio, D., Takeda, J., Kosugi, M., Hisamoto, H., Sasaki, S., Komatsu, H., & Suzuki, K. 2006. PH-independent fluorescent chemosensor for highly selective lithium ion sensing. *Anal. Chem.* 79(3): 1237–1242.
9. Rochat, S., & Severin, K. 2010. Fluorescence sensors for lithium ions and small peptides. *Chimia* 64(3): 150.
10. Minta, A., & Tsien, R. Y. 1989. Fluorescent indicators for cytosolic sodium. *J. Biol. Chem.* 264(32): 19449–19457.
11. Martin, V. V., Rothe, A., & Gee, K. R. 2005. Fluorescent metal ion indicators based on benzoannelated Crown Systems: A green fluorescent indicator for intracellular sodium ions. *Bioorg. Med. Chem. Lett.* 15(7): 1851–1855.
12. de Silva, A. P., Gunaratne, H. Q., Gunnlaugsson, T., & Nieuwenhuizen, M. 1996. Fluorescent switches with high selectivity towards sodium ions: Correlation of ion-induced conformation switching with fluorescence function. *Chem. Commun.* 16: 1967–1968.
13. Poronik, Y. M., Clermont, G., Blanchard-Desce, M., & Gryko, D. T. 2013. Nonlinear optical chemosensor for sodium ion based on Rhodol chromophore. *J. Org. Chem.* 78(23): 11721–11732.
14. Yamada, K., Nomura, Y., Citterio, D., Iwasawa, N., & Suzuki, K. 2005. Highly sodium-selective fluoroionophore based on conformational restriction of oligoethyleneglycol-bridged biaryl boron–dipyrromethene. *J. Am. Chem. Soc.* 127(19): 6956–6957.
15. Lu, D., He, L., Wang, Y., Xiong, M., Hu, M., Liang, H., Huan, S., Zhang, X. B., Tan, W. 2017. Tetraphenylethene derivative modified DNA oligonucleotide for in situ potassium ion detection and imaging in living cells. *Talanta* 167: 550–556.
16. Schwarze, T., Riemer, J., Holdt, & H. J. 2018. A ratiometric fluorescent probe for K+ in water based on a phenylaza 18-crown-6 lariat ether. *Eur. J. Chem.* 24(40): 10116–10121.
17. Xia, W.-S., Schmehl, R. H., & Li, C.-J. 1999. A highly selective fluorescent chemosensor for K+ from a bis-15-crown-5 derivative. *J. Am. Chem. Soc.* 121(23): 5599–5600.
18. Xia, W.-S., Schmehl, R. H., & Li, C.-J. 2000. A fluorescent 18-crown-6 based luminescence sensor for lanthanide ions. *Tetrahedron* 56(36): 7045–7049.

19. He, H., Li, H., Mohr, G., Kovacs, B., Werner, T., & Wolfbeis, O. S. 1993. Novel type of ion-selective fluorosensor based on the inner filter effect: An optrode for potassium. *Anal. Chem.* 65(2): 123–127.

20. Lin, Q., & Buccella, D. 2018. Highly selective, red emitting BODIPY-based fluorescent indicators for intracellular Mg^{2+} imaging. *J. Mater. Chem. B* 6(44): 7247–7256.

21. Dong, Y., Li, J., Jiang, X., Song, F., Cheng, Y., & Zhu, C. 2011. Na+ triggered fluorescence sensors for Mg^{2+} detection based on a coumarin Salen moiety. *Org. Lett.* 13(9): 2252–2255.

22. Kang, J., Kang, H. K., Kim, H., Lee, J., Song, E. J., Jeong, K.-D., Kim, J. 2013. Fluorescent chemosensor based on bispicolylamine for selective detection of magnesium ions. *Supramol. Chem.* 25(2): 65–68.

23. Liu, Z., Xu, H., Song, C., Huang, D., Sheng, L., & Shi, R. 2011. A simple fluorescent chemosensor for Mg^{2+} based on C=N isomerization with highly selectivity and sensitivity. *Chem. Lett.* 40(1): 75–77.

24. Tsien, R. Y. 1980. New calcium indicators and buffers with high selectivity against magnesium and protons: Design, synthesis, and properties of prototype structures. *Biochem. Int.* 19(11): 2396–2404.

25. Haugland, R. P. 2015. Ethidium homodimer-1 (ED-1). *Handbook of Fluorescent Dyes and Probes*, First Edition. R. W. Sabnis. pp. 199–202.

26. Takesako, K., Sasamoto, K., Ohkura, Y., Hirose, K., & Iino, M. 1997. Low-affinity fluorescent indicator for intracellular calcium ions. *Anal. Commun.* 34(12): 391–392.

27. He, H., Jenkins, K., & Lin, C. 2008. A fluorescent chemosensor for calcium with excellent storage stability in water. *Anal. Chim. Acta* 611(2): 197–204.

28. Kim, H. M., Kim, B. R., Hong, J. H., Park, J.-S., Lee, K. J., & Cho, B. R. 2007. A two-photon fluorescent probe for calcium waves in living tissue. *Angew. Chem. Int. Ed. Engl.* 46(39): 7445–7448.

29. Kim, H. M., Kim, B. R., An, M. J., Hong, J. H., Lee, K. J., & Cho, B. R. 2008. Two-photon fluorescent probes for long-term imaging of calcium waves in live tissue. *Eur. J. Chem.* 14(7): 2075–2083.

30. Mohan, P. S., Lim, C. S., Tian, Y. S., Roh, W. Y., Lee, J. H., & Cho, B. R. 2009. A two-photon fluorescent probe for near-membrane calcium ions in live cells and tissues. *Chem. Commun.* 36: 5365–5367.

31. Dong, X., Yang, Y., Sun, J., Liu, Z., & Liu, B.-F. 2009. Two-photon excited fluorescent probes for calcium based on internal charge transfer. *Chem. Commun.* 26: 3883–3885.

32. Matsui, A., Umezawa, K., Shindo, Y., Fujii, T., Citterio, D., Oka, K., & Suzuki, K. 2011. A near-infrared fluorescent calcium probe: A new tool for intracellular multicolour Ca^{2+} imaging. *Chem. Commun.* 47(37): 10407–10409.

33. Kawakami, J., Kimura, H., Nagaki, M., Kitahara, H., & Ito, S. 2004. Intramolecular exciplex formation and complexing behavior of 1-(2-naphthalenecarboxy)-n-(p-substituted benzenecarboxy)oxaalkanes as fluorescent chemosensors for calcium and barium ions. *J. Photochem. Photobiol. A Chem.* 161(2–3): 141–149.

34. Nakahara, Y., Kida, T., Nakatsuji, Y., & Akashi, M. 2004. A novel fluorescent indicator for Ba^{2+} in aqueous micellar solutions. *Chem. Commun.* 2004(2): 224–226.

35. Nakahara, Y., Kida, T., Nakatsuji, Y., & Akashi, M. 2005. Fluorometric sensing of alkali metal and alkaline earth metal cations by novel photosensitive monoazacryptand derivatives in aqueous micellar solutions. *Org. Biomol. Chem.* 3(9): 1787–1794.

36. Licchelli, M., Orbelli Biroli, A., & Poggi, A. 2006. A prototype for the chemosensing of Ba^{2+} based on self-assembling fluorescence enhancement. *Org. Lett.* 8(5): 915–918.

4 Lanthanide Metal Ions Detection
Mechanism and Applications

Sruthi Guru and Indu Tucker Sidhwani

4.1 INTRODUCTION

Lanthanide or lanthanoid ions constitute 15 elements from lanthanum ($Z = 57$) to lutetium ($Z = 71$), present in the form of metal complexes, functional materials, and various mineral assemblages. These elements together with scandium ($Z = 21$) and yttrium ($Z = 39$) constitute the rare-earth elements (REEs) (Figure 4.1). Throughout this chapter, REE is used to signify lanthanides. Lanthanides occur in nature in the form of phosphate or carbonate minerals in ores (bastnäsite, monazite, and xenotime), clays, coal ash, and acid mine drainage [1]. The extraction and isolation of lanthanides from these sources involve separation from more abundant metal ions, such as Fe, Mn, Cu, and Al [2]. These challenges have been overcome via advancement in extraction and chromatographic methods leading to formation of lanthanide-based small molecules, and supramolecular assemblies such as metal–organic frameworks (MOFs) [3–9]. The study of lanthanide-based materials has gained momentum in the last few decades with their applications as LEDs in lighting and display devices, permanent magnets in wind turbines and electric car batteries, critical components of lasers and phosphors, and agents in medical imaging (Table 4.1) [10,11]. As a result of an extensive use of lanthanides, they have become a cause of major concern with respect to their availability, sustainability, recyclability, and their

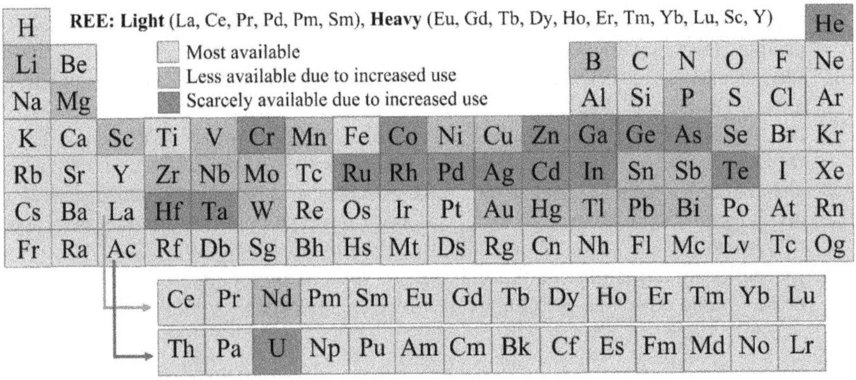

FIGURE 4.1 Availability of elements in the Earth's crust.

DOI: 10.1201/9781003352372-4

TABLE 4.1
Lanthanide Elements: Applications and Availability

Applications	Rare-Earth Element	Availability/Scarcity
Cell phones	La, Pr, Nd, Eu, Gd, Tb, Dy	Most available: La, Ce, Pr, Sm, Eu, Gd, Tb, Ho, Er, Tm, Yb, Lu
Hybrid engines	La, Nd, Dy	Limited availability: Nd
Alloy manufacture	La, Ce, Y	Scarcity: Dy
Magnets	Pr, Sm, Gd	Synthetic: Pm
Permanent magnets	Tb, Dy	
Catalyst in petroleum refining	Ce, Lu, Nd	
LEDs and display devices	Eu, Tb	
Storage and communication devices	Nd, Pr, Er, Yb	
Lasers	Ho, Yb, Nd	
Phosphors	Tb, Er	
Medical imaging	Tm, Yb, Nd	
Luminescent probes	Eu, Sm, Tb, Dy	

role in environmental pollution. Hence, lanthanide ion detection has become a major step toward remedying this effect. Certain properties that impede their detection are insolubility of lanthanides, and similar physical properties with other lanthanide elements [1,12]. This chapter addresses these challenges and how these have been overcome to aid in mining, recycling, and separations of lanthanides, so that their potential is realized, and their broader applications can be discovered.

4.2 NEED AND SCOPE FOR DETECTION OF LANTHANIDE IONS

As can be seen from Table 4.1, there are various applications of lanthanide-based functional materials as critical components in medical diagnostics, electronics, clean energy, automotive, aerospace, and environmental applications. Optimum concentration is critical to derive the best outcome from these promising lanthanide ions which have a direct impact on the performance diagnostics of the device. Metallic elements such as Li, Mg, Sc, V, Cr, Mn, Co, Ni, Cu, Zn, Ga, Sr, Y, Zr, Nb, Mo, Ru, Rh, Pd, Ag, Cd, In, Sn, Nd, Dy, Hf, Ta, W, Os, Ir, Pt, Au, Hg, Tl, Pb, Bi, and the lanthanides (Nd, Dy) form a group of energy critical elements (ECEs). These elements are critical components of clean, renewable energy applications, but are scarcely available or have limited availability in the Earth's crust; their extraction from respective ores entails huge cost and energy. Therefore, sustainable practices involving green energy techniques are employed for acquiring these ECEs. The common problems faced in conventional technologies, especially involving rare-earth metals for separation from mineral sources, are similar ionic radii and chemical properties. Counter current solvent–solvent extraction processes used to be widely adopted method until recently, which is a scalable and efficient process producing highly pure REs from any ore composition. However, uncertain selectivity afforded by this process due to unknown molecular speciation and a large amount of caustic waste produced during the process made this technique less popular. Green technologies used for separation of REs aim at minimizing solvent use, having controllable molecular speciation for tuning selectivity, and focus on recycling these critical metals instead of extraction from mineral resources [13].

4.3 PROPERTIES OF LANTHANIDE IONS

4.3.1 OXIDATION STATES AND COORDINATION PROPERTIES

Lanthanide ions (Ln^{III}) having electronic configurations of $[Xe]4f^n$ ($n = 0$ to 14) possess a rich variety of electronic levels of the order $14!/n!(14-n)!$, resulting in interesting optical properties [14–16]. The lanthanides are most commonly found in +III oxidation state (Ln^{III}), though in case of cerium and europium, Ce^{IV} and Eu^{II} are dominant under normal physiological conditions [1,17–19]. Due to the greater proximity of 4f orbitals to nucleus radial probability distribution of 4f orbitals is closer to the nucleus than 6s and 5p orbitals, there is poor shielding of 4f orbitals from increasing nuclear charge. This poor shielding effect causes reduction in ionic radius, termed as lanthanide contraction, especially for lanthanide ions having a coordination number of 8, from 116 pm (La^{III}) to 98 pm (Lu^{III}) [1,19–21]. This feature has been useful for the application of Pr^{III}-based complexes as probes for calcium binding sites in protein, as the ionic radius of Pr^{III} and Ca^{II} is similar, 1.12 Å for coordination number 8 [22,23].

The bonding in Ln^{III} complexes is largely ionic in character, without the significant impact of a poor shielding effect of 4f electrons. This ionic bonding provides flexible, sterically driven coordination geometries, which enable Ln(III) ions to adopt higher coordination numbers 8–12. Ln^{III} ions act as Lewis acids and their hydroxide and phosphate salts have low solubilities at pH = 7 (with K_{sp} of the order of 10^{-30}, and solubilities of 0.1–1 pM) [24].

4.3.2 STEREOCHEMICAL REQUIREMENTS

New and improved separation processes are required for REEs, as they have similar ionic radii and coordination numbers/geometries [25]. In supramolecular tetrahedral cage complexes, lanthanide extraction and separation is done by metal ion self-sorting, based on the multivalent supramolecular cooperativity of the lanthanide complexes [26]. Tris(tridentate) ligands having similar coordination moieties, the scaffold rigidity as well as the geometry greatly influence the structural stability. Li et al. hypothesize that ligands having moderate rigidity operate like levers among tetrahedral vertices with metal centers. This results in a small distortion in the coordination geometry of one metal center, and this distortion gets transferred to the other three vertices resulting in a greater energy barrier than that in a perfectly symmetrical tetrahedral complex. The same effect is not seen in a flexible tetrahedron; hence, Li et al. concluded that ligand rigidity increases the mechanical-coupling effect in a complex and plays a role in the cooperative enhancement of stability of the complex and the selectivity of the metal ion. It is found that ligands having ideal coordination sites for lanthanide separation in a supramolecular system coordinate with all the 14 lanthanide elements, while at the same time having different binding affinities toward each lanthanide element [27]. When there is a higher chelating affinity among the coordinating moieties and the metal ions, the structural stability of the complex improves; however, there is a negative impact on metal ion selectivity. This is because it results in kinetically trapped assembly structures, thereby hindering transformation into thermodynamically favored product. To break the kinetically stable structures, higher temperatures and/or longer reaction times may be required, which may affect the feasibility of practical applications. For practical applications, both the factors, efficient chelating ability of lanthanide ions, and moderate binding affinity should be considered. Hooley's group reported an acylhydrazone-phenolate–based bis(tridentate) ligand having anionic coordination sites. This ligand binds with lanthanide ions to form Ln2L3 complexes in DMSO. Its binding affinity toward smaller metals is kinetically favored, whereas its binding affinity toward larger metals is thermodynamically favored [28]. However, it was found that Ln_2Ln_3 complexes could not dissociate in DMSO even after 20 h, whereas pcam-based $Ln_4(L_1)_4$ complexes dissociated readily in DMSO, owing to weaker binding affinity between ligands and metal ions and reached the thermodynamic equilibrium on a minute timescale. For feasible lanthanide separation, it is required that the supramolecular assemblies reach thermodynamic equilibrium rapidly on a minute timescale. Use of ancilliary ligands (ALs) helps in enhancing the anion binding in lanthanide-based sensors [29]. Herein, first the AL binds to the Ln depending upon its affinity toward that Ln, reduces solvent quenching, and makes a binding site available for the target analyte. The ALs (e.g., DO_2A, that is, 1,4,7,10-tetraazacyclododecane-1,7-bisacetate) are chosen depending on their stability of the Ln-AL complex, denticity (number of donor groups of AL binding to the Ln) and the size/geometry of the binding site.

4.3.3 LUMINESCENT PROPERTIES

Lanthanides Ln^{III}, barring La^{III} ($4f^0$) and Lu^{III} ($4f^{14}$) exhibit luminescent f–f emissions covering the entire spectrum of electromagnetic radiations. In the visible region of the spectra, Eu^{3+}, Tb^{3+}, Sm^{3+}, and Tm^{3+} emit red, green, orange, and blue radiations, respectively; whereas Pr^{3+}, Nd^{3+}, Sm^{3+}, Dy^{3+}, Ho^{3+}, Er^{3+}, Tm^{3+}, and Yb^{3+} emit in the near-infrared region, and Ce^{3+} shows a broadband emission at 370–410 nm as a result of 5d–4f transition [15,17] (Figure 4 2).

The 4f–4f transitions of lanthanide ions are Laporte forbidden as they are well shielded by filled $5s^25p^6$ subshells [30]. This impacts the absorption efficiency of these transitions on direct photoexcitation. To overcome this low absorption efficiency of the 4f–4f transitions, an effect termed as "antenna effect" or luminescence sensitization is introduced by means of a strong absorbing chromophore that sensitizes Ln^{3+} ions [16,30,31]. The steps leading to the much acclaimed and profited antenna effect of Ln^{3+} ions are elucidated in Figure 4.3, and explained as follows:

i. The organic ligands containing chromophores absorb light upon direct photoexcitation.
ii. The energy of excitation is transferred to Ln^{III} excited state via intramolecular energy transfer process.

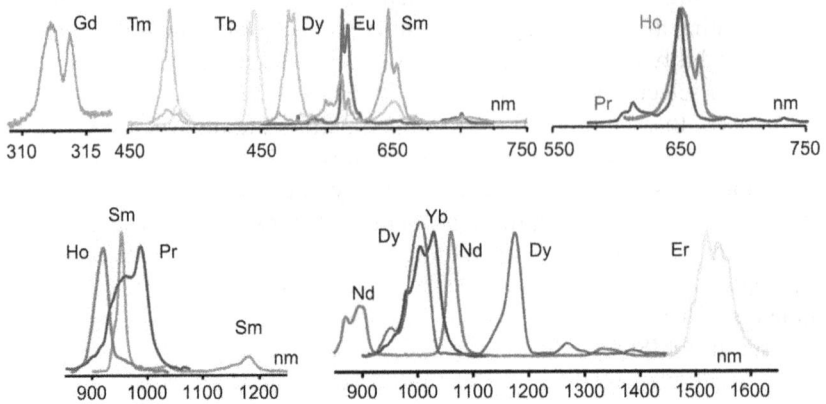

FIGURE 4.2   Emission spectra of Ln³⁺ ions. (Source: Bünzli [17].).

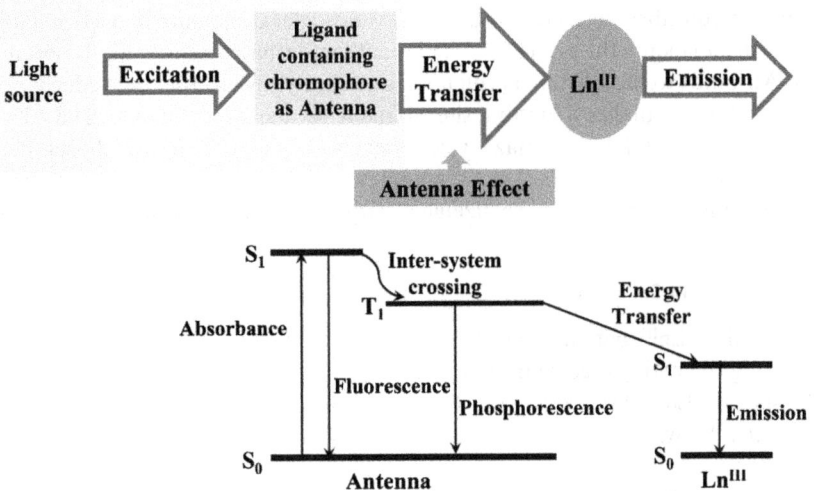

FIGURE 4.3   Antenna effect or lanthanide sensitization in Ln^III using ligands containing chromophore.

iii. The Ln³⁺ ion shows luminescence spectra by virtue of a radiative pro-
    cess, which increases the luminescence quantum yield of Ln³⁺ in nor-
    mal conditions at room temperature. For MOFs as ligands for Ln^III ions,
    luminescence is further enhanced due to nullification of the solvent
    quenching and self-quenching characteristic of Ln^III ions. Thus, the
    strong luminescence property of LnMOFs is successfully utilized in
    designing chemical sensors.

4.3.4 PARAMAGNETIC PROPERTIES

The paramagnetic properties of REE are because of the presence of unpaired 4f electrons. Due to shielding of 4f orbitals by the completely filled 5s and 5p orbitals, the magnetic behavior of Ln^{3+} is less affected by the coordination of Ln^{3+} with other ions in comparison to 3d transition metals. The magnetic properties can be considered as that of a free Ln^{3+} ion, signifying that REEs have unquenched orbital momentum because of the presence of the 4f orbitals in the inner core which prevents the crystal field from quenching the orbital momentum [32]. The value of magnetic moments is significantly different from the spin-only values due to the strong spin–orbit coupling. The maximum number of unpaired electrons is 7, in Gd^{3+}, with a magnetic moment of 7.94 B.M., but the largest magnetic moments, at 10.4–10.7 B.M., are exhibited by Dy^{3+} and Ho^{3+}. However, in Gd^{3+}, all the electrons have parallel spin and this property is important for the use of gadolinium complexes as contrast reagent in magnetic resonance imaging (MRI) scans. Almost all lanthanides exhibit paramagnetic behavior except La, Yb, and Lu (having no unpaired f electrons). Some of the lanthanides become ferromagnetic at sufficiently low temperatures, especially Gd, Tb, Dy, Er, Tu, and Y (e.g., Gd^{3+} becomes ferromagnetic at room temperature), while the other become antiferromagnetic (e.g., Ho) [33]. These magnetic properties empower Ln to be used as permanent magnets in magnetic storage disks (e.g., Nd (mostly), Pr, Nd, Sm, Gd, Tb, Dy)) [34] and in form of effective shift or relaxation agents in MRI applications (Tm^{3+} and Dy^{3+} as ^{23}Na shift reagents and Gd^{3+} contrast and spin relaxation reagent) [35].

4.4 DETECTION AND SEPARATION TECHNIQUES

The detection and separation techniques using chemical sensors, luminescent sensors using ratiometric detectors, and electrochemical techniques are discussed based on the coordination properties, stereochemical requirements, and fluorescent and electrochemical properties of lanthanide ions.

4.4.1 CHEMICAL TECHNIQUES

The most commonly used technique for separation of lanthanides is solvent extraction. The main principle that underlies this technique is that when a ligand or extractant is dissolved in a water-immiscible organic solvent which is in contact with an aqueous solution of heavy and light rare-earth (HRE and LRE) metals, the RE ions coordinate with acidic extractants [di-(2-ethylhexyl)phosphoric acid (DEHPA or HDEHP), 2-ethylhexylphosphonic acid mono-2-ethylhexyl ester (HEH [EHP]), tributyl phosphate (TBP), naphthenic acid, or versatic acid] at the liquid–liquid interface and move into the organic phase (Figure 4.4).

Once the mixture settles in a mixer settler, selective partitioning of the REs between the two phases can be induced based on binding affinities of the extractant with the Ln ion and based on Lewis acidity trends [36]. Figure 4.4 shows a flow chart of extraction, scrubbing, and stripping processes occurring inside a mixer settler. There are multiple repetitions to ensure all the REs (heavier ones in the organic

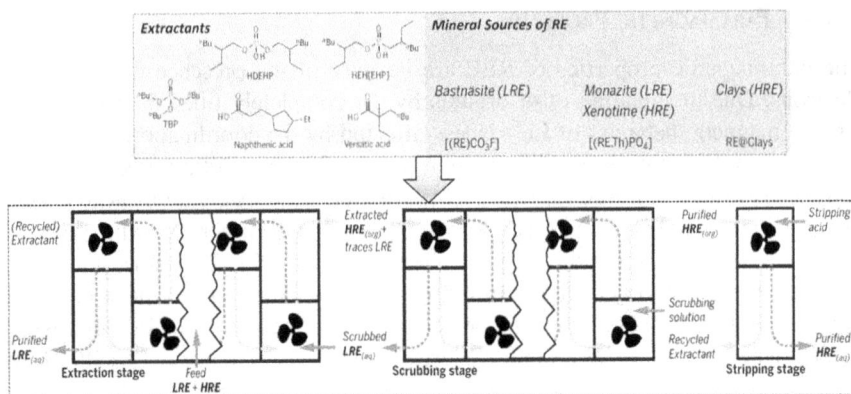

FIGURE 4.4 Commonly used extractants, mineral sources of RE and solvent extraction process (extraction, scrubbing, and stripping stages) in an industrial mixer settler. (Source: Cheisson and Schelter [36].)

phase and lighter ones in the aqueous phase) are extracted completely. There are various parameters that need to be optimized to ensure selectivity of extractant that reduces the number of iterations and makes the process efficient. Actinide separation by solvent extraction processes also uses phosphorus-based extractants. Among these extractants, TBP was found to be the most powerful and efficient extractant for Ce^{IV} and Ln^{III} ions [36].

4.4.1.1 Mechanism Involved in Chemical Techniques

Li et al. showed that the pcam-based ditopic and tritopic ligands have greater selectivity toward lanthanide ions due to their moderate chelating affinity in thermodynamically favored complexes. The cooperative enhancement of tris(tridentate) ligands toward selective lanthanide elements for forming tetrahedral cage complexes enables these ligands to act as good industrial extractants for lanthanide separation. To further prove their hypothesis, Li et al. synthesized L4–6 complexes containing hydrophobic alkyl groups at the periphery to provide improved phase separation. The tests for lanthanide separation using these complexes showed that di-dodecanamine groups (L6) provided dispersed well in CHCl3 for liquid–liquid extraction. The structure of the complex was found to be intact from ^1H NMR spectra, and the lanthanide metal contents in aqueous and organic phases were measured by ICP-MS. The ratio of distribution coefficient of lanthanide ion in the two phases is the separation factor of that lanthanide ion and is expressed as $S_{Ln(a)/Ln(b)} = D_{Ln(a)} / D_{Ln(b)}$, where distribution coefficient $D_{Ln(a)}$ is expressed as $D_{Ln(a)}$ $D_{Ln(a)} = [Ln(a)]_{aq} / [Ln(a)]_{org}$). Poor water stability of the core of the tetrahedral cage complex affects the separation factor, and this problem can be solved by using more stable tetrahedral frameworks in extractants for lanthanide separation. These stable supramolecular complexes can also be used for actinides as well along with lanthanides separation in radioactive waste treatment and recycling of these REEs having similar oxidation states, chemical properties, and ionic radii [25].

4.4.2 FLUORESCENT TECHNIQUES

Dolai et al. synthesized a novel TTA-generated C-dots fluorescent sensor that can sense lanthanide and actinide. The C-dots were embedded in a porous silica aerogel matrix. The encapsulation of C-dots was done without any changes in the structural and physical properties of the aerogel matrix and embedded C-dots. This hybrid framework was used for selective and sensitive fluorescence sensing of UO_2^{2+}, Sm^{3+}, and Eu^{3+} ions by the group. The sensing influence was studied by fluorescence emission spectra recorded on aqueous solutions of UO_2^{2+}, Sm^{3+}, and Eu^{3+} ions to the TTA−C-dot/aerogel F as shown in Figure 4.5. A significant red shift from 445 to 480 nm (excitation at 400 nm) was observed for UO_2^{2+} addition and fluorescence quenching was observed for Sm^{3+} and Eu^{3+}. When similar studies were carried out using glucose C-dot aerogel, no significant change was observed. Hence, they concluded that the changes in shift and intensities in the spectra were due to the interaction of the TTA units on the C-dots with UO_2^{2+}, Sm^{3+}, and Eu^{3+} [37].

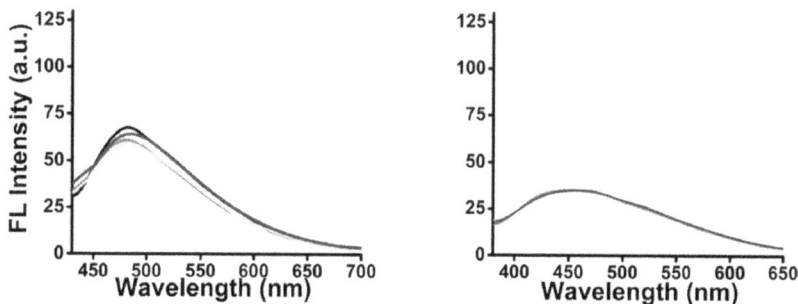

FIGURE 4.5 Fluorescence emission spectra following the addition of UO_2^{2+}, Sm^{3+}, and Eu^{3+} to (a) TTA−C-dot−aerogel and (b) glucose derivative-C-dot−aerogel. (Source: Dolai et al. [37].)

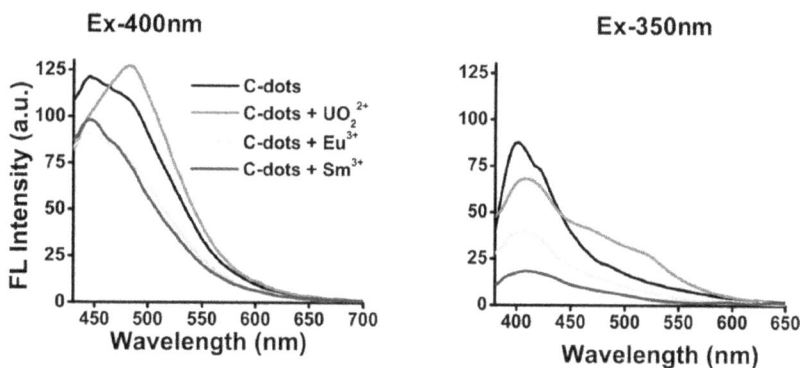

Lanthanide ion detection and separation in fluorescent ratiometric sensors is also gaining interest among researchers, especially in those sensors having multiple metal emission centers [38–40]. Two or more lanthanide ions possessing similar ionic radii showing similar chemical behavior can be doped into the same crystallographically equivalent metal positions of MOFs, thereby resulting in two or more metal emission sites at the respective lanthanide center [39]. Heffern et al. formed strong Eu^{3+} and Tb^{3+} ion emission centers, emitting red and green light, respectively. This emission is clearly visible to the eye and is used for the fabrication of luminescence materials [40]. In another study, Cui et al. optimized the Eu^{3+}/Tb^{3+} ratio of the precursors in the synthesis to form mixed-lanthanide MOF temperature sensors, $Eu_{0.0069}Tb_{0.9931}$-DMBDC and $Tb_{0.9}Eu_{0.1}$-PIA (where H2DMBDC is 2,5-dimethoxy-1,4-benzenedicarboxylate and H2PIA is 5-(pyridin-4-yl)isophthalic acid). These Ln-based MOF ratiometric sensors rely on characteristic emissions from Tb^{3+} and Eu^{3+} ions [41,42]. Zhou et al. studied a series of $[Eu_{2x}Tb_{2(1-x)}(FDA)_3]$ (H2FDA: furan-2,5-dicarboxylic acid) systems as ratiometric sensors for organic mixtures. In these mixtures, the emission of $[Eu_{2x}Tb_{2(1-x)}(FDA)_3]$ can be chemically modified from red to orange, yellow, and green light by changing the ratio of Eu^{3+}/Tb^{3+} (Figure 4.6a). The strong emissions in the visible spectra suggest a role of FDA^{2-} as an efficient antenna for sensitization of both Eu^{3+} and Tb^{3+} ions [43]. A similar strategy was adopted by Zhang et al. to form a series of zeolite-like MOFs (ZMOFs) with a combination of Tb^{3+} and Eu^{3+} ions. Among the series of complexes, $Eu_{0.6059}Tb_{0.3941}$-ZMOF showed excellent selectivity for lysophosphatidic acid, a biomarker for ovarian cancer [44]. In another study, Zhang et al. introduced a mixture of Ln^{3+} ions into anionic porous MOFs via ion exchange. Nonluminescent Bio-MOF-1 was modified by ion exchange of Tb^{3+} and Eu^{3+} cations, resulting in Tb/Eu@bio-MOF-1 material which emitted orange-red light due to emissions from both Tb^{3+} and Eu^{3+} ions (Figure 4.6b). In this case, dual emission results from (a) energy transfer takes place from organic ligand biphenyl-4,4′-dicarboxylic acid to Ln^{3+} ions and (b) metal-to-metal energy transfer

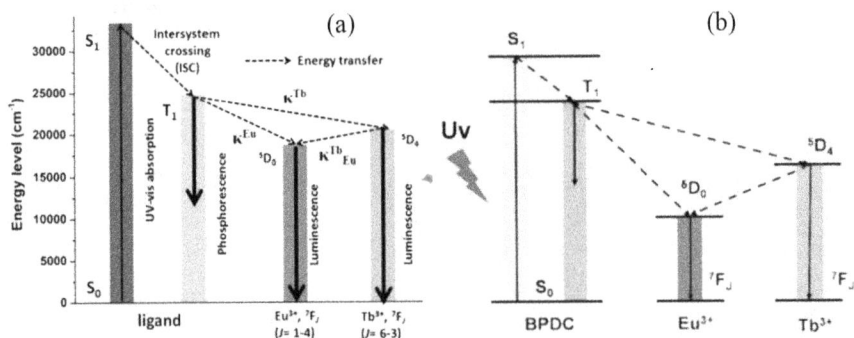

FIGURE 4.6 (a) Ligand–Ln^{3+} energy transfer processes of $[Eu_{2x}Tb_{2(1-x)}(FDA)_3]$ (S_0, S_1, and T_1 are the ground state, the singlet state, and the triplet state of the ligand, respectively). (Source: Zhou et al. [43].) (b) Energy transfer from the T_1 state of biphenyl-4,4′-dicarboxylic acid to Ln^{3+} ions and the metal-to-metal energy transfer from Tb^{3+} to Eu^{3+}. (Source: Zhang et al. [45].)

from Tb^{3+} ion to the Eu^{3+} ion [45]. Dual emission from Ln-MOF can be obtained from the metal centers and organic ligands on partial energy transfer of the light absorbed by the ligand to the lanthanide center. Wang et al. used this concept to show dual emission of europium-based MOF $[Eu_2(L)_3(H_2O)_2(DMF)_2]\cdot16H_2O$ (H_2L: 1,4-bis(5-carboxy-1H-benzimidazole-2-yl)benzene)) from both organic linkers and lanthanide centers [46]. Similarly, Yang et al. showed dual emission of europium-based MOF with 5-boronoisophthalic acid (5-bop) as organic ligand. In this case, the emission from the ligand and from Eu^{3+} ions was due to insufficient intermolecular energy transfer from ligand to Eu^{3+} ions [47].

Although many researchers have MOF-based ratiometric sensors with multiple centers for detection of lanthanide ions, there are many challenges to be overcome. These include (1) poor water stability of MOF-based sensors resulting in low recyclability, (2) controlling luminophore moieties for efficient ratiometric sensing functions, (3) improving detection limit for practical and complex systems under real-time fluctuations in temperature/pressure/concentration of analytes, and (4) proper establishment of sensing mechanism at each emission center [39].

4.4.2.1 Mechanism Involved in Fluorescent Techniques

The detection of an analyte containing Ln^{3+} ions by a luminescent sensor occurs by either one or a combination of photophysical mechanisms [48]:

a. **Inter/intramolecular charge transfer (ICT):** In this mechanism, the fluorescent sensors contain a single molecule having both electron-donating and electron-withdrawing groups which are in conjugation with each other. This results in push–pull of π-electron system in the excited state of these groups. Now, when a Ln^{3+} ion interacts with the electron-donating group of the fluorophore, the latter's electron donation ability reduces, and a blue shift (lower wavelength) is observed in the absorption spectra. Similarly, interaction of Ln^{3+} with the electron-withdrawing group of the fluorophore enhances the electron-withdrawing character of that group, thus rendering stability to the fluorophore system and showing a red shift (higher wavelength) in the absorption spectrum (Figure 4.7a).

b. **Energy transfer (ET):** This mechanism can be either electronic energy transfer (EET) where electron is transferred via resonance, and fluorescence resonance energy transfer (FRET), where energy gets transferred. The mode of transfer depends on the orientation and distance between energy acceptor and energy donor molecules in compounds having multiple chromophores. In fluorescence resonance energy transfer, energy transfer takes place from the excited-state donor group to either single state or to triplet state of the acceptor group. For a distance below 10 Å between the acceptor and the donor, Dexter-type ET commences. For a distance of 10–100 Å, Förster-type ET takes place (Figure 4.7b). Förster-type ET is a radiationless transfer of energy released from an excited molecule to a molecule which in turn gets excited and creates an exciton. Dexter-type ET involves a fluorescence quenching mechanism and transfer of electron from an excited molecule (donor) to another excited molecule (acceptor).

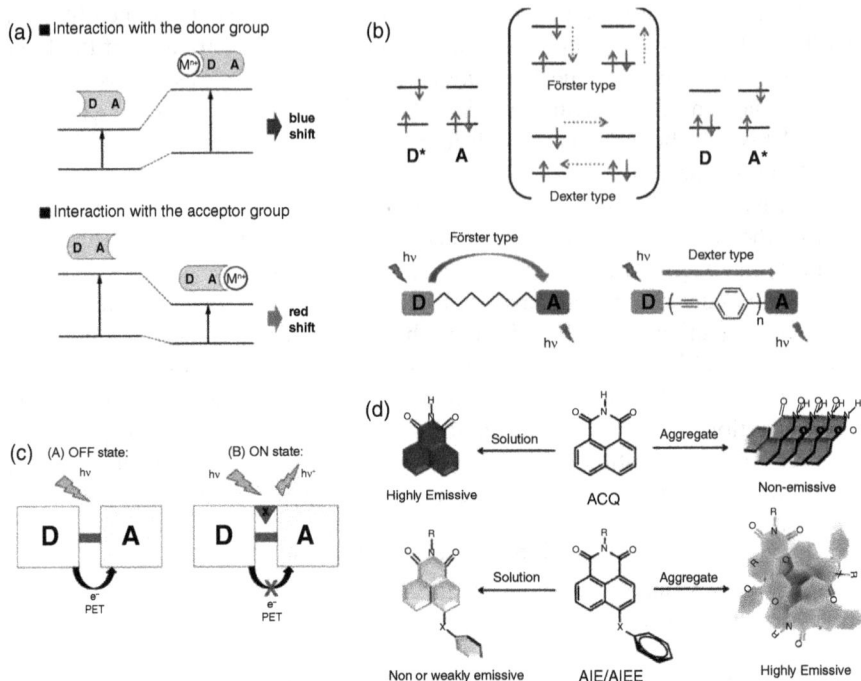

FIGURE 4.7 Schematic representation of photophysical mechanisms: (a) inter/intramolecular charge transfer (ICT); (b) energy transfer (ET); (c) photo-induced electron transfer (PET); and (d) aggregation-induced emission (AIE). (Source: Gawas et al. [48].)

    c. **Photo-induced electron transfer (PET):** In PET sensors, donor is linked to the acceptor via a non-conjugated bridge. Upon excitation, the fluorophore group accepts or donates electron to the ground state acceptor which is present in the same molecule. Fluorescence quenching normally takes place in this mechanism, which can be rectified by using a guest analyte to deactivate the PET mechanism (Figure 4.7c).

    d. **Aggregation-induced emission (AIE) or aggregation-caused quenching:** In this mechanism, the fluorescence quenching is observed for an organic fluorophore when it is present in aggregate form, whereas some non-fluorescent organic molecule in solution starts emitting upon. Thus, aggregation-induced enhanced emission is observed (Figure 4.7d).

4.4.3 ELECTROCHEMICAL TECHNIQUES

Most of the REEs have a prevalent trivalent oxidation state; while some exist in divalent (Eu^{II}) or tetravalent (Ce^{IV}) states in ambient conditions [36]. In solution state, almost all rare-earth metals are reduced to their divalent state (58), and due to high reactivity, REs are cannot be separated in the solution state. In the solid state, however, the redox behavior of REs can be exploited for their separation [49]. In industrial processes, for the extraction of cerium, the ores undergo oxidative roasting to form Ce^{IV} salts, which are removed as insoluble $Ce(OH)_4$ or CeO_2 residue. The same cannot be said for europium, as it is present in small quantities in mineral resources and cannot be efficiently reduced early in the separation process. Way back in 1979, Donohue identified the photoredox redox characteristics of lanthanides as a potential tool for REE separations [50]. He successfully separated cerium from lanthanide mixture by photooxidation of Ce^{3+} in acidic aqueous solution and precipitation of Ce^{4+} with iodate.

Electrophoretic separation is a very old and established technique for lanthanide separation from its complexes. The first reported use of this technique was done by Kendell and his colleagues in 1920s where "ion migration method" was used to separate pairs of the lanthanide ions from their complexes [51]. This was followed by Konstantinov et al. [52] who conducted separations of REs in an arrangement of discontinuous electrolytes without gel. They also showed the importance of complexing agent (acetate) and pH on RE separation and managed to separate a mixture of 12 REs within 9–10 h. These works along with many others established capillary isotachophoresis (ITP) as a powerful analytical separation technique [53–55]. Principles of ITP are employed in many modern separation schemes of REs, and the first practically feasible arrangement for capillary isotachophoretic method was suggested by Nakatsuka et al. [56]. Another most popular separation technique is capillary zone electrophoresis (CZE), which was first established by Foret et al. [57], where complete separation of lanthanide ions is achieved within a few minutes. Other less popular techniques include micellar electrokinetic capillary chromatography, capillary gel electrophoresis, ion-exchange electrokinetic chromatography, capillary isoelectric focusing, affinity capillary electrophoresis, capillary electrochromatography, or separations by capillary electrophoresis on microchips [58–61]. In most of the electrophoretic techniques, hydroxyisobutyric acid is the most suitable complexing agent that improves the separations of lanthanide metal ions.

Electrorefining process is another way of employing redox behavior of lanthanide ions. Krishnamurthy et al. used the reported Molycorp process for extraction of Eu^{3+}. In this process, after concentration by solvent extraction, an aqueous feed containing a substantial amount of Eu^{3+} was reduced over a Zn(Hg) column and precipitated as $Eu^{II}(SO_4)$, while the other REs remained in the solution [62]. In 2015, Van Gerven and his colleagues reported excellent photoseparation of Eu from non-equimolar binary Eu/Y doped concentrations in red lamp phosphors by photochemical reduction of Eu(III) [63]. Another factor apart from redox activity of REEs that is useful for devising suitable separation techniques is their Lewis acidity, as was demonstrated by Cheisson et al. [36]. As the Lewis acidity of REs increases, it can affect the irreversible oxidative electron transfer rate of a coordinated redox-active ligand [64]. Fang et al. demonstrated how selectively oxidized RE complexes could be separated by filtration. The electrokinetics involved in the separation helped the group construct a model for predicting separation factors based on the cyclic voltammograms of the isolated RE complexes. More recently, Yang et al. were successful in recovering REs from rare-earth permanent magnets (e.g., Neodymium–Iron–Boron) using an electrorefining technique in a molten fluoride electrolyte [65]. The separation rates of Nd and Pr in the molten fluoride electrolyte increased with increasing current. By electrolysis, the RE ions moved to the cathode, leaving the porous Fe_2B alloy and metallic Fe in solution.

One disadvantage of the electrorefining process is the use of cyclic electrolysis, which leads to a low current efficiency. For overcoming this disadvantage, Wang et al. showed separation of separation of U and lanthanides on Al cathodes using Li metal as a sacrificial anode [66]. They used an electrochemical technique as well as electronic absorption spectroscopy to study the U species separation. After the potentiometric deposition of the ions, the recovery U reached 99% with the main products as Al_3U and Al_4U. Simultaneously, successful separation of U(III) from La (III) and Sm (II) was achieved with separation factors $SF_{U/La} = 251$ and $SF_{U/Sm} = 4662$, respectively (Figure 4.8).

This study by Wang et al. is of great significance to improve current efficiency in the pyrochemical processing of spent nuclear fuel using molten salt electrolysis. Pyrochemical processing involves high-temperature actinides or lanthanide separation using an electrochemical method. The recovery of uranium, reuse of transuranic elements as a nuclear fuel, and reduction of nuclear waste are some advantages of pyrochemical processing over other nuclear spent fuel treatment techniques. By this technique, actinide ions dissolved in LiCl-KCl molten salts at 450°C can be recovered as pure actinide metals at the cathode in the electrorefining step of pyrochemical processing. Also, the lanthanide species dissolved in the LiCl-KCl molten salts plays a significant role in metal purification during the electrorefining step.

4.4.3.1 Mechanisms Involved in Electrochemical Techniques

In electrophoresis techniques employed in CZE and ITP, the lanthanide separation takes place based on the different electrophoretic mobilities of the separated Ln^{III} ions. There is a very negligible difference, however, in the range of 72.3–67.0×10^{-5} cm/V s from La^{3+} to Lu^{3+}, which is not detectable by even the most effective CE separation technique. Therefore, other factors are employed to enhance the ion mobility among lanthanide ions to facilitate their separation. Lanthanide contraction, whereby the decrease in ionic radius of Ln^{III} with increasing atomic number, occurs due to the poor shielding effect of f-orbitals. This causes increased

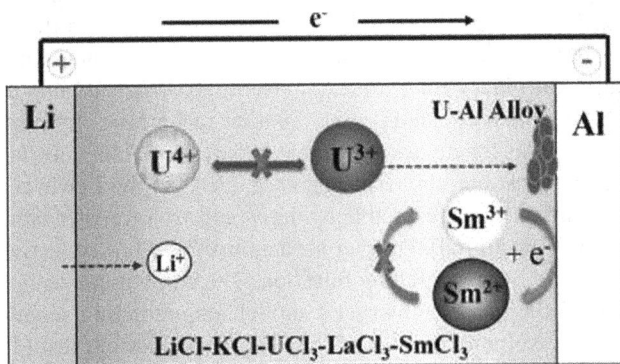

FIGURE 4.8 Separation of U and lanthanides on Al cathodes using Li metal as a sacrificial anode. (Source: Wang et al. [66].)

strength of cation-anion, ion-dipole, and ion-induced dipole interactions, leading to stronger bonds between Ln ions and ligand donor groups [4]. Therefore, the role of complexing agents for Ln-complex formation in both on-capillary (labile complexes) and precapillary complexation (inert complexes) is significant. It was found that when for labile systems, the complexing agent is added to BGE, Ln^{3+} forms complexes with complexing ligand (L^-) as $Ln^{3+} + nL^- \rightleftharpoons LnL_n^{(3-n)+}$. These complexing agents are anions of weak organic acids that undergo a dissociation/protonation equilibrium: $HL \rightleftharpoons H^+ + L^-$. The species of lanthanide ion complexes present depends on the total concentration of complexing agent and the pH. The effective mobility (μ_{eff}) of the lanthanide ion is expressed as the weighted average of the mobilities of the individual species [67]

$$\mu_{eff} = \sum_{i=0}^{N} \mu_{LnL_i} \chi_{LnL_i} = \mu_{Ln^{3+}} \chi_{Ln^{3+}} + \mu_{LnL_1} \chi_{LnL_1} + \mu_{LnL_2} \chi_{LnL_2} + \cdots + \mu_{LnL_N} \chi_{LnL_N}$$

For electrorefining process, the electrochemical reactions occurring at the cathode are $Ln(III) + e^- \rightarrow Ln(II)$. Kuznetsov and Gaune-Escard [68] studied the electrochemistry of redox process of lanthanide ions (Eu^{III}) in NaCl-KCl melt at 973–1,123 K on glassy carbon working electrode. $SmCl_3$, $EuCl_3$, and $YbCl_3$ also show similar electrochemical behavior in molten salt electrolytes NaCl-KCl, KCl, and CsCl. In these molten alkali chlorides, the cathodic peak current (for electroreduction) is directly proportional to $v^{1/2}$, where v is the polarization rate, and concentration of $LnCl_3$ (melt). In contrast, the peak potential is independent of the polarization rate up to $v = 0.1$ V/s and concentration of $LnCl_3$ (melt) [68]. On potentiostatic electrolysis at the peak cathodic potential, there was no formation of any solid phase at the electrode or any electrode transformation observed. Hence, the electrode process $Ln(III) + e^- \rightarrow Ln(II)$ is controlled by the mass transfer after this stage and produces reduced form (Ln^{II}) that is soluble in the melt. The diffusion coefficient of $Ln(II)$ is obtained based on the electroreduction of $Ln(III)$ to $Ln(II)$ using the Delahay equation expressed as $i_p = 0.496nFc_0 AD^{1/2}(\alpha n_\alpha Fv/RT)^{1/2}$, where i_p is the cathodic peak current, A is the electrode area, c_0 is the bulk concentration of species, D is the diffusion coefficient, v is the scan rate, α is the transfer coefficient, and $n\alpha$ is the number of electrons transferred for the reduction process in the rate determining step. For the mass transfer control process, $i_d = nFcD / \delta$, where i_d is the diffusion limiting current density, n is the number of moles of electrons, F is Faraday's constant, c is the bulk concentration of species in the melt, D is the diffusion coefficient, and δ is the diffusion layer thickness [68]. The activation energies for diffusion are calculated by $-\Delta E_{act} / 2.303R = \partial \log D / \partial(1/T)$. It is found that as the Ln oxidation state increases, the diffusion coefficients decrease whereas the activation energies for diffusion increase [69]. Also, due to lanthanide contraction, the diffusion coefficients decrease from Sm to Yb. The stability of the Ln complex is higher for smaller Ln ionic radius, so diffusion coefficients decrease whereas the activation energies for diffusion increase with faster movement of the diffusing species. The formal standard redox potentials are expressed as $E^*_{Ln^{III}/Ln^{II}} = \left(E_p^C + E_p^A\right)/2 + RT/F \ln\left(D_{ox}/D_{red}\right)^{1/2}$. Typically, lanthanide chlorides in alkali chloride melts are $LnCl_6^{3-}$ and $LnCl_4^{2-}$ [70].

Their relative stability depends on the ionic radius of Ln ion. Further, the separation coefficients (θ) of REEs having the same oxidation state can be found from their formal standard electrode potentials: $\ln\theta = (nF/RT)\left(E_1^* - E_2^*\right)$ [69].

4.4.4 BIOLOGICAL TECHNIQUES

Lanthanides have been employed in biological applications, as they are fairly abundant (La–Nd - ~10–70 ppm which is similar to Cu and Zn, while Pm–Lu are 10–100 times rarer) [1,71]. Also, the higher charge-to-radius ratio of lanthanides is comparable to common biological Lewis acids having Ca^{II}, Mg^{II}, and divalent first-row transition metal ions. This enables them to act as robust Lewis acid catalysts, and their catalytic efficiency is more prominent owing to their higher coordination number, flexible coordination geometry, and/or redox-inertness [72]. Finally, the similar size and functionalities of lanthanides with Ca^{II}, Fe^{III}, and Mg^{II} help in providing an evolutionary connection to analogous biological pathways involving these metals and offering selective recognition based on their distinct coordination chemistry [73,74]. Pol et al. reported an essential dependence of methanotrophic bacterium on lanthanides resulting from replacement Ca^{2+} by Ln^{3+} ion (La, Ce, Pr, and Nd) in the active site of the methanol dehydrogenase (MDH) enzyme to act as a cofactor [75]. They isolated the microorganism from volcanic mudpot water, which had an acidic pH of 1–2 and contained 2–3 μM of lanthanides (La, Ce, and Nd), thus confirming bioavailability of REs in acidic environments. They showed that substitution of lanthanides as cofactors to MDH enzyme increased the catalytic activity of pyrroloquinoline quinone-dependent MDH in methanotrophs and methylotrophs. Since then, many RE-dependent bacteria have been found and the research in this field is progressing well [36,76,77]. Recently, Cotruvo et al. have shown that there are direct analogies between lanthanides and Ca^{II} and Fe^{III} with respect to acquisition, transport, utilization, and storage and coordination chemistry. They demonstrated that protein isolated from RE-dependent bacteria was able to bind with REs with 10^8-fold selectivity over Ca. They derived protein lanmodulin from its Ca analog called calmodulin, with some structural differences [78].

4.4.4.1 Mechanism Involved in Biological Techniques

REEs can be recovered from solid phase by their dissolution in aqueous solution, from which their selective uptake can be done. This uptake can be done by selective bioabsorption or bioaccumulation using microbial cells [79]. The mechanisms for REE recovery involve biochemical processes, such as redoxolysis, acidolysis, and complexolysis (Figure 4.9a) [77].

The redoxolysis reaction is a two-step process as shown below:

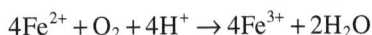

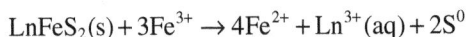

$$4Fe^{2+} + O_2 + 4H^+ \rightarrow 4Fe^{3+} + 2H_2O$$

$$LnFeS_2(s) + 3Fe^{3+} \rightarrow 4Fe^{2+} + Ln^{3+}(aq) + 2S^0$$

There is dissolution of the mineral happening when Fe^{2+} in the mineral gets converted into Fe^{3+}, and there is electron transfer from the mineral to the microbe.

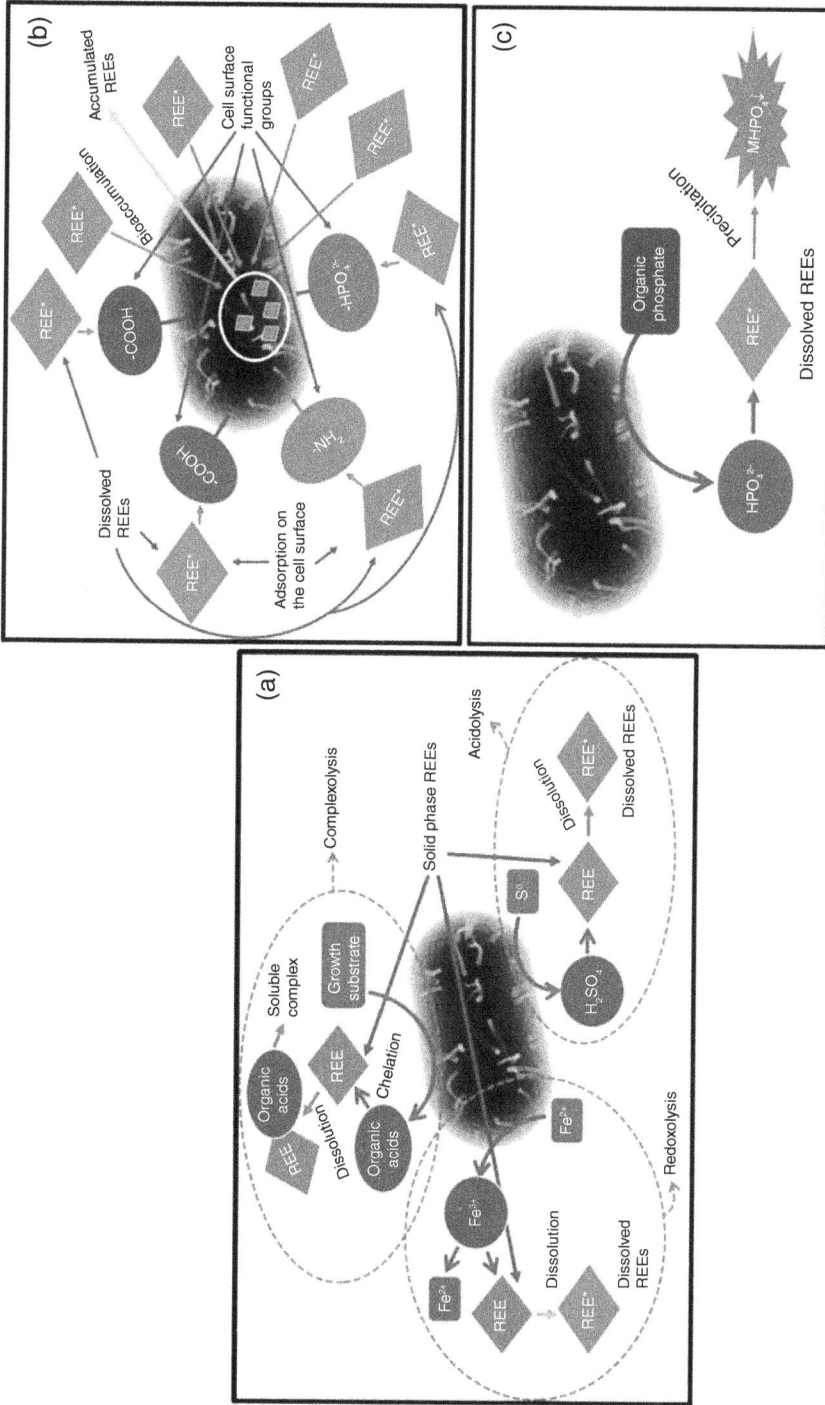

FIGURE 4.9 Mechanism of RE biorecovery by microorganisms: (a) mobilization of REE, (b) biosorption and bioaccumulation, and (c) bioreduction and bioprecipitation. (Source: Dev et al. [77].)

The reactions involving the acidolysis process mediated by both sulfur-oxidizing and phosphate-solubilizing bacteria are as follows:

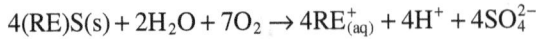

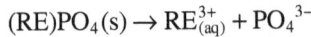

$$4(RE)S(s) + 2H_2O + 7O_2 \rightarrow 4RE^+_{(aq)} + 4H^+ + 4SO_4^{2-}$$

$$(RE)PO_4(s) \rightarrow RE^{3+}_{(aq)} + PO_4^{3-}$$

The sulfur-oxidizing bacteria (e.g., *Acidithiobacillus ferrooxidans*, *Acidithiobacillus thiooxidans*, *Alicyclosbacillus disulfidooxidans*, and *Sulfobacillus acidophilus*) oxidize sulfide to sulfuric acid, thereby causing dissolution of RE. The phosphate-solubilizing bacteria cause dissolution of phosphate ions as well as RE, and these include *Acetobacter chroococcum*, *Bacillus circulans*, *Cladosporium herbarum*, *Bradyrhizobium japonicum*, *Enterobacter agglomerans*, *Pseudomonus chlorora-phis*, and *Rhizobium leguminosarum*. The organic acids secreted by these phosphate-solubilizing bacteria are gluconic, citric, acetic, oxalic, succinic, 2-ketogluconic, and malic acids, depending on the organic carbon employed for their growth [80]. For example, the strain *Acinetobacter* sp. secretes gluconic acid using glucose as a carbon source, whereas it secretes malic acid when mannitol is used as the carbon source [81].

In the process of complexolysis, the organic acids, such as succinic, fumaric, lactic, malic, citric, oxalic, and acetic acid, produced by the sulfate-oxidizing or phosphate-solubilizing bacterial strains form an REE-organic acid complex formation, followed by extraction of the respective REE [82]. Chelating agents such as desferri-oxamine have been used to expedite stable complex formation between the REE and secreted organic acid [83]. Antonick et al. reported the use of spent medium having gluconic acid produced by *Gluconobacter oxydans* to extract Y, Sm, Yb, Eu, Nd, and Ce from synthetic phosphogypsum via formation of gluconate–REE complex [84].

Biosorption is becoming a popular technique for recovery of REEs due to high efficiency of recovery, low sludge production, and ease of operation [85]. The adsorption of REEs on microbial cell surface can occur by electrostatic interactions, ion exchange, complex formation (via carboxyl, amine, and hydrophosphate groups), and precipitation [86,87]. Maleke et al. showed the presence of PO_4^{3-}, $C-PO_3^{2-}$, COOH, and CO groups on the cell surface of *Thermus scotoductus*, which were responsible for specific adsorption of Eu through electrostatic interactions [88]. The specificity of functional groups toward REEs can be seen as shown in Table 4.2. FTIR analysis identifies the functional groups, while X-ray Absorption Spectroscopy shows the specificity of functional groups toward REEs, and Extended X-ray Adsorption Fine Structure analysis shows the role of pH in binding of RE to functional groups on cell's surface [88–91]. A pH of 3–4 favors binding of RE to $-PO_4^{3-}$, while pH of 6–7 favors RE binding to COOH groups. It is also found that lighter REEs (e.g., La and Nd) are adsorbed efficiently at $pH > 4$, whereas medium and heavy REEs (e.g., Gd) get adsorbed strongly at $pH < 4$ [89].

Bioaccumulation involves the intracellular uptake of REEs that have been adsorbed on the cell surface by bioadsorption. The mechanism of bioaccumulation is schematically presented in Figure 4.9b. After adsorption, a complex present in the lipid membrane bilayer of the cell translocates and imports the REEs into intracellu-lar space [88]. During this metabolically active process, the REEs are sequestered by

TABLE 4.2

Specificity of Different Functional Groups on Microbial Cell Surface toward REEs

Functional Group on Cell Surface	Specificity Toward REE	Reference
-PO_4^{3-}	La, Nd, Ce	[89]
-COOH and PO_4^{3-}	Gd, Yb, Er, and Sm	[89,92]
PO_4^{3-}, C-PO_3^{2-}, COOH, and CO	Eu	[88]

the proteins and peptide ligands present in the intracellular space, thereby enabling selective recovery of REEs from a mixed metal solution [93,94]. Some examples of bioaccumulation shown by microorganisms include that of Ce and Nd by *Bacillus cereus* [95]. In this case, exposure to Ce enhanced the expression of the -COOH group on cellular surface. In another example, *Arthrobacter luteolus* eliminates the low solubility of REEs in lipids by formation of lipid-soluble REE–siderophore complex, which moves across the cell membrane [96]. Ochsner et al. used TonB-dependent transporter to transport this lanthanide–siderophore complex by proton motive force in *Methylobacterium extorquens* [97].

Bioprecipitation is a process whereby dissolved REEs can be recovered using microorganisms [98]. Inorganic phosphates are released during microbial metabolism as byproducts along with the phosphate precipitate of REE [99,100].

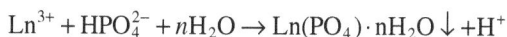

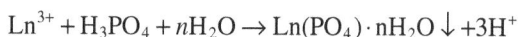

$$Ln^{3+} + HPO_4^{2-} + nH_2O \rightarrow Ln(PO_4) \cdot nH_2O \downarrow + H^+$$

$$Ln^{3+} + H_3PO_4 + nH_2O \rightarrow Ln(PO_4) \cdot nH_2O \downarrow + 3H^+$$

The mechanism of REE bioprecipitation is shown in Figure 4.9c, where enzyme phosphatase present inside the periplasm and extracellular polymeric substances of *Serratia* sp. interacts to form inorganic phosphate (Pi) and enable precipitation of REEs [101]. It was reported that recovery of Nd (>90%) and Eu (>85%) was possible using phosphate-based precipitation process involving *Serratia* sp. immobilized on polyurethane foam [102]. Similarly, recovery of Dy as DyPO$_4$ on the fungi *Penidiella* sp. cell surface at pH 2.5 was reported by Horiike et al. [103]. pH is an important factor that influences solubility of the REE-phosphate precipitate. For instance, pH of 5 reduces the precipitation of La as LaPO$_4$ by *Citrobacter* sp. mediated phosphorylation [104]. Bioprecipitation enables specific precipitation of REEs, thus allowing selective recovery of specific REE. Feng et al. showed how *B. megaterium* caused selective dissolution of heavy REEs (Gd, Dy, Er, Ho, and Tm) and precipitation of light REEs (La, Ce, Pr, Nd, and Sm) as their phosphate salts [105].

4.4.5 PHYSICAL AND THERMAL TECHNIQUES

Separation of lanthanides and REEs can be done by physical and thermal separation methods that can adopt two approaches: (1) pre-consumer recycling (recovery of REEs from manufacturing scrap) and (2) post-consumer recycling (recovery from

REE-bearing discarded consumer products) [106–108]. In the case of metals, recycling process involves (1) cleaning and collection, (2) pre-processing and physical disassembly, and (3) processing and pyrometallurgy (smelting, incineration, combustion, and pyrolysis) [106,107].

Gueroult et al. showed an alternate route to pyrometallurgical process and explored the potential of plasma mass separation using NdFeB magnets [109]. The composition of these magnets is Nd (28%–35%); Pr, Tb, Dy, and Gd: (0%–10%); B (1%–2%); Ni (0%–15%); Al, V, Nb (0%–1%); Fe (≤68%) as shown in Figure 4.10 [110]. Gueroult et al. were successful in recycling Nd from NdFeB magnets, thereby establishing the process for separation of REE from non-REE elements, but require further separation processes to separate one REE from another.

Figure 4.11a shows the flowchart for the plasma separation process. On sending the output stream from Stage 1 from single pass again through the plasma filter (Stage 2) and varying the mass threshold, it is possible to separate heavy REE (Stages 3H) from light REE (Stage 3L), followed by advanced separation processes to further separate one REE from another. Figure 4.11b shows the plot between input stream containing multiple REEs versus output stream containing a mixture of REEs corresponding to single pass in the plasma filter for different filter width [narrow: A, intermediate: B, wide: C] [109].

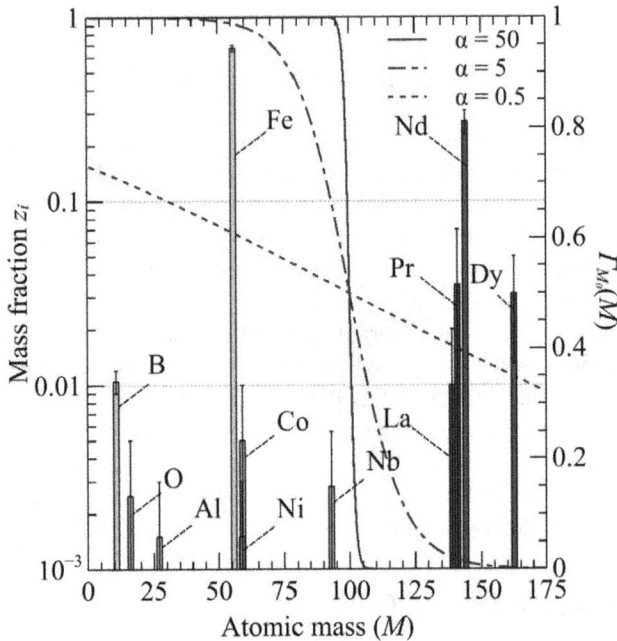

FIGURE 4.10 Composition of rare-earth permanent magnet NdFeB waste. (Source: Gueroult et al. [109] and Firdaus et al. [110].)

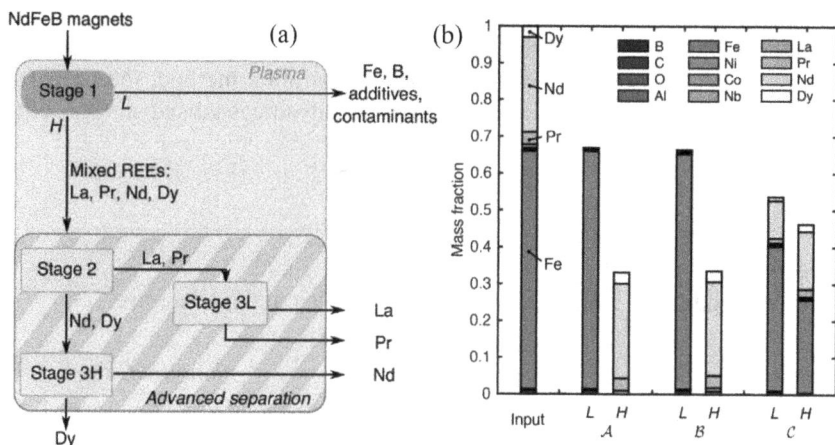

FIGURE 4.11 (a) Flowchart of plasma separation process. (b) Input stream versus mass fraction plot for each separated stream L (left stack) and H (right stack) for different filter width [narrow: *A*, intermediate: *B*, wide: *C*]. (Source: Gueroult et al. [109].)

4.4.5.1 Mechanism Involved in Plasma Separation Technique

Variation of the mass threshold enhances the potential of plasma separation process, especially in the case of NdFeB magnets. For addressing mass discrimination of REEs, analytical filter function is defined as $\Gamma_{M_0}(M) = \left[1 + \tanh\left(\alpha[1 - M/M_0]\right) \right]/2$ where α is the reciprocal of the filter width, M is the element atomic mass, and M_0 is the filter threshold mass (100 amu). For Case A, where filter width is very narrow (an ideal filter), $\alpha = 50$ and $\Gamma_{M_0}(M)$ is a step function. For Case B, where filter width is broader but comparable to the mass gap between non-REEs and REEs, $\alpha = 5$ and $\Gamma_{M_0}(M)$ is variable. For case C, where filter width is wide, $\alpha = 0.5$ and $\Gamma_{M_0}(M)$ is almost linear over the mass range taken. The mass fractions x_i, y_i, and z_i of element i with mass M_i in the heavy, light, and input stream containing all j elements, respectively, are expressed as

$$x_i(\text{heavy}) = \frac{z_i \left[1 - \Gamma_{M_0}(M_i) \right]}{\sum_j z_j \left[1 - \Gamma_{M_0}(M_j) \right]} \quad \text{and} \quad y_i(\text{light}) = \frac{z_i \Gamma_{M_0}(M_i)}{\sum_j z_j \Gamma_{M_0}(M_j)}$$

The separation factor $\left(\beta_i \right)$ and separation efficiency $\left(\eta_i \right)$ are defined as

$$\beta_i = \frac{x_i(1 - z_i)}{z_i(1 - x_i)} \quad \text{and} \quad \eta_i = \frac{\theta x_i}{z_i}$$

where θ is the ratio of output (heavy) stream to input stream [109].

4.5 CHOICE OF DETECTION TECHNIQUES BASED ON POINT OF APPLICATION

The preferred choice of technique, the detection mechanism, and parameters ensuring efficient and sensitive detection depend on various factors, such as available sources, distribution, and leaching, as well as extraction background, extraction theory, extraction technology, separation efficiency, and adsorption capacity [111-112]. Table 4.3 summarizes the separation method, the properties of lanthanides exploited by that method, and the advantages/disadvantages of the method used [106,113,114].

TABLE 4.3
Choice of Separation Method Based on Properties of Lanthanides

Separation Method	Parameters Influencing Separation Capacity	Prominent Features
Fractionation crystallization [115]	Solubility of nitrate/sulfate/borate complexes of REE, variable oxidation state of REE	Large number of steps involved
Solvent extraction [116]	Solubility of metal ions in two non-miscible solvents	Purification and separation required at different extraction stages
Liquid/gas chromatography [117]	Centrifugal mixer speed, sedimentation time	Used with various aqueous solutions with simple scaling improvements
Liquid–liquid separation [118]	Solubility, pH sensitivity, selective enrichment behavior of ions	Group separation of elements into light, intermediate, and heavy element streams
Organophosphorus extraction [119]	Solubility, pH, precipitation as phosphate salts	Good separation factor, especially for light REE [di-(2-ethylhexyl) phosphinic acid (P227/P229)]; Susceptibility toward emulsification needs to be controlled
Amine extraction [120]	Complexation of REE with quaternary ammonium $R_4N^+X^-$	X^- transfers ions to organic phase of the complex anion, which is formed by complexation with RE
Extraction by diglycolamides [121]	Affinity of REE toward amide skeleton; distribution ratios and equilibrium constants of REE toward length of amide group alkyl chain	Glucosamine and N,N', N'-tetraoctyl-3-oxoprenediamide have a strong affinity toward Ln^{3+}
Ionic liquid extraction [122]	Solubility, pH, complexing, dilution factor	Selective and sustainable extraction
Adsorption-desorption separation [123]	Chelating agent, desorption efficiencies, desorbing agent concentration, the solid–liquid phase ratio, contact time	Efficient and economical for extraction of REE from solution
Ion-exchange extraction [124]	Resins, eluents, and pH	No clarity in the separation of equal amounts of REE at higher concentration
High-efficiency chelation ion chromatography (HPCIC) [125]	Formation/dissociation of surface ligand functionalized stationary phase complexes	Suitable for distinct separation of REE without losing the organic part of the ligand

(Continued)

TABLE 4.3 (*Continued*)
Choice of Separation Method Based on Properties of Lanthanides

Separation Method	Parameters Influencing Separation Capacity	Prominent Features
Solid extraction through magnetic adsorbent [126]	External magnetic field, leaching efficiency of REE	Leaching behavior of REE needs to be controlled to improve recovery rate
Membrane separation [127]	Pore size of membrane, adsorbing capacity, ionic radii of REE	Shape and size can be adjusted for selective and targeted recovery of REE; high membrane replacement cost
Plasma separation [109]	Ionization energy of REE, filter width	Minimal waste and environment-friendly
Separation using Microfluidic device [128]	Phase separation time in REE extraction, solvent dilution	High separation efficiency of heavy REE
Fluorescence detection followed by extraction [129]	Light absorptive properties, Concentration of fluorophore, temperature of solution medium, pH, solvent, ionic radii of REE	High recovery and efficiency
Electrophoresis [58]	Ligand concentration, pH, ionic radii of REE	Separation among Ln^{3+} is difficult due to negligible differences in electrophoretic mobilities of Ln^{3+}
Electrochemical separation [130]	Leaching behavior of REE, type of electrolyte, electrode assembly, applied voltage	High separation efficiency of small particles
Electrodialysis [131]	Electric potential, ion-exchange membrane	Low pressure and high purity of REEs obtained
Electrorefining [69]	Current, separation rate of REE with respect to migration of metal ion toward cathode	High purity of extracted REE
Electrowinning [132]	Electrorefining process, solid-state electro transport	Ultrapure REEs with low sludge formation
Phytoremediation [133]	Hyperaccumulating properties of crops/leaves/shoot	Environment-friendly, but non-selective separation of REE
Biosorption [134]	Binding capacity, ionic concentration in aqueous solution	High removal efficiency of REE
Biometallurgy [135]	Binding capacity of REE with chelating complex on microbial cell surface, transport of REE across cell membrane	Low temperature and energy required, tendency of microbial contamination
Pyrometallurgy [136]	Temperature of blast furnace, separation time, removal of slag	High energy consumption, but high efficiency of REE recovery
Hydrometallurgy [137]	redox potential, leaching kinetic behavior, particle size, pH, temperature, agitation, sludge removal	High generation of sludge, but high efficiency of REE recovery

4.6 CRITICAL EXAMINATION AND SANITY CHECKS OF DETECTION TECHNIQUES

Some of the detection and separation techniques mentioned in Table 4.3 are critically examined in this section using some case examples covering recent progress made by researchers in this field, and the checkpoints that they had come across in adopting the specific technique.

Akbar et al. in 2021 analyzed the aqueous chemistry of polydentate tripodal ligand 5,50,50′-(Cyclohexane-1,3,5-triyltris(oxy))tris(methylene)triquinolin-8-ol (CYTOM5OX) with Eu and Tb [129]. These lanthanides form symmetrical tripodal ligand when 8-hydroxyquinoline is used as the chelating ligand. By luminescence spectra, Akbar et al. were able to affirm the lanthanide species and their corresponding formation constants. Upon titration, it was found that 1:1 complex solutions of Eu and Tb ions formed complex with chelator at a pH of 1.8–10.2. From the results of luminescence titrations, plots of fluorescence versus pH for these Eu and Tb complexes were drawn. For Eu(CYTOM5OX) system, five complexes [[Eu(H$_3$L)]$^{3+}$, [Eu(H$_2$L)]$^{2+}$, [Eu(HL)]$^+$, EuL, and [EuL(OH)]$^-$], were found from optimized fit. For the Tb(CYTOM5OX) system, on optimization of fluorescence-pH plot, Tb(H$_4$L)$^{4+}$, Tb(H$_3$L)$^{3+}$, Tb(H$_2$L)$^{2+}$, and Tb(HL), TbL and TbL$_2$, and one hydroxo species were obtained. The stability constants are determined by spectrometric batch titrations using EDTA in the pH range 10.2–2.1 and by substituting equilibrium constants corresponding to Ln-EDTA, K$_a$ of EDTA in

$$LnY + L \rightleftharpoons LnL + Y \quad K = \frac{[LnL][Y]}{[LnY][L]} = \frac{\beta_{110}^{LnL}}{\beta_{110}^{LnY}}$$

The value of log β_{110} for Eu (33.28) and Tb (34.19) complexes were obtained. The results tend to be misleading since relative affinities of the lanthanide ions for a ligand Y are compared based on log K values, whereas the competition of lanthanide ions with H$^+$ for the ligand Y is not considered. The pM values provide information about the relative sequestering ability of the ligands under some defined conditions. Hence, the pLn for CYTOM5OX chelates was calculated to determine LnY stability for Eu and Tb ions and were between 21 and 22. Figure 4.12 presents the plot between pLn and pH (2–9), and it is observed that sequestering ability of ligand Y increases with the increase in the atomic number of lanthanide ions. Higher stability constants indicate soluble and thermodynamically stable complexes in aqueous solution. It was found that Tb[CYTOM5OX] has a longer excited-state lifetime than its Eu counterpart. The increase in coordinated water molecules for a complex reduce the luminescence lifetimes.

Areti et al. showed how a hydrophilic glucopyranosyl conjugate (L) shows 75-fold enhanced fluorescence (green) with La^{3+} among the other Ln^{3+} ions in the HEPES buffer (pH = 7.4) with a minimum detection limit of 140 nM (16 ± 2 ppb) for La^{3+} (Figure 4.13) [138]. The glucosyl-amino conjugate of 8-hydroxyquinoline (L) has optimum water solubility, biological compatibility, metal ion binding core, and absorbs and emits visible light. Thus, complex of L with La^{3+} can be used for reversible sensing of La^{3+} and be used for cell imaging applications. Figure 4.13a and b

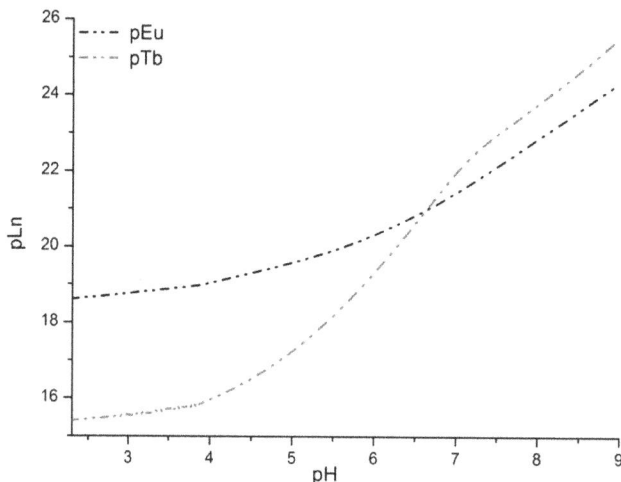

FIGURE 4.12 Plot of pLn versus pH for CYTOM5OX, pLn calculated for Eu^{3+} and Tb^{3+} when $[Ln^{3+}] = 10^{-5}$ M, [CYTOM5OX] = 10^{-5} M. (Source: Akbar et al. [129].)

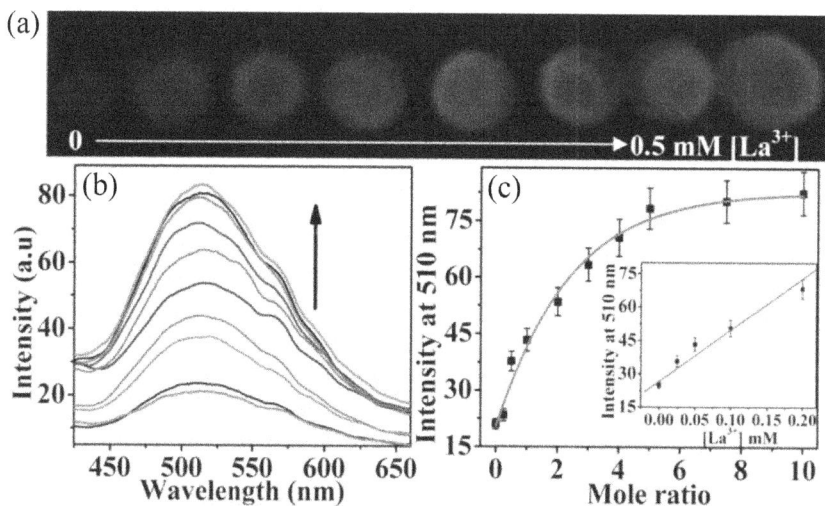

FIGURE 4.13 (a) Samples used in the titration on Whatman cellulose filter paper under $\lambda = 365$ nm. (b) Fluorescence spectra on titration of L ($\lambda_{exc} = 360$ nm) with La^{3+} on Whatman cellulose filter paper. (c) Intensity vs. $[La^{3+}]/[L]$ mole ratio at $\lambda = 510$ nm (inset: linear concentration region in plot of intensity vs. $[La^{3+}]$ for the receptor L). (Source: Areti et al. [138].)

shows the sensitivity of L toward concentration of La^{3+} on Whatman filter paper strips in luminescence titration at an excited wavelength of 360 nm. As the concentration of La^{3+} increased, L shows increased emission intensity at 510 nm (Figure 4.13c). Based on concentration dependence studies, the detection limit of L toward La^{3+} on Whatman cellulose paper strips is $10 \pm 1\,\mu M$ (1.3 ppm).

Byracki et al. reported the changes in fluorescence intensity upon complexation of Ln^{3+} (especially La^{3+}, Gd^{3+}, Tb^{3+}, Dy^{3+}, Er^{3+}, and Yb^{3+}) complexation with calix [4] azacrown ether based Bodipy probe [139]. This probe was highly selective toward Yb^{3+}. The binding of Ln^{3+} to the Bodipy probe was found to depend on the presence of two proximal $CONH_2$ groups in the probe, while two -COOH groups of the crown ether ring are selective toward lanthanides having very similar ionic radii, rather than those with larger or smaller radii. The complex mechanism is influenced by concentration of fluorophore, temperature of solution medium, pH, solvent, and metal ionic radii.

Bekiari et al. used a molecule having diazostilbene chromophore and a benzo-15-crown-5 ether moiety as an efficient fluorescent probe for Ln^{3+} ions (Nd^{3+}, Pr^{3+}, Ce^{3+}) depending on the suitable ionic radii of Ln^{3+} which can fit into the cavity of crown ether to form stable and functional complexes. The order of ionic radius (in pm) is In (96) < Er (103) < Tb (106) < Eu (109) < Nd (112) < Pr (113) < Ce (115) [140].

Schelter et al. showed that proligand H_3TriNO_x ($[(2\text{-tBuNOH})C_6H_3CH_2]_3N$) forms complexes with rare-earth metals and enables separations of simple mixtures of rare-earth metal salts (especially Nd/Dy and Eu/Y for rare-earth recycling) [141]. Prof. Schelter has been recognized with the US EPA Green Chemistry Challenge: 2017 Academic Award for his pioneering work on a simple and targeted approach for recycling of REE from consumer goods (batteries, cellphones, etc.) using tailored metal complexes with organic ligands, such as tris(2-tert-butylhydroxylaminato)benzylamine (H_3TriNO_x). An electron-donating derivative of the proligand H_3TriNO_xR ($[(2\text{-tBuNOH})C_6H_3CH_2]_3N$; R = 5-Ome) improved the separation efficiency by influencing the electronic and physical properties of the complex. Schelter et al. probed the effect of substituents on the proligand by using electron-donating and -withdrawing groups along the aryl-backbone (R = 4-tBu, 5-Ph, 4-CF$_3$) to form new proligands. These proligands coordinated to Nd and Dy by protonolysis reactions, and dimerization equilibrium constants as well as molar solubility of the resulting complexes were determined. It was found that greater electron donation of the aryl-substituents increased the extent of dimerization of the Nd complexes. Hence, Nd/Dy mixture was separated based on dimerization equilibrium and solubility differences. The study was extended to other binary mixtures of REs, such as Eu/Y. Hence, proligands having $TriNO_x^{3-}$ framework can be modified to target specific RE separations.

4.7 LANTHANIDES-ACTINIDE CHEMISTRY AND THEIR SEPARATION TECHNIQUES

As we progress in the separation techniques of lanthanides, the separation of actinides [Ac (Z = 89) to Lr (Zr = 103)] follows similar techniques, owing to the similar chemical behavior of lanthanides (4f) and actinides (5f) elements. The separation of Ln from An is very important in the spent nuclear fuel obtained during the nuclear fuel cycle [142] difficult due to their similar chemical properties, and researchers are trying out new and efficient ways for their separation. The transuranic elements (e.g., Np, Am, and Cu) are formed in small amounts on activation of U and Po in a nuclear reactor, and these comprise minor actinides (MAs). These MAs have long half-lives, ranging from 10^2 to 10^6 years, and pose environmental hazard due to their migration from the repository to the environment as a result of heat generated inside the nuclear fuel reactor. Therefore, there is need to convert these MAs into short-lived or stable nuclides by burning (or transmutation) them a high energy/high flux reactor or in accelerator-driven subcritical system in high neutron flux. For recycling these MAs in the transmutation fuel cycle, separation of An^{3+} from Ln^{3+} is required for preventing problems in fuel fabrication and during neutron irradiation processes. Lanthanides when present in large amounts in the spent fuel do not form solid solutions with actinides and separate in different phases, especially at higher temperatures [143]. The MAs concentrate in these phases, leading to non-uniform heat distribution inside the fuel matrix on irradiation, which results in further problems. In addition to this, there are high chances of neutron poisoning effect of some lanthanide nuclides. These concerns emphasize the separation of the lanthanides from the actinides before the transmutation of the actinides. Some of the successful separation processes are TALSPEAK (Trivalent Actinide Lanthanide Separations by Phosphorus-reagent Extraction from Aqueous Complexes) process using aminopolycarboxylic acid developed by Weaver and Kappelmann [144]; TRUEX (TransUranium EXtraction) employing carbamoylmethyl phosphine oxide (CMPO) in nirtic acid; TeRtiary AMine EXtraction (TRAMEX) process, using tertiary amine in highly concentrated chloride media (e.g., 1M LiCl); [145]; recovery of yttrium and lanthanides by addition of a flocculant (like polyacrylamide) to the phosphoric acid, followed by removal of uranium from the precipitate left behind [146]; Ln/An using separation di-ethylenetriamine-penta acetic acid [147]; use of lipophilic/hydrophilic "N," "O," and/or "S" donor heteropolycyclic ligands as the actinide selective ligand [148]. However, significant progress in Ln/An separation still needs to be made in finding suitable ligands having favorable features with respect to fast kinetics, high selectivity over all Ln^{3+} ions, high stability (thermal, radiolytic, in low pH conditions, and in hydrolysis reactions), good separation efficiency, solubility in organic diluents, and ability to undergo complete incineration to reduce the secondary waste generation, among others. Also, green extraction methods like membrane extraction, ionic liquid-based extraction, and solid phase extraction need to be further refined for dealing with large quantities of nuclear spent fuel and radioactive wastes.

4.8   CONCLUSIONS AND FUTURE SCOPE

Lanthanides comprising 15 elements from lanthanum (Z = 57) to lutetium (Z = 71) are focused upon in this chapter and their properties that influence their application are discussed, followed by their detection and separation techniques. Their oxidation states [+2, +3 (most preferred), +4], coordination properties (8–12 coordination number) and flexible, sterically driven coordination geometries form the basis for their catalytic behavior. The rich variety of electronic levels of the order $14!/n!(14-n)!$ result in interesting optical properties like antenna effect in lanthanide-based functional materials, which aids their use as luminescent sensors. The paramagnetic behavior of lanthanides (especially Gd, Tb, Dy, Ho, Er, Tu, and Y) helps in their use as magnetic storage disks and in MRI applications. However, these wide-ranging applications have reduced their availability and can pose a serious problem in the future, especially in applications which cannot do without the use of lanthanides. Hence, detection, separation, and recycling of lanthanide ions from their functional materials is necessary for maintaining continuous supply of lanthanides for their use in medical diagnostics, electronics, clean energy, automotive, aerospace, and environmental applications. For this purpose, stereochemistry involved in lanthanide-based complexes and properties (similar oxidation state, ionic radii, coordination number, geometries) of lanthanide elements that pose as impediments to their separation (either from other elements or among themselves) from functional materials are discussed briefly before moving onto the detection and separation techniques. To delve into the details of lanthanide detection and separation techniques (chemical, fluorescent, electrochemical, biological, physical, and thermal), their critical aspects, the mechanism involved, and parameters influencing separation efficiency are discussed with the help of research works from prominent researchers in the field of lanthanide separation. The preferred choice of detection and separation techniques, critical examination, and sanity checks of these techniques for ensuring efficient and sensitive detection are comprehensively described to aid researchers working in this field. A preview to Ln^{3+}/Ac^{3+} separation techniques is also given to understand the importance of Ln^{3+} separation from An^{3+} in the fuel matrix before the transmutation of MAs in the nuclear spent fuel.

The research in the recovery of REEs from their functional materials is rapidly growing over the last decade. There is significant transformation and progress in the separation techniques employed for REE separation from more energy-intensive methods (e.g., hydrometallurgy, pyrometallurgy, and solvent extraction) to greener and sustainable methods (e.g., electrochemical, ionic liquid-based extraction, solid phase extraction, plasma separation, and phytomining). Further research for promising environmental-friendly REE separation/recovery techniques, which can be employed to handle multiple feedstocks and large amounts of electronic/nuclear waste, is required. This requirement can be met using (1) hyperaccumulator plant species which can collect large concentrations of lanthanides (e.g., *Phytolacca acinosa*) in conjunction with microorganisms used for biosorption of Ln, (2) variable magnetic fields to exploit paramagnetic and ferromagnetic properties of Ln, (3) tuning the Ln-ligand binding affinities for effective extraction using ancilliary ligands (e.g., DO_2A, that is, 1,4,7,10-tetraazacyclododecane-1,7-bisacetate), and (4) ensuring high purity of recovered REE using carbon-based materials (e.g., oxidized multiwalled nanotubes).

REFERENCES

1. Contruvo, J. A. The chemistry of lanthanides in biology: Recent discoveries, emerging principles, and technological applications. *ACS Cent. Sci.* 2019; 5: 1496–1506.
2. Erickson, B. Rare-earth recovery: U.S. efforts to extract valuable elements from coal waste surge. *Chem. Eng. News* 2018; 96(28): 29–33.
3. Vogt, C., Klunder, G. Separation of metal ions by capillary electrophoresis - Diversity, advantages, and drawbacks of detection methods. *Fresenius J. Anal. Chem.* 2001; 370: 316–331.
4. Nash, K. L., Jensen, M. P. Analytical-scale separations of the lanthanides: A review of techniques and fundamentals. *Sep. Sci. Technol.* 2001; 36(5-6): 1257–1282.
5. Pourjavid, M. R., Norouzi, P., Ganjali, M. R., Nemati, A., Zamani, H. A., Javaheri, M. Separation and determination of medium lanthanides: A new experiment with use of ion-exchange separation and fast Fourier transform continuous cyclic voltammetry. *Int. J. Electrochem. Sci.* 2009; 4: 1650–1671.

6. Bogart, J. A., Cole, B. E., Boreen, M. A., Lippincott, C. A., Manor, B. C., Carroll, P. J., Schelter, E. J. Accomplishing simple, solubility-based separations of rare earth elements with complexes bearing size-sensitive molecular apertures. *Proc. Natl. Acad. Sci. USA*. 2016; 113: 14887–14892.

7. Fang, H., Cole, B. E., Qiao, Y., Bogart, J. A., Cheisson, T., Manor, B. C., Carroll, P. J., Schelter, E. J. Electro-kinetic separation of rare earth elements using a redox-active ligand. *Angew. Chem. Int. Ed.* 2017; 56: 13450–13454.

8. Li, X.-Z., Zhou, L.-P., Yan, L.-L., Dong, Y.-M., Bai, Z.-L., Sun, X.-Q., Diwu, J., Wang, S., Bünzli, J.-C., Sun, Q.-F. A supramolecular lanthanide separation approach based on multivalent cooperative enhancement of metal ion selectivity. *Nat. Commun.* 2018; 9: 547.

9. Gao, H. Y., Peng, W. L., Meng, P. P., Feng, X. F., Li, J. Q., Wu, H. Q., Yan, C. S., Xiong, Y. Y., Luo, F. Lanthanide separation using size-selective crystallization of Ln-MOFs. *Chem. Commun.* 2017; 53: 5737–5739.

10. Haxel, G. B., Hedrick, J. B., Orris, G. J. Rare earth elements -Critical resources for high technology. U.S.G.S. Fact Sheet 2002, 087–02.

11. Kostelnik, T. I., Orvig, C. Radioactive main group and rare earth metals for imaging and therapy. *Chem. Rev.* 2019; 119: 902–956.

12. Firsching, F. H., Brune, S. N. Solubility products of the trivalent rare-earth phosphates. *J. Chem. Eng. Data* 1991; 36: 93–95.

13. Higgins, R. F., Ruoff, K. P., Kumar A., Schelter, E. J. Coordination chemistry-driven approaches to rare earth element separations. *Acc. Chem. Res.* 2022; 55(18): 2616–2627.

14. Bunzli, J. C., Piguet, C. Taking advantage of luminescent lanthanide ions. *Chem. Soc. Rev.* 2005; 34: 1048–1077.

15. Eliseeva, S. V., Bunzli, J. C. Lanthanide luminescence for functional materials and bio-sciences. *Chem. Soc. Rev.* 2010; 39: 189–227.

16. Zhao, S.-N., Wang, G., Poelman, D., Van Der Voort, P. Luminescent lanthanide MOFs: A unique platform for chemical sensing. *Materials* 2018; 11(4): 572.

17. Bünzli, J.-C. G. (2016). Lanthanide luminescence: From a mystery to rationalization, understanding, and applications. In: Jean-Claude G. Bünzli (ed.) *Handbook on the Physics and Chemistry of Rare Earths*, vol. 50, pp. 141–176, Elsevier: Amsterdam, The Netherlands.

18. Lumpe, H., Pol, A., Op den Camp, H. J. M.; Daumann, L. J. Impact of the lanthanide contraction on the activity of a lanthanidedependent methanol dehydrogenase - a kinetic and DFT study. *Dalton Trans.* 2018; 47: 10463–10472.

19. Cotton, S. A., Raithby, P. R. Systematics and surprises in lanthanide coordination chemistry. *Coord. Chem. Rev.* 2017; 340: 220–231.

20. Shannon, R. D. Revised effective ionic radii and systematic studies of interatomic distances in halides and chalcogenides. *Acta Crystallogr. Sect. A: Cryst. Phys., Diffr. Theor. Gen. Crystallogr.* 1976; 32: 751–767.

21. Seitz, M., Oliver, A. G., Raymond, K. N. The lanthanide contraction revisited. *J. Am. Chem. Soc.* 2007; 129: 11153–11160.

22. Horrocks, W. D., Sudnick, D. R. Lanthanide ion probes of structure in biology. Laser-induced luminescence decay constants provide a direct measure of the number of metal-coordinated water molecules. *J. Am. Chem. Soc.* 1979; 101: 334–340.

23. Evans, C. H. (1990). Biochemistry of the Lanthanides. New York: Plenum Press.

24. Firsching, F. H., Brune, S. N. Solubility products of the trivalent rare-earth phosphates. *J. Chem. Eng. Data* 1991; 36: 93–95.

25. Li, X.-Z., Zhou, L.-P., Yan, L.-L., Dong, Y.-M., Bai, Z.-L., Sun, X.-Q., Diwu, J., Wang, S., Bünzli, J.-C., Sun, Q.-F. A supramolecular lanthanide separation approach based on multivalent cooperative enhancement of metal ion selectivity. *Nat. Commun.* 2018; 9(1): 1–10.

26. Yan, L.-L., et al. Stereocontrolled self-assembly and self-sorting of luminescent europium tetrahedral cages. *J. Am. Chem. Soc.* 2015; 137: 8550–8555.

27. Zebret, S., Besnard, C., Bernardinelli, G., Hamacek, J. Thermodynamic discrimination in the formation of tetranuclear lanthanide helicates. *Eur. J. Inorg. Chem.* 2012; 2012: 2409–2417.

28. Johnson, A. M., Young, M. C., Zhang, X., Julian, R. R., Hooley, R. J. Cooperative thermodynamic control of selectivity in the self-assembly of rare earth metal-ligand helices. *J. Am. Chem. Soc.* 2013; 135: 17723–17726.

29. Cable, M. L., Kirby, J. P., Gray, H. B., Ponce, A. Enhancement of anion binding in lanthanide optical sensors. *Acc. Chem. Res.* 2013; 46(11): 2576–2584. doi:10.1021/ar400050t.

30. Bünzli, J. C. G., Eliseeva, S. V. (2010). Basics of lanthanide photophysics. In: Hänninen, P., Härmä, H. (eds.) *Lanthanide Luminescence*. Springer Series on Fluorescence, vol. 7. Berlin: Springer.

31. Parker, D. Luminescent lanthanide sensors for pH, pO2 and selected anions. *Coord. Chem. Rev.* 2000; 205: 109–130.

32. Sorace, L., Gatteschi, D. (2015). Electronic structure and magnetic properties of lanthanide molecular complexes. In: Layfield, R. A., Murugesu, M. (eds.) *Lanthanides and Actinides in Molecular Magnetism*, vol. 2, pp. 1–26, Wiley-VCH Verlag GmbH & Co. KGaA: Boschstr. 12, 69469 Weinheim, Germany.

33. Cullity, B. D., Graham, C. D. (2011). *Introduction to Magnetic Materials*. Hoboken, NJ: John Wiley & Sons; Schüßler-Langeheine, C., Weschke, E., Ott, H., Grigoriev, A. Y., Möller, A., Meier, R., ... Kaindl, G. Magnetic effects in the band structure of ferromagnetic and antiferromagnetic lanthanide-metal films. *J. Electron Spectrosc. Relat. Phenom.* 2001; 114-116: 795–799. doi:10.1016/s0368-2048(00)00316-9.

34. Bünzli, J.- C. G. (2000). Lanthanides. In: Jean-Claude G. Bu¨ Nzli (ed.) *Kirk-Othmer Encyclopedia of Chemical Technology*, pp. 1–43, Wiley: Chichester, U.K.

35. Corsi, D. M., Platas-Iglesias, C., van Bekkum, H., Peters, J. A. Determination of paramagnetic lanthanide(III) concentrations from bulk magnetic susceptibility shifts in NMR spectra. *Magn. Reson. Chem.* 2001; 39(11): 723–726.

36. Cheisson, T., Schelter, E. J. Rare earth elements: Mendeleev's bane, modern marvels. *Science* 2019; 363(6426): 489–493.

37. Dolai, S., Bhunia, S. K., Zeiri, L., Paz-Tal, O., Jelinek, R. Thenoyltrifluoroacetone (TTA)-carbon dot/aerogel fluorescent sensor for lanthanide and actinide ions. *ACS Omega* 2017; 2(12): 9288–9295.

38. Li, Q., Sun, K., Chang, K., Yu, J., Chiu, D. T., Wu, C., Qin, W. Ratiometric luminescent detection of bacterial spores with terbium chelated semiconducting polymer dots. *Anal. Chem.* 2013; 85(19): 9087–9091.

39. Wu, S., Min, H., Shi, W., Cheng, P. Multicenter metal-organic framework-based ratiometric fluorescent sensors. *Adv. Mater.* 2019; 32: 1805871.

40. Heffern, M. C., Matosziuk, L. M., Meade, T. J. Lanthanide probes for bioresponsive imaging. *Chem. Rev.* 2014; 114: 4496.

41. Cui, Y. J., Xu, H., Yue, Y. F., Guo, Z. Y., Yu, J C., Chen, Z. X., Gao, J. K., Yang, Y., Qian, G. D., Chen, B. L. A luminescent mixed-lanthanide metal-organic framework thermometer. *J. Am. Chem. Soc.* 2012; 134: 3979.

42. Rao, X. T., Song, T., Gao, J. K., Cui, Y. J., Yang, Y., Wu, C. D., Chen, B. L., Qian, G. D. A highly sensitive mixed lanthanide metal-organic framework self-calibrated luminescent thermometer. *J. Am. Chem. Soc.* 2013; 135: 15559.

43. Zhou, J., Li, H., Zhang, H., Li, H., Shi, W., Cheng, P. A bimetallic lanthanide metal-organic material as a self-calibrating color-gradient luminescent sensor. *Adv. Mater.* 2015; 27(44): 7072–7077.

44. Zhang, S. Y., Shi, W., Cheng, P., Zaworotko, M. J. A mixed-crystal lanthanide zeolite-like metal-organic framework as a fluorescent indicator for lysophosphatidic acid, a cancer biomarker. *J. Am. Chem. Soc.* 2015; 137: 12203.

45. Zhang, Y., Li, B., Ma, H., Zhang, L., Zheng, Y. Rapid and facile ratiometric detection of an anthrax biomarker by regulating energy transfer process in bio-metal-organic framework. *Biosens. Bioelectron.* 2016; 85: 287–293.

46. Wang, D. B., Tan, Q. H., Liu, J. J., Liu, Z. L. A stable europium metal-organic framework as a dual-functional luminescent sensor for quantitatively detecting temperature and humidity. *Dalton Trans.* 2016; 45: 18450.

47. Yang, Z. R., Wang, M. M., Wang, X. S., Yin, X. B. Boric-acid-functional lanthanide metal-organic frameworks for selective ratiometric fluorescence detection of fluoride ions. *Anal. Chem.* 2017; 89: 1930.

48. Gawas, P. P., Ramakrishna, B., Veeraiah, N., Nutalapati, V. Multifunctional hydantoins: Recent advances in optoelectronics and medicinal drugs from Academia to the chemical industry. *J. Mater. Chem. C* 2021; 9(46): 16341–16377.

49. Fieser, M. E., et al., Structural, spectroscopic, and theoretical comparison of traditional vs recently discovered Ln2+ ions in the [K (2.2. 2-cryptand)][(C5H4SiMe3) 3Ln] complexes: the variable nature of Dy2+ and Nd2+. *J. Am. Chem. Soc.* 2015; 137: 369–382.

50. Donohue, T. Photochemical separation of cerium from rare earth mixtures in aqueous solution. *Chem. Phys. Lett.* 1979; 61: 601–604.

51. Kendall, J., Crittenden, E. D., The separation of isotopes, *Proc. Natl. Acad. Sci. USA* 1923; 9: 75; Kendall, J., Separations by the ionic migration method, *Science* 1928; 67: 163–167.

52. Konstantinov, B. P., Lyadov, N. S., Oshurkova, O. V., Separation of Rare-Earth Elements by Ionic Mobilities, *Zh. Prikl. Khim.* 1972; 5: 963–969.

53. Everaerts, F. M., Beckers, J. L., Verheggen, T. P. E. M. (1975). Isotachophoresis. *Theory, Instrumentation and Applications*, Amsterdam: Elsevier, pp. 1–4.

54. Gebauer, P., Bocek, P. Recent progress in capillary isotachophoresis, *Electrophoresis* 2002; 23: 3858–3864.

55. Janoš, P. Analytical separations of lanthanides and actinides by capillary electrophoresis, *Electrophoresis* 2003; 24: 1982–1992.

56. Nakatsuka, I., Taga, M., Yoshida, H. Separation of lanthanides by capillary tube isotachophoresis using complex-forming equilibria, *J. Chromatogr.* 1981; 205: 95–102.

57. Foret, F., Fanali, S., Nardi, A., Bocek, P. Capillary zone electrophoresis of rare earth metals with indirect UV absorbance detection, *Electrophoresis* 1990; 11: 780–783.

58. Janoš, P. Analytical separations of lanthanides and actinides by capillary electrophoresis. *Electrophoresis* 2003; 24(1213): 1982–1992.

59. Collins, G. E., Lu, Q. Radionuclide and metal ion detection on a capillary electrophoresis microchip using LED absorbance detection. *Sens. Actuators B* 2001; 76: 244–249.

60. Timerbaev, A. R. Recent progress in capillary electrophoresis of metal ions. *Electrophoresis* 2000; 21: 4179–4191.

61. Liu, B.-F., Liu, L.-F., Cheng, J.-K. Analysis of inorganic cations as their complexes by capillary electrophoresis. *J. Chromatogr. A* 1999; 834: 277–308.

62. Krishnamurthy, N., Gupta, C. K. (2016). Extractive Metallurgy of Rare Earths, Boca Raton, FL: CRC Press.

63. Van den Bogaert, B., Havaux, D., Binnemans, K., Van Gerven, T. Photochemical recycling of europium from Eu/Y mixtures in red lamp phosphor waste streams. *Green Chem.* 2015; 17: 2180–2187.

64. Fang, H., Cole, B.E., Qiao, Y., Bogart, J.A., Cheisson, T., Manor, B.C., Carroll, P.J. Schelter, E.JElectro-kinetic Separation of Rare Earth Elements Using a Redox-Active Ligand. *Angew. Chem. Int. Ed.* 2017; 56: 13450–13454.

65. Yang, Y., Lan, C., Guo, L., An, Z., Zhao, Z., Li, B. Recovery of rare-earth element from rare-earth permanent magnet waste by electro-refining in molten fluorides. *Sep. Purif. Technol.*, 2019; 233: 116030.

66. Wang, D.-D., Liu, Y.-L., Yang, D.-W., Zhong, Y.-K., Han, W., Wang, L., Chai, Z.-F., Shi, W.-Q. Separation of uranium from lanthanides (La, Sm) with sacrificial Li anode in LiCl-KCl eutectic salt. *Sep. Purif. Technol.* 2022; 292: 121025.

67. Timernaev, A. R. Strategies for selectivity control in capillary electrophoresis of metal species. *J. Chromatogr. A* 1997; 792: 495–518.

68. Kuznetsov, S. A., Gaune-Escard, M. Redox electrochemistry and formal standard redox potentials of the Eu (III)/Eu (II) redox couple in an equimolar mixture of molten NaCl/KCl. *Electrochim. Acta* 2001; 46: 1101.

69. Kuznetsov, S. A., Gaune-Escard, M. (2005). Electrochemistry and electrorefining of rare earth metals in chloride melts. In: Proceedings of 7-th International Symposium on Molten Salts Chemistry and Technology, vol. 2.

70. Yamana, H., Fujii, T., Shirai, O. (2003). *International Symposium on Ionic Liquids in Honour of* Mar*celle Gaune-Escard*, Carry le Rouet, France.

71. Rumble, J. R. (2019). CRC Handbook of Chemistry and Physics, 100th ed. (Internet Version 2019); Boca Raton, FL: CRC Press/Taylor & Francis.

72. Perrin, D. D. (1982). Ionisation Constants of Inorganic Acids and Bases in Aqueous Solution, 2nd ed.; New York: Pergamon Press.

73. Martinez-Gomez, N. C.; Vu, H. N.; Skovran, E. Lanthanide chemistry: From coordination in chemical complexes shaping our technology to coordination in enzymes shaping bacterial metabolism. *Inorg. Chem.* 2016; 55: 10083−10089.

74. Picone, N., Op den Camp, H. J. M. Role of rare earth elements in methanol oxidation. *Curr. Opin. Chem. Biol.* 2019; 49: 39−44.

75. Pol, A., Barends, T. R. M., Dietl, A., Khadem, A. F., Eygensteyn, J., Jetten, M. S. M., Op den Camp, H. J. M. Rare earth metals are essential for methanotrophic life in volcanic mudpots. *Environ. Microbiol.* 2014; 16: 255–264.

76. McSkimming, A., Cheisson, T., Carroll, P. J., Schelter, E. J. *J. Am. Chem. Soc.* 2018; 140: 1223–1226; Picone, N., Op den Camp, H. J. M. Role of rare earth elements in methanol oxidation. *Curr. Opin. Chem. Biol.* 2018; 49: 39–44.

77. Dev, S., Sachan, A., Dehghani, F., Ghosh, T., Briggs, B. R., Aggarwal, S. Mechanisms of biological recovery of rare-earth elements from industrial and electronic wastes: A review. *Chem. Eng. J.* 2020; 397: 124596.

78. Cotruvo Jr., J. A., Featherston, E. R., Mattocks, J. A., Ho, J. V., Laremore, T. N. Lanmodulin: a highly selective lanthanide-binding protein from a lanthanide-utilizing bacterium. *J. Am. Chem. Soc.* 2018; 140: 15056–15061.

79. Sethurajan, M., van Hullebusch, E. D., Nancharaiah, Y. V. Biotechnology in the management and resource recovery from metal bearing solid wastes: Recent advances, *J. Environ. Manage* 2018; 211: 138–153.

80. do Carmo, T. S., Moreira, F. S., Cabral, B. V., Dantas, R. C. C., de Resende, M. M., Cardoso, V. L., Ribeiro, E. J. Phosphorus recovery from phosphate rocks using phosphate-solubilizing bacteria, *Geomicrobiol. J.* 2019; 36: 195–203.

81. Nahas, E. (2007). Phosphate solubilizing microorganisms: Effect of carbon, nitrogen, and phosphorus sources. In: Velázquez, E., Rodríguez-Barrueco, C. (Eds.), First International Meeting on Microbial Phosphate Solubilization. Developments in Plant and Soil Sciences, vol. 102, Dordrecht: Springer, pp. 111–115.

82. Qu, Y., Li, H., Wang, X., Tian, W., Shi, B., Yao, M., Zhang, Y. Bioleaching of major, rare earth, and radioactive elements from red mud by using indigenous chemoheterotrophic bacterium Acetobacter sp. *Minerals* 2019; 9: 67.

83. Kraemer, D., Kopf, S., Bau, M. Oxidative mobilization of cerium and uranium and enhanced release of "immobile" high field strength elements from igneous rocks in the presence of the biogenic siderophore desferrioxamine B. *Geochim. Cosmochim. Acta* 2015; 165: 263–279.

84. Antonick, P. J., Hu, Z., Fujita, Y., Reed, D. W., Das, G., Wu, L., Shivaramaiah, R., Kim, P., Eslamimanesh, A., Lencka, M. M., Jiao, Y., Anderko, A., Navrotsky, A., Riman, R. E. Bio- and mineral acid leaching of rare earth elements from synthetic phosphogypsum. *J. Chem. Thermodyn.* 2019; 132: 491–496.

85. Das, N., Das, D. Recovery of rare earth metals through biosorption: An overview. *J. Rare Earths* 2013; 31: 933–943.

86. Andrès, Y., Gérente, C. (2011). Removal of rare earth elements and precious metal species by biosorption. In: Kotrba, P., Mackova, M., Macek, T. (Eds.), *Microbial Biosorption of Metals*, Dordrecht: Springer, pp. 179–196.

87. Zhuang, W.-Q., Fitts, J. P., Ajo-Franklin, C. M., Maes, S., Alvarez-Cohen, L., Hennebel, T. Recovery of critical metals using biometallurgy. *Curr. Opin.Biotechnol.* 2015; 33: 327–335.

88. Maleke, M., Valverde, A., Vermeulen, J.-G., Cason, E., Gomez-Arias, A., Moloantoa, K., Coetsee-Hugo, L., Swart, H., van Heerden, E., Castillo, J. Biomineralization and bioaccumulation of europium by a thermophilic metal resistant bacterium. *Front. Microbiol.* 2019; 10: 81.

89. Ngwenya, B. T., Mosselmans, J. F. W., Magennis, M., Atkinson, K. D., Tourney, J., Olive, V., Ellam, R. M. Macroscopic and spectroscopic analysis of lanthanide adsorption to bacterial cells. *Geochim. Cosmochim. Acta* 2009; 73: 3134–3147.

90. Takahashi, Y., Yamamoto, M., Yamamoto, Y., Tanaka, K. EXAFS study on the cause of enrichment of heavy REEs on bacterial cell surfaces. *Geochim. Cosmochim. Acta* 2010; 74: 5443–5462.

91. Ohnuki, T., Jiang, M., Sakamoto, F., Kozai, N., Yamasaki, S., Yu, Q., Tanaka, K., Utsunomiya, S., Xia, X., Yang, K., He, J. Sorption of trivalent cerium by a mixture of microbial cells and manganese oxides: Effect of microbial cells on the oxidation of trivalent cerium. *Geochim. Cosmochim. Acta* 2015; 163: 1–13.

92. Jiang, M., Ohnuki, T., Utsunomiya, S. Biomineralization of middle rare earth element samarium in yeast and bacteria systems. *Geomicrobiol. J.* 2018; 35: 375–384.

93. Lederer, F. L., Curtis, S. B., Bachmann, S., Dunbar, W. S., MacGillivray, R. T. A. Identification of lanthanum-specific peptides for future recycling of rare earth elements from compact fluorescent lamps. *Biotechnol. Bioeng.* 2017; 114: 1016–1024.

94. Merroun, M. L., Ben Chekroun, K., Arias, J. M., González-Muñoz, M. T. Lanthanum fixation by Myxococcus xanthus: Cellular location and extracellular polysaccharide observation. *Chemosphere* 2003; 52: 113–120.

95. Challaraj Emmanuel, E. S., Vignesh, V., Anandkumar, B., Maruthamuthu, S. Bioaccumulation of cerium and neodymium by Bacillus cereus isolated from rare earth environments of Chavara and Manavalakurichi, India, Indian. *J. Microbiol.* 2011; 51: 488–495.

96. Emmanuel, E. C., Ananthi, T., Anandkumar, B., Maruthamuthu, S. Accumulation of rare earth elements by siderophore-forming Arthrobacter luteolus isolated from rare earth environment of Chavara, India. *J. Biosci.* 2012; 37: 25–31.

97. Ochsner, A. M., Hemmerle, L., Vonderach, T., Nüssli, R., Bortfeld-Miller, M., Hattendorf, B., Vorholt, J. A. Use of rare-earth elements in the phyllosphere colonizer Methylobacterium extorquens PA1. *Mol. Microbiol.* 2019; 111: 1152–1166.

98. Liang, X., Gadd, G. M. Metal and metalloid biorecovery using fungi. *Microb. Biotechnol.* 2017; 10: 1199–1205.

99. Feng, M.-H., Ngwenya, B. T., Wang, L., Li, W., Olive, V., Ellam, R. M. Bacterial dissolution of fluorapatite as a possible source of elevated dissolved phosphate in the environment. Geochim. *Cosmochim. Acta* 2011; 75: 5785–5796.

100. Crocket, K. C., Hill, E., Abell, R. E., Johnson, C., Gary, S. F., Brand, T., Hathorne, E. C. Rare earth element distribution in the NE Atlantic: Evidence for benthic sources, longevity of the seawater signal, and biogeochemical cycling. *Front. Mar. Sci.* 2018; 5. doi: 10.3389/fmars.2018.00147.

101. Macaskie, L. E., Yong, P., Paterson-Beedle, M., Thackray, A. C., Marquis, P. M., Sammons, R. L., Nott, K. P., Hall, L. D. A novel non line-of-sight method for coating hydroxyapatite onto the surfaces of support materials by biomineralization. *J. Biotechnol.* 2005; 118: 187–200.
102. Deplanche, K., Merroun, M. L., Casadesus, M., Tran, D. T., Mikheenko, I. P., Bennett, J. A., Zhu, J., Jones, I. P., Attard, G. A., Wood, J., Selenska-Pobell, S., Macaskie, L. E. Microbial synthesis of core/shell gold/palladium nanoparticles for applications in green chemistry. *J. R. Soc. Interface* 2012; 9: 1705–1712.
103. Horiike, T., Kiyono, H., Yamashita, M. (2018). Dysprosium biomineralization by penidiella sp. strain T9. In: Endo, K., Kogure, T., Nagasawa, H. (eds.) *Biomineralization*, Singapore: Springer, pp. 251–257.
104. Tolley, M. R., Strachan, L. F., Macaskie, L. E. Lanthanum accumalation from acidic solutions using a Citrobacter sp. immobilized in a flow-through bioreactor. *J. Ind. Microbiol.* 1995; 14: 271–280.
105. Feng, M.-H., Ngwenya, B. T., Wang, L., Li, W., Olive, V., Ellam, R. M. Bacterial dissolution of fluorapatite as a possible source of elevated dissolved phosphate in the environment. *Geochim. Cosmochim. Acta* 2011; 75: 5785–5796.
106. Opare, E. O., Struhs, E., Mirkouei, A. A comparative state-of-technology review and future directions for rare earth element separation. *Renew. Sust. Energ. Rev.* 2021; 143, 110917. doi: 10.1016/j.rser.2021.110917.
107. Kaya, M. (2018). 3-current WEEE recycling solutions. In: Vegli'o, F., Birloaga, I. (eds.) *Waste Electrical and Electronic Equipment Recycling*, Sawston: Woodhead Publishing, pp. 33–93.
108. Reck, B. K., Graedel, T. E. Challenges in metal recycling. *Science* 2012; 337: 690–695.
109. Gueroult, R., Rax, J.-M., Fisch, N. J. Opportunities for plasma separation techniques in rare earth elements recycling. *J. Clean Prod.* 2018; 182: 1060–1069.
110. Dudarko, O., Kobylinska, N., Kessler, V., Seisenbaeva, G. Recovery of rare earth elements from NdFeB magnet by mono- and bifunctional mesoporous silica: Waste recycling strategies and perspectives. *Hydrometallurgy* 2022; 210: 105855.
111. Firdaus, M., Akbar Rhamdhani, M., Durandet, Y., John Rankin, W., McGregor, K. Review of high-temperature recovery of rare earth (Nd/Dy) from magnet waste. *J. Sustain. Metall.* 2016; 2(4): 276–295.
112. Liu, T., Chen, J. Extraction and separation of heavy rare earth elements: A review, Sep. *Purif. Technol.* 2021; 276: 119263.
113. Traore, M., Gong, A., Wang, Y., Qiu, L., Bai, Y., Zhao, W., Liu, Y., Chen, Y., Liu, Y., Wu, H., Li, S., You, Y. Research progress of rare earth separation methods and technologies. *J. Rare Earths* 2023; 41(2): 182–189..
114. Ambaye, T. G., Vaccari, M., Castro, F. D., Prasad, S., Rtimi, S. Emerging technologies for the recovery of rare earth elements (REEs) from the end-of-life electronic wastes: a review on progress, challenges, and perspectives. *Environ. Sci. Pollut. Res.* 2020; 27(29): 36052–36074.
115. Zhu, D. M., Chen, Q. W., Qiu, T. H., Zhao, G. F., Fang, X. H., Optimization of rare earth carbonate reactive-crystallization process based on response surface method. *J. Rare Earths* 2021; 39: 98.
116. Nascimento, M., Valverde, B. M., Ferreira, F. A., Gomes, R. D. C., Soares, P. S. M. Separation of rare earths by solvent extraction using DEHPA. *Rev. Esc. Minas.* 2015; 68(4): 427.
117. Chen, B., He, M., Zhang, H., Jiang, Z., Hu, B. Chromatographic techniques for rare earth elements analysis. *Phys. Sci. Rev.* 2017; 2. doi: 10.1515/psr-2016-0057.
118. Neves, H. P., Ferreira, G. M. D., Ferreira, G. M. D., de Lemos, L. R., Rodrigues, G. D., Leão, V. A., et al. Liquid-liquid extraction of rare earth elements using systems that are more environmentally friendly: Advances, challenges and perspectives. *Sep. Purif. Technol.* 2022; 282: 120064.

119. Das, S., Behera, S. S., Murmu, B. M., Mandal, D., Samantray, R., Parhi, P. K., et al. Extraction of scandium(III) from acidic solutions using organo-phosphoric acid reagents: A comparative study. *Sep. Purif. Technol.* 2018; 202(III): 248.
120. Liu, J., Huang, K., Wu, X. H., Liu, H. Z. Enrichment of low concentration rare earths from leach solutions of ion-adsorption ores by bubbling organic liquid membrane extraction using N1923. *ACS Sustain. Chem. Eng.* 2017; 5(9): 8070.
121. Peng, X. J., Su, J. H., Li, H., Cui, Y., Lee, J. Y., Sun, G. X. Theoretical elucidation of rare earth extraction and separation by diglycolamides from crystal structures and DFT simulations. *J. Rare Earths.* 2021; 39(7): 858.
122. Li, R., Marion, C., Espiritu, E. R. L., Multani, R., Sun, X. Q., Waters, K. E. Investigating the use of an ionic liquid for rare earth mineral flotation. *J. Rare Earths.* 2021; 39(7): 866.
123. Allam, E. M., Lashen, T. A., El-Enein, S. A. A., Hassanin, M. A., Sakr, A. K., Cheira, M. F., et al. Rare earth group separation after extraction using sodium diethyldithiocarbsamate/polyvinyl chloride from lamprophyre dykes leachate. *Materials (Basel).* 2022; 15(3): 1211.
124. He, Y., Tang, S. W., Yin, S. H., Li, S. W. Research progress on green synthesis of various high-purity zeolites from natural material-kaolin. *J. Clean Prod.* 2021; 306: 127248.
125. Dicholkar, D. D., Kumar, P., Kaur Heer, P., Gaikar, G. V., Kumar, S., Natarajan, R. Synthesis of N,N,N,N'-Tetraoctyl-3-oxapentane-1,5-diamide (TODGA) and its steam thermolysis-nitrolysis as a nuclear waste solvent minimization method. *Ind. Eng. Chem. Res.* 2013; 52(7): 2457.
126. Gueroult, R., Fisch, N. J. Practical considerations in realizing a magnetic centrifugal mass filter. *Phys. Plasmas* 2012; 19: 122503–122506. doi:10.1063/1.4771674.
127. Hu, Y., Florek, J., Larivi'ere, D., Fontaine, F.-G., Kleitz, F. Recent advances in the separation of rare earth elements using mesoporous hybrid materials. *Chem. Rec.* 2018; 18: 1261–76. doi:10.1002/tcr.201800012.
128. Kolar, E., Catthoor, R. P. R., Kriel, F. H., Sedev, R., Middlemas, S., Klier, E., et al. Microfluidic solvent extraction of rare earth elements from a mixed oxide concentrate leach solution using Cyanex(r) 572. *Chem. Eng. Sci.* 2016; 148: 212–218. doi: 10.1016/j.ces.2016.04.009.
129. Akbar, R., Baral, M., Kanungo, B. K. Spectroscopic, photophysical, solution thermodynamics and computational study of europium and terbium complexes with a flexible quinolinol-based symmetric tripodal chelator. *Spectrochim. Acta A Mol. Biomol. Spectrosc.* 2021; 247: 119124.
130. Maes, S., Zhuang, W.-Q., Rabaey, K., Alvarez-Cohen, L., Hennebel, T. Concomitant leaching and electrochemical extraction of rare earth elements from monazite. *Environ. Sci. Technol.* 2017; 51: 1654–1661. doi: 10.1021/acs. est.6b03675.
131. Marder, L., Bernardes, A. M., Zoppas Ferreira, J. Cadmium electroplating wastewater treatment using a laboratory-scale electrodialysis system. *Separ. Purif. Technol.* 2004; 37: 247–255. doi: 10.1016/j.seppur.2003.10.011.
132. Xiaowei, Z., Zhiqiang, W., Dehong, C., Zongan, L., Ruiying, M., Shihong, Y. Preparation of high purity rare earth metals of samarium, ytterbium and thulium. *Rare Met. Mater. Eng.* 2016; 45: 2793–2797.
133. van der Ent, A., Baker, A. J. M., Reeves, R. D., Chaney, R. L., Anderson C. W. N., Meech, J. A., et al. Agromining: farming for metals in the future? *Environ. Sci. Technol.* 2015; 49: 4773–80. doi: 10.1021/es506031u.
134. Park, D. M., Reed, D. W., Yung, M. C., Eslamimanesh, A., Lencka, M. M., Anderko, A., Fujita, Y., Riman, R. E., Navrotsky, A., Jiao, Y. Bioadsorption of rare earth elements through cell surface display of lanthanide binding tags. *Environ. Sci. Technol.* 2016; 50: 2735e2742. doi: 10.1021/acs.est.5b06129.

135. Hocheng, H., Chakankar, M., Jadhav, U. (2017). Biohydrometallurgical Recycling of Metals from Industrial Wastes. Boca Raton, FL: CRC Press.
136. Yoon, H.-S., Kim, C.-J., Chung, K. W., Lee, J.-Y., Shin, S. M., Lee S.-J., et al. Leaching kinetics of neodymium in sulfuric acid of rare earth elements (REE) slag concentrated by pyrometallurgy from magnetite ore. *Kor. J. Chem. Eng.* 2014; 31: 1766–72.
137. Tunsu, C., Petranikova, M., Gergorić, M., Ekberg, C., Retegan, T. Reclaiming rare earth elements from end-of-life products: a review of the perspectives for urban mining using hydrometallurgical unit operations. *Hydrometallurgy* 2015; 156: 239–58. doi: 10.1016/j.hydromet.2015.06.007.
138. Areti, S., Bandaru, S., Teotia, R., Rao, C. P. Water-Soluble 8-hydroxyquinoline conjugate of amino-glucose as receptor for La3+ in HEPES buffer, on whatman cellulose paper and in living cells. *Anal. Chem.* 2015; 87(24): 12348–12354.
139. Bayrakcı, M., Kursunlu, A. N., Güler, E., and Ertul, Ş. A new calix [4] azacrown ether based boradiazaindacene (Bodipy): Selective fluorescence changes towards trivalent lanthanide ions. *Dyes Pigm.* 2013; 99(2): 268–274.
140. Bekiari, V., Judeinstein, P., Lianos, P. A sensitive fluorescent sensor of lanthanide ions. *J. Lumin.* 2003; 104(1-2): 13–15.
141. Cole, B. E., Cheisson, T., Nelson, J. J. M., Higgins, R. F., Gau, M. R., Carroll, P. J. and Schelter, E. J. Understanding molecular factors that determine performance in the rare earth (TriNOx) separations system. *ACS Sustain. Chem. Eng.* 2020; 8(39): 14786–14794.
142. Leoncini, A., Huskens, J., Verboom, W. Ligands for f-element extraction used in the nuclear fuel cycle. *Chem. Soc. Rev.* 2017; 46: 7229; Veliscek-Carolan, J.: Separation of actinides from spent nuclear fuel: A review. *J. Hazard. Mater.* 2016; 318: 266.
143. Christiansen, B., Apostolidis, C., Carlos, R., Courson, O., Glatz, J. P., Malmbeck, R., Pagliosa, G., Römer, K., Serrano-Purroy, D. Advanced aqueous reprocessing in P&T strategies: Process demonstrations on genuine fuels and targets. *Radiochim. Acta* 2004; 92: 475.
144. Weaver, B., Kappelmann, F. A. 1964. Talspeak, A new method of separating americium and curium from the lanthanides by extraction from an aqueous solution of an aminopolyacetic acid complex with a monoacetic organophosphate or phosphonate. ORNL-3559.
145. Stevenson, C., Mason, E., Gresky, A. Progress in nuclear energy. Series III. Vol. 4 *Process Chemistry*, (1970) 3423.
146. Allaedini, G., Zhang, P. Treatment of phosphoric acid sludge for rare earths recovery II: Effect of sonication and flocculant solution temperature on settling rate. *Sep. Sci. Technol.* 2018; 1–11. doi: 10.1080/01496395.2018.1536715.
147. Tachimori, S., Sato, A., Nakamura, H.: Separation of transplutonium and rare-earth elements by extraction with di-isodecyl phosphoric acid from DTPA solution. *J. Nucl. Sci. Technol.* 1979; 16: 434.
148. Bhattacharyya, A., Mohapatra, P. K. Separation of trivalent actinides and lanthanides using various 'N', 'S'and mixed 'N, O'donor ligands: A review. *Radiochim. Acta* 2019; 107(9-11): 931–949.

5 Optimized Dyes for Actinide Metal Ions Detection and Applications

Kavya Bhakuni, Sruthi Guru, and Indu Tucker Sidhwani

5.1 INTRODUCTION TO ACTINIDES

The modern periodic table has given a special place at the bottom to the f-block elements, which include lanthanides (Ln) and actinides (An). Though their IUPAC names are lanthanoids and actinoids, the names lanthanides and actinides are also allowed. The lanthanides consist of elements from atomic no. 57–72. When a newly added electron enters in 4f orbitals, there is a decrease in the size of the atoms due to lanthanide contraction. Actinides include elements with atomic numbers 89–103 and are characterized by the gradual filling of the 5f orbitals. The first actinide to be discovered was uranium by extraction from U_3O_8 ore (also known as pitchblende) by Klaproth in 1789 and named after the planet Uranus. Thorium was discovered by Berzelius in 1829 from monazite, which is the principal ore of thorium. It is named after "Thor" the Scandinavian God of thunder and weather. The three naturally occurring isotopes of thorium, namely, Th 232, Ur 235, and U 238, have long half-lives of the order of billions of years and are present in significant amounts on the Earth's surface. These are described as primordial isotopes. Actinium and Protactinium are also found in small amount in the Earth and are formed as the decay product of ^{235}U and ^{238}U. Microscopic amount of plutonium is also reported and is formed because of neutron capture by uranium (Cotton 1991, Petrucci et al. 2007). Most of the actinides after uranium are synthetic products of the 20th century (Figure 5.1) and are obtained through different processes involving the bombardment of lighter elements. Actinide series derived its name from the first element Actinium (Cotton 1991, Petrucci et al. 2007, Morss et al. 2006).

Interestingly, lanthanides inherently coexist with actinides such as thorium and uranium as well as other metal ions in minerals and are relatively abundant in the Earth's crust. The primary mineral of thorium is called monazite. It is a phosphate ore with substantial lanthanide content (Morss et al. 2006). Besides, the isotopes of thorium are radioactive and thorium-232 is the most stable with a half-life of 14,000,000,000 years. This isotope comprises almost 100% of the Earth's naturally

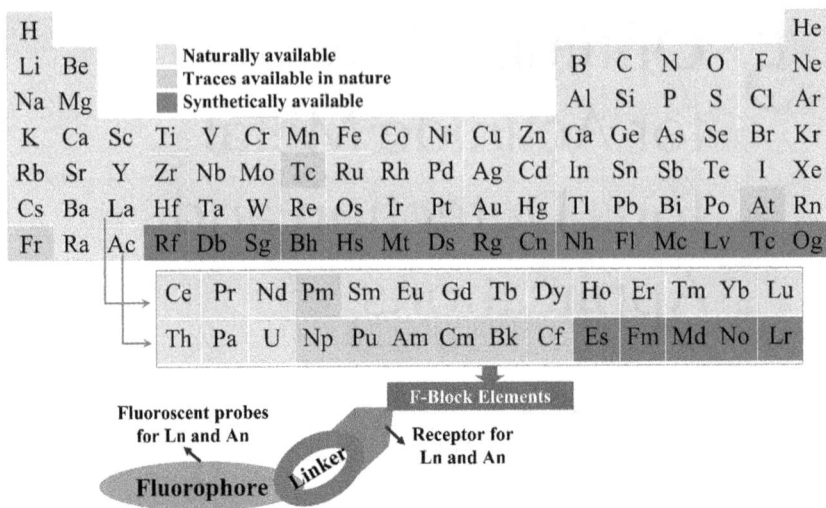

FIGURE 5.1 Periodic table depicting natural, trace, and artificial elements; and small-molecule-based fluorescent probes for f-block metal ions.

occurring thorium. Protactinium is silvery-white and it is among the ten least abundant elements and is found in uranium ores (pitchblende) at very minimal concentrations. Uranium is found in the ores, such as pitchblende, uraninite, and autunite. In their natural form, actinide metal ions may get attached to rock surfaces in contact with the aqueous phase. Moreover, actinide ions can form colloidal-sized particles which are waterborne particles having at least one dimension as 1–1000 nm (Silva and Nitsche 1995; Zänker and Hennig 2014).

5.2   GENERAL PROPERTIES

Actinides exhibit a larger number of oxidation states than lanthanides because of a small energy gap among 5f, 6d, and 7f orbitals. Hence, in actinides, 5f orbitals participate in bonding to a greater extent than the 4f orbitals in lanthanides. The most stable oxidation state for actinides is the +3 oxidation state. Actinides also show higher oxidation states such as +4, +5, and +7. They show a greater tendency to form complexes than lanthanides due to their high effective nuclear charge and higher charge density. This increases their binding affinity towards ligands as they can easily accept lone pair of electrons from the ligands. The overall charge of an actinide complex can differ broadly (Silva and Nitsche 1995). In acidic solutions, trivalent and tetravalent actinide ions are found in simple hydrated ions An^{3+} and An^{4+} (Katz et al. 1986). However, in higher oxidation states, actinides form oxygenated species in solution known as actinyl ions AnO_2^+ and AnO_2^{+2}. These actinyl ions are exceptionally stable and have a symmetrical and almost-linear geometry (Choppin 1988). Besides, most of the actinides form oxides with different oxidation states, with M_2O_3 being one of the most common oxides, where M is one of the actinide elements (Cotton 1991, Petrucci et al. 2007, Morss et al. 2006). Additionally, the hydroxide, carbonate, sulfate, phosphate, and fluoride form insoluble complexes with the actinides, whereas chloride and nitrate compounds of actinides are soluble. Apart from this, the most significant natural ligands are humic and fulvic acids, which are abundant in natural water resources (MacCarthy and Suffet 1988). Other significant ligands that tend to form a weak complex with actinides are phenols, amine, and hydroxyl groups (Silva and Nitsche 1995). Furthermore, several actinides in free and combined forms, such as hydrides, carbides, and alloys, may ignite at room temperature and may spread radioactive contaminates (Cotton 1991, Petrucci et al. 2007, Morss et al. 2006). Consequently, actinides in their colloidal form demonstrate differently with diverse migration activities. Thus, depending on the precipitation and colloid formation conditions, these processes may increase or retard the migration of the actinide ions. Additionally, the oxidation state of actinides is a crucial property based on which processes such as precipitation, complexation, sorption, and colloid formation vary substantially from one oxidation state to another. Actinide cations in the solution phase demonstrate four types of reactions that are highly significant in their detection and analysis, namely, precipitation, complexation, sorption, and colloid formation (Silva and Nitsche 1995). If the concentration of actinides in the solution exceeds the solubility product constant of the certain actinide salt, precipitation occurs. Several inorganic and organic ligands can form complexes with the actinide ions in the solution.

Especially, actinide ions such as Am^{3+} and Cm^{3+} are deemed as one of the most dangerous elements due to their severe and long-lasting radioactivity with a half-life $(T_{1/2})$ reported as 103–104 years (Choppin 1988). Consequently, actinides, which comprise key radionuclides in the nuclear energy cycle, show both radiotoxicity and chemotoxicity, due to which they require special management.

5.3   APPLICATIONS OF ACTINIDES

The most common and well-known element is uranium, which is used as a nuclear fuel. The radioactivity in actinides plays a major role in the chemistry and composition of particles in crystals. Besides, actinides are critical elements in understanding the nuclear chemistry and are a great source of energy such as nuclear power. Additionally, elements such as plutonium and uranium are used as fuels and in nuclear weapons. One of the oxides of thorium (ThO_2) has been used as an incandescent gas mantle since the 1880s (Cotton 1991, Petrucci et al. 2007, Morss et al. 2006). The 238-plutonium isotope was used as a power source in the Apollo-12 Lunar and for other satellites for a week. The 241-Americium isotope is used in smoke detectors as the ionizing source. Most importantly, uranium is the primary fuel in nuclear reactors (Cotton 1991, Petrucci et al. 2007, Morss et al. 2006). Additionally, uranium is also utilized in staining dyes for ceramic products as well as biological samples. Apart from this, actinides have attracted substantial attention lately. They perform crucial roles in various significant fields, such as material science (Pallares and Abergel 2020; Jia et al. 2019) metallurgy (Gu 2018), catalysis (Abney et al. 2017, Arnold et al. 2015), manufacturing (Kennedy et al. 2017), and biomedical engineering (Kuncewicz et al. 2019; Wilson et al. 2019; Dam et al. 2007). Moreover, actinides and lanthanides both have contributed significantly to clinical applications, majorly in imaging and therapeutic applications. Some of the highly advanced imaging techniques such as positron emission tomography and single-photon imaging employ radionuclide-based contrast agents that use actinides (Kostelnik and Orvig 2019). Such techniques are frequently used for the diagnosis of pathologies that are difficult to diagnose by traditional procedures like magnetic resonance imaging. Besides, actinides are also for targeted alpha particle therapy to treat tumours, as alpha radiation initiates confined cell death through permanent DNA damage (Durakovic 2016; CDC 2015; Eidson 1994).

5.4   HEALTH HAZARDS DUE TO ACTINIDES

Unfortunately, there are various kinds of health hazards associated with actinides due to their high radioactivity. The direct health impact due to nuclear explosions involves acute radiation syndrome (ARS) and injury due to blast and external radiation (Ramzaev et al. 2013). More specifically, ARS is a consequence of extremely high levels of radiation exposure for a short period of time with immediate effects on the haematopoietic system and radiosensitive undifferentiated cells (Brugge 2005; Landauer 2002). Moreover, the long-term effects of radiation comprise both somatic and genetic adverse influences. It is a well-known fact that uranyl salts and their complex can diffuse in blood, thus leading to damage to vital organs, such as kidneys, livers, and bones (Franic 2005; Kozai et al. 2015). This kind of long-term exposure can occur due to radioactive fallout and/or internal contamination due to war or industrial disasters. These types of scenarios are catastrophic as radioactive particulates or radiations can spread worldwide due to the movement of the wind. This kind of phenomenon was observed for the first time as the black rain after Hiroshima and Nagasaki bombings (Ramzaev et al. 2013; Kozai et al. 2015). Lately, due to the Fukushima nuclear disaster, widespread contamination by radioactive fallout was observed in the Mediterranean basin and the grasslands of Siberia (Ramzaev et al. 2013; Franic 2005). These radioactive materials can alter the cellular structure as well as function varying from morphological to transgenerational genetic effects (Brugge et al. 2005; Landauer 2002). This demonstrates the potentially irreversible consequences of chronic internal radiation exposure on the genetic pool of the Earth's population. All these observations reveal the potential irrevocable outcomes of prolonged internal radiation exposure on the genomic pool of the Earth's inhabitants. Overall, despite numerous applications, actinides are inherent heavy metal ions, and high intake or long-term contact with these ions can lead to adverse effects and irreversible damage to human health. Furthermore, these hazardous actinides are inevitably released into the environment due to various industrial activities, such as uranium mining, testing and disintegration of nuclear-powered satellites, nuclear reactor accidents, and nuclear weapons production (Kozai et al. 2015; Selvakumar et al. 2018). Thus, the overutilization of radioactive actinides in various industries has led to the pollution of natural resources, such as water, soil, and the atmosphere.

Until now, a variety of classical techniques comprising inductively coupled plasma atomic emission spectroscopy, plasma-mass spectrometry, X-ray fluorescence spectroscopy, and resonance light scattering are being used for quantitative evaluation of actinides (Li et al. 2016). Nevertheless, these approaches are expensive and often involve complicated instrument panels, mass samples, and time-exhausting practices. In some cases, various problems are encountered, such as inadequate selectivity, minimal precision, and bad accuracy. Thus, the development of innovative methods for the selective detection of actinides is currently challenging and is generating a lot of attention. The classic examples of metal organic frameworks (Li et al. 2019), functionalized nanoparticles (Mondal and Yarger 2019), quantum dots (Dolai et al. 2017), biosensors (Chen et al. 2015), and multivariate pattern analysis (Roy et al. 2018) were effectively utilized for precise determination of specific f-block elements. Amongst all these methods, sensors, specifically, chemosensors are one of the easiest methods that generate a promising visual and physical response due to selective binding with actinides. Moreover, contrary to traditional techniques, the use of a chemosensor for the detection of actinides presents a low-cost method with higher selectivity and sensitivity. Chemical sensing refers to continuous monitoring of the presence of a chemical species. This sensing technique uses fluorescence emission to signal a molecular recognition event. It was first demonstrated during the early 1980s when Tsien reported the synthesis of the first chemosensor as a Ca indicator (Tsien 1980). These Ca indicators are based on calcium ion chelate receptors, covalently linked to simple aromatic rings or other dyes as chromophores. Fluorescent sensors are divided into two groups, fluorescent biosensors and fluorescent chemosensors. Even though biosensors represent an important area in sensing, they fall outside the scope of this chapter. Here, we will focus our attention only on chemosensors. Additionally, this field is so broad, so we focus our attention on dyes and new trends in the development of fluorescent sensors, such as the fabrication of nanosensors.

5.5 DETECTION TECHNIQUES

Diverse approaches have been developed for the detection of actinides that includes employing potentiometric membrane sensors, nanoparticles, fluorescent molecular dyes, conjugated polymer sensors, and so on. Fluorescent dyes are defined as compounds which both absorb and emit strongly in the visible region, and which own their potential for applications due to their fluorescent properties (Lourdes et al. 2007). Particularly, numerous nanoparticle-based assays have been advanced as fast and cost-effective methods for actinide ions quantification. For instance, the quenching of CdSe/CdS quantum dot emission due to the presence of uranium was utilized for the development of an assay with a limit of detection (LOD) of 74.5 ppb (Singhal et al. 2017). Furthermore, Singh and coworkers created a gold nanoparticle encapsulated hydrogel for the portable detection of uranyl ions. Apart from this, DNAzymes were also used to identify uranyl ions employing gold nanowires through surface-enhanced Raman spectroscopy (He et al. 2020). Besides, fluorescent sensors for actinides in the form of UO_2^{2+} were also proposed by employing a carbon dot (C-dot)–aerogel hybrid prepared through in situ carbonization of 2-thenoyltrifluoroacetone (TTA) (Dolai et al. 2017). Figure 5.2A shows the concentration-dependent shifts of the fluorescence emission ($\lambda_{ex} = 400$ nm) when UO_2^{2+}, Sm^{3+}, or Eu^{3+} are added. Only UO_2^{2+} induced a significant fluorescence shift even when present in low concentrations. Figure 5.2B shows fluorescence quenching ($\lambda_{ex} = 350$ nm and $\lambda_{emission} = 350$ nm) vs. ion concentrations. This shows that metal ion concentrations

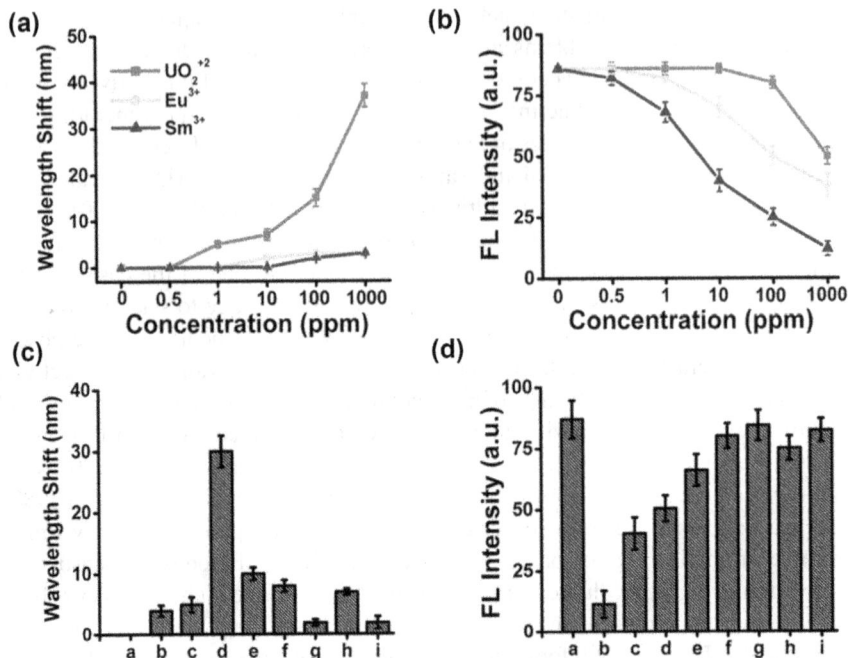

FIGURE 5.2 (A) Shifts of the fluorescence peak (λe_x 400 nm/λ_{em} 445 nm) and (B) intensity (λ_{ex} 350 nm/λ_{em} 405 nm) of TTA–C-dot–aerogel upon addition of metal ions. Comparison of the (C) fluorescence shifts (λ_{ex} 400 nm/ λ_{em} 445 nm) and (D) fluorescence intensities (λ_{ex} 350 nm/λ_{em} 405 nm) on adding 100 ppm of (a) control TTA–C-dot/aerogel, (b) Sm^{3+}, (c) Eu^{3+}, (d) UO_2^{2+}, (e) Ce^{3+}, (f) Nd^{3+}, (g) Gd^{3+}, (h) Cd^{2+}, and (i) Pb^{2+} (Dolai et al. 2017).

are directly related to C-dots' fluorescence quenching, thereby indicating that ion adsorption within the aerogel pores is responsible for the fluorescence quenching effect. The most pronounced fluorescence quenching was by Sm^{3+}, followed by UO_2^{2+} and Eu^{3+} ions at higher concentrations. Figure 5.2C and 5.2D shows the fluorescence shifts with ($\lambda_{ex} = 400$ nm and $\lambda_{emission} = 445$ nm) and intensities with ($\lambda_{ex} = 350$ nm and $\lambda_{em} = 405$ nm) for different metal ions at a concentration of 100 ppm: f different metal ions: (a) control TTA–C-dot/aerogel, (b) Sm^{3+}, (c) Eu^{3+}, (d) UO_2^{2+}, (e) Ce^{3+}, (f) Nd^{3+}, (g) Gd^{3+}, (h) Cd^{2+}, and (i) Pb^{2+}.

Liu et al. (2019) proposed a detailed method using a luminescent metal organic framework-based probe for extremely sensitive and selective detection of Th^{4+} contamination in natural freshwater media. Interestingly, sorbent biopolymers such as cellulose, chitin/chitosan, and alginate demonstrate reasonable competencies and marginal selectivity for actinides. Nevertheless, the backbones of such polymers can be effortlessly functionalized by fusion of various biomass or ion-exchangers can increase their viability for actinide sorption (Abd El-Magied et al. 2017). For instance, functionalized alginate gels and hydrogels have been demonstrated to be robust and effective at eliminating uranium from aqueous solution (Gok and Aytas 2009). Furthermore, chemosensors are considered an easy-to-monitor tag. They give a promising optical and physical reaction upon selective binding with a certain metal ion or analyte that leads to the formation of optical, electrochemical, or fluorescent chemosensors. In contrast to the above-mentioned methods, the use of chemosensory materials as a detecting agent presents a great deal of advantages. It is a low-cost and regional examination along with being highly selective and sensitive. More specifically, fluorescent chemosensors have outstanding advantages for real-time visualization.

A chemosensor typically comprises of a fluorophore, a binding site, and a linker to combine the two fractions if necessary. There can be multiple identification mechanisms for selective recognition that may include photoinduced electron transfer (PET), photoinduced charge transfer, Förster resonance energy transfer, inter/intramolecular charge transfer (ICT); energy transfer (ET), and aggregation-induced emission (AIE) (Gawas et al. 2021). These fluorescent mechanisms were discussed in detail in Chapter 4. Some of the prime strategies to construct fluorescent chemosensors are demonstrated in Figure 5.3. The popular approach is the sensor that attaches to metal ions through non-covalent interactions; thus, the process is reversible (Figure 5.3a) (Fang and Dehaen 2021). An alternative strategy involves an irreversible chemical transformation induced by a target metal ion eventually leads to the formation of a new product (Figure 5.3b). Lately, the supramolecular assembly of small molecules has exhibited a favourable approach for the detection of analytes. These combinations are both driven by a targeted metal ion and further employed as a fluorescent chemosensor for the targeted analyte (Figure 5.3c), which causes detectable fluorescence changes.

(a) Strategy I: Noncovalent interactions (reversible)

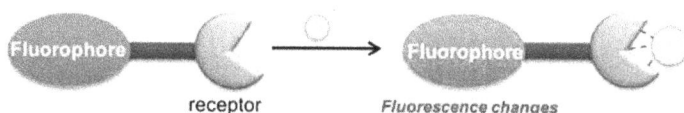

(b) Stratege II: Reaction-based process (irreversible)

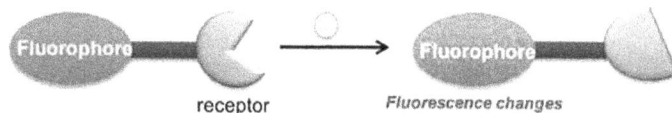

(c) Stratege III: Supramolecular assembly

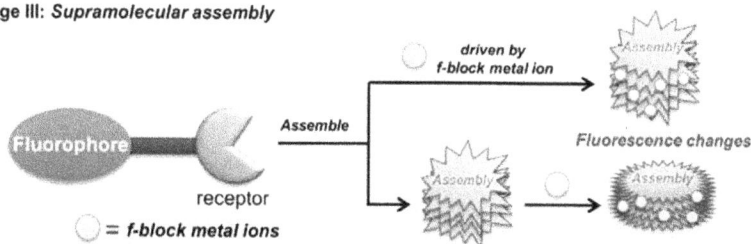

FIGURE 5.3 Prime strategies to construct a fluorescent chemosensor (Fang and Dehaen, 2021).

In this chapter, we have reviewed the development of a specific kind of chemosensors, namely, dyes for the detection of actinide metal ions. The following sections are dedicated to the molecular structures of dyes along with their sensing mechanisms. However, before proceeding towards the detailed mechanism and applications of dyes for actinide detection, we will be discussing the basics of dyes.

5.5.1 DYES FOR F-BLOCK METAL DETECTION

Dyes are well-known chemical products that are widely utilized in the textile industry, food industry, and cosmetic and printing industry (Bauer et al. 2001; Ganesh et al. 1994). Chemically, dyes are organic compounds that contain specific groups called chromophores and auxochromes. An azo dye is depicted in Figure 5.4 which contains the linkage R−N=N−R′ in its structure, with R being an aryl group and R′ being a substituted aryl group. The colour of the dye is due to the chromophore group whereas the auxochromes are either electron-withdrawing or -donating side groups that assist the chromophore by shifting the adsorption or by enhancing the intensity of absorption (Christie 2001). Some major chromophores are azo (−N=N−), carbonyl (−C=O), methine (−CH=), nitro (−NO₂), and quinoid groups. The auxochromes can be either acids or bases, for example, amine (−NH₃), carboxyl (−COOH), sulfonate (−SO₃H), and hydroxyl (−OH). Moreover, dyes can be categorized into various groups such as indigoid dyes or xanthene dyes depending on their different structures and applications (Lima et al. 2007).

Apart from the use of dyes in various industrial processes, they find important applications in analytical chemistry as well. Dyes readily form a distinctly coloured complex with the metal ion in aqueous media. The formation of the metal–dye complex depends on various factors such as the number of ligands in the dye, the coordination number of the metal, and the electron-donating ligand or ion. Therefore, the metal–dye complex formed may be largely divided into two groups: the 1:1 metal complex and the 1:2 metal complex (Chavan 2011). These complexes have shown distinct colours in wavelength range that extends over the entire spectrum permitting their application in spectrometry. With the application of UV–Visible spectroscopy, the absorption spectra of sample solutions can be recorded, and the concentration of metal ions can be determined. Chemically, dyes interact effortlessly with metal ions with the help of heteroatoms S, N, and O (Kilincarslan et al. 2007; Mahapatra et al. 1990). Thus, these compounds can chelate with several metal ions to form a specific

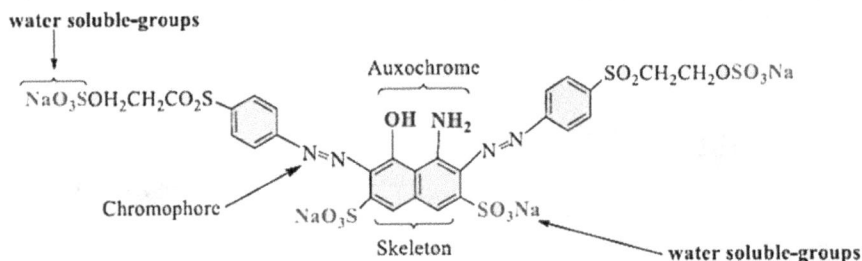

FIGURE 5.4 Structure of an Azo dye.

metal–dye complex. This property of the dyes makes them an ideal candidate for detection and quantitative analysis in analytical chemistry. Most famously, they are used as an indicator in titrimetric analysis. They also find application in spectrophotometric analysis. Overall, the dyes (more especially azo dyes) are applied successfully as spectrophotometric chemosensors.

Consequently, various novel spectrophotometric methods are developed for the quantitative determination of metal ions using different dyes (Ashok and Sharma 2009). A simple technique was developed by Bonishko et al. (2008) and Rydchuk et al. (2009) for the determination of osmium (IV) ions using the dye named Congo Red and Orange G, respectively. Furthermore, the complexation of metal and dyes modulates the photophysical and colouristic properties of dyes. As a result, with the formation of the metal–dye complex, there occurs a substantial decline in the absorption band of the dyes along with the advent concomitantly of a new absorption band with distinct absorbance. Moreover, the formation of metal–dye complexes is affected by numerous factors such as pH, temperature, solvents, and ionic strengths along with the concentration and structure of the dye (Goftar et al. 2014).

Apart from this, dyes are also utilized in electrochemistry particularly in metal ions detection (Wang et al. 2008). Interestingly, electrochemical analysis is acknowledged to be an efficient method for industrial process control and environmental monitoring (Gupta et al. 2011). Various research works have stipulated that certain azo dyes are electrochemically reactive. Subsequently, these dyes can undergo reduction or oxidation on the different electrodes. As a result, this electrochemical conduct permits the detection of dye by voltammetric technique. Furthermore, the detection of trace metal ions can be conducted, which is based on the decline of dyes oxidation/reduction peak after complex formation (Coulibaly et al. 2012).

5.5.1.1   Different Dyes for Actinide Detection

As discussed in the previous section, dyes work as chemosensors for metal detection. Additionally, they can also be utilized for the detection of radioactive metal ions that fall in the actinide group. These can be applied to recognize target metal ions in the environment as well as organisms. Furthermore, dyes as sensors have easy operation, are rapid, and favour on-site detection, with good recyclability. This section covers various dyes that find applications for actinide detection and their optimized conditions.

Azo dyes such as 1,8-dihydroxynaphthalene-3,6-disulphonic acid-2,7-bis[(azo-2)-phenylarsonic acid (Azo C1) (Collins and Lu 2001) and 2-(5-bromo-2-pyridylazo)-5-diethylaminophenol (Azo C2) (Gray et al. 2001) having a typical colorimetric trait can be utilized for the detection of actinide ions. The structure of the above-mentioned dyes is illustrated in Figure 5.5. Interestingly, Savvin was the first to synthesize Azo C1 and is very sensitive to uranyl-ion concentration as a free indicator in aqueous solution (Savvin 1959; Kuznetsov et al. 1960). The LOD was 7.4×10^{-6} mol L^{-1}. Thus, the Azo C1 was employed for colorimetric detection of uranyl ions in ores, building materials, seawater, and plant samples (Kiriyama and Kuroda 1974; Nakashima et al. 1992; Greene et al. 2005; Jauberty et al. 2013). Furthermore, the Azo C2 solution changes the colour from yellow to purple when it forms a 1:1 complex with uranium (Das et al. 2010).

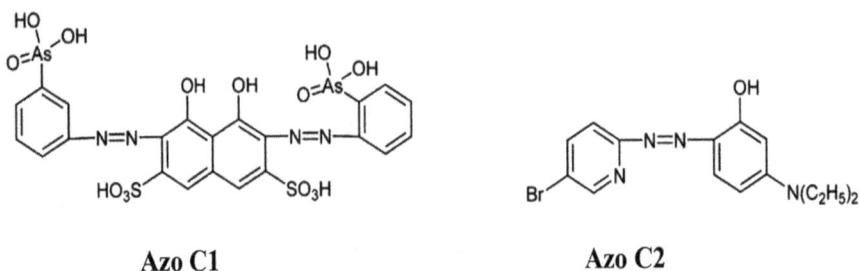

Azo C1   **Azo C2**

FIGURE 5.5 Important azo dyes used for actinide detection: 1,8-dihydroxynaphthalene-3,6-di-sulphonic acid-2,7-bis[(azo-2)-phenylarsonic acid (Azo C1) and 2-(5-bromo-2-pyridylazo)-5-di-ethylaminophenol (Azo C2) (Fang and Dehaen 2021).

Nevertheless, Azo C2 demonstrates reactions with some metal ions, such as Cu^{2+}, Co^{2+}, Ni^{2+}, Cd^{2+}, and Zn^{2+}. However, this can be removed by the addition of 1,2-cyclo-hexylenedinitrilotetraacetic acid or triethanolamine as masking agents. Furthermore, modifications were carried out in Azo C2, which has led to improved selectivity and sensitivity for UO_2^{2+} (Wen et al. 2018). However, both Azo C1 and Azo C2 suffer from drawbacks. For example, Azo C2 has a low selectivity for the actinides and displays similar reactivity for Th (IV) and the trivalent lanthanides. Thus, to prevent recognition of the lanthanides rather than actinides, a pre-purification step to remove the lanthanides is essential. Additionally, this dye is not soluble in water, thus, only minor colour changes are observed upon combining with metal. Hence, there was a need for better actinide-sensing dyes.

Consequently, studies demonstrated that expanded porphyrin complexes exhibited remarkable colour changes upon binding with metal. In this regard, hexaphyrin (1.0.1.0.0.0) or isoamethyrin shows a very distinct colour change in the presence of uranium as depicted in Figure 5.6 (Sessler et al. 2001). In this specific case, it was noted that the molar absorptivity amplifies by a factor of five in the presence of a metal cation (Sessler et al. 2006). This has led us to contemplate that isoamethyrin might act as an exceptional actinide sensor. Moreover, Sessler et al. showed that isoamethyrin permits the naked-eye detection of various actinides such as uranyl, neptunyl, and

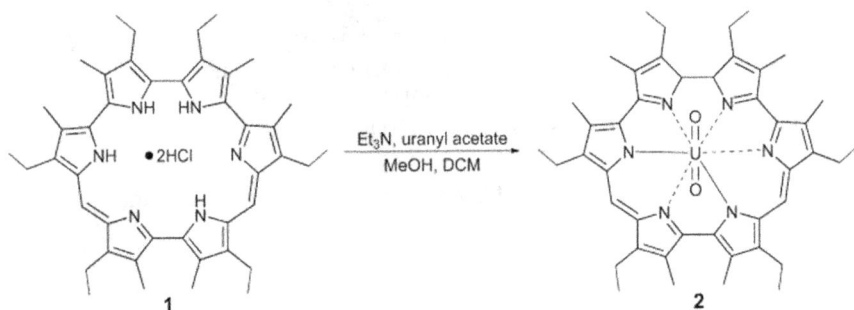

FIGURE 5.6 Representation of reaction between uranyl acetate and hexaphyrin (1.0.1.0.0.0) which ultimately leads to oxidation of the ring (Sessler et al. 2004).

plutonyl cations under aptly selected solution phase conditions. Moreover, research carried out with uranyl acetate in a 1:1 mixture of methanol and dichloromethane (v/v) signifies a detection limit of less than 28 ppb (Sessler et al. 2004). Additionally, the progress of isoamethyrin and its related species as prospective actinide sensors is ongoing at a fast pace.

The colour changes observed upon addition of Np(VI) or Pu(VI) cations to 1:1 (v/v) MeOH–CH_2Cl_2 solutions of isoamethyrin. Regrettably, because of restrictions related to extremely radioactive species, Sessler et al. were not able to achieve the expected molar absorptivities of these complexes. In particular, the addition of Pu(VI) or Np(VI) to an isoamethyrin and Et_3N solution generates an instant colour transformation. Whereas, in the case of uranium, it requires 24 h to show a substantial colour change.

Interestingly, it was also discovered that natural flavonoids can also be used as fluorescent sensors for actinide detection (Figure 5.7) (Fang and Dehaen 2021). In this regard, the first natural fluorescent probe for Th^{4+} can be traced back to the flavonoid morin (Figure 5.7a). This flavonoid formed a stable yellow-coloured complex with Th^{4+} in a 1:2 (metal:ligand) ratio (Milkey and Fletcher 1957). Afterwards, two other flavonoids, namely, quercetin (Figure 5.7b) (Guilbault 1990) and 3-hydroxyflavone (Figure 5.7c) (Bottei and D'Alessio 1967) were too proposed to act as fluorescent analyses for the identification of Th^{4+} under particular conditions. Subsequently, in vitro monitoring of natural Th^{4+} in urine was demonstrated using 3-hydroxyflavone (Kalaiselvan et al. 2011). Nevertheless, the above-mentioned flavonoid probes were found to suffer from interference of UO_2^{2+} and lanthanide ions. Along with this, they also lacked adequate selectivity.

Liu et al. in 2022 designed a new flavonoid fluorescent probe HTPAF with triphenylamine functional group. This probe HTPAF had a large Stokes shift of $\lambda = 110$ nm and could selectively detect UO_2^{2+} in $DMSO:H_2O$ (v/v = 5:95) at pH = 4.5 without the presence of general metal ions and specific metal ions (e.g., Al^{3+}, Cu^{2+}, Fe^{3+}, and Th^{4+}). On adding UO_2^{2+}, fluorescence at 474 nm was quenched. Also, there was no interference in the presence of metal ions (Pb^{2+}, Cd^{2+}, Cr^{3+}, Fe^{3+}, Co^{2+}, Th^{4+}, La^{3+}, etc.), especially Cu^{2+} and Al^{3+}.

Apart from these flavonoids, phytochemicals, such as curcumin and esculin can act as effective sensors for UO_2^{2+} with a fluorescence quenching response due to the formation of a 1:2 (metal:ligand) complex. The detection was identified by the colour transition from bright yellow to orange in an acetic acid buffer solution (pH = 4.0)

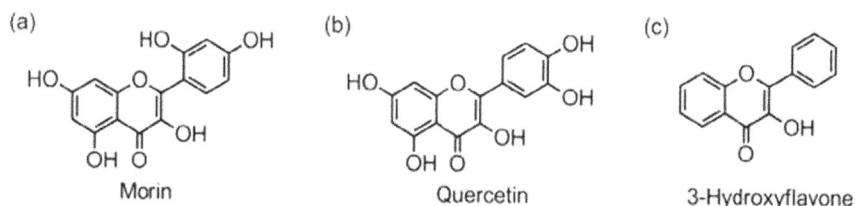

(a) Morin (b) Quercetin (c) 3-Hydroxyflavone

FIGURE 5.7 Biochemical structures of different natural flavonoid-based fluorescent probes for actinide detection (Fang and Dehaen 2021).

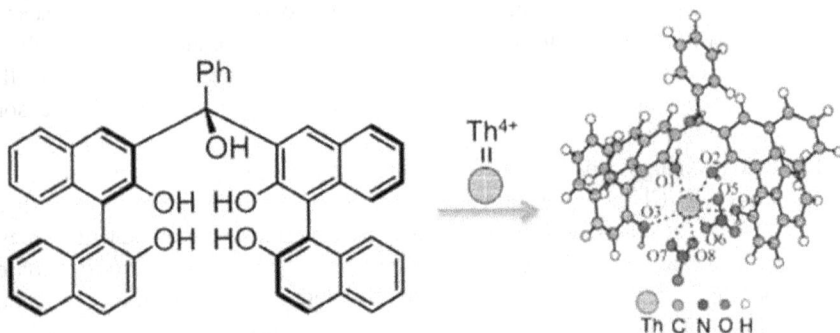

FIGURE 5.8   BINOL-based fluorescence sensor for Th⁴⁺ (Wen et al. 2013).

with an LOD of 3.7×10^{-6} M (Zhu et al. 2016). Angel and coworkers discovered that UO_2^{2+} can be detected through fluorescence quenching of a commonly seen fluorescein analogue dye, calcein, in a barbituric acid buffer (pH = 4.0) solution with a detection limit of 60 nM (Nivens et al. 2002).

Later, a novel fluorescence sensor 1,10-bis-2-naphthol (BINOL) was developed by Wen et al. for the discerning identification of Th⁴⁺ in a fluorescence quenching manner (Wen et al. 2013). The chemical structure of a BINOL-based fluorescent sensor is depicted in Figure 5.8. When one equivalent of Th⁴⁺ was added to the CH_3OH-H_2O (1:1, v/v) solution of BINOL, the strong emission at 385 nm was severely quenched, whereas the screening of other alkalis, alkaline earth, and transition and f-block metal ions showed only a marginal quenching response. Thus, BINOL demonstrated high sensitivity as well as selectivity for Th⁴⁺ with a LOD of 6.0×10^{-7} M. From a mechanistic viewpoint, DFT calculations established that all the hydroxyl groups from phenols and two nitrate anions were taking part in binding with Th⁴⁺, thus leading to the formation of an octahedral coordination complex. Successive studies confirmed BINOL as a reliable chemosensor for Th⁴⁺ detection in the surface water samples.

Fascinatingly, in contrast to all these dyes, Luo et al. (2001) for the first time reported the AIE effect of compounds that are faintly emissive or non-emissive in solution but display fluorescence upon aggregation or in the solid state. This can be attributed to the repression of a non-radiative route linked with the restriction of intramolecular motion (RIMs). In consideration of this, 2,6-pyridinedicarboxylic acid (PDA) functionalized TPE derivative (Figure 5.9) was proposed as the earliest instance of fluorescent probe for Th⁴⁺ recognition by virtue of the AIE effect (Wen et al. 2016). This unique compound displays almost no fluorescence in CH_3OH-H_2O (7:3, v/v) solution. However, the addition of Th⁴⁺ into this media turned ON the fluorescence with a distinct emission peak at around 475 nm. Moreover, this Turn-ON fluorescent reaction for Th⁴⁺ was complemented by a clear colour conversion from colourless to yellow green; thus enabling a speedy detection of Th⁴⁺ by the naked eye. The subsequent binding ratio of Th⁴⁺ was determined to be 1:1 with an LOD of 1.67×10^{-7} M. Furthermore, researchers also reported the synthesis of a series of novel benzo[a]xanthene-aminoquinoline conjugates (Figure 5.10), and their application as

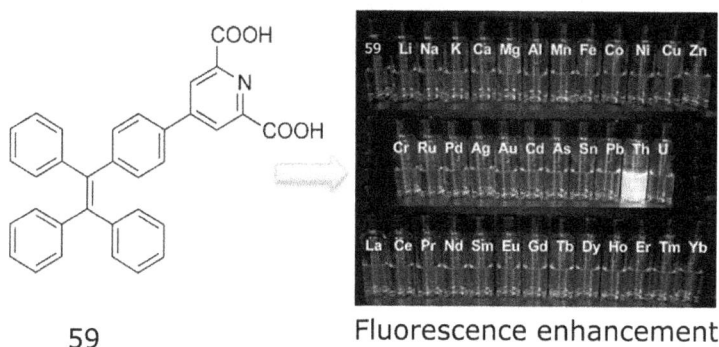

FIGURE 5.9   2,6-Pyridinedicarboxylic acid functionalized TPE derivative for turn-on fluorescence recognition of Th⁴⁺ (Wen et al. 2016).

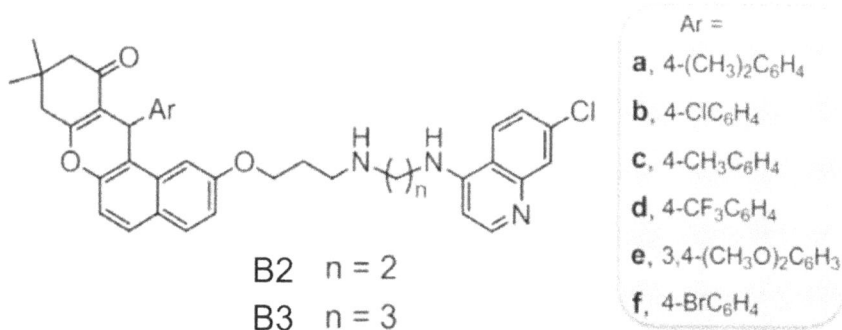

FIGURE 5.10   Chemical structures of benzo[a]xanthene-incorporated aminoquinoline conjugates B2 and B3 (Fang and Dehaen 2021).

fluorescent chemosensors for the selective detection of Th^{4+} (Singh et al. 2015; Fang and Dehaen 2021). The compound B2-a is a characteristic example of the ligands B2–B3 that all possess an analogous elementary skeletons. Even though almost no variation was observed in the absorption spectra of B2-a in methanol upon addition of Th^{4+}. However, an intense emission band at 374 nm was highly quenched with a quenching efficiency of 93%. This indicates that Th^{4+} only effects the excited state of B2-a.

Lin et al. in 2019 detected uranyl ion using a novel tetraphenyl ether (TPE)-based fluorescent sensor. They made use of AIE property of TPE and incorporated salophen moiety into the AIE-active tetraphenylethene moiety. Fluorescence titrimetric studies reveal that this sensor was able to detect uranyl ions in a simple one-step assay with LOD as low as 0.039 µM.

Furthermore, in 2008, a new class of compounds called Pillararenes was introduced (Ogoshi et al. 2008; Cao et al. 2009). These were macrocyclic hosts, possessing a symmetrical pillar-shaped structural design with the linker of methylene bridges at the 2,5-positions. Fascinatingly, due to their unique structure, simplistic preparation, and attractive host–guest properties, they have been investigated in various research fields and have become a vital part of supramolecular chemistry (Ogoshi et al. 2016; Fang et al. 2020). Later on, Fang et al. (2015) reported pillararene-based fluorescent chemosensors for the discriminatory detection of metal ions. Particularly, pillar[5] arene (Figure 5.11) tethered with triazole-linked 8-oxyquinolines at only one rim was reported as a subsequent fluorescence investigator for Th^{4+} followed by F^- (Fang et al. 2015). Upon addition of one equivalent of various metal ions into CH_3CN-H_2O (9:1, v/v) solution containing pillar[5]arene, only Th^{4+} produced almost complete quenching of the distinctive strong emission band at 390 nm.

Another promising nominee for actinide detection (particularly UO_2^{2+}) was found to be cyclic peptides (Lebrun et al. 2014). Due to their unique binding properties, Wang et al. described a phosphorylated cyclic peptide CP2 to have high selectivity and sensitivity for the fluorescent identification of UO_2^{2+} (Figure 5.12). Thus, it exemplifies the first instance of fluorescent sensors created from the functionalization of cyclic peptides for identifying UO_2^{2+} (Yang et al. 2015; Zhang et al. 2019). Initially, cyclic peptide CP1 was investigated for UO_2^{2+} sensing. It was reported that the fluorescence emission maxima at 360 nm of cyclic peptide CP1 got quenched with the addition of UO_2^{2+}. Nevertheless, CP1 was unsuccessful in achieving selectivity for UO_2^{2+} as additional metal ions such as Cu^{2+}, Nd^{3+}, and Th^{4+} might possibly cause the same fluorescence quenching effect. Thus, in order to enhance the selectivity, a

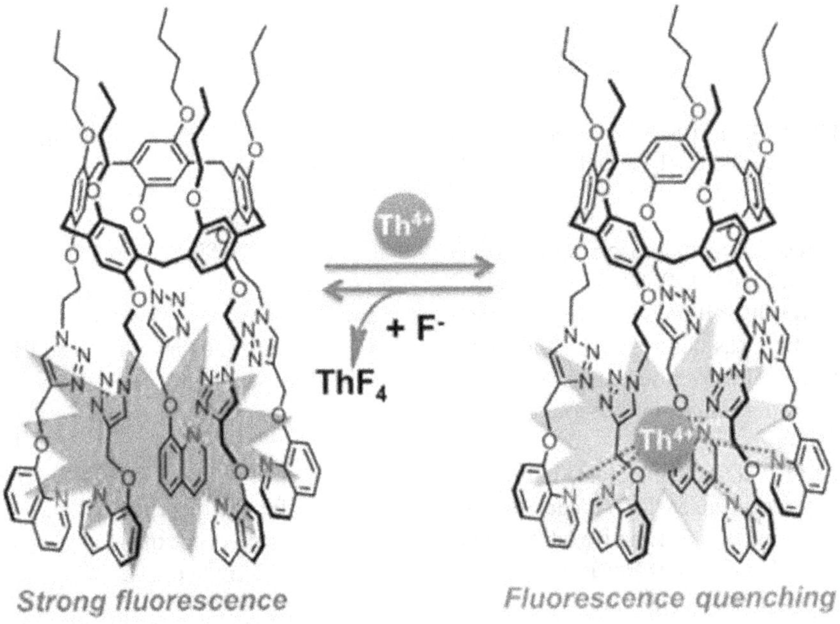

FIGURE 5.11 A non-symmetric pillar[5]arene tethered with triazole-linked 8-oxyquinolines at only one rim as a sequential fluorescence probe for Th^{4+} followed by F^- (Fang and Dehaen 2021).

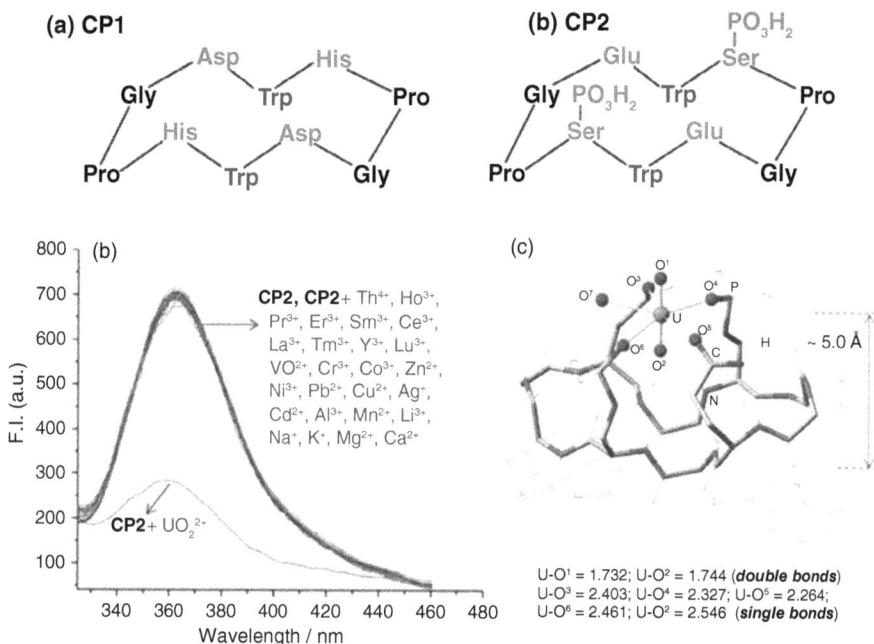

FIGURE 5.12 (a) Chemical structures of cyclic peptide (CP1) and phosphorylated cyclic peptide (CP2); (b) fluorescent responses of CP2 to various metal ions (1 equiv.) at MES buffer (20 mM, pH = 6.0) solution; (c) DFT calculated UO_2^{2+}-CP2 structure (Yang et al. 2015).

new cyclic peptide CP2 functionalized with two phosphoryl groups was constructed. It was found that the presence of one equivalent of UO_2^{2+} noticeably quenched the emission of CP2 at 360 nm with a quenching efficiency of 75%. In contrast, the general transition metals, alkali metals, alkaline earth metals, and almost all f-block metal ions, except the radioactive ones, demonstrated minimal fluorescence quenching. The LOD and stoichiometry of CP2 for UO_2^{2+} were 0.36 µM and 1:1, respectively. Besides, the binding constant CP2 between and UO_2^{2+} was much greater than that of CP1. Furthermore, DFT calculations also backed the 1:1 binding ratio, as suggestive of five oxygen atoms (two from P=O of phosphate, two from carboxylate of Glu, and one from H_2O) involved in coordination with the equatorial plane of UO_2^{2+}.

A Turn-ON resonance fluorescence recognition of UO_2^{2+} was achieved by a di-tetradentate macrocyclic (DTM) ligand via coordination of UO_2^{2+} with the DTM–Eu^{3+} complex to form a heterobinuclear complex (Figure 5.13) (Wang et al. 2018). When compound DTM chelated UO_2^{2+} or Eu^{3+} alone, the in situ generated complex only showed very weak resonance fluorescence at 358 nm in tris-HCl buffer solution (pH = 7.6). By contrast, simultaneously binding UO_2^{2+} and Eu^{3+} with DTM granted the resulting formed a heterobinuclear complex with strong resonance fluorescence owing to the cation–cation interaction between UO_2^{2+} and Eu^{3+}. The sequence of adding Eu^{3+} into the solution of DTM and then adding UO_2^{2+} offered a stronger resonance fluorescence signal. A linear relationship between the resonance fluorescence intensity

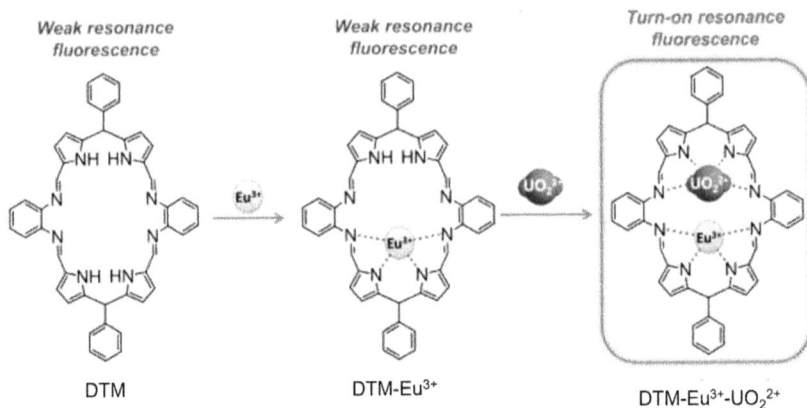

FIGURE 5.13 Turn-ON resonance fluorescence recognition of UO_2^{2+} via coordination of UO_2^{2+} with a DTM–Eu^{3+} complex (Wang et al. 2018).

of 91-Eu^{3+}-UO_2^{2+} and the concentration of UO_2^{2+} from 0.008 to 1.2 µM was obtained, from which the LOD of UO_2^{2+} was estimated to be 2.0×10^{-9} M. However, this protocol was suitable to be carried out only in a narrow range of 7.2–7.8, and Fe^{3+} and Cu^{2+} may cause an obstacle for the selective sensing of UO_2^{2+}. Interestingly, this system was successfully applied to the quantification of UO_2^{2+} in environmental water samples.

Overall, several investigations can, therefore, be carried out by synthesizing new dye as a ligand for metal complexation. The study of physicochemical characteristics of new dyes and their electrochemical behaviour and complexation as well as polymerization of dye and the thermodynamic study of the complexation of metal cations in poly-dye/poly film dye and their application still need extensive research.

Providing used nuclear fuel is particularly challenging because of high and long-term radioactivity of An, especially plutonium (Salvatores and Palmiotti 2011). Nuclear waste is proportioned into its different components, and An is transmuted into less hazardous isotopes. Lanthanides, however, need to be detected and removed prior to transmutation of actinides because they quench the transmutation process through neutron processing. Separation of actinides is complicated due to their similar physical and chemical properties with lanthanides (Nash 1993). Significant efforts are currently being pursued to develop efficient processes.

Some of the successful separation processes are TALSPEAK (Trivalent Actinide Lanthanide Separations by Phosphorus-reagent Extraction from Aqueous Complexes) process using aminopolycarboxylic acid developed by Weaver and Kappelmann (1964) TRUEX (TransUranium EXtraction) employing carbamoylmethyl phosphine oxide (CMPO) in nitric acid; TeRtiary AMine EXtraction (TRAMEX) process, using tertiary amine in highly concentrated chloride media (e.g., 1M LiCl) (Stevenson et al. 1970); recovery of yttrium and lanthanides by the addition of a flocculant (like polyacrylamide) to the phosphoric acid, followed by removal of uranium from the precipitate left behind (Allaedini and Zhang 2018); Ln/An using separation di-ethylenetriamine-penta acetic acid (Tachimori et al. 1979); use of lipophilic/hydrophilic

"N," "O," and/or "S" donor heteropolycyclic ligands as the actinide selective ligand (Bhattacharyya and Mohapatra 2019). However, significant progress in Ln/An separation still needs to be made in finding suitable ligands having favourable features with respect to fast kinetics, high selectivity over all Ln^{3+} ions, high stability (thermal, radiolytic, in low pH conditions, and in hydrolysis reactions), good separation efficiency, solubility in organic diluents, ability to undergo complete incineration to reduce the secondary waste generation, among others. Also, green extraction methods like membrane extraction, ionic liquid-based extraction, and solid phase extraction need to be further refined for dealing with large quantities of nuclear spent fuel and radioactive wastes.

5.5.1.2 Conclusions and Future Outlook

The selective identification and detection of actinide ions develop into an imperative requirement due to their high toxicity and strong radiotoxicity. Herein, we reviewed the advancements carried out in dyes as chemosensors for selective identification and detection of actinide ions. However, despite the above-stated instances suggesting the increasing research interests in this specific field, these detection of actinides by fluorescent chemosensors have slower progress than other frequently seen s-block and d-block metal ions, anions, neutral, etc. Consequently, broad prospects and challenges persist in this evolving research field.

Furthermore, considering the potential applications, very few efforts of fluorescent probes have been made to selectively detect specific actinide ions in environmental analysis, particularly the ore samples and spent nuclear fuels. Very limited examples are capable of sensing and detecting these f-block cations in live-cell imaging. Meanwhile, the selective recognition of certain actinide cations in specific cellular organelles and live animal models (e.g., zebrafish and mice) has also not been addressed. More importantly, uncovering the selective fluorescence detection events for actinides that are implicated in pathological processes (e.g., cancers, disease diagnostics, and clinical analysis) ought to be quite promising. Undeniably, great accomplishments could be anticipated soon with continuous and collaborative efforts poured into this research area.

REFERENCES

Abd El-Magied, M.O., A.A. Galhoum, A.A. Atia, A.A. Tolba, M.S. Maize, T. Vincent and E. Guibal (2017) Cellulose and chitosan derivatives for enhanced sorption of erbium (III). *Colloids Surf. A* 529: 580–593.

Abney, C.W., R.T. Mayes, T. Saito, and S. Dai (2017) Materials for the recovery of uranium from seawater. *Chem. Rev.* 117: 13935–14013.

Allaedini, G. and P. Zhang (2018) Treatment of phosphoric acid sludge for rare earths recovery II: Effect of sonication and flocculant solution temperature on settling rate. *Sep. Sci. Technol.* 54: 1–11.

Arnold, P.L., M.W. McMullon, J. Rieb, and F.E. Kuhn (2015) C-H bond activation by f-block complexes. *Angew. Chem. Int. Ed.* 54: 82–100.

Ashok, K. and S. Sharma (2009) Spectrophotometric trace determination of iron in food, milk, and tea samples using a new bis-azo dye as analytical reagent. *Food Anal. Methods* 2: 221–225.

Bauer, C., P. Jacques, and A. Kalt (2001) Photooxidation of an azo dye induced by visible light incident on the surface of TiO2. *J. Photochem. Photobiol.* 140: 87–92.

Bhattacharyya, A., and P.K. Mohapatra (2019) Separation of trivalent actinides and lanthanides using various 'N', 'S' and Mixed 'N, O' donor ligands: A Review. *Radiochim. Acta* 107: 931–949.

Bonishko, O.S., T.Y. Vrublevska, O.Z. Zvir, O.P. Dobryanska (2008) Spectrophotometric of osmium (IV) ions in intermetallic compounds. *Mater. Sci.* 44: 248–253.

Bottei, R.S. and A.S. D'Alessio (1967) Fluorimetric determination of thorium with flavonol. *Anal. Chim. Acta* 37: 405–409.

Brugge, D., J.L. de Lemos, B. Oldmixon (2005) Exposure pathways and health effects associated with chemical and radiological toxicity of natural uranium: A review. *Rev. Environ. Health* 20: 177–194.

Cao, D., Y. Kou, J. Liang, Z. Chen, L. Wang, and H. Meier (2009) A facile and efficient preparation of pillararenes and a pillarquinone. *Angew. Chem. Int. Ed.* 48: 9721–9723.

Center for Disease Control and Prevention (March 2015) Emergency Preparedness and Response: Acute Radiation Syndrome. Available at https://emergency.cdc.gov/radiation/ars.asp. Accessed 23 Mar 2015.

Chavan, R.B. (2011) Environmentally friendly dyes. In: *Handbook of Textile and Industrial Dyeing Principles*, Ed. M. Clark, Sawston, UK: Woodhead Publishing Limited, pp. 515–561.

Chen, Q., J. Zuo, J. Chen, P. Tong, X. Mo, L. Zhang, and J. Li (2015) A label-free fluorescent biosensor for ultratrace detection of Terbium (III) based on structural conversion of G-quadruplex DNA mediated by ThT and Terbium (III). *Biosens. Bioelectron.* 2: 326–331.

Choppin, G.R. (1988) Chemistry of actinides in the environment. *Radiochim. Acta* 43: 82–83.

Christie, R. (2001) *Colour Chemistry*. Cambridge, UK: The Royal Society of Chemistry.

Collins, G.E., and Q. Lu (2001) Microfabricated capillary electrophoresis sensor for uranium (VI). *Anal. Chim. Acta* 436: 181–189.

Cotton, S. (1991) *Lanthanides and Actinides*, Houndsmill, Basingstoke, Hampshire and London: Macmillan Education Company Ltd.

Coulibaly, M., Mureşan, L.M., Popescu, I.C. (2012) Detection of Cu(II) using its reaction with indigo carmine and differential pulse voltammetry. *Stud. Chem. J.* 3: 65–72.

Dam, H.H., D.N. Reinhoudt, and W. Verboom (2007) Multicoordinate ligands for actinide/lanthanide separations. *Chem. Soc. Rev.* 36: 367–377.

Das, S.K., C.S. Kedari, and S.C. Tripathi (2010) Spectrophotometric determination of trace amount of uranium (VI) in different aqueous and organic streams of nuclear fuel processing using 2-(5-Bromo-2-Pyridylazo-5-Diethylaminophenol). *J. Radioanal. Nucl. Chem.* 285: 675–681.

Dolai, S., Bhunia, S.K., Zeiri, L., Paz-Tal, O., and R. Jelinek (2017) Thenoyltrifluoroacetone (TTA)-carbon dot/aerogel fluorescent sensor for lanthanide and actinide ions. *ACS Omega* 2: 9288–9295.

Durakovic, A. (2016) Medical effects of internal contamination with actinides: Further controversy on depleted uranium and radioactive warfare. *Environ. Health Prev. Med.* 21: 111–117.

Eidson, A.F. (1994) The effect of solubility on inhaled uranium compound clearance: A review. *Health Phys.* 67: 1–14.

Fang, Y., and W. Dehaen (2021) Small-molecule-based fluorescent probes for f-block metal ions: A new frontier in chemosensors. *Coord. Chem. Rev.* 427: 213524.

Fang, Y., Y. Deng, and W. Dehaen (2020) Tailoring pillararene-based receptors for specific metal ion binding: From recognition to supramolecular assembly. *Coordin Chem. Rev.* 415: 213313.

Fang, Y., C. Li, L. Wu, B. Bai, X. Li, Y. Jia, W. Feng, and L. Yuan (2015) A non-symmetric pillar [5] arene based on triazole-linked 8-oxyquinolines as a sequential sensor for thorium (IV) followed by fluoride ions. *Dalton Trans.* 44: 14584–14588.

Franic, Z. (2005) Estimation of the adriatic sea water turnover time using Fallout90Sr as a radioactive tracer. *J. Marine Sys.* 57: 1–12.

Ganesh, R., G.D. Boardman, and D. Michelson (1994) Fate of azo dyes in sludges. *Water Res.* 28: 1367–1376.

Gawas, P.P., B. Ramakrishna, N. Veeraiah, and V. Nutalapati (2021) Multifunctional hydantoins: Recent advances in optoelectronics and medicinal drugs from Academia to the chemical industry. *J. Mater. Chem. C* 9(46): 16341–16377.

Goftar, M.K., K. Moradi, and N.M. Kor (2014) Spectroscopic studies on aggregation phenomena of dyes. *Eur. J. Exp. Biol.* 4: 72–81.

Gok, C., and S. Aytas (2009) Biosorption of Uranium (VI) from aqueous solution using calcium alginate beads. *J. Hazard. Mater.* 168: 369–375.

Gray, H.N., B. Jorgensen, L. Donald, and A.K. McClaugherty (2001) Smart polymeric coatings for surface decontamination. *Ind. Eng. Chem. Res.* 40: 3540–3546.

Greene, P.A., C.L. Copper, D.E. Berv, J.D. Ramsey, and G.E. Collins (2005) Colorimetric detection of uranium(VI) on building surfaces after enrichment by solid phase extraction. *Talanta* 66: 961–966.

Gu, Z. (2018) History review of nuclear reactor safety. *Ann. Nucl. Energy* 120: 682–690.

Guilbault, G.G. (1990) *Practical Fluorescence*, 2nd Ed. New York: Marcel Dekker Inc., CRC Press.

Gupta, V.K., M.R. Ganjali, P. Norouzi, H. Khani, A. Nayak, and S. Agarwal (2011) Electrochemical analysis of some toxic metals by ion-selective electrodes. *Crit. Rev. Anal. Chem.* 41: 282–313.

He, X., X. Zhou, W. Liu, Y. Liu, and X. Wang (2020) Flexible DNA hydrogel SERS active biofilms for conformal ultrasensitive detection of uranyl ions from aquatic products. *Langmuir* 36: 2930–2936.

Jauberty, L., N. Drogat, J.L. Decossas, V. Delpech, V. Gloaguen, and V. Sol (2013) Optimization of the arsenazo-III method for the determination of uranium in water and plant samples. *Talanta* 115: 751–754.

Jia, J.H., Q.W. Li, Y.C. Chen, J.L. Liu, and M.L. Tong (2019) Luminescent single-molecule magnets based on lanthanides: Design strategies, recent advances and magneto-luminescent studies. *Coord. Chem. Rev.* 378: 365–381.

Kalaiselvan, S., A.R. Lakshmanan, and V. Meenakshisundaram (2011) In vitro monitoring of natural thorium in urine using fluorimeter. *Talanta* 87: 80–84.

Katz, J.J., G.T. Seaborg, L.R. Morss (1986) *The Chemistry of the Actinide Elements*, New York: Chapman and Hall, pp. 1133–1146.

Kennedy, Z.C., D.E. Stephenson, J.F. Christ, T.R. Pope, B.W. Arey, C.A. Barrett, and M.G. Warner (2017) Enhanced anti-counterfeiting measures for additive manufacturing: Coupling lanthanide nanomaterial chemical signatures with blockchain technology. *J. Mater. Chem. C* 5: 9570–9578.

Nash, K.L. (1993) A review of the basic chemistry and recent developments in trivalent f-elements separations. *Solvent Extr. Ion Exch.* 11: 729–768.

Kilincarslan, R., E. Erdem, H. Kocaokutgen (2007) Synthesis and spectral characterization of some new azo dyes and their metal complexes. *Transit. Met. Chem.* 32: 102–106.

Kiriyama, T., and R. Kuroda (1974) Ion-exchange separation and spectrophotometric determination of zirconium, thorium, and uranium in silicate rocks with arsenazo III. *Anal. Chim. Acta* 71: 375–381.

Kostelnik, T.I. and C. Orvig (2019) Radioactive main group and rare earth metals for imaging and therapy. *Chem. Rev.* 119: 902–956.

Kozai, N., S. Suzuki, N. Aoyagi, F. Sakamoto, T. Ohnuki (2015) Radioactive fallout cesium in sewage sludge ash produced after the fukushima daiichi nuclear accident. *Water Res.* 68: 616–626.

Kuncewicz, J., J.M. Dabrowski, A. Kyzioł, M. Brindell, P. Łabuz, O. Mazuryk, W. Macyk, and G. Stochel (2019) Perspectives of molecular and nanostructured systems with d- and f-block metals in photogeneration of reactive oxygen species for medical strategies. *Coord. Chem. Rev.* 398: 113012.

Kuznetsov, V.I., S.B. Savvin, and V.A. Mikhailov (1960) Progress in the analytical chemistry of uranium, thorium, and plutonium. *Russ. Chem. Rev.* 29: 243–267.

Landauer, M.R. (2002) Radiation-induced performance decrement. *Mil. Med.* 16: 128–30.

Lebrun, C., M. Starck, V. Gathu, Y. Chenavier, and P. Delangle (2014) Engineering short peptide sequences for uranyl binding. *Chem. Eur. J.* 20: 16566–16573.

Li, M., G. Ren, F. Wang, Z. Li, W. Yang, D. Gu et al. (2019) Two metal-organic zeolites for highly sensitive and selective sensing of Tb^{3+}. *Inorg. Chem. Front.* 6: 1129–1134.

Li, W., J.T. Mayo, D.N. Benoit, L.D. Troyer, Z.A. Lewicka, B.J. Lafferty, et al. (2016) Engineered superparamagnetic iron oxide nanoparticles for ultra-enhanced uranium separation and sensing. *J. Mater. Chem. A* 4: 15022–15029.

Lima, E.C., B. Royer, J.C.P. Vaghetti, J.L. Brasil (2007) Adsorption of Cu(II) on Araucaria angustifolia Wastes: Determination of the optimal conditions by statistic design of experiments. *J. Hazard. Mater.* 140: 211–220.

Lin, N., W. Ren, J. Hu, B. Gao, D. Yuan, X. Wang, and J. Fu. (2019) A novel tetraphenylethene-based fluorescent sensor for uranyl ion detection with aggregation-induced emission character. *Dyes Pigm.* 166: 182–188.

Liu, B., W. Cui, J. Zhou, and H. Wang (2022) A novel triphenylamine-based flavonoid fluorescent probe with high selectivity for uranyl in acid and high water systems. *Sensors* 22: 6987.

Liu, W., X. Dai, Y. Wang, L. Song, L. Zhang, D. Zhang, et al. (2019) Ratiometric monitoring of thorium contamination in natural water using a dual-emission luminescent europium organic framework. *Environ. Sci. Technol.* 53: 332–341.

Lourdes, B.-D., D.N. Reinhoudt, and M. Crego-Calama (2007) Design of fluorescent materials for chemical sensing. *Chem. Soc. Rev.* 36: 993–1017.

Luo, J., Z. Xie, J.W. Lam, L. Cheng, H. Chen, C. Qiu et al. (2001) Aggregation-induced emission of 1-Methyl-1, 2, 3, 4, 5–Pentaphenylsilole. *Chem. Commun.* 18: 1740–1741.

MacCarthy, P. and I.H. Suffet (Eds.) (1988) Aquatic humic substances. *Advances in Chemistry*, Washington: American Chemical Society. doi: 10.1007/978-94-009-5095-5_11.

Mahapatra, B.B., N.P. Ajith Kumar, P.K. Bhoi (1990) Polymetallic complexes. Part-XXX. Complexes of Cobalt-, Nickel-, Copper-, Zinc-, Cadmium- and Mercury(II) with doubly-tridentate chelating azo-dye ligand. *J. Ind. Chem. Soc.* 67: 800–802.

Milkey, R.G., and M.H. Fletcher (1957) A fluorimetric study of the thorium-morin system. *J. Am. Chem. Soc.* 79: 5425–5435.

Mondal, P., and J.L. Yarger (2019) Colorimetric dual sensors of metal ions based on 1, 2, 3-Triazole-4, 5-dicarboxylic acid-functionalized gold nanoparticles. *J. Phys. Chem. C* 123: 20459–20467.

Morss, L.R., N.M. Edelstein, and J. Fuger (2006) *The Chemistry of the Actinide and Transactinide Elements*, Dordrecht: Springer.

Nakashima, T., K. Yoshimura, and T. Taketatsu (1992) Determination of uranium(VI) in seawater by ion-exchanger phase absorptiometry with arsenazo III. *Talanta* 39: 523–527.

Nivens, D.A., Y. Zhang, and S.M. Angel (2002) Detection of uranyl ion via fluorescence quenching and photochemical oxidation of calcein. *J. Photochem. Photobiol. A* 152: 167–173.

Ogoshi, T., S. Kanai, S. Fujinami, T.-A. Yamagishi, and Y. Nakamoto (2008) Para-bridged symmetrical pillar [5] Arenes: Their lewis acid catalyzed synthesis and host-guest property. *J. Am. Chem. Soc.* 130: 5022–5023.

Ogoshi, T., T.A. Yamagishi, and Y. Nakamoto (2016) Pillar-shaped macrocyclic hosts Pillar[n] arenes: New key players for supramolecular chemistry. *Chem. Rev.* 116: 7937–8002.

Pallares, R.M., and R.J. Abergel (2020) Transforming lanthanide and actinide chemistry with nanoparticles. *Nanoscale* 12: 1339–1348.

Petrucci, R.H., W.S. Harwood, G.E. Herring and J. Madura (2007) *General Chemistry: Principles and Modern Applications*, Upper Saddle River, NJ: Pearson.

Ramzaev, V., A. Barkovsky, Y. Goncharova, A. Gromov, M. Kaduka, I. Romanovich (2013) Radiocesium fallout in the grasslands on sakhalin, kunashir and shikotan islands due to fukushima accident. *J. Environ. Radioact.* 118: 128–42.

Roy, B., T.S. Roy, S.A. Rahaman, K. Das, and S. Bandyopadhyay (2018) A minimalist approach for distinguishing individual lanthanide ions using multivariate pattern analysis. *ACS Sensors* 3: 2166–2174.

Rydchuk, M., T. Vrublevska, O. Korkuna, and M. Volchak (2009) Application of orange G as a complexing reagent in spectrophotometric determination of osmium(IV). *Anal. Chem.* 54: 1051–1063.

Salvatores, M. and G. Palmiotti (2011) Radioactive waste partitioning and transmutation within advanced fuel cycles: Achievements and challenges. *Prog. Part. Nucl. Phys.* 66: 144–166.

Savvin, S.B. (1959) Photometric determination of thorium and uranium with the reagent arsenazo III. *Akad. Nauk SSSR Biochem. Sect. (Engl. Transl.)* 127: 1231–1234.

Selvakumar, R., G. Ramadoss, M.P. Menon, K. Rajendran, P. Thavamani, R. Naidu, M. Megharaj (2018) Challenges and complexities in remediation of uranium-contaminated soils: A review. *J. Environ. Radioact.* 192 (2018): 592–603.

Sessler, J.L., D. Seidel, A.E. Vivian, V. Lynch, B.L. Scott, and D.W. Keogh (2001) Hexaphyrin (1.0. 1.0. 0.0): An Expanded porphyrin ligand for the actinide cations uranyl (UO^{22+}) and neptunyl (NpO^{2+}). *Angew. Chem. Int. Ed.* 40: 591–594.

Sessler, J.L., P.J. Melfi, and D. Seidel, A.E.V. Gorden, D.K. Ford, P.D. Palmer, and C.D. Tait (2004) Hexaphyrin (1.0.1.0.0.0). A new colorimetric actinide sensor. *Tetrahedron* 60: 11089–11097.

Sessler, J.L., P.J. Melfi, and G. Pantos (2006) Uranium complexes of multidentate N-donor ligands. *Coordin. Chem. Rev.* 250: 816–843.

Silva, R.J. and H. Nitsche (1995) Actinide environmental chemistry. *Radiochim. Acta* 70: 377–396.

Singh, H., J. Sindhu, and J.M. Khurana (2015) Synthesis of novel fluorescence xanthene-aminoquinoline conjugates, determination of dipole moment and selective fluorescence chemosensor for Th^{4+} ions. *Opt. Mater.* 42: 449–457.

Singhal, P., S.K. Jha, B.G. Vats, and H.N. Ghosh (2017) Electron-transfer-mediated uranium detection using quasi-type II core-shell quantum dots: Insight into mechanistic pathways. *Langmuir* 33: 8114–8122.

Stevenson, C., E. Mason, and A. Gresky (1970) A Progress in nuclear energy. Series III. *Proc. Chem.*, vol. 4, Pergamon Press.

Tachimori, S., A. Sato, and H. Nakamura (1979) Separation of transplutonium and rare-earth elements by extraction with di-isodecyl phosphoric acid from DTPA solution. *J. Nucl. Sci. Technol.* 16: 434.

Tsien, R.Y. (1980) New Calcium indicators and buffers with high selectivity against magnesium and protons: Design, synthesis, and properties of prototype structures. *Biochem.* 19: 2396–2404.

Wang, J., X. Xiao, B. He, M. Jiang, C. Nie, Y.-W. Lin, and L. Liao (2018) A novel resonance fluorescence chemosensor based on the formation of heterobinuclear complex with a di-tetradentate macrocyclic ligand and europium (III) for the determination of uranium (VI). *Sens. Actuat. B: Chem.* 262: 359–364.

Wang, Y., H. Xu, J. Zhang, and G. Li (2008) Electrochemical sensors for clinic analysis. *Sensors* 8: 2043–2081.

Weaver, B., and F.A. Kappelmann (1964) TALSPEAK: a new method of separating americium and curium from the lanthanides by extraction from an aqueous solution of an aminopolyacetic acid complex with a monoacidic organophosphate or phosphonate. No. ORNL-3559. Oak Ridge National Lab.(ORNL), Oak Ridge, TN (United States)..

Wen, J., L. Dong, J. Tian, T. Jiang, Y.Q. Yang, Z. Huang, et al. (2013) Fluorescent BINOL-based sensor for thorium recognition and a density functional theory investigation. *J. Hazard. Mater.* 263: 638–642.

Wen, J., L. Dong, S. Hu, W. Li, S. Li, X. Wang (2016) Fluorogenic thorium sensors based on 2,6-pyridinedicarboxylic acid-substituted tetraphenylethenes with aggregation-induced emission characteristics. *Chem. Asian J.* 11: 49–53.

Wen, J., S. Li, Z. Huang, W. Li, and X. Wang (2018) Colorimetric detection of Cu^{2+} and UO_2^{2+} by mixed solvent effect. *Dyes Pigm.* 152: 67–74.

Wilson, R.J., N. Lichtenberger, B. Weinert, and S. Dehnen (2019) Intermetalloid and heterometallic clusters combining p-Block (Semi)Metals with d- or f-block metals. *Chem. Rev.* 119: 8506–8554.

Yang, C.-T., J. Han, M. Gu, J. Liu, Y. Li, Z. Huang, et al. (2015) Fluorescent recognition of uranyl ions by a phosphorylated cyclic peptide. *Chem. Commun.* 51: 11769–11772.

Zänker, H. and Hennig, C. (2014) Colloid-borne forms of tetravalent actinides: A brief review. *J. Contamin. Hydrol.* 157: 87–105.

Zhang, Y., Y. Deng, N. Ji, J. Zhang, C. Fan, T. Ding et al. (2019) A rationally designed flavone-based ESIPT fluorescent chemodosimeter for highly selective recognition towards fluoride and its application in live-cell imaging. *Dyes Pigm.* 166: 473–479.

Zhu, J.H., X. Zhao, J. Yang, Y.T. Tan, L. Zhang, S.P. Liu, Z.F. Liu, and X.L. Hu (2016) Selective colorimetric and fluorescent quenching determination of uranyl ion via its complexation with curcumin. *Spectrochim. Acta A* 159: 146–150.

6 Functionalized Fluorescent Tags for Protein Labeling in Biotechnology

Deepshikha Verma, Rita Mahapatra, and Poonam Mishra

6.1 INTRODUCTION

The simplest unit of life, the cell, is a complex network of various biomolecules embodied in an aqueous environment encircled by a membranous structure. These biomolecules such as proteins, carbohydrates, and lipids are essential for the structural and functional integrity of the cell. Studying these molecules in vivo has been vital in determining their specific function concerning cell metabolic processes. In vitro, these can be detected using various qualitative and quantitative methods like biochemical tests such as Bradford (Kruger 1994), Molisch's test (Foulger 1931), and Saponification (Brown and Marnett 2011), or more advanced methods like microscopy imaging like fluorescence recovery after photobleaching (Qin et al. 2011), CD spectroscopy (Greenfield 2004), etc. Also, identification of protein can be achieved by chromatographic or electrophoretic methods followed by western blotting with tagged antibodies with either chromogenic or fluorogenic tags. Further, recognition and analysis of these biomolecules in vivo have been a humongous task for scientists using these indirect methods of labeling proteins. Moreover, the complexity of the protein network necessitates specific labeling to identify and characterize the protein of interest (POI) from the rest of the protein pool of a cell. With the advent of proteomics and genomics, available protein data is enormous, analysis of which by the conventional methods become cumbersome. To study these proteins for their quantitation, existence in a multi-protein complex, and role in a biological process, it is tagged with a certain known small molecule known as tags. Based on their chemical nature, these tags can be chemical or biological entities like FLAG-tag (Knappik and Plückthun 1994), HA-tag (Field et al. 1988) V5-tag (Hanke et al. 1995), or full protein of small size like green fluorescent protein (GFP) (Shimomura, Johnson, and Saiga 1962; Sattarzadeh et al. 2015) are few to name.

This chapter will first explain different protein tags available in two categories: (1) fluorescent protein tags and (2) non-fluorescent tags. Subsequently, the application of these protein tags in the recognition of proteins across the various model organisms, viz., microbes, plants, and animals will be discussed.

DOI: 10.1201/9781003352372-6

6.2 FUNCTIONALIZED FLUORESCENT TAGS

6.2.1 FLUORESCENT PROTEIN TAGS

Fluorophores or probes that can be used to decipher the occurrence and dynamics of proteins are known as fluorescent tags. These are the molecules that can absorb light of a specific wavelength and emit light of a specific light. This specificity depends on their chemical structure, the presence of a solvent, and its target protein (Rashidian, Dozier, and Distefano 2013). Different known probes fluorescein, rhodamine, and GFPs are used regularly for labeling proteins in vitro or in vivo. Among these, fluorescein and rhodamine are examples of highly efficient organic dyes used for tagging or labeling antibodies and other proteins. Fluorescein isothiocyanate and tetramethylrhodamine-5-(and 6)-isothiocyanate are the commercial derivatives of fluorescein (excitation/emission wavelength: 494/520 nm) and rhodamine (excitation/emission wavelength: 541/572 nm) that are frequently used as bioconjugates covalently linked to proteins with their side chain or sulfhydryl group for protein tagging (Weiss 1999; Hu et al. 2021). Apart from this, there are certain naturally occurring fluorescent proteins like GFP found in many coelenterates like *Aequorea victoria*, *Phialidium*, and *Obellia* (Morin and Hastings 1971). This protein of 238 amino acids and a molecular weight of 27 kDa has an absorption in blue light at 395 nm and emission at 509 nm (Yang, Moss, and Phillips 1996). This GFP coding gene was first successfully cloned by Prasher (Prasher et al. 1992), followed by its expression in a heterologous model organism like *E. coli* by Chalfie et al. (1994). Further, several other mutants like YFPs, RFPs, and BFPs were created by mutagenesis that can have an emittance of different colors such as yellow, red, and blue, respectively (Patterson et al. 1997). In year 2008, Martin Chalfie, Osamu Shimomura, and Roger Y. Tisen were awarded the Nobel Prize in Chemistry for the discovery and development of the GFP.

The use of these fluorescent proteins as fusion tags at the N-terminal or C-terminal of the protein has been a pivotal tool for protein chemists and molecular biologists to determine their localization, quantitation, and interactions with other proteins or DNA in vivo across the different model organisms as discussed in the later part of this chapter.

6.2.2 NON-FLUORESCENT TAGS OF PROTEINS

Non-fluorescent tags of proteins involve the fusion of small peptides of varying lengths to the recombinant proteins to modify their sequence to enhance affinity tag-based purification (Arnau et al. 2011), increase solubility and yield (Walls and Loughran 2011), and detection via antibodies through epitope tags like biotin, c-myc, or FLAG (Dundas, Demonte, and Park2013; Lai et al. 2023; Sasaki et al. 2012). These tagged proteins are created by the in-frame fusion of polypeptide coding DNA sequence at either the N-terminus or C-terminus, based on the requirement, present in an expression vector to a DNA sequence that encodes the target protein using various PCR-based techniques or cloning methods as shown in Figure 6.1 (Gama and Breitwieser 2002; Raran-Kurussi and Waugh 2017).

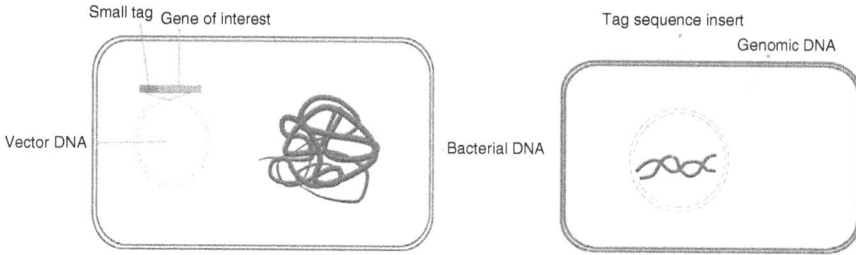

FIGURE 6.1 Protein labels can be attached by cloning where the gene of interest is added to the vector having the tag of interest or by inserting the tag DNA sequence into the gene of interest.

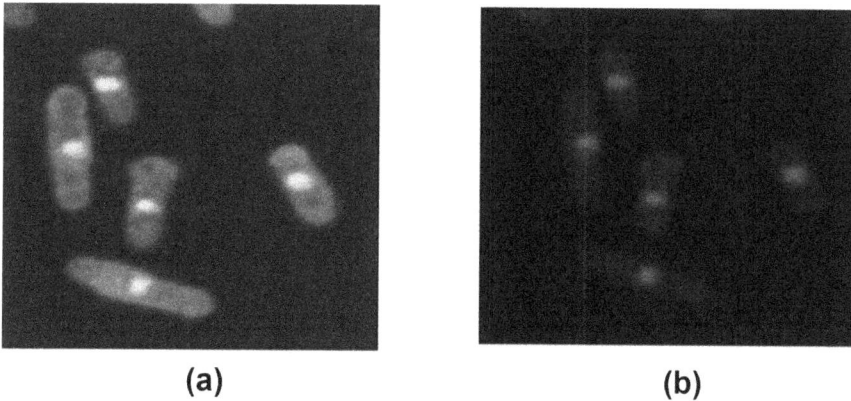

(a)   (b)

FIGURE 6.2 Labeling of Cut4 with GFP (A) and (B) DAPI as expressed and visualized under the fluorescence microscope.

Further, to remove these small tags, a short, specific peptide spacer linker is placed between the tag and the target protein which is cleaved by the use of TEV (Tobacco etch virus) protease or enterokinase (Raran-Kurussi et al. 2017). Some of the commonly known and widely used non-fluorescent tags are as follows (Figure 6.2):

6.2.2.1 Poly-Histidine (His)$_6$ and Maltose Binding Protein (MBP) Tag

These are the most widely used tags for the detection and purification of the proteins expressed in heterologous prokaryotic systems like *E. coli*. These tags can be used individually or placed in tandem positions as a dual tag of His6-MBP. Once the protein is expressed and purified using these tags, the next step is to remove the tags from the recombinant proteins as these tags may or may not interfere with the proper folding and achieving the right conformation for its function. Tev protease effectively identifies the amino acid sequence ENLYFQ/G and cleaves between Q and G to remove the tags placed (Raran-Kurussi et al. 2017).

6.2.2.2 c-Myc Tag

With the development of a murine antibody against the c-Myc protein in 1985 (Evan et al. 1985), a novel label for tagging the protein was available to the scientific community. This c-Myc epitope is 11-amino acid-long tag EQKLISEEDL and has been widely used for different experiments of protein chemistry and cell biology like immunoprecipitation, western blotting, confocal microscopy, and flow cytometry. This tag has been used for protein study across different cells varying from microbial cells of prokaryotic or eukaryotic origin (Dennler et al. 2015; Bowen et al. 2021).

6.2.2.3 FLAG-Tag

The FLAG-tag fusion is also another method for labeling proteins. This is a small peptide of eight amino acids DYKDDDDK, which can be used as a single repeat or in 3X FLAG peptide form (Hopp et al. 1988; Einhauer and Jungbauer 2001). Identification of FLAG peptide is carried out by antibody M1. Proteins can be labeled at the N- or C-terminus with FLAG peptide. Again, it can be used across various cell types, ranging from microbial cells like bacteria (Einhauer and Jungbauer 2001), yeast (Noguchi et al. 2008), or mammalian cells (Valdez-Sinon et al. 2020). This tag is used for immunodetection, co-immunoprecipitation, and ELISA in different cells.

6.3 APPLICATION OF PROTEIN TAGS

6.3.1 IN MICROBIAL CELLS

The use of different tags has been proven vital in studying the localization, quantitation, and interaction studies for protein–protein or protein–DNA through co-localization, etc. The best-known method of labeling proteins with specific small tags is through recombinant DNA technology by fusion of tags either at the N- or C-terminal of protein while cloning the gene in a commercially available vector that can be expressed in bacterial or yeast cells as mentioned in Table 6.1.

TABLE 6.1
Details of the Different Protein Tags Available for Microbial Protein Labeling

Serial No.	Name of Protein Tag	Size (kDa)	Host Organism	References
1	Poly-Histidine tag	0.84	Bacteria, yeast	Hochuli, Döbeli, and Schacher (1987)
2	Poly-Arginine Tag	0.8	Bacteria, yeast	Smith et al. (1984)
3	Maltose Binding Protein (MBP)	40	Bacteria	Waugh (2016)
4	Glutathione S-transferase (GST)	26	Bacteria, Yeast	Schäfer et al. (2015)
5	FLAG	1	Yeast	Hopp et al. (1988)
6	c-Myc	1	Yeast	Terpe (2003)
7	Haemagglutinin (HA)	1	Yeast	Terpe (2003)

TABLE 6.2
Use of Different Vector-Based Labeling of Protein at the Genomic DNA Level

S. No.	Name of the Protein Tag	Vector Used	References
1	c-Myc	pFA6a-MX6 based	Bähler et al. (1998),
2	FLAG	vectors system	Gadaleta et al. (2013),
3	GFP		Noguchi et al. (2008),
4	HA		Atkins and Izant (1995)

Another method, which is preferably used in single-cell eukaryotes like yeast for example, in *Schizosaccharomyces pombe*, a well-known model organism for studying cell cycle-related events, for which Paul Nurse in 2001 was awarded the Nobel Prize, is tagging of protein with 13xMyc (13Myc), 5xFLAG (5FLAG), and 3xHA (3HA) at the genomic DNA level with the help of plasmid-based method as shown in Table 6.2 with the use of tag-specific primers of 20–100 nucleotides (Gadaleta et al. 2013; Bähler et al. 1998).

The tagged proteins are expressed using a host cell protein expression system. Their localization can be detected directly if the protein has a fluorescent tag using a fluorescence microscope (Image 2). Further, their quantitation can be carried out by fluorescence spectroscopy or protein-specific expression-based cell sorting can be done by fluorescence-associated cell sorting.

On the other hand, the proteins with non-fluorescent tags can also be used for all the above studies with the help of different antibodies in an indirect detection method involving antibodies generated against those small tags or epitopes. Thus, usage of either fluorescent or non-fluorescent tags can be used to deduce much information.

6.3.2 Tagged Protein Fluorescence in Animal Models

Live-cell imaging has been an essential technique for understanding microbial functioning in complex surfaces (Faria, Joao, and Jordao 2015). Microbial biofilms have been known to consist of various different populations of bacterium such as planktonic and persisters (Kostakioti, Hadjifrangiskou, and Hultgren 2013). These different populations especially persisters are shown to be more resistant to biocides and antibiotics. Multiple auto-fluorescent proteins such as red fluorescent proteins (DsRed) along with enhanced green fluorescent protein (eGFP) can be applied to assess different cell populations in the biofilms. *Pseudomonas putida* tagged with mCherry acted as a genetic tool to co-localize and visualize single cells in microbial pathogenicity study (Lagendijk et al. 2010)

Whole-body imaging using fluorescent protein tags has also been shown to quantitatively track the growth of the tumor and metastasis, expression of genes, and bacterial pathogenesis. Since whole-body imaging is a non-invasive and sensitive method, it is more popular but it has cons too. One of the major drawbacks is interference through skin autofluorescence. However, the use of proper filters can minimize the interference of autofluorescence to some extent. Van Zyl and co-workers (2015) used mCherry integrated *Lactobacillus plantarum* and *Enterococcus mundtii* to study the protein expression and metabolism in the intestinal tract of mice using an in vivo imaging system (IVIS). IVIS is an advanced fluorescence imaging system that has the capability to illuminate in vivo fluorescent sources. Using IVIS, neutrophil dual cell populations can be synchronously monitored in a live mouse injected with cells labeled with far-infra infrared lipophilic dyes (Leslie et al. 2021). Dual color imaging has also proved to be useful in differentially labeling cancer in cytoplasm and nucleus and associated fluorescent color imaging forms a potent tool to elucidate the mechanism of cancer cell migration and nucleus deformation (Yang, Moss, and Phillips 1996; Yamauchi et al. 2005). In recent years, *C. elegans* has emerged as an ideal model to study host–pathogen interactions at the system level using fluorescent probes (FPs). For instance, *Mycobacterium nematophilum* was found to induce morphological changes in the nematode (Hodgkin, Kuwabara, and Corneliussen 2000; Gravato-Nobre et al. 2005). For optical imaging in live *C. elegans*, self-associated split FPs have been recently proposed. It is an ideal genetically encoded fluorescent protein probe set up to study the protein–protein interaction (PPI) in live *C. elegans*. Recently, CRISPR-mediated fluorescent labeling of *C. elegans* has been developed (Goudeau et al. 2020); it is considered to be the most invaluable tool for studying infectious diseases. Another very interesting in vivo model which deserves a mention here is zebrafish model. Zebrafish shares high similarity to humans and around 70% of all human genes share homologs in zebrafish. Thus, very well suited to study immune pathways and cell types using fluorescent protein probes and live imaging (Goudeau et al. 2020). Zebrafish is a model of choice for *Shigella flexneri* and *Mycobacterium abscesses* to study host responses against diseases. Small-molecule trackers routinely being used in zebrafish model are Lysotracker Red, fluorescent tyramine-conjugated chalcones such as fluorescent chemosensor (DiCesare and Lakowicz 2002), and Bezochalcone fluorescent probe (HAB) for target-induced fluorogenesis (Colucci-Guyon et al. 2019).

Microscopy and flow cytometry are effective alternative methods to monitor single-defined cell populations in live animals. Dynamics of the individual cells can be studied with IVM (intravital microscopy) and CD8+ T cells interaction with dendritic cells can be studied in the lymph nodes using EGFP fluorescence.

A flow cytometer detects the cell size and density based on the visible light scatter. In addition, visible light flow cytometers have fluorescent light sources. When a cell expressing associated fluorophore comes in contact with fluorescent light, it emits a signal which is captured by detectors. An experiment of flow cytometry can have fluorescent colors ranging from 2 to 30 measuring different targets. Cell labeling methods for fluorescent proteins are very widely used in animal studies (Verma et al. 2019). Florescent labeling helped in the sorting of a FoxP3+GFP+DTR+ (Thy1.2) cell population for adoptive transfer studies.

The fluorescent proteins have an advantage over synthetic dyes of expression inside the cells but sometimes the size of the fluorescent protein itself becomes a disadvantage (size more than 25 kDa) and can compromise the expression of co-expressed protein or biomolecule. Also, unwanted autofluorescence can increase the background while recording fluorescence. The difference in brightness and stability is also an issue, like white blue and cyan are not as bright as GFP. Different variants are available to combat these issues. For multiplexed applications, tandem dyes were nothing but fluorescent proteins covalently linked with synthetic molecules.

6.3.3 PROTEIN LABELING AND DETECTION IN PLANT

Proteins in plant cells constitute highly heterogeneous populations due to their functional diversity but present at relatively low concentrations. Proteins made up of polypeptides with various molecular sizes, complexes (e.g. "clusters" or "modules" of interacting molecules that carry out cellular functions), spatial and time-dependent concentrations (e.g. proteins in the nucleus for transcription or in the mitochondrion for energy regeneration), and charge (pI ranges from 3 to 12). Proteins present in compartments like the cytosol or distinct organelles like the mitochondrion or plastid to highly hydrophobic proteins embedded within the different cell membranes are some aspects of this complexity (Tanz et al. 2013). As a consequence, a multi-step procedure is often necessary to extract subsets of specific proteins, to accomplish that protein-based conjugates are valuable constructs for a broad-spectrum application. In the field of biotechnology for the development of biosensors, the immobilization of proteins on solid surfaces has been used extensively recyclable catalysts and protein microarrays (Zhang et al. 2018). For the identification of protein produced in plants can be identified by labeled proteins. The labeling can be classified into two categories: in vivo metabolic labeling and in vitro chemical modification. Metabolic labeling exploits biosynthetic incorporation of isotopically labeled nutrients or amino acids into proteins (Gygi et al. 1999; Oda et al. 1999). This part of the chapter gives an overview of protein labeling and tagging techniques which can be used for plant protein detection or PPI.

6.3.3.1 GFP Tagging

Fluorescent protein (FP) tagging approaches are widely used to determine the subcellular location of plant proteins (Tanz et al. 2013). The most commonly used tags are the GFP, the yellow fluorescent protein, and the cyan fluorescent protein, which provide an additional handle for protein localization studies (Cristea et al. 2005). The discovery and cloning of the fluorescent protein from the bioluminescent jellyfish *Aequorea victoria* have revolutionized studies in cell biology by enabling the dynamic monitoring of protein localization in the living cell using fluorescent microscopy (Prasher et al. 1992). Because of their intrinsic fluorescence ability and minimal toxicity, fluorescent proteins have been widely used as non-invasive markers in many living organisms. Extensive mutagenesis screens had been carried out and numerous GFP variants with distinct fluorescence characteristics had been generated (Shaner, Steinbach, and Tsien 2005). GFP is a naturally occurring, globular, and soluble acidic FP. The primary structure of GFP consists of a monomer composed of 238 amino acid residues, with a molecular weight of 27–30 kDa, the maturation of the chromophore requires oxygen (Yang, Moss, and Phillips 1996). GFP absorbs blue light or UV light and emits green fluorescence, with the main excitation peak at 395 nm, the lowest excitation peak at 475 nm, and the emission peak at 509 nm (Misteli and Spector 1997). The development of genetically encoded FPs and the increasing capability of software for image acquisition and analysis has enabled in vivo studies of protein functions and processes. Genetically encoded FPs are at the core of a variety of approaches to probe PPI in living cells (Lalonde et al. 2008). Verticillium wilt caused by the soil-borne fungal pathogen *Verticillium dahliae* in cotton (*Gossypium hirsutum*) is a widespread and destructive disease. Zhang et al. (2013) have obtained a GFP labeled *V. dahliae* strain (TV7) by transforming GFP into defoliating strain V991. GFP is extremely stable and can easily withstand the high-temperature treatment.

6.3.3.2 Labeling of Proteins with Fluorophores

Fluorophores aid in identifying their cellular localization, compared to the GFP, the small size of these organic molecules makes them less likely to perturb the native structure and function of the protein. Table 6.3 gives an overview of fluorophores used in protein labeling.

6.3.3.3 Förster Resonance Energy Transfer (FRET)

PPI has been extensively studied using FRET measured by fluorescence lifetime imaging microscopy (FLIM). However, implementing this technology to detect protein interactions in living multicellular organisms at single-cell resolution and under native conditions is still difficult to achieve. FRET analysis is an excellent approach to study in planta PPI dynamics in real-time. Long et al. (2018) described the optimization of the labeling conditions to detect FRET-FLIM in living plants.

6.3.3.4 Bimolecular Fluorescence Complementation (BiFC)

Ghosh et al. (2000) demonstrated that a fluorophore in *E. coli* when such as GFP could be split into N- and C-terminal fragments, which on their own are non-fluorescent. Upon interaction, the fused fragments (now in close proximity) can refold, mature, and fluoresce (Ghosh, Hamilton, and Regan 2000). The method was later adapted for use in mammalian and plant systems (Hu et al. 2021; Bracha-Drori et al. 2004; Walter et al. 2004). BiFC has two major limitations: (1) the irreversible complementation of the split fluorescent protein fragments with an estimated half-life time of 10 years precludes the analysis of interaction dynamics (but not necessarily their frequency); and (2) the tendency of most split fragments to spontaneously reassemble autonomously if unhindered (Magliery et al. 2005; Miller et al. 2015) have coupled BiFC analysis with flow cytometry for analysis, sorting (fluorescence-activated cell sorting), or both. Flow cytometry is a technique used to examine a population of particles or cells in free suspension, one by one, typically by collecting and analyzing fluorescence and scattered light (Oda et al. 1999).

6.3.3.5 Split-Luciferase Complementation Assay

As plants are not bioluminescent organisms, split-luciferase complementation is a very sensitive detection system, giving high signal-to-noise ratios. From bacteria and fungi to vertebrates, from land-dwelling hexapods to marine mollusks and fishes, bioluminescence has evolved convergently in as many as 40 different species (Haddock, Moline, and Case 2010). The terms luciferase and luciferin are generic and are used for all classes of light-producing enzymes; their respective substrates are usually preceded by the species name. Different luciferase systems are used exclusively in planta, where the luciferases from the North American firefly (*Photinus pyralis*) and the sea pansy (*Renilla reniformis*) are utilized. Advantages of the split-luciferase

TABLE 6.3
Overview of Fluorophore Labeling for Protein–Protein Interaction in Plants

Methods	Expression System	References
FRET-SE	Fluorescence	Piehler (2005), Zacharias et al. (2002)
BiFC	Fluorescence	Magliery et al. (2005)
CoIP	Immunostaining	Magliery et al. (2005)
Split-Luciferase	Bioluminescence	Fujikawa and Kato (2007)
Renilla Luciferase	*Arabidopsis thaliana* protoplast	Fujikawa and Kato (2007)
Renilla Luciferase	*Nicotiana benthamiana* leaf extract	Lund et al. (2015)
Firefly Luciferase	*Arabidopsis thaliana* protoplast	Chen et al. (2008)
Firefly Luciferase	*A. thaliana* protoplast *N. benthamiana* leaf extract	Lai et al. (2023)

complementation assay include flexible temporal control with a fast turnover and/or reversibility of probe assembly, thereby allowing for highly dynamic PPI analysis in planta. Table 6.3 provides an overview of the luciferase used in planta.

6.3.3.6 Planta CoIP

CoIP was adapted from column affinity chromatography (or affinity purification; Cuatrecasas, Wilchek, and Anfinsen 1968). Thus, affinity purification is principally based on molecular recognition between a matrix-bound molecule and the target molecule in an extract that is presented to it. CoIP is a chimera between an in vivo and in vitro system. Correctly labeled an *ex vivo* technique, CoIP is often used as an alternative method to one of the above systems, allowing the detection of interaction in specific tissues or developmental stages, under experimentally determined conditions, and/or in different genetic backgrounds. The combination of CoIP and MS analysis is a particularly powerful screening tool for discovering novel interactors or establishing interaction maps of POI (Miernyk and Thelen 2008).

6.3.3.7 Enzymatic Protein Labeling

GFP is bulkier than enzymatic labeling. Enzymatic labeling has numerous applications including the creation of clinically relevant conjugates with polymers, cytotoxins, or imaging agents (Zhang et al. 2018).

There are enzymes such as

1. Peptidase – sortase A and subtiligase
2. Transferase – Microbial transglutaminase, N-myristoyl transferase, farnesyl transferase, and phosphopentethiinyl transferase
3. Ligase – Tubulin tyrosine ligase, lipoic acid ligase, and biotin ligase
4. Oxidoreductases – Formyl glycine-generating enzyme

6.4 CONCLUSION

This chapter summarizes the available protein tags with their known application across the varying species and known kingdoms. In near future, new technologies should be developed to facilitate large-scale studies of tagged proteins with a prospect on their dynamics and spatial and temporal resolutions of interactions in selected cell types and even whole plants or multicellular organisms.

REFERENCES

Arnau, J., C. Lauritzen, G.E. Petersen, and J. Pedersen. 2011. "Reprint of: Current strategies for the use of affinity tags and tag removal for the purification of recombinant proteins." *Protein Expression and Purification*, September. doi: 10.1016/j.pep.2011.08.013.

Atkins, D., and J.G. Izant. 1995. "Expression and analysis of the green fluorescent protein gene in the fission yeast schizosaccharomyces pombe." *Current Genetics* 28 (6): 585–88. doi: 10.1007/BF00518173.

Bähler, J., J.Q. Wu, M.S. Longtine, N.G. Shah, A. McKenzie, A.B. Steever, A. Wach, P. Philippsen, and J.R. Pringle. 1998. "Heterologous modules for efficient and versatile PCR-based gene targeting in schizosaccharomyces pombe." *Yeast (Chichester, England)* 14 (10): 943–51. doi: 10.1002/(SICI)1097-0061(199807)14:10<943::AID-YEA292>3.0.CO;2–Y.

Bowen, J., J. Schneible, K. Bacon, C. Labar, S. Menegatti, and B.M. Rao. 2021. "Screening of yeast display libraries of enzymatically treated peptides to discover macrocyclic peptide ligands." *International Journal of Molecular Sciences* 22 (4): 1634. doi:10.3390/ijms22041634.

Bracha-Drori, K., K. Shichrur, A. Katz, M. Oliva, R. Angelovici, S. Yalovsky, and N. Ohad. 2004. "Detection of protein-protein interactions in plants using bimolecular fluorescence complementation." *The Plant Journal* 40 (3): 419–27. doi:10.1111/j.1365-313X.2004.02206.x.

Brown, H.A., and L.J. Marnett. 2011. "Introduction to lipid biochemistry, metabolism, and signaling." *Chemical Reviews* 111 (10): 5817–20. doi: 10.1021/cr200363s.

Chalfie, M., Y. Tu, G. Euskirchen, W.W. Ward, and D.C. Prasher. 1994. "Green fluorescent protein as a marker for gene expression." *Science (New York, N.Y.)* 263 (5148): 802–5. doi: 10.1126/science.8303295.

Chen, H., Y. Zou, Y. Shang, H. Lin, Y. Wang, R. Cai, X. Tang, and J.-M. Zhou. 2008. "Firefly luciferase complementation imaging assay for protein-protein interactions in plants." *Plant Physiology* 146 (2): 323–24. doi: 10.1104/pp.107.111740.

Colucci-Guyon, E., A.S. Batista, S.D.S. Oliveira, M. Blaud, I.C. Bellettini, B.S. Marteyn, K. Leblanc, P. Herbomel, and R. Duval. 2019. "Ultraspecific live imaging of the dynamics of zebrafish neutrophil granules by a histopermeable fluorogenic benzochalcone probe." *Chemical Science* 10 (12): 3654–70. doi: 10.1039/c8sc05593a.

Cristea, I.M., R. Williams, B.T. Chait, and M.P. Rout. 2005. "Fluorescent proteins as proteomic probes." *Molecular & Cellular Proteomics: MCP* 4 (12): 1933–41. doi: 10.1074/mcp.M500227-MCP200.

Cuatrecasas, P., M. Wilchek, and C.B. Anfinsen. 1968. "Selective enzyme purification by affinity chromatography." *Proceedings of the National Academy of Sciences* 61 (2): 636–43. doi: 10.1073/pnas.61.2.636.

Dennler, P., L.K. Bailey, P.R. Spycher, R. Schibli, and E. Fischer. 2015. "Microbial transglutaminase and C-Myc-Tag: A strong couple for the functionalization of antibody-like protein scaffolds from discovery platforms." *ChemBioChem* 16 (5): 861–67. doi: 10.1002/cbic.201500009.

DiCesare, N., and J.R. Lakowicz. 2002. "Chalcone-analogue fluorescent prfobes for saccharides signaling using the boronic acid group." *Tetrahedron Letters* 43 (14): 2615–18. doi: 10.1016/s0040-4039(02)00312-x.

Dundas, C.M., D. Demonte, and S. Park. 2013. "Streptavidin-biotin technology: Improvements and innovations in chemical and biological applications." *Applied Microbiology and Biotechnology* 97 (21): 9343–53. doi: 10.1007/s00253-013-5232-z.

Einhauer, A., and A. Jungbauer. 2001. "The FLAGTM peptide, a versatile fusion tag for the purification of recombinant proteins." *Journal of Biochemical and Biophysical Methods* 49 (1–3): 455–65. doi: 10.1016/S0165-022X(01)00213-5.

Evan, G.I., G.K. Lewis, G. Ramsay, and J.M. Bishop. 1985. "Isolation of monoclonal antibodies specific for human C-Myc proto-oncogene product." *Molecular and Cellular Biology* 5 (12): 3610–16. doi: 10.1128/mcb.5.12.3610-3616.1985.

Faria, S., I. Joao, and L. Jordao. 2015. "General overview on nontuberculous mycobacteria, biofilms, and human infection." *Journal of Pathogens* 2015: 809014. doi: 10.1155/2015/809014.

Field, J., J. Nikawa, D. Broek, B. MacDonald, L. Rodgers, I.A. Wilson, R.A. Lerner, and M. Wigler. 1988. "Purification of a RAS-responsive adenylyl cyclase complex from saccharomyces cerevisiae by use of an epitope addition method." *Molecular and Cellular Biology* 8 (5): 2159–65. doi: 10.1128/mcb.8.5.2159-2165.1988.

Foulger, J.H. 1931. "The use of the molisch (α-naphthol) reactions in the study of sugars in biological fluids." *Journal of Biological Chemistry* 92 (2): 345–53. doi: 10.1016/S0021-9258(18)76522-8.

Fujikawa, Y., and N. Kato. 2007. "Technical advance: Split luciferase complementation assay to study protein-protein interactions in arabidopsis protoplasts: Protein-protein interactions in living plant cells." *The Plant Journal* 52 (1): 185–95. doi: 10.1111/j.1365-313 X.2007.03214.x.

Gadaleta, M.C., O. Iwasaki, C. Noguchi, K. Noma, and E. Noguchi. 2013. "New vectors for epitope tagging and gene disruption in *Schizosaccharomyces Pombe*." *BioTechniques* 55 (5): 257–63. doi: 10.2144/000114100.

Gama, L., and G.E. Breitwieser. 2002. "Generation of epitope-tagged proteins by inverse polymerase chain reaction mutagenesis." In *In Vitro Mutagenesis Protocols*, edited by J. Braman, 77–83. Totowa, NJ: Humana Press. doi: 10.1385/1-59259-194-9:077.

Ghosh, I., A.D. Hamilton, and L. Regan. 2000. "Antiparallel leucine zipper-directed protein reassembly: Application to the green fluorescent protein." *Journal of the American Chemical Society* 122 (23): 5658–59. doi: 10.1021/ja994421w.

Goudeau, J., M. Samaddar, K. Adam Bohnert, and C. Kenyon. 2020. "Addendum: A lysosomal switch triggers proteostasis renewal in the immortal C. Elegans germ lineage." *Nature* 580 (7802): E5. doi: 10.1038/s41586-020-2108-0.

Gravato-Nobre, M.J., H.R. Nicholas, R. Nijland, D. O'Rourke, D.E. Whittington, K.J. Yook, and J. Hodgkin. 2005. "Multiple genes affect sensitivity of caenorhabditis elegans to the bacterial pathogen microbacterium nematophilum." *Genetics* 171 (3): 1033–45. doi: 10.1534/genetics.105.045716.

Greenfield, N.J. 2004. "Circular dichroism analysis for protein-protein interactions." *Methods in Molecular Biology (Clifton, N.J.)* 261: 55–78. doi: 10.1385/1-59259-762-9:055.

Gygi, S.P., Y. Rochon, B.R. Franza, and R. Aebersold. 1999. "Correlation between protein and MRNA abundance in yeast." *Molecular and Cellular Biology* 19 (3): 1720–30. doi: 10.1128/MCB.19.3.1720.

Haddock, S.H.D., M.A. Moline, and J.F. Case. 2010. "Bioluminescence in the sea." *Annual Review of Marine Science* 2 (1): 443–93. doi: 10.1146/annurev-marine-120308-081028.

Hanke, T., D.F. Young, C. Doyle, I. Jones, and R.E. Randall. 1995. "Attachment of an oligopeptide epitope to the C-terminus of recombinant SIV Gp160 facilitates the construction of SMAA complexes while preserving CD4 binding." *Journal of Virological Methods* 53 (1): 149–56. doi: 10.1016/0166-0934(95)00003-d.

Hochuli, E., H. Döbeli, and A. Schacher. 1987. "New metal chelate adsorbent selective for proteins and peptides containing neighbouring histidine residues." *Journal of Chromatography* 411: 177–84. doi: 10.1016/s0021-9673(00)93969-4.

Hodgkin, J., P.E. Kuwabara, and B. Corneliussen. 2000. "A novel bacterial pathogen, microbacterium nematophilum, induces morphological change in the nematode C. Elegans." *Current Biology: CB* 10 (24): 1615–18. doi: 10.1016/s0960-9822(00)00867-8.

Hopp, T.P., K.S. Prickett, V.L. Price, R.T. Libby, C.J. March, D.P. Cerretti, D.L. Urdal, and P.J. Conlon. 1988. "A short polypeptide marker sequence useful for recombinant protein identification and purification." *Bio/Technology* 6 (10): 1204–10. doi: 10.1038/nbt1088-1204.

Hu, G., M. Zhong, J. Zhao, H. Gao, L. Gan, H. Zhang, S. Zhang, and J. Fang. 2021. "Fluorescent probes for imaging protein disulfides in live organisms." *ACS Sensors* 6 (3): 1384–91. doi: 10.1021/acssensors.1c00049.

Knappik, A., and A. Plückthun. 1994. "An improved affinity tag based on the FLAG peptide for the detection and purification of recombinant antibody fragments." *BioTechniques* 17 (4): 754–61.

Kostakioti, M., M. Hadjifrangiskou, and S.J. Hultgren. 2013. "Bacterial biofilms: Development, dispersal, and therapeutic strategies in the dawn of the postantibiotic era." *Cold Spring Harbor Perspectives in Medicine* 3 (4): a010306. doi: 10.1101/cshperspect.a010306.

Kruger, N.J. 1994. "The bradford method for protein quantitation." In *Basic Protein and Peptide Protocols*, edited by John M. Walker, 32:9–16. Totowa, NJ: Humana Press. doi: 10.1385/0-89603-268-X:9.

Lagendijk, E.L., S. Validov, G.E.M. Lamers, S. de Weert, and G.V. Bloemberg. 2010. "Genetic tools for tagging gram-negative bacteria with MCherry for visualization in vitro and in natural habitats, biofilm and pathogenicity studies." *FEMS Microbiology Letters* 305 (1): 81–90. doi: 10.1111/j.1574-6968.2010.01916.x.

Lai, R., W. Li, Z. Xu, W. Liu, Q. Zeng, W. Lin, J. Jiang, J. Lai, and C. Yang. 2023. "A robust method for identification of plant SUMOylation substrates in a library-based reconstitution system." *Plant Communications* 100573. doi: 10.1016/j.xplc.2023.100573.

Lalonde, S., D.W. Ehrhardt, D. Loqué, J. Chen, S.Y. Rhee, and W.B. Frommer. 2008. "Molecular and cellular approaches for the detection of protein-protein interactions: Latest techniques and current limitations: The detection of protein-protein interactions." *The Plant Journal* 53 (4): 610–35. doi: 10.1111/j.1365-313X.2007.03332.x.

Leslie, J., S.M. Robinson, F. Oakley, and S. Luli. 2021. "Non-invasive synchronous monitoring of neutrophil migration using whole body near-infrared fluorescence-based imaging." *Scientific Reports* 11 (1): 1415. doi: 10.1038/s41598-021-81097-8.

Long, Y., Y. Stahl, S. Weidtkamp-Peters, W. Smet, Y. Du, T.W.J. Gadella, J. Goedhart, B. Scheres, and I. Blilou. 2018. "Optimizing FRET-FLIM labeling conditions to detect nuclear protein interactions at native expression levels in living arabidopsis roots." *Frontiers in Plant Science* 9: 639. doi: 10.3389/fpls.2018.00639.

Lund, C.H., J.R. Bromley, A. Stenbæk, R.E. Rasmussen, H.V. Scheller, and Y. Sakuragi. 2015. "A reversible renilla luciferase protein complementation assay for rapid identification of protein-protein interactions reveals the existence of an interaction network involved in xyloglucan biosynthesis in the plant golgi apparatus." *Journal of Experimental Botany* 66 (1): 85–97. doi: 10.1093/jxb/eru401.

Magliery, T.J., C.G.M. Wilson, W. Pan, D. Mishler, I. Ghosh, A.D. Hamilton, and L. Regan. 2005. "Detecting protein–protein interactions with a green fluorescent protein fragment reassembly trap: Scope and mechanism." *Journal of the American Chemical Society* 127 (1): 146–57. doi:10.1021/ja046699g.

Miernyk, J.A., and J.J. Thelen. 2008. "Biochemical approaches for discovering protein-protein interactions: Biochemical approaches for discovering protein-protein interactions." *The Plant Journal* 53 (4): 597–609. doi: 10.1111/j.1365-313X.2007.03316.x.

Miller, K.E., Y. Kim, W.-K. Huh, and H.-O. Park. 2015. "Bimolecular fluorescence complementation (BiFC) analysis: Advances and recent applications for genome-wide interaction studies." *Journal of Molecular Biology* 427 (11): 2039–55. doi: 10.1016/j.jmb.2015.03.005.

Misteli, T., and D.L. Spector. 1997. "Applications of the green fluorescent protein in cell biology and biotechnology." *Nature Biotechnology* 15 (10): 961–64. doi: 10.1038/nbt1097-961.

Morin, J.G., and J.W. Hastings. 1971. "Energy transfer in a bioluminescent system." *Journal of Cellular Physiology* 77 (3): 313–18. doi: 10.1002/jcp.1040770305.

Noguchi, C., M.V. Garabedian, M. Malik, and E. Noguchi. 2008. "A vector system for genomic FLAG epitope-tagging in schizosaccharomyces pombe." *Biotechnology Journal* 3 (9–10): 1280–85. doi: 10.1002/biot.200800140.

Oda, Y., K. Huang, F.R. Cross, D. Cowburn, and B.T. Chait. 1999. "Accurate quantitation of protein expression and site-specific phosphorylation." *Proceedings of the National Academy of Sciences of the United States of America* 96 (12): 6591–96. doi: 10.1073/pnas.96.12.6591.

Patterson, G.H., S.M. Knobel, W.D. Sharif, S.R. Kain, and D.W. Piston. 1997. "Use of the green fluorescent protein and its mutants in quantitative fluorescence microscopy." *Biophysical Journal* 73 (5): 2782–90. doi: 10.1016/S0006-3495(97)78307-3.

Piehler, J. 2005. "New Methodologies for Measuring Protein Interactions in Vivo and in Vitro." *Current Opinion in Structural Biology* 15 (1): 4–14. doi: 10.1016/j.sbi.2005.01.008.

Prasher, D.C., V.K. Eckenrode, W.W. Ward, F.G. Prendergast, and M.J. Cormier. 1992. "Primary structure of the aequorea victoria green-fluorescent protein." *Gene* 111 (2): 229–33. doi: 10.1016/0378-1119(92)90691-h.

Qin, K., C. Dong, G. Wu, and N.A. Lambert. 2011. "Inactive-state preassembly of G(q)-coupled receptors and G(q) heterotrimers." *Nature Chemical Biology* 7 (10): 740–47. doi: 10.1038/nchembio.642.

Raran-Kurussi, S., S. Cherry, D. Zhang, and D.S. Waugh. 2017. "Removal of affinity tags with TEV protease." *Methods in Molecular Biology (Clifton, N.J.)* 1586: 221–30. doi: 10.1007/978-1-4939-6887-9_14.

Raran-Kurussi, S., and D.S. Waugh. 2017. "Expression and purification of recombinant proteins in escherichia coli with a His6 or Dual His6-MBP tag." In *Protein Crystallography*, edited by A. Wlodawer, Z. Dauter, and M. Jaskolski, 1607:1–15. Methods in Molecular Biology. New York: Springer. doi: 10.1007/978-1-4939-7000-1_1.

Rashidian, M., J.K. Dozier, and M.D. Distefano. 2013. "Enzymatic labeling of proteins: Techniques and approaches." *Bioconjugate Chemistry* 24 (8): 1277–94. doi:10.1021/bc400102w.

Sasaki, F., T. Okuno, K. Saeki, L. Min, N. Onohara, H. Kato, T. Shimizu, and T. Yokomizo. 2012. "A high-affinity monoclonal antibody against the flag tag useful for g-protein-coupled receptor study." *Analytical Biochemistry* 425 (2): 157–65. doi: 10.1016/j.ab.2012.03.014.

Sattarzadeh, A., R. Saberianfar, W.R. Zipfel, R. Menassa, and M.R. Hanson. 2015. "Green to red photoconversion of GFP for protein tracking in vivo." *Scientific Reports* 5 (1): 11771. doi: 10.1038/srep11771.

Schäfer, F., N. Seip, B. Maertens, H. Block, and J. Kubicek. 2015. "Purification of GST-tagged proteins." *Methods in Enzymology* 559: 127–39. doi: 10.1016/bs.mie.2014.11.005.

Shaner, N.C., P.A. Steinbach, and R.Y. Tsien. 2005. "A guide to choosing fluorescent proteins." *Nature Methods* 2 (12): 905–9. doi: 10.1038/nmeth819.

Shimomura, O., F.H. Johnson, and Y. Saiga. 1962. "Extraction, purification and properties of aequorin, a bioluminescent protein from the luminous hydromedusan, aequorea." *Journal of Cellular and Comparative Physiology* 59 (3): 223–39. doi: 10.1002/jcp.1030590302.

Smith, J.C., R.B. Derbyshire, E. Cook, L. Dunthorne, J. Viney, S.J. Brewer, H.M. Sassenfeld, and L.D. Bell. 1984. "Chemical synthesis and cloning of a poly(Arginine)-coding gene fragment designed to aid polypeptide purification." *Gene* 32 (3): 321–27. doi: 10.1016/0378-1119(84)90007-6.

Tanz, S.K., I. Castleden, I.D. Small, and A.H. Millar. 2013. "Fluorescent protein tagging as a tool to define the subcellular distribution of proteins in plants." *Frontiers in Plant Science* 4: 214. doi: 10.3389/fpls.2013.00214.

Terpe, K. 2003. "Overview of tag protein fusions: From molecular and biochemical fundamentals to commercial systems." *Applied Microbiology and Biotechnology* 60 (5): 523–33. doi: 10.1007/s00253-002-1158-6.

Valdez-Sinon, A.N., A. Gokhale, V. Faundez, and G.J. Bassell. 2020. "Protocol for immuno-enrichment of FLAG-tagged protein complexes." *STAR Protocols* 1 (2): 100083. doi: 10.1016/j.xpro.2020.100083.

Van Zyl, W. F. (2015). Fluorescence and bioluminescence imaging of the intestinal colonization of Enterococcus mundtii ST4SA and Lactobacillus plantarum 423 in mice infected with Listeria monocytogenes EGde (Doctoral dissertation, Stellenbosch: Stellenbosch University).

Verma, D., M. Stapleton, J. Gadwa, K. Vongtongsalee, A.R. Schenkel, E.D. Chan, and D. Ordway. 2019. "Mycobacterium avium infection in a C3HeB/FeJ mouse model." *Frontiers in Microbiology* 10: 693. doi: 10.3389/fmicb.2019.00693.

Walls, D., and S.T. Loughran. 2011. "Tagging recombinant proteins to enhance solubility and aid purification." *Methods in Molecular Biology (Clifton, N.J.)* 681: 151–75. doi: 10.1007/978-1-60761-913-0_9.

Walter, M., C. Chaban, K. Schütze, O. Batistic, K. Weckermann, C. Näke, D. Blazevic, et al. 2004. "Visualization of protein interactions in living plant cells using bimolecular fluorescence complementation." *The Plant Journal* 40 (3): 428–38. doi: 10.1111/j.1365-313X.2004.02219.x.

Waugh, D.S. 2016. "Crystal structures of MBP fusion proteins." *Protein Science: A Publication of the Protein Society* 25 (3): 559–71. doi: 10.1002/pro.2863.

Weiss, S. 1999. "Fluorescence spectroscopy of single biomolecules." *Science* 283 (5408): 1676–83. doi: 10.1126/science.283.5408.1676.

Yamauchi, K., M. Yang, P. Jiang, N. Yamamoto, M. Xu, Y. Amoh, K. Tsuji, et al. 2005. "Real-time in vivo dual-color imaging of intracapillary cancer cell and nucleus deformation and migration." *Cancer Research* 65 (10): 4246–52. doi: 10.1158/0008-5472.CAN-05-0069.

Yang, F., L.G. Moss, and G.N. Phillips. 1996. "The molecular structure of green fluorescent protein." *Nature Biotechnology* 14 (10): 1246–51. doi: 10.1038/nbt1096-1246.

Zacharias, D.A., J.D. Violin, A.C. Newton, and R.Y. Tsien. 2002. "Partitioning of lipid-modified monomeric GFPs into membrane microdomains of live cells." *Science* 296 (5569): 913–16. doi:10.1126/science.1068539.

Zhang, Y., K.-Y. Park, K.F. Suazo, and M.D. Distefano. 2018. "Recent progress in enzymatic protein labelling techniques and their applications." *Chemical Society Reviews* 47 (24): 9106–36. doi: 10.1039/c8cs00537k.

7 Fluorescent Polymer-Based Molecular Recognition and Application

Sonkeshwar Sharma and Mohammad Asif Ali

7.1 INTRODUCTION OF FLUORESCENT POLYMER

Luminescence is a phenomenon in which a substance emits visible, ultraviolet, or infrared light in the optical range at a given temperature.[1] It is caused by a substance's chemical change, electrical energy, or other intrinsic crystal stress without heating. Fluorescence and phosphorescence are good examples of luminescence. However, other types of luminescence include hemi-, bio-, tribo-, and thermo-luminescence. Fluorescence belongs to photoluminescence. In this phenomenon, the molecules absorb electromagnetic radiation and become excited to the first excited singlet state (absorption spectra). The electromagnetic radiation is released into the optical range from the first excited singlet state's lowest vibrational state to the ground state. The latter process is known as fluorescence. The time of the fluorescence process is of the order of 10^{-1} s.[2] The definition of fluorescence depends on the type of molecules, crystals, and particles, such as organic molecules,[3] nano-crystalline semiconductors,[4] and metal nanoparticles,[5] and varies accordingly as the process of emissions for different polymers has different and complex mechanisms. Materials with luminescence behavior are fluorescent materials and have recently been in high demand. For this, many researchers have tried to explore certain substances with fluorescent behavior, such as silica particles,[6] glass,[7] gold surfaces,[5] quantum dots,[8] and carbon dots.[9] Some researchers developed their interest in polymers with luminescence properties, known as fluorescent polymers. These polymers open a new door with unique characteristics and share dependencies with other fluorescent materials. Fluorescent polymers have advantages over small fluorescent molecules, such as their enhanced signal response even after perturbation due to the cooperative conformational effects of its chain segments[10] and their superior visco-elastic and mechanical properties. With these unique characteristics, fluorescent polymers are widely used to manufacture new devices.[11,12] Like other fluorescent materials, these polymers show a similar widespread use in areas such as sensing[13] and imaging,[14] optoelectronics,[15] fluorescent bioprobes,[16] molecular imaging,[17] photodynamic treatments,[18] OLEDs,[19] storage data security,[20] encryption,[21] anti-counterfeiting materials,[22] and other fields.[23,24] Furthermore, this chapter discusses the utility of fluorescent polymers, how they

DOI: 10.1201/9781003352372-7

129

perform luminescence, the reasons behind their luminescence, the types of fluorescent polymers, their mechanistic approach, and their effects on the environment, etc. Therefore, due to their optical or luminescence behavior, it is widely used in molecular recognition selectively.

7.2   TYPES OF FLUORESCENT POLYMER

Fluorescent polymers contain fluorophosphoric side chains, which have the capability of luminescence, making them an important polymeric material. This side chain mostly determines the utility of the polymer, and by making a slight change in the side chain, its luminescence behavior can be changed. Depending on their structural development, fluorescent polymers are of different types and show a variety of applications. Some fluorescent polymers that have conjugated polymeric chains and show semi-conductive behaviors are known as conjugated fluorescent (CF) polymer, and non-conjugative polymeric chain polymers are termed non-conjugated fluorescent (NCF) polymer. Aggregation-induced emission (AIE) polymers are a particular type of polymer and have advantages over traditional fluorescent polymers. Due to their fine-tuned morphology, unique structural presence, and composition, AIEs show a wide range of chemo/biosensing, imaging, and theranostics applications. Over the past decades, many fluorescent polymers have been developed to meet the needs of people.[25] Modulating the photophysical properties of fluorescent polymers via changes in environmental stimuli is a growing area of research nowadays, and lights, the medium's pH, pressure in the environment, heat energy, an electrical or magnetic environment, and a chemical environment are examples of environmental stimuli.[26,27] Mechanoresponsive luminescent materials change their emission colors from these external factors. The types of fluorescent polymers depend on how they are manufactured and the type of precursor used. The need to develop a variety of fluorescent polymers is a driving force, as a single polymer cannot be used for various applications. The above-mentioned fluorescent polymers, i.e., CF polymers, NCF polymers, and AIE polymers, are widely used in various applications and are manufactured accordingly. Many fluorescent polymers are being tested at the nanoscale, called fluorescent-based nanomaterials. It is being said that they have excellent qualities. Therefore, nanotechnology opens a new window in the field of luminescent materials. Many nanomaterials with luminescence properties have been developed, and many more with enhanced properties are coming in the near future. Some fluorescent dyes are also reported in several research publications to make fluorescent polymers and are mentioned below (Table 7.1).

7.3   MECHANISTIC ACTION OF FLUORESCENT POLYMER-BASED MOLECULAR RECOGNITION: PH-RESPONSIVE FLUORESCENT POLYMER

Fluorescent polymer is a real-time detector for drug systems because it shows its fluorescence behavior at different pHs.[34] Dai et al. (2022) experimented with the antibiotic drug on bacterial biofilm with a branched polymer, i.e. [poly (MBA-AEPZ)-AEPZ-NA]

TABLE 7.1

Overview of Fluorescent Molecules, Their Reactive Groups, Types of Polymerizations, and Their Uses

Fluorescent Molecules	Reactive Groups	Type of Polymerization	Uses	References
	Methacryoyl-	Atom-transfer radical polymerization	Monitoring of encapsulation of Doxorubicin	Chen and Wu[29]
	Styryl-	Reversible addition-fragmentation chain transfer polymerization	Cell imaging study	Ma et al.[30]
	Azide-	Copper-catalyzed azide-alkyne cycloaddition	Bacteria detection and inhibition	Li et al.[31]

(Continued)

TABLE 7.1 (Continued)
Overview of Fluorescent Molecules, Their Reactive Groups, Types of Polymerizations, and Their Uses

Fluorescent Molecules	Reactive Groups	Type of Polymerization	Uses	References
	Amino-	Ring-opening polymerization	Bio-imaging of cancer cell	Xue et al.[32]
	Styryl-	Free radical polymerization	Supramolecular hyperbranched polymer	Li et al.[33]

Adopted and modified from Reference[28]. Copyright 2021, John Wiley and Sons.

and found a good result; it realizes effective real-time biofilm detection, eradicating biofilm, and realize fluorescence imaging-guided treatment of bacterial biofilm infections. Biofilms are multicellular aggregates of bacteria that are enclosed in extracellular polymeric materials. The biofilm's presence restricts the antibiotic drug's penetration effect and triggers the bacteria's tolerance to the antibiotic, making it challenging to eliminate the bacteria. The branched polymer can overcome bacterial resistance, and the fluorescence image shows a real-time visualization of the biofilm by guided infection control. The positively charged branched polymer can transport antibiotics into the biofilm by effectively penetrating the bacterial biofilm. The positively charged polymer neutralizes the anionic character of the bacterial biofilm and distorts its structural integrity. This neutralizes bacterial resistance and allows antibiotics to transfer to the biofilm easily. The designed polymers show a weak fluorescence emission intensity under physiological conditions (pH 7.4), but their emission changes to green light emission within the local biofilm microenvironment (pH 5.5) and shows a real-time visualization of bacterial biofilms. This experiment has been done positively on live zebrafish (Figure 7.1).

7.3.1 THE FLUORESCENT POLYMER AS A SELF-REPORTING SENSOR AGENT

Polymers are associated with multiple monomers, and the associated mechanisms for polymerization are of different types, such as addition or chain growth polymerization and condensation polymerization or step-growth polymerization.[35] Estupinan

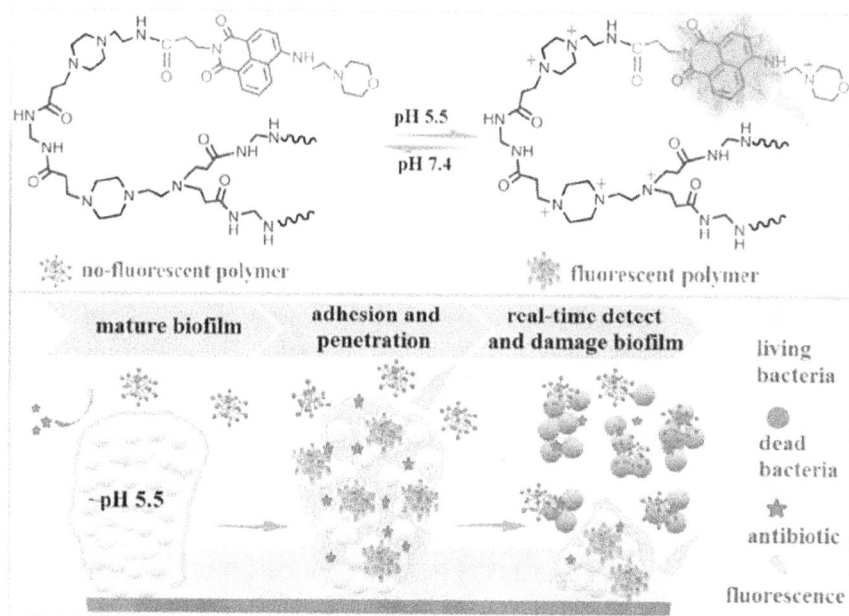

FIGURE 7.1   pH-responsive green fluorescent polymers, synergistic action with antibiotics for real-time biofilm imaging and elimination. (Reproduced with permission from reference.[34] Copyright 2022, American Chemical Society.)

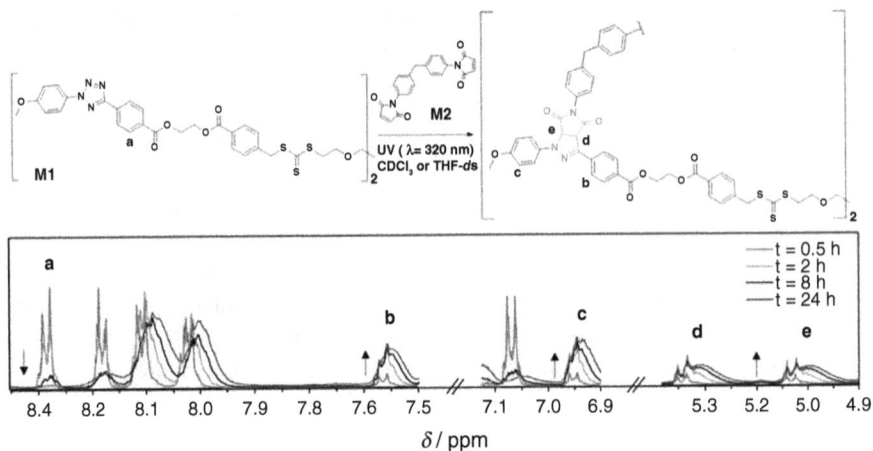

FIGURE 7.2 Condensation polymerization of two monomers, i.e., tetrazole RAFT agent M1 and bismaleimide M2. The ^{1}H NMR spectra (CDCl$_3$) demonstrate monomer conversion and evolution of the signals used to determine the pyrazoline yield. (Reproduced with permission from reference.[35] Copyright 2017, American Chemical Society.)

et al. introduced self-reporting step-growth fluorescent polymer and their methodology. The introduced polymer follows the photo-induced nitrile imine-mediated tetra-zole-ene cycloaddition process to access the kinetics of step-growth polymerization. The step-growth polymerization occurs with two monomers containing the tetrazole moiety and the active dialkenes group to produce the fluorescent pyrazoline-containing polymer. The number of ligation points in polymerization is directly proportional to the fluorescent emission of step-growth polymerization. Hence the reported polymer and the mechanism is the perfect example of the self-reporting sensor system (Figure 7.2).

The mechanism discussed above shows how the fluorescent polymer behaves when used for molecular recognition. These are only a few instances demonstrating how effective fluorescent polymers are as instruments in the medical and therapeutic professions. It helps analyze the process of the reaction during polymerization and also helps in assessing real-time drug interactions in the biomedical field. Drug interactions with biological cells can be easily studied using fluorescent polymers; therefore, this fluorescent polymer is also used in cancer therapy.

7.4 ENVIRONMENTAL IMPACTS AND ITS BIO-DEGRADABILITY

Degradable and biodegradable materials are very much discussed these days. They have wide applications in agriculture, industrial food, medical, and environmental protection.[36–39] Degradable polymeric substances can be converted into smaller molecular fragments, such as oligomers, and monomers, due to natural or industrial conditions after decomposition.[40] Biodegradability is a subclass of degradability, and enzymes or bacteria are responsible for this process.[41] Recently, many researchers have tried to develop degradable or biodegradable materials to protect the environment from waste materials because once the materials are used, they must be converted into valuable biomass.[42,43] Many researchers found results but not good enough. When researchers investigated the degradability or biodegradability of polymeric materials, they found that many materials are readily converted to micro or nano form, but the conversion of micro or nano molecules to valuable biomass takes much longer.[44,45] Micro or nano polymeric materials are more dangerous than they initially as they are the hidden type of waste pollution, causing a more negative impact on the environment.[46] Fluorescent polymers have wide applications in the medical field, and there are many different ways people take these materials into their bodies during treatment or at the time of medical dosing. Once its role in our body is over, it passes directly into the environment in the form of waste materials. Here it has a minimal shelf life, so it must be quickly converted into valuable biomass, but there are many limitations in the degradation of the polymeric material.

In most cases, fluorescent polymers have a brief life span. During polymer formation, their functional groups are primarily involved in the polymerization and are not available to support the fragmentation or degradation process.[47] Therefore, many fluorescent polymers have an extended period for their biomass conversion. When scientists or researchers identified these problems, many products were banned for this reason.[48] Many researchers tried to add some external oxidative additives to enhance the polymer degradation tendency, but this compromised their usefulness.[49,50] These external oxidative additives have a distinctly negative effect on the environment; many green oxidative additives have also been developed,[51] but the problem persists. So far, degradability or biodegradability has been ensured in industrial conditions, but this is not a permanent solution. Once natural degradation occurs, compromising the polymer's durability and usefulness is a significant problem, and researchers are still trying to solve this problem. Recently, many polymers have been developed and used extensively, but they have a concise shelf life, so they play a significant role in causing environmental pollution. To avoid these scenarios, researchers try to develop biodegradable polymers or enhance the degradation of existing polymers by adding some oxidative additives. Hence it is important to know whether the synthetic polymer is degradable or not, what is its shelf life and how much negative impact it has on the environment if not handled properly. These are the only parameters by which people protect the environment through human activities.

7.5   APPLICATION OF FLUORESCENT POLYMER

Fluorescent polymers have broad applications in medical imaging, drug delivery, discrimination of explosives in water,[52] and theranostics. Polymeric nanoparticles are an emerging field nowadays and are used in many applications. Due to their high surface area interaction, nanotechnology opens up a wide range of opportunities in the polymer field, such as sensors,[53,54] catalysts,[55] energy conservation and storage.[56] The fluorescent polymer is also applicable to analyze the calcium ion concentration and sense the intracellular temperature during the oxidative phosphorylation process and is of great help in neurodegenerative disease.[57] Weng et al.[58] reported an in-situ monitor growth process of fluorescent polymers based on an angle-scanning-based surface plasmon coupled emission approach. This polymerization follows an electrochemically mediated atom-transfer radical mechanism. Fluorescent polymers are used for detecting explosive materials and discrimination of explosives in water, for which the fluorescent polymer sensor assay has been developed.[52] Fluorescent polymers are used as advanced fluorescent polymer probes for site-specific labeling of proteins in living cells using Halo tag technology.[59] Using nanotechnology, researchers have successfully designed and developed a fluorescence resonance energy transfer-mediated multicolored and photo-switchable fluorescent polymer nanoparticle capable of tunable emission properties. The developed polymer has excellent potential for multiplex bioanalysis. For these two dyes are used as monomers, namely, 4-ethoxy-9-allyl-1,8-naphthalimide and allyl-(7-nitro-benzo [1,2,5] oxidazol-4-yl)-amine at different ratio combinations.[60] A reported assay has been developed for detecting DNA hybridization via fluorescent polymer super quenching.[61] β-glucosidase is associated with many diseases, and it is important to analyze its activity effectively, so the researchers developed a fluorescent polymer nanoparticle to determine selective β-glucosidase activity. The method involves the formation of a water-soluble fluorescent polymer nanoparticle as a result of cross-link polymerization of polyethyleneimine molecules with the hydrolyzed product of hydroquinone, which is formed after hydrolysis of the β-arbutin substrate with the help of β-glucosidase.[62] Researchers have developed a protocol to detect DNA methylation that is useful in cancer therapy. The method is very homogeneous, convenient, and sensitive, using an optically amplifying cationic conjugated polymer (ccp, poly((1,4-p henylene)-2,7-[9,9-bis(6'-N,N,N-trimethyl ammonium)-hexyl fluorene] dibromide)).[63] Bu et al.[64] reported a photochemically fluorescent color-tunable system consisting of photoresponsive conjugated polymer nanospheres. They synthesized as photoresponsive conjugated polymers with red, green, and blue (RGB) fluorescence with photoisomeric dithienylethene moieties used as a side chain. Fluorescent polymer nanoparticles doped with ionic liquid-like salts of a cationic dye (octadecyl rhodamine B) with a bulky hydrophobic counter-ion (fluorinated tetraphenylborate) that serves as a spacer minimizing dye aggregation and self-quenching is reported.[65] As a result, a new class of nanomaterials for sensing, imaging, and light harvesting is now possible. Aoki et al.[66] reported a note on near-infrared fluorescent nanoparticles of low-bandgap p-conjugated polymer for in vivo molecular imaging and suggested that dye-doped polymer nanoparticles of the low-band gap polymer are a promising candidate for in vivo molecular probes.

All discussed publications show that fluorescent polymers have wide applications in various fields and have promising potential in medicine. Fluorescent polymers are used as indicators in many qualitative and quantitative analyses, such as it is used to identify and discriminate explosives in water. It is also a real-time indicator in many biological reactions, sensing, diagnostics, imaging, and organic electronic devices.

7.6 CONCLUSION AND FUTURE PERSPECTIVES

This book chapter focuses on the research carried out over the years, particularly on the design, types of fluorescent polymers, their mechanistic action as molecular recognition, their applications, and their effects on the environment. In the early stages, researchers became interested in fluorophores containing polymers because they have distinct photoluminescence, tunable emission colors, and biocompatibility, despite their commercial viability at the time was not good. The electrochromic characteristics of polymers can be employed to enhance the performance of signal transduction systems in sensors and biosensors and their design. Conjugated polymers have undoubtedly been used to fabricate flexible optoelectronic devices and phototransistors due to their softness, flexibility, and lightweight. However, electrochemical applications and the creation of portable electrochemical sensors have received less attention. Thus, tunable characteristics of polymers, such as band gap tuning, can be advantageous in electrochemical applications. Moreover, although some conjugated frameworks have been used to deliver pharmaceuticals such as doxorubicin and bio-imaging, their biological applications are still in their infancy and have much scope for future research. Some factors, including renewable resources or bioavailability, biocompatibility, solubility in aquatic and biological conditions, and low cytotoxicity, are very important to consider when designing fluorescent materials. Fluorescent materials have achieved extraordinary success over the past decades, resulting in additional research and discoveries in this area with enormous potential. Developing and preparing new monomers to make innovative fluorescent polymers with new properties and functionalities is desirable to meet people's future expectations. Nanotechnology is playing a significant role in this regard, and many fluorescent nanomaterials or polymers with excellent properties are being developed, and many more will be done in the near future. Therefore, nanopolymers or nanomaterials have a broad scope in future research.

REFERENCES

1. Braslavsky, S.E. 2007. Glossary of terms used in photochemistry, (IUPAC Recommendations 2006). *Pure and Applied Chemistry*, 79(3): 293–465.
2. Kapoor, K. L., ed. 2011. *Textbook of Physical Chemistry*, Quantum Chemistry and Molecular Spectroscopy, 4th Volume. Haryana, India: Macmillan.
3. McNaught, A.D., and A. Wilkinson. 1997. *Compendium of chemical terminology* (Vol. 1669). Oxford: Blackwell Science..
4. Jin, Y., and X. Gao. 2009. Plasmonic fluorescent quantum dots. *Nature Nanotechnology*, 4(9): 571–576.
5. Huang, C.C., Z. Yang, K.H. Lee, and H.T. Chang. 2007. Synthesis of highly fluorescent gold nanoparticles for sensing mercury (II). *Angewandte Chemie*, 119(36): 6948–6952.
6. El-Safty, S.A., M.A. Shenashen, M. Ismael, M. Khairy, and M.R. Awual. 2013. Mesoporous aluminosilica sensors for the visual removal and detection of Pd (II) and Cu (II) ions. *Microporous and Mesoporous Materials*, 166: 195–205.
7. Crego-Calama, M., and D.N. Reinhoudt. 2001. New materials for metal ion sensing by self-assembled monolayers on glass. *Advanced Materials*, 13(15): 1171–1174.
8. Yadav, N.,R. P. Gaikwad, V. Mishra, and M. Gawande. 2023. Synthesis and photocatalytic Applications of functionalized carbon quantum dots. *Bulletin of the Chemical Society of Japan*, 95(11): 1638–1679.
9. Yadav, N., Mudgal, D. and V. Mishra. 2023. In-situ synthesis of ionic liquid-based-carbon quantum dots as fluorescence probe for hemoglobin detection. *Analytica Chimica Acta*, 1272 341502.
10. Ahumada, G., and M. Borkowska. 2022. Fluorescent polymers conspectus. *Polymers*, 14(6): 1118.

11. Minotto, A., I. Bulut, A.G. Rapidis, G. Carnicella, M. Patrini, E. Lunedei, H.L. Anderson, and F. Cacialli. 2021. Towards efficient near-infrared fluorescent organic light-emitting diodes. *Light: Science & Applications*, 10(1): 1–10.

12. Park, J.M., S.H. Nam, K.I. Hong, Y.E. Jeun, H.S. Ahn, and W.D. Jang. 2021. Stimuli-responsive fluorescent dyes for electrochemically tunable multi-color-emitting devices. *Sensors and Actuators B: Chemical*, 332: 129534.

13. Weldeab, A.O., L. Li, S. Cekli, K.A. Abboud, K.S. Schanze, and R.K. Castellano. 2018. Pyridine-terminated low gap π-conjugated oligomers: Design, synthesis, and photophysical response to protonation and metalation. *Organic Chemistry Frontiers*, 5(21): 3170–3177.

14. Yamagishi, H., T. Matsui, Y. Kitayama, Y. Aikyo, L. Tong, J. Kuwabara, T. Kanbara, M. Morimoto, M. Irie, and Y. Yamamoto. 2021. Fluorescence switchable conjugated polymer microdisk arrays by cosolvent vapor annealing. *Polymers*, 13(2): 269.

15. Li, Q., W.C. Zhang, C.F. Wang, and S. Chen. 2015. In situ access to fluorescent dual-component polymers towards optoelectronic devices via inhomogeneous biphase frontal polymerization. *RSC Advances*, 5(124): 102294–102299.

16. Liu, Y., Y.M. Wang, W.Y. Zhu, C.H. Zhang, H. Tang, and J.H. Jiang. 2018. Conjugated polymer nanoparticles-based fluorescent biosensor for ultrasensitive detection of hydroquinone. *Analytica Chimica Acta*, 1012: 60–65.

17. Lichon, L., C. Kotras, B. Myrzakhmetov, P. Arnoux, M. Daurat, C. Nguyen, D. Durand, K. Bouchmella, L.M.A. Ali, J.O. Durand, and S. Richeter. 2020. Polythiophenes with cationic phosphonium groups as vectors for imaging, siRNA delivery, and photodynamic therapy. *Nanomaterials*, 10(8): 1432.

18. Hu, L., Z. Chen, Y. Liu, B. Tian, T. Guo, R. Liu, C. Wang, and L. Ying. 2020. In vivo bioimaging and photodynamic therapy based on two-photon fluorescent conjugated polymers containing dibenzothiophene-S, S-dioxide derivatives. *ACS Applied Materials & Interfaces*, 12(51): 57281–57289.

19. Li, C., Y. Xu, Y. Liu, Z. Ren, Y. Ma, and S. Yan. 2019. Highly efficient white-emitting thermally activated delayed fluorescence polymers: Synthesis, non-doped white OLEDs and electroluminescent mechanism. *Nano Energy*, 65: 104057.

20. Jiang, J., P. Zhang, L. Liu, Y. Li, Y. Zhang, T. Wu, H. Xie, C. Zhang, J. Cui, and J. Chen. 2021. Dual photochromics-contained photoswitchable multistate fluorescent polymers for advanced optical data storage, encryption, and photowritable pattern. *Chemical Engineering Journal*, 425: 131557.

21. Wang, M., K. Jiang, Y. Gao, Y. Liu, Z. Zhang, W. Zhao, H. Ji, T. Zheng, and H. Feng. 2021. A facile fabrication of conjugated fluorescent nanoparticles and micro-scale patterned encryption via high resolution inkjet printing. *Nanoscale*, 13(34): 14337–14345.

22. Abdollahi, A., H. Alidaei-Sharif, H. Roghani-Mamaqani, and A. Herizchi, 2020. Photoswitchable fluorescent polymer nanoparticles as high-security anticounterfeiting materials for authentication and optical patterning. *Journal of Materials Chemistry C*, 8(16): 5476–5493.

23. Gou, Z., X. Zhang, Y. Zuo, and W. Lin. 2019. Synthesis of silane-based poly (thioether) via successive click reaction and their applications in ion detection and cell imaging. *Polymers*, 11(8): 1235.

24. Ostos, F.J., G. Iasilli, M. Carlotti, and A. Pucci. 2020. High-performance luminescent solar concentrators based on poly (Cyclohexylmethacrylate)(PCHMA) films. *Polymers*, 12(12): 2898.

25. Baker, S.N. and G.A. Baker. 2010. Luminescent carbon nanodots: Emergent nanolights. *Angewandte Chemie International Edition*, 49(38): 6726–6744.

26. Mishra, V., S.H.. Jung, H.M. Jeong, and H.I.. Lee. 2014. Thermoresponsive ureidoderivatized polymers: The effect of quaternization on UCST properties.. *Polymer Chemistry*, 5(7): 2411–2416.

27. Nandi, M., B. Maiti, K. Srikanth, and P. De. 2017. Supramolecular interaction-assisted fluorescence and tunable stimuli-responsiveness of L-phenylalanine-based polymers. *Langmuir*, 33(40): 10588–10597.

28. Ban, Q., Y. Li, and S. Wu. 2022. Self-fluorescent polymers for bioimaging. *View*, 3(2): 20200135.

29. Chen, J.I. and W.C. Wu. 2013. Fluorescent polymeric micelles with aggregation-induced emission properties for monitoring the encapsulation of doxorubicin. *Macromolecular Bioscience*, 13(5): 623–632.

30. Ma, C., Q. Ling, S. Xu, H. Zhu, G. Zhang, X. Zhou, Z. Chi, S. Liu, Y. Zhang, and J. Xu. 2014. Preparation of biocompatible aggregation-induced emission homopolymeric nanoparticles for cell imaging. *Macromolecular Bioscience*, 14(2): 235–243.

31. Li, Y., H. Yu, Y. Qian, J. Hu, and S. Liu. 2014. Amphiphilic star copolymer-based bimodal fluorogenic/magnetic resonance probes for concomitant bacteria detection and inhibition. *Advanced Materials*, 26(39): 6734–6741.

32. Xue, Y., W. Liang, Y. Li, Y. Wu, X. Peng, X. Qiu, J. Liu, and R. Sun. 2016. Fluorescent pH-sensing probe based on biorefinery wood lignosulfonate and its application in human cancer cell bioimaging. *Journal of Agricultural and Food Chemistry*, 64(51): 9592–9600.

33. Li, H., W. Chen, F. Xu, X. Fan, T. Liang, X. Qi, and W. Tian. 2018. A color-tunable fluorescent supramolecular hyperbranched polymer constructed by pillar [5] arene-based host-guest recognition and metal ion coordination interaction. *Macromolecular Rapid Communications*, 39(10): 1800053.

34. Dai, X., Q. Xu, L. Yang, J. Ma, and F. Gao. 2022. pH-responsive fluorescent polymer-drug system for real-time detection and in situ eradication of bacterial biofilms. *ACS Biomaterials Science & Engineering*, 8(2): 893–902.

35. Estupinan, D., T. Gegenhuber, J.P. Blinco, C. Barner-Kowollik, and L. Barner. 2017. Self-reporting fluorescent step-growth raft polymers based on nitrile imine-mediated tetrazole-ene cycloaddition chemistry. *ACS Macro Letters*, 6(3): 229–234.

36. Edlund, U., M. Hakkarainen, S. Karlsson, Y. Liu, E. Ranucci, M. Ryner, M.S. Lindblad, K.M. Stridsberg, and I.K. Varma. 2003. *Degradable Aliphatic Polyesters* (Vol. 157). Heidelberg: Springer.

37. Kyrikou, I. and D. Briassoulis. 2007. Biodegradation of agricultural plastic films: A critical review. *Journal of Polymers and the Environment*, 15(2): 125–150.

38. Seppälä, J.V., A.O. Helminen, and H. Korhonen. 2004. Degradable polyesters through chain linking for packaging and biomedical applications. *Macromolecular Bioscience*, 4(3): 208–217.

39. Tawagi, E., T. Ganesh, H.L.M. Cheng, and J.P. Santerre. 2019. Synthesis of degradable-polar-hydrophobic-ionic co-polymeric microspheres by membrane emulsion photopolymerization: In vitro and in vivo studies. *Acta Biomaterialia*, 89: 279–288.

40. Göpferich, A. 1996. Mechanisms of polymer degradation and erosion. *Biomaterials*, 17(2): 103–114.

41. He, Y., K. Xie, P. Xu, X. Huang, W. Gu, F. Zhang, and S. Tang. 2013. Evolution of microbial community diversity and enzymatic activity during composting. *Research in Microbiology*, 164(2): 189–198.

42. Calabrò, P.S. and M. Grosso. 2018. Bioplastics and waste management. *Waste Management*, 78: 800–801.

43. Perotto, G., L. Ceseracciu, R. Simonutti, U.C. Paul, S. Guzman-Puyol, T.N. Tran, I.S. Bayer, and A. Athanassiou. 2018. Bioplastics from vegetable waste via an eco-friendly water-based process. *Green Chemistry*, 20(4): 894–902.

44. Attallah, O.A., M. Mojicevic, E.L. Garcia, M. Azeem, Y. Chen, S. Asmawi, and M. Brenan Fournet. 2021. Macro and micro routes to high performance bioplastics: Bioplastic biodegradability and mechanical and barrier properties. *Polymers*, 13(13): 2155.

45. Rahman, M.H. and P.R. Bhoi. 2021. An overview of non-biodegradable bioplastics. *Journal of Cleaner Production*, 294: 126218.
46. Lett, Z., A. Hall, S. Skidmore, and N.J. Alves. 2021. Environmental microplastic and nanoplastic: Exposure routes and effects on coagulation and the cardiovascular system. *Environmental Pollution*, 291: 118190.
47. Boaen, N.K. and M.A. Hillmyer. 2005. Post-polymerization functionalization of polyolefins. *Chemical Society Reviews*, 34(3): 267–275.
48. Rochman, C.M., S.M. Kross, J.B. Armstrong, M.T. Bogan, E.S. Darling, S.J. Green, A.R. Smyth, and D. Veríssimo. 2015. Scientific evidence supports a ban on microbeads. *Environmental Science & Technology*, 49: 10759–10761.
49. Abdelmoez, W., I. Dahab, E.M. Ragab, O.A. Abdelsalam, and A. Mustafa. 2021. Bio- and oxo-degradable plastics: Insights on facts and challenges. *Polymers for Advanced Technologies*, 32(5): 1981–1996.
50. Hadiyanto, H., A. Khoironi, I. Dianratri, K. Huda, S. Suherman, and F. Muhammad. 2022. Biodegradation of oxidized high-density polyethylene and oxo-degradable plastic using microalgae Dunaliella salina. *Environmental Pollutants and Bioavailability*, 34(1): 469–481.
51. da Luz, J.M.R., M.D.C.S. da Silva, L.F. dos Santos, and M.C.M. Kasuya. 2019. Plastics polymers degradation by fungi. In *Microorganisms. BoD-Books on Demand* (261–270). Blumenberg, M., Shaaban, M. and Elgaml, A. eds., 2020. London, UK: IntechOpen.
52. Woodka, M.D., V.P. Schnee, and M.P. Polcha. 2010. Fluorescent polymer sensor array for detection and discrimination of explosives in water. *Analytical Chemistry*, 82(23): 9917–9924.
53. Dararatana, N., F. Seidi, and D. Crespy. 2018. pH-sensitive polymer conjugates for anticorrosion and corrosion sensing. *ACS Applied Materials & Interfaces*, 10(24): 20876–20883.
54. Hou, J., M. Li, and Y. Song. 2018. Recent advances in colloidal photonic crystal sensors: Materials, structures and analysis methods. *Nano Today*, 22: 132–144.
55. Jaroonwatana, W., T. Theerathanagorn, M. Theerasilp, S. Del Gobbo, D. Yiamsawas, V. D'Elia, and D. Crespy. 2021. Nanoparticles of aromatic biopolymers catalyze CO_2 cycloaddition to epoxides under atmospheric conditions. *Sustainable Energy & Fuels*, 5(21): 5431–5444.
56. Arico, A.S., P. Bruce, B. Scrosati, J.M. Tarascon, and W. Van Schalkwijk. 2011. Nanostructured materials for advanced energy conversion and storage devices. *Materials for Sustainable Energy: A Collection of Peer-Reviewed Research and Review Articles from Nature Publishing Group*, 148–159. doi: 10.1038/nmat1368.
57. Qiao, J., Y.H. Hwang, D.P. Kim, and L. Qi. 2020. Simultaneous monitoring of temperature and Ca^{2+} concentration variation by fluorescent polymer during intracellular heat production. *Analytical Chemistry*, 92(12): 8579–8583.
58. Weng, Y.H., L.T. Xu, M. Chen, Y.Y. Zhai, Y. Zhao, S.K. Ghorai, X.H. Pan, S.H. Cao, and Y.Q. Li, 2019. In situ monitoring of fluorescent polymer brushes by angle-scanning based surface plasmon coupled emission. *ACS Macro Letters*, 8(2): 223–227.
59. Berki, T., A. Bakunts, D. Duret, L. Fabre, C. Ladavière, A. Orsi, M.T. Charreyre, A. Raimondi, E. Van Anken, and A. Favier. 2019. Advanced fluorescent polymer probes for the site-specific labeling of proteins in live cells using the HaloTag technology. *ACS Omega*, 4(7): 12841–12847.
60. Chen, J., P. Zhang, G. Fang, P. Yi, F. Zeng, and S. Wu, 2012. Design and synthesis of FRET-mediated multicolor and photoswitchable fluorescent polymer nanoparticles with tunable emission properties. *The Journal of Physical Chemistry B*, 116(14): 4354–4362.
61. Kushon, S.A., K.D. Ley, K. Bradford, R.M. Jones, D. McBranch, and D. Whitten. 2002. Detection of DNA hybridization via fluorescent polymer superquenching. *Langmuir*, 18(20): 7245–7249.

62. Liu, J., H. Bao, C. Liu, F. Wu, and F. Gao. 2019. "Turn-on" fluorescence determination of β-glucosidase activity using fluorescent polymer nanoparticles formed from polyethylenimine cross-linked with hydroquinone. *ACS Applied Polymer Materials*, 1(11): 3057–3063.

63. Feng, F., L. Liu, and S. Wang. 2010. Fluorescent conjugated polymer-based FRET technique for detection of DNA methylation of cancer cells. *Nature Protocols*, 5(7): 1255–1264.

64. Bu, J., K. Watanabe, H. Hayasaka, and K. Akagi. 2014. Photochemically colour-tuneable white fluorescence illuminants consisting of conjugated polymer nanospheres. *Nature Communications*, 5(1): 1–8.

65. Reisch, A., P. Didier, L. Richert, S. Oncul, Y. Arntz, Y. Mély, and A.S. Klymchenko, 2014. Collective fluorescence switching of counterion-assembled dyes in polymer nanoparticles. *Nature Communications*, 5(1): 1–9.

66. Aoki, H., J.I. Kakuta, T. Yamaguchi, S. Nitahara, and S. Ito. 2011. Near-infrared fluorescent nanoparticle of low-bandgap π-conjugated polymer for in vivo molecular imaging. *Polymer Journal*, 43(11): 937–940.

8 Biomolecular Fluorescent Probes for Hazardous Ion Detection and Application

Darshankumar Prajapati, Parul Shrivastava,
Shilpi Thakur, and Poonam Mishra

8.1 INTRODUCTION

Metals in their respective cation and anion forms are present everywhere in living or non-living species. Metal ions play various structural and functional roles including metabolism, osmotic regulation, catalysis, and cell signaling in all living systems (Li et al., 2022; Park et al., 2020). Moreover, they are useful for various industrial purposes, viz. manufacturing of electronic devices and diagnosis of diseases due to their high strength, a good conductor of heat and electricity, lightness, boiling, and melting point properties (Chowdhury et al., 2018).

Some of the metals like sodium (Na), potassium (K), calcium (Ca), and magnesium (Mg) play an important role in the maintenance of fluid and electrolyte balance, muscle contraction, and cell signaling including neurotransmission and enzyme function (Guliani et al., 2023, Zheng et al., 2020; Carter et al., 2014). Similarly, transition metals such as iron (Fe), zinc (Zn), and copper (Cu) act as a component of proteins and enzymes, which has a role in many vital biological processes such as oxygen transport, energy production, neurotransmission, regulation of gene expression, and synthesis of essential molecules (Zheng et al., 2020). Even though some of them are essential for sustaining life (like iron, zinc, or cobalt), their concentration in organisms must be maintained within a permissible range. They can be hazardous at higher concentrations (John De Acha et al., 2019; Goldhaber, 2003). On the other hand, heavy metals such as cadmium (Cd), lead (Pb), arsenic (As), and mercury (Hg) are detrimental even at low concentration (Xie at al., 2017), showing a close association with cancer or neurodegenerative diseases (Bhatti et al., 2009; Huff et al., 2007).

The environment has been polluted by the presence of living or non-living substances (e.g., pathogenic microbes, sewage discharge, industrial waste, animal defecation, and heavy metals). Heavy metal pollution is becoming a global environmental concern due to speedy industrialization (Fu and Wang, 2011; Arora et al., 2008). Furthermore, heavy metal ions such as lead (Pb^{2+}), mercury (Hg^{2+}), arsenic (As^{2+}), cadmium (Cd^{2+}), and chromium (Cr^{2+}) are usually characterized by

DOI: 10.1201/9781003352372-8

non-degradability, high toxicity, carcinogenicity, and persistence in the environment for longer periods and bioaccumulation in living organisms (Singh et al., 2021; Vardhan et al., 2019).

Therefore, the detection of these metal ions at low concentrations is a matter of priority for environmental protection as well as for disease prevention. To detect and quantify these ions, highly sensitive and selective methods are required (Yadav et al., 2022, 2023). There are a number of techniques for the detection of metal ions reported by the researcher such as spectroscopic detection techniques, electrochemical detection techniques, and optical sensor detection techniques. Optical sensor detection techniques present several attractive features such as the ease of integration in microfluidic platforms (John De Acha et al., 2019; Kuswandi et al., 2007) and the capability of monitoring hazardous environments (Aiestaran et al., 2009). Among the optical sensors, fluorescent ones have gained more attention in recent years compared to other analytical methods, because of their high specificity, low detection limits, fast response time, in situ detection, and technical simplicity (Nan et al., 2021). There has been a remarkable growth in the use of small-molecule fluorescent probes in the biological sciences.

Their working principle of fluorescent probes is based on the change in the fluorescence properties of probe before and after specific binding of probe and analyte which leads to the change in the fluorescence signal. The change in the fluorescence signal is further detected, realizing the detection of different analytes in the given samples (Li et al., 2022; Xu et al., 2016). For determination of metal ions, the sensor must consist of two components: a fluorescent carrier and an ionic carrier, which may be independent species or covalently linked on a molecule (Li et al., 2022). It is of great significance in chemistry, biology, and agriculture to search for biomolecular fluorescent probes with high selectivity and specificity for metal-ion detection (Xu and Xu, 2016). Some examples of biomolecular probes are nucleic acid-based fluorescent probes, amino acid and peptide-based fluorescent probes, etc. Biomolecular fluorescent probes exhibit enormous application for the detection of hazardous metal ions.

This chapter summarizes the various biomolecular probes used for the detection of hazardous metal ions in the biological and environmental samples. Basically, this chapter is divided into three kinds of materials: the first section focuses on the different types of hazardous metal ion and their hazardous effects; the second one is dedicated to the detection of these hazardous ions using different biomolecular fluorescent probe techniques; and the third one is application of fluorescent probes. We also try to summarize the different mechanisms of fluorescent sensing from the literature.

8.2 HAZARDOUS IONS AND ITS HAZARDOUS EFFECTS

There are different metals in their elemental or compound form that have beneficial or hazardous effects on various biological systems. Here, in this review, we emphasize some of the metal ions and their hazardous effects such as iron, zinc, copper, cadmium, mercury, lead, chromium, aluminum, cobalt, nickel, arsenic, and platinum-group elements (PGEs).

8.2.1 IRON (Fe²⁺ OR Fe³⁺)

Iron is considered the most essential metal in living cells. It plays a vital role in the living system like electron transfer processes, regulates the activities of enzymes, and stabilizes the structure of complex biological compounds such as hemoglobin (Elmas Karuk et al., 2020; Powers et al., 2019). Although iron is essential for all living cells, elevated levels of it cause health hazard damage to essential tissues such as the liver and kidneys (Conway and Henderson, 2019; Nakamura et al., 2019). Therefore, there is a requirement to find sensitive methods to detect Fe^{3+} in medical, environmental, and industrial samples.

8.2.2 ZINC (Zn²⁺)

Zinc plays a crucial role in gene transcription, regulation of metalloenzymes, neural signal transmission, and apoptosis (Voegelin et al., 2005; Frederickson and Bush, 2001). According to the U.S. Food and Nutritional Board, the safe range of Zn^{2+} in human body is from 2 mg (0–5 years) to 15 mg (adults). Elevated levels of zinc cause neurological diseases like Parkinson's and Alzheimer's disease (Koh et al., 1996).

8.2.3 COPPER (Cu²⁺)

Copper is the third most abundant transition element after iron and zinc. It plays a vital role in different physiological processes in living cells such as cellular energy generation, oxygen transport and activation and signal transduction as well as it acts as a catalytic co-factor in many metalloenzymes (Jung et al., 2009; Zhou et al., 2009; Gaggelli et al., 2006). Accumulation of copper leads to serious neurodegenerative diseases like Alzheimer's disease (Hao et al., 2020), Indian Childhood Cirrhosis (Hahn et al., 1995), Indian Prion Disease (Brown, 2001), and Menkes and Wilson diseases (Waggoner et al., 1999).

8.2.4 ALUMINUM (Al³⁺)

Aluminum is not essential metal for biological systems, but it is drawing the attention of researchers due to its potential toxicity and its application in automobiles, chemical drugs, packaging materials, electrical equipment, machineries, food additives, water purification, etc. (Chowdhury et al., 2018; Soni et al., 2001). It is supposed to play a significant role in neurodegenerative diseases like Parkinson's and Alzheimer's diseases (Nayak, 2002). Elevated levels of aluminum lead to the brain and kidney problems (Yousef et al., 2005).

8.2.5 CADMIUM (Cd²⁺)

Cadmium is a human carcinogen, and it is very hazardous to human beings. It also pollutes the environment severely. Ingestion of any substantial amount of cadmium causes poisoning and damage to the liver and kidneys. In Japan, itai-itai is a fatal disease caused by cadmium (Chowdhury et al., 2018). So, it is important to detect cadmium at trace level.

8.2.6 Chromium (Cr^{3+}, Cr^{4+}, and Cr^{6+})

Chromium, in its trivalent oxidation state (Cr^{3+}), plays a vital role in the nucleic acids, fats, proteins, and carbohydrates metabolism by activation of some enzymes and stabilizes protein and nucleic acid (Li et al., 2011; Mao et al., 2007). Chromium in its trivalent oxidation state (Cr^{3+}) is less toxic than its higher hexavalent (Cr^{6+}) and tetravalent oxidation (Cr^{4+}) states for human (Bagchi et al., 2002). Increased level of chromium more than the sufficiency acts upon human body by binding to deoxyribonucleic acid (DNA). It also affects the cellular structures and components. Some anthropogenic uses of chromium like alloying, plating, textile dyeing, and pigmenting and tanning of animal hide cause severe environmental pollution (Zhou et al., 2008; Zayed and Terry, 2003). Hence, detection of chromium by simple and quick methods is very important.

8.2.7 Nickel (Ni^{2+})

Nickel is one of the essential transition metals for living organisms as it plays an important role in metabolism and biosynthesis. It also has many industrial importance. Nickel has some hazardous effects on gastrointestinal systems, kidneys, and blood making the bivalent metal chemically toxic (Stafilov, 2000). So, the detection of this metal ion is an important purpose.

8.2.8 Cobalt (Co^{2+})

Cobalt is an essential metal for eukaryotic and prokaryotic organisms (Barceloux and Barceloux, 1999). It also plays a vital role in the nutrition of plants and animals. Vitamin B12, which contains cobalt as metal ion, plays a vital role in the synthesis of DNA, red blood cells formation, and nervous system protection (Okamoto and Eltis, 2011). Moreover, as a redox active metal ion, excessiveness of cobalt can cause serious health hazards on human body such as cardiomyopathy, hyperglycemia, dermatitis, and cancer (Raux et al., 2000; Kobayashi and Shimizu, 1999).

8.2.9 Mercury (Hg^{2+})

Among the heavy metal ions, mercury is the most widely used metal in industrial and agricultural processes. It is the third potential toxic element after lead and arsenic in nature (Mishra et al., 2023; Rice et al., 2014). Mercury coming from the industry continues to be dumped into soil and waterways, which further circulate into the ecosystem and eventually bioaccumulated in the human being causing adverse effects to human health and causes severe environmental pollution (Wang et al., 2008). Mercury pollutants can enter the body through the skin, respiratory tract, and gastrointestinal tract causing severe kidney and stomach diseases because of their affinity to thiol groups in proteins and enzymes (Shuai et al., 2021; Zhao et al., 2020; Kim et al., 2012). Mercury in its all forms has the potential to cause health hazards since it could accumulate in the vital organs such as the brain and kidneys, which leads to diseases that cannot be treated (Elmas Karuk et al., 2020; Camaschella, 2017). Mercury pollution is a global concern because of its severe effect on ecology and human health.

8.2.10   Lead (Pb²⁺)

Lead is not categorized under a biological important metal, but it is useful for car batteries, pigments, ammunition, cable sheathing, weights for lifting, lead crystal glass, and radiation protection (Aksuner et al., 2011). The exposure to high levels of lead may cause multiple disorders in humans by affecting the renal, gastrointestinal, hematological, cardiovascular, reproductive, and nervous systems (Singh et al., 2021; Engwa et al., 2019). Also, the long-term exposure of lead may cause DNA damage, cell apoptosis, cell cycle arrest, interference in signal transduction pathways, and genomic instability such as genotoxicity (Wallace and Djordjevic, 2020; Wang and Shi, 2001).

8.2.11   Arsenic (As³⁺ and As⁵⁺)

Arsenic is a teratogenic and carcinogenic toxic element, which causes serious skin problems, neurodegenerative disorders, and cardiovascular disease (Wang et al., 2012). The organic form of arsenic is less toxic than its inorganic form. In ground water, it mainly exists as arsenate (As^{5+}) and arsenite (As^{3+}). The detection of this hazardous and fatal heavy metal in its trace amount is of the greatest concern.

8.2.12   Platinum-Group Elements (PGEs)

PGEs, including platinum (Pt), palladium (Pd), rhodium (Rh), and ruthenium (Ru), have been widely applied in different industrial processes because of their specific physical and chemical properties (Li et al., 2013). PGE are used as efficient catalysts, which induce powerful transformations to synthesize drug molecules (Sore et al., 2012; Wu et al., 2011a). Among all PGEs, palladium can be transported to biological materials and accumulated in the food chain which may result in potential health hazards (Merget and Rosner, 2001). PGE has been associated with asthma, nausea, increased hair loss, increased spontaneous abortion, dermatitis, and other serious health problems in humans (Bencs et al., 2003). Palladium can also bind to thiol-group containing amino acids, some proteins, and other macromolecules and biochemicals (vitamin B6) which perhaps damages various cellular processes (Li et al., 2013).

8.3   DETECTION OF HAZARDOUS METAL IONS

Metals and their respective ions mentioned in the above portion are found to be omnipresent. They can be found biological and non-biological, living and non-living species. Some of the metals viz. copper (Cu), zinc (Zn), and iron (Fe) are very important for numerous enzymatic as well as biological processes that occur in any living being including humans (Chowdhury et al., 2018). Simultaneously, metals are also useful in industries due to their striking characteristics such as high strength, light weight, good conductor of electricity and heat, and melting and boiling points. Manufacturing of electronic gadgets is one of the widespread applications of metals (Naghdi et al., 2018). They are also applicable in the medical field as a medicinal agent and for the purpose of diagnosis. On the other hand, the presence of metals like lead (Pb), mercury (Hg), and cadmium (Cd) imposes a serious threat to health. Due to an increase in anthropogenic activities, the release of such hazardous metals and their ions into waterbodies has been tremendously increased and has put the ecosystem and environment at high risk (Nagajyothi et al., 2018). Being toxic, non-degradable, and having the ability to bioaccumulate and biomagnify, heavy metal ions have been a continuous contributor to water pollution, thereby imposing a serious threat to human health. Through water, these metal ions reach the soil and are eventually absorbed by the plant, which finally reach animals and humans (Zhu et al., 2019a; Kahlon et al., 2018; Ayangbenro and Babalola, 2017, Mallampati et al., 2013). Due to such serious environmental problems as well as threats to human health risk and well-being, contaminated niches, especially aquatic ecosystems, must be made free of metal ions.

Removal of metal ions from the contaminated niches is one of the important tasks, but prior to that a suitable technique for the detection and quantification should be developed so that one can estimate the level of pollution and think of suitable methods for remediation purposes. Thus, researchers have been forced to develop suitable techniques for the detection of metal-ion contaminants in various sources.

Detection of metal ions is aided by an ion detector or a probe, which is an instrument or a device designed to sense the occurrence of metal ions in its surroundings and sometimes these probes are also useful for the purpose of quantification. However, a detection method should be fast, sensitive, accurate, economically affordable, and environmentally green. There are a number of techniques for the detection of metal ions, but a single technique for the detection of a large variety of metal ions is rather missing. The techniques used for the detection of metal ions can be categorized into three categories: spectroscopic detection techniques, electrochemical detection techniques, and optical detection techniques.

8.3.1 SPECTROSCOPIC METHODS FOR DETECTION

Spectroscopic techniques are versatile, i.e., they can be employed for determination of a variety of metal ions simultaneously with very low concentration detection limits. Spectroscopic detection of metal ions includes highly accurate and sensitive techniques such as atomic absorption spectroscopy (AAS), inductively coupled plasma mass spectroscopy (ICP-MS), X-ray fluorescence (XRF) spectrometry, inductively coupled plasma-optical emission spectrometry (ICP-OES), high-resolution surface plasmon resonance spectroscopy, and neutron activation analysis (Gong et al., 2016; Wang et al., 2015; Sitko et al., 2015; Losev et al., 2015; Pöykiö and Perämäki, 2003). However, these techniques require trained personnel to work with and they are even quite expensive.

8.3.1.1 Atomic Absorption Spectroscopy (AAS)

AAS consists of a primary light source (a hollow cathode lamp or an electrode-less discharge lamp of the same element, which is to be analyzed); plasmas, flames, or electro-thermal atomizers for produce gas-phase atoms or ions for analysis; a monochromator; a detector (solid-state detectors or photomultiplier tubes); and an electronic display system (Figure 8.1). In this method, a specific wavelength is imparted on metal ions, which will in turn absorb the energy during excitation from the ground

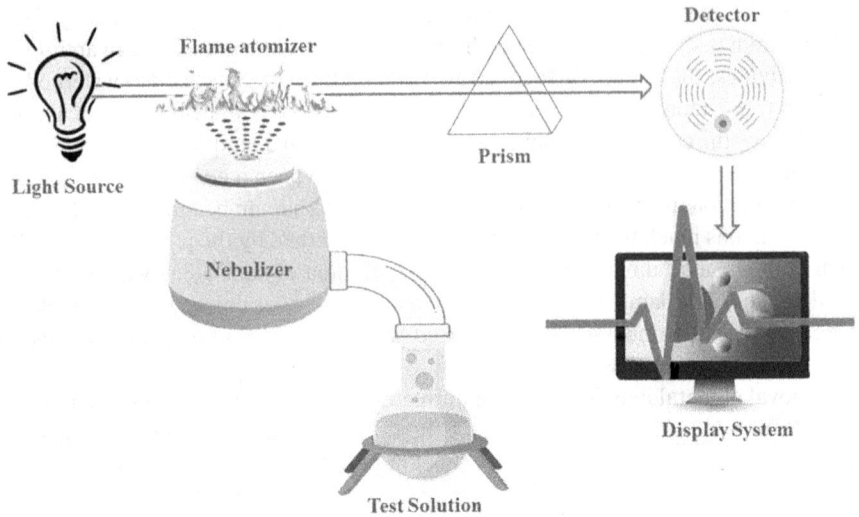

FIGURE 8.1 A block diagram of single-beam atomic absorption spectrophotometry for detection of hazardous metal ions.

state to the excited state. This absorbed energy is measured directly proportional to the concentration of metal ions present in the sample.

Shirkhanloo et al. (2011) reported the presence of hazardous metal ions like copper (Cu^{2+}), lead (Pb^{2+}), and cadmium (Cd^{2+}) in water using a Flame atomic absorption spectroscopic technique, where the concentrations were found to be 2, 3, and 0.2 $\mu g/dm^3$, respectively. Detection of mercury metal ions (Hg^{2+}) can be performed by modified AAS equipped with flow injection mercury systems. Moreover, graphite furnace AAS (GF-AAS) was used for the detection of metal ions such as cadmium (Cd^{2+}), lead (Pb^{2+}), arsenic (As^{3+}), copper (Cu^{2+}), and mercury (Hg^{2+}) with detection limits of 0.014, 0.49, 0.28, 0.19, and 0.061 ppm, respectively (Nie et al., 2008).

8.3.1.2 X-Ray Fluorescence Spectroscopy (XRF)

In XRF spectroscopy, metal ions are irradiated with X-rays or gamma rays which lead to ionization. Such high energy radiation bombardment can even eject electrons from inner orbitals of K or L shell (Figure 8.2a). These vacant places are filled by electrons from higher energy shells which are complemented by the photon emission or fluorescence, which is eventually measured. As every element possesses a distinctive set of energy levels, each element generates its own specific fluorescence spectrum, and thus it becomes easy to determine the elemental composition. Figure 8.2b represents a block diagram of XRF spectroscopy composed of a source of X-rays, sample slot, collimator, fluorescence detector, and output system.

XRF technique has been used to screen and identify toxic elements in various FDA-regulated products (Palmer et al., 2009) and many other such sources which are summarized in Table 8.1.

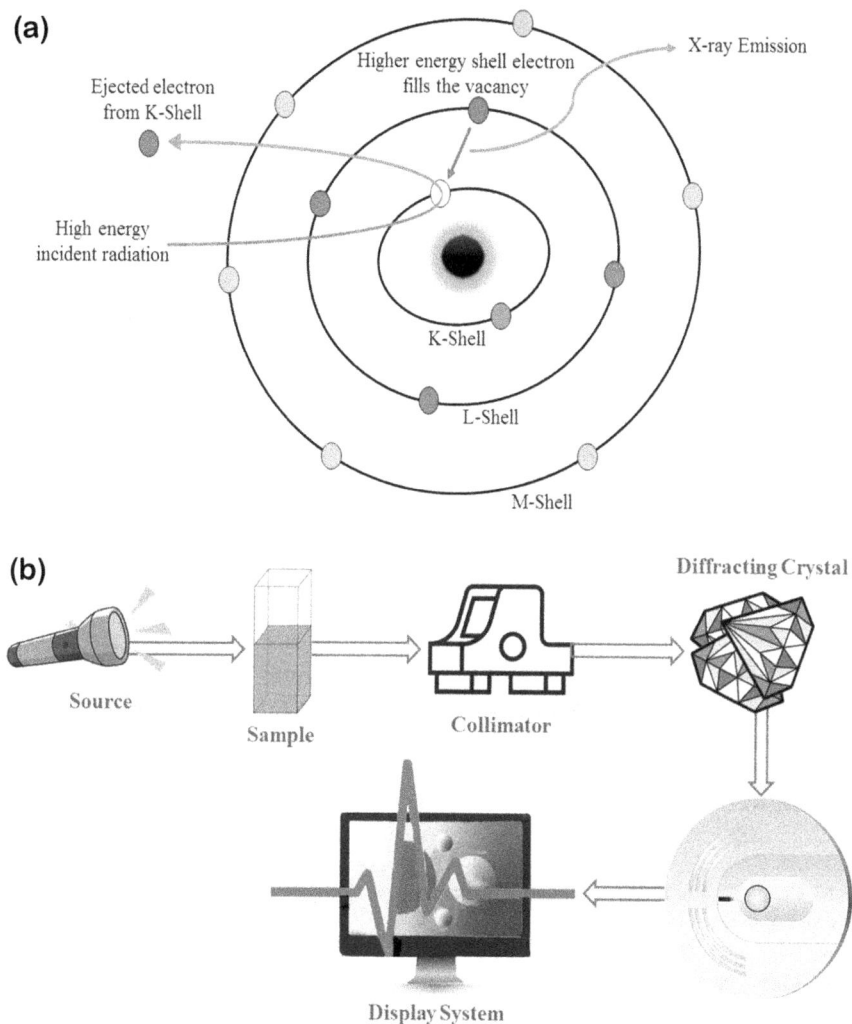

FIGURE 8.2 (a) Schematic representation of high energy X-rays or gamma rays mediated ejection of electron and K-capture. (b) A block diagram of X-ray fluorescence spectroscopy (Malik et al., 2019).

8.3.2 ELECTROCHEMICAL METHODS FOR DETECTION

Electrochemical methods are comparatively cheaper, user-friendly, and reliable. These techniques involve simple procedures for analysis of samples contaminated with metal ions. Moreover, these methods offer the advantage of having a very short analytical time as compared to other spectroscopic techniques (Malik et al., 2019). However, these electrochemical methods have lower sensitivity and broader limits of detection (LODs) as compared to other optical and spectroscopic methods. In addition to this, for a variety of metal ions, these methods usually require modifications to improve the detection limits and sensitivity (Bansod et al., 2017).

TABLE 8.1

Detection of Hazardous Metal Ions Using Analytical Spectroscopic Techniques

Metal Ion(s)	Sample / Source	Spectroscopic Technique	References
Cd, Co, Cu, Ni, Pb	Water and food samples	AAS[a]	Gouda et al. (2023)
Cd, Co, Cr, Pb, Ni, Cu, Mn, Zn	*Oryza sativa* L.	AAS	Wasim et al. (2019)
Pb, Cd, Cu	Environmental and biological samples	AAS	Zhao et al. (2019)
Pb, Cd, As, Hg, Mn, and Zn	Environmental samples	AAS	Li et al. (2018)
As, Cd, Cr, Cu, Ni, Zn, Pb	Soils	XRFS[b]	Taha (2017)
Cd	Brown rice	AFS[c]	Hafuka et al. (2017)
Mn, Pb, Cr, Cd	Kulufo River, Arbaminch, Gamo Gofa, Ethiopia	AAS	Tsade (2016)
Pb, Cd, Cr, Cu, Ni	Chinese tea	GF-AAS[d]	Zhong et al. (2016)
As, Pb, U	Aerial parts of *Origanum sipyleum* L.	XRFS	Durmuşkahya et al. (2016)
Pt	Rocks	GF-AAS	Odonchimeg et al. (2016)
Pb	Water samples	AFS	Beltrán et al. (2015)
Cu, Pb	Aqueous solution	XRFS	Hutton et al. (2014)
Cr, Ni, Cu, Zn, Hg, Pb	Fish tissues	XRFS	Zarazúa et al. (2014)
Pb, Sb, Al, As	Tube wells of District Pishin, Balochistan, Pakistan	AAS	Tareen et al. (2014)
Pb, Cd, Zn, Ni, Cr, Mn, Fe	Water and therapeutic mud	AAS	Radulescu et al. (2014)
Hg	Muscle samples of fish	GF-AAS	Moraes et al. (2013)
Cr, Ni, Cu, Zn, Zr, Rb, Y, Ba, Pb, Sr, Ga, V, Nb	Surface soil samples	XRFS	El-Bahi et al. (2013)
Fe, Ni, Mn, Cu, Zn, Pb	Coastal seawaters	XRFS	Peng et al. (2012)
As, Cr, Cu, Ni, Pb, V, Zn	Soil	XRFS	Ene et al. (2010)

[a] Atomic absorption spectroscopy.
[b] X-ray fluorescence spectroscopy.
[c] Atomic fluorescence spectroscopy.
[d] Graphite furnace atomic absorption spectroscopy.

The occurrence of any metal ion in the water sample usually creates a change in electrical parameters in the electrochemical setup such as voltage, current, charge, electrochemical impedance, and electroluminescence (Cui et al., 2015). These techniques can be categorized into amperometric, potentiometric, coulometric, impedance measurement, voltammetric, and electrochemiluminescent techniques, based on the changes made in the electrical signal due to the presence of metal ions.

8.3.2.1 Amperometric Electrochemical Techniques

Amperometry is a potentiostatic technique. In this method, the solution containing electroactive metal ions is subjected to a fixed potential difference, applied between working and reference electrodes. As a result of reduction of the reduction of present metal ions, the flow of very small currents is produced, which is recorded as a function of time. Such experiments are known as amperometry techniques. By this technique, any metal ion is detected among various electroactive species (Sun et al., 2022).

8.3.2.2 Potentiometric Electrochemical Techniques

Potentiometric detection techniques are based on the measurement of electromotive force without applying any electric currents. These techniques are generally used for the quantitative detection of ions in solutions. Moreover, it is quite effective in the detection of metal ions due to its several advantages such as short response time, low cost, high selectivity, and broad range of response (Aragay and Merkoci, 2012). However, this technique survives with certain disadvantages like higher detection limits and lower sensitivity. Some attempts have been made to overcome these issues by using electrodes modified with metal nanoparticles and carbon nanotubes and were found promising (Düzgün et al., 2011; Bakker and Pretsch, 2008).

8.3.2.3 Voltammetric Electrochemical Techniques

Voltammetry is the most frequently and commonly used technique for the detection of metal ions, among electrochemical methods. In these techniques, current is measured at different applied potentials to obtain a current–voltage curve. This is a widely used technique for detection of metal ions due to its higher accuracy, lower limit of detections, and increased sensitivity. There are various forms of voltammetry with a common basic principle measuring the current by varying the potential. Different forms of voltammetry are cyclic voltammetry, pulse voltammetry, stripping voltammetry, square-wave voltammetry, linear sweep voltammetry, and differential pulse anodic stripping voltammetry. The selective detection of Hg^{2+} was attempted by employing cyclic voltammetry using gold nanoparticles-thiol functionalized reduced graphene oxide-modified glassy carbon electrode (GCE/rGO-SH/Au nanoparticles) (Devi et al., 2018). Cumulative detection of copper, lead, and cadmium was carried out by employing differential pulse voltammetry with the help of a carbon paste electrode modified with hexagonal mesoporous silica (HMS)-immobilized quercetin (HMS-Qu/CPE) (Xia et al., 2010). Square-wave anodic stripping voltammetry equipped with cubic and octahedral Fe_3O_4 nanocrystal-modified electrodes has been performed to detect metal ions such as Zn^{2+}, Cd^{2+}, Pb^{2+}, Cu^{2+}, and Hg^{2+} (Yao et al., 2014). Table 8.2 summarizes the detection of metal ions using various electrochemical methods aided with modified electrodes and detection limits.

8.3.3 OPTICAL METHODS FOR DETECTION

Some phenomena like absorption, reflection, or luminescence can be used as optical methods of metal-ion detection. Ionophores, specific indicator dyes, capillary-type devices, optical fibers, integrated optics, etc. are usually used for the optical detection of metal ions. These conventional spectrophotometric and electroanalytical

TABLE 8.2

Detection of Hazardous Metal Ions Using Various Electrochemical Techniques

Metal Ion(s)	Modification of Electrode	Electrochemical Technique	Detection Limit (mol/L)	References
Hg	Screen-printed electrode modified with zirconium antimonate ionophore	PmET[a]	5×10^{-8}	Aglan et al. (2018)
Hg	Graphene-modified glassy carbon electrode	CV[b]	1×10^{-9}	Talat et al. (2018)
Hg	Pt/CeO2/urease-modified electrode	AmET[c]	1.8×10^{-8}	Gumpu et al. (2017)
Cd	Bi/glassy carbon electrode (Bi/GCE)	SWASV[d]	NA	Zhao et al. (2017)
Cd	A boron-doped diamond electrode modified by 0.5 mM p-aminomethyl benzoic acid	ASV[e]	1.8×10^{-9}	Innuphata and Chootoa (2017)
Cd, Co, Ni, As, Cr, Pb	Glucose oxidase-functionalized cobalt oxide-modified glassy carbon electrode	AmET	50*	Mugheri et al. (2016)
Hg	Hydroxyapatite (HA) nanoparticle-modified glassy carbon electrode (GCE)	SWV[f]	1.4×10^{-7}	Kanchana et al. (2015)
Hg	Indium tin oxide (ITO) electrodes modified by gold nanoparticles (Au nanoparticles)	LSV[g]	1×10^{-6}	Ratner and Mandler (2015)
Pb	PVC-based carboxymethyl cellulose Sn(IV) phosphate composite membrane electrode	PmET	1×10^{-6}	Inamuddin et al. (2015)
Cd	PVC-based polyaniline Sn(IV) silicate composite cation exchanger ion-selective membrane electrode	PmET	1×10^{-7}	Naushad et al. (2015)
Hg	Modified palm shell activated carbon paste electrode based on Kryptofix®5*	PmET	1.0×10^{-7}	Ismaiel et al. (2012)
Cd	Multiwalled carbon nanotubes functionalized by dithizone-modified electrode	PmET	1.0×10^{-7} mol	Karimi et al. (2012)
Pb	Zirconium(IV) iodosulphosalicylate-based electrode	PmET	4.0×10^{-6}	Rahman and Rahman (2012)
Pb	Polypyrrole-modified electrode	PmET	7.0×10^{-7}	Mazloum-ardakani et al. (2012)
Pb	Graphitic carbon modified with 4-amino salicylic acid	CV & DPASV[h]	9×10^{-8}	Kempegowda and Malingappa (2012)
Pb	A solid paraffin-based carbon paste electrode modified with 2-aminothiazole-functionalized silica gel	ASV	7.3×10^{-9}	Silva et al. (2011)

(Continued)

TABLE 8.2 (Continued)
Detection of Hazardous Metal Ions Using Various Electrochemical Techniques

Metal Ion(s)	Modification of Electrode	Electrochemical Technique	Detection Limit (mol/L)	References
Hg	Thiourea-functionalized nanoporous silica-modified carbon paste electrode	PmET	7×10^{-8}	Javanbakht et al. (2009)
Hg	Poly(vinylferrocenium) (PVF(+))-modified platinum electrode	AmET	5×10^{-10}	Celebi et al. (2009)
Pb,	Single carbon fiber electrode	AmET	1.3×10^{-6}	Li et al. (2007)

[a] Potentiometric electrochemical technique.
[b] Cyclic voltammetry.
[c] Amperometric electrochemical technique.
[d] Square-wave anodic stripping voltammetry.
[e] Anodic stripping voltammetry.
[f] Square-wave voltammetry.
[g] Linear sweep voltammetry.
[h] Differential pulse anodic stripping voltammetry.

techniques are efficient to detect trace amount of metal ions which can be successfully employed. However, these techniques are time-consuming, costly, and quite complex to understand and thus an easier and cheaper technique was required. Many classes of fluorescent probes have been constructed to detect a wide variety of analytes through various emission mechanisms focusing on classes of analyte (Kaur et al., 2016; Evans and Beer, 2014), receptor or transducer architecture (Alreja and Kaur, 2016; Ding et al., 2015), and other aspects of fluorescence sensing (Schäferling, 2012; Formica et al., 2012).

8.3.4 MECHANISMS OF FLUORESCENCE SENSING

Fluorescence spectroscopy can be applied effectively for the sensing and detection of metal ions if a fluorophore can be used as a transducer in a sensor or a probe. In free (unbound) and bound states, the fluorophores reveal different optical properties by some different photophysical mechanisms (Figure 8.3). This makes chemosensing a faster, sensitive, and inexpensive process. The known fluorescence sensing mechanisms are: (1) photoinduced electron transfer (PET), (2) intramolecular charge transfer (ICT), (3) Förster resonance energy transfer (FRET), and one more recently developed (4) aggregation-caused quenching (ACQ) or aggregation-induced emission (AIE) (Wu et al., 2011b; De Silva et al., 1997; Silva et al., 1992).

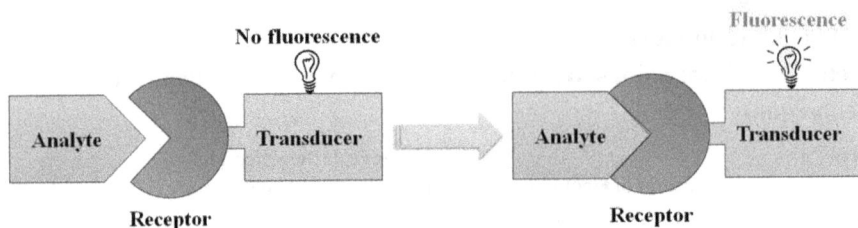

FIGURE 8.3 A general concept of fluorescence generation.

8.3.4.1 Photoinduced Electron Transfer (PET)

A reactive fluorescent sensor essentially possesses a binding site (receptor) and a photon-interaction site (fluorophore) (Bryan et al., 1989). In this process, from HOMO (Highest Occupied Molecular Orbital) of the receptor, an electron of a photo-excited fluorophore is transferred to the energetically closed HOMO of the fluorophore. Now, as the HOMO of fluorophore is completely occupied, the excited electron of LUMO (Lowest Occupied Molecular Orbital) cannot be transferred to the HOMO; however, it is back donated to the HOMO of the receptor displaying the fluorescence quenching (the "off" state). On the other hand, the receptor donates its electron to the cation once cation binds, so the reduction potential of the receptor is increased which in turn lowers the energy of the receptor HOMO than that of fluorophore HOMO. Thus, the PET process is constrained and excited electrons of the fluorophore are transferred to its ground state exhibiting fluorescence emission (the "on" state) (Valeur and Leray, 2000).

8.3.4.2 Intramolecular Charge Transfer (ICT)

In some cases, fluorescence is also emitted by ICT of chemosensors. ICT sensing probes are the combination of electron donor and electron acceptor groups within a conjugated π (pie) system having both receptor and fluorophore together. Upon excitation, an electron redistribution occurs from e⁻ donor to e⁻ acceptor, and as a result dipole moment is created within the molecule. Now, when an analyte binds to the molecule, the dipole moment may be increased or decreased depending on the electronic relationship between receptor and fluorophore as well as the nature of analyte. Now, upon analyte binding, if there is reduced conjugation and destabilization of excited state; there will be a decrease in molar absorptivity with a blue shift in absorbance and emission of fluorescence on reduction of dipole moment. On the other hand, if there is increased conjugation and enhanced stabilization of excited state, there will be an increase in molar absorptivity with red shift in absorbance and emission of fluorescence on enhanced dipole moment (Alreja and Kaur, 2016). There are a few other charge transfer processes including Metal Ligand Charge Transfer and Twisted Internal Charge Transfer (Hush and Reimers, 1998; Grabowski, 1979).

8.3.4.3 Förster Resonance Energy Transfer (FRET)

A non-radiative energy transfer occurs from an excited donor fluorophore to an appropriate energy acceptor via long-range dipole–dipole interaction (Yuan et al., 2013). However, in this fluorescence quenching method, the excited electron is not able to return to its ground state, but the acceptor gets excited and fluorophore exhibits a far red-shifted emission. The distance between the donor and the acceptor must be between 10 and 100 Å to generate a considerable spectral overlap between the absorption profile of acceptor and the emission profile of donor.

8.3.4.4 Aggregation-Caused Quenching (ACQ) or Aggregation-Induced Emission (AIE)

In ACQ, fluorescence intensity of aggregated fluorophores reduces because of the formation of less fluorescent species like pyrene (Hong et al., 2009). On the contrary, in AIE, enhanced fluorescence on aggregation of fluorophores has also been exhibited by molecules such as hexaphenylsilol (Luo et al., 2001). In fact, such molecules possess some groups as "rotors" that undergo rotation in dilute solution. Now, as this rotation is a non-radiative process, in dilute solution such molecules display no fluorescence emission (Wang et al., 2015; Hu et al., 2014). On the other hand, in concentrated solution fluorescence is enhanced due to the elimination of non-radiative rotation.

8.3.5 Biomolecular Fluorescent Probes

Fluorescent probes (sensors) enable highly sensitive and highly selective detection of many target analytes including metal ions (Juskowiak, 2011; Thompson, 2005). These may include chemical sensors as well as biomolecular probes. Various biomolecular probes have been described below.

8.3.5.1 Nucleic Acid-Based Fluorescent Probes

Recently, DNA probe technology has been an area of interest in research and innovation. With the help of this technology, a variety of sensitive and selective fluorescent probes have been constructed with nucleic acid as bioreceptors. These probes possess numerous applications including the detection of metal ions (Marti et al., 2007; Liu et al., 2009). Generally, such probes are synthetic RNA or DNA molecules having a specific sequence for a specific target molecule comprising a reporter group which can be monitored with the help of fluorescence spectroscopy. Nucleic acid chemistry is well-understood, and they can be easily incorporated into probe oligonucleotides, using the commercial synthesizers. Due to the characteristic of specific target–probe interactions like hybridization and molecular recognition, such probes label their target molecule by binding to them. Based on the mechanism of recognition (probe–target interactions), oligonucleotide biosensors can be categorized as (1) hybridization probes and (2) aptamer affinity probes. The recognition interactions of the hybridization probes are generally reliant on complementarity between probe and target analyte molecules. Hybridization probes are commonly employed in various fields where detection or monitoring of metal ions is required (Liu et al., 2022; Zhan et al., 2016). Aptamer affinity probes can bind non-nucleic acid molecules that act as three-dimensional bioreceptors. There are a variety of applications of aptamer affinity probes including the detection of metal ions (Li and Yu, 2009). In consideration of the degree of automation and system integration, the aptamer-based detection and subsequent analysis can be classified: aptamer-based assay, aptasensor, and lab-on-a-chip (Amaya-González et al., 2013).

Detection of metal ions is generally accomplished using functional nucleic acid (FNA)-based fluorescent probes. Nucleic acids, whose functions are beyond the conventional applications as genetic material, are known as functional nucleic acids (Liu et al., 2009). These FNAs can be natural or artificial. The former one includes ribozymes and riboswitches, whereas the latter one incorporates aptamers, ribozymes, and DNAzymes. FNA-based probes showed great potential in detection of metal ions as they are highly efficient, highly sensitive, cheaper, and easy to operate (Zhan et al., 2015). Metal ion-specific DNAs are specific sequences of nucleic acid that can selectively bind with metal ions and create strong metal-nitrogenous base complexes. Such most remarkable DNAs are thymine (T)-rich DNA that binds Hg^{2+} to form T-Hg^{2+}-T mismatch (Wu et al., 2011c) and cytosine (C)-rich DNA that binds Ag^+ to form C-Ag^+-C mismatch (Zhan et al., 2012) (Figure 8.4). According to Torigoe et al. (2010), the T-Hg^{2+}-T mismatches are more stable than the natural adenine-thymine (A-T) base pairing. Moreover, researchers have also reported GT-rich oligonucleotides that specifically recognize arsenite (As^{3+}) (Liang et al., 2013); a Cu^{2+} selective ssDNA and modified streptavidin (SA) aptamer selective to Pt^{2+} (Cai et al., 2015).

Based on the construction of the FNA sensors, they can be classified into six types viz. microchip, microfluidic paper-based analytical devices (µPADs), microfluidic lab-on-a-chip (LOC) system, lateral flow dipstick, personal glucose meter (PGM), and disc-based analytical platform. Table 8.3 describes the performance of various types of FNA-based metal-ion detectors.

These six types of sensors have the following properties: (1) aptamers or FNAs that have been designed for target metal ions are in limited number; (2) FNA-based heavy metal-ion sensors of the LOC category detect mainly Pb^{2+}, Hg^{2+}, UO_2^{2+}, and Cu^{2+} and researchers are still in search for the sensors for metal ions viz. Zn^{2+}, As^{3+}, Pd^{2+}, Cd^{2+}, and Ag^+; (3) these probes are easy to operate, portable, and reliable; and (4) nanopolymers, carbon nano-crystals, noble metals (Au/AgNPs), metal compounds (MBs), etc. play an important role as carrier or as signal reporter or catalyst (Zhan et al., 2016).

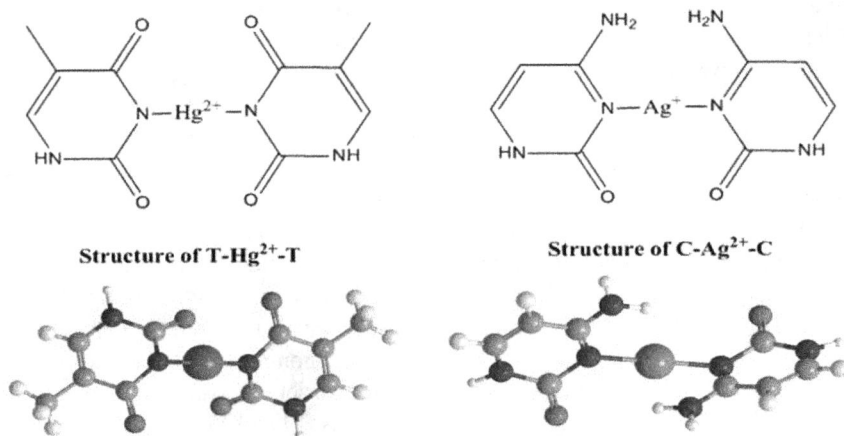

Structure of T-Hg^{2+}-T   **Structure of C-Ag^{2+}-C**

FIGURE 8.4 Two-dimensional (2D) and three-dimensional (3D) structures of T-Hg^{2+}-T and C-Ag^+-C.

TABLE 8.3

Comparison of Different FNA-Based Metal Ion Detection Methods

Detection Method	Metal Ion(s)	Response Time	LODs	References
Microfluidic LOC system	Pb^{2+}	5 min	500 nM	Shaikh et al. (2005)
		-	11 nM	
		Overnight	-	Dalavoy et al. (2008)
Lateral flow dipstick	Cu^{2+}	20 min	10 nM	Fang et al. (2010)
	Hg^{2+}	10 min	5 nM	Duan and Guo (2012)
		10 min	3 nM	Yang et al. (2012)
		~1 h	1 pM	Chen et al. (2014b)
	Pb^{2+}	~2.5 h	10 pM	Chen et al. (2013)
Microchip	Hg^{2+}	-	8.6 nM	Du et al. (2012)
		~30 min	10 nM	Lee and Mirkin (2008)
	Pb^{2+}	~1 h	2 ppb	Zuo et al. (2009)
	Cu^{2+}	~1 h	0.6 ppb	
Personal glucose meter	Pb^{2+}	~2.5 h	1 pM	Zhang et al. (2015a)
		1.5 h	5 nM	Xiang and Lu (2013)
		40 min	1 pM	Fu et al. (2013)
	UO$_2^{2+}$	-	9.1 nM	Xiang and Lu (2011)
Disc-based analytical platform	Hg^{2+}/ Pb^{2+}	~2 h	0.5 nM	Zhang et al. (2015c)
Microfluidic paper-based analytical device	Hg^{2+}	30 min	50 nM	Chen et al. (2014a)
	Pb^{2+}	50 min	10 pM	Zhang et al. (2013)
	Hg^{2+}		0.2 nM	

Apart from this, catalytic DNAs or DNAzymes are DNA enzymes that possess efficient structure identification abilities and catalytic activities. The first ever Pb^{2+}-dependent DNAzyme was isolated by Breaker and Joyce (1994), which had the ability to cleave RNA. The 8–17 RNA-cleaving DNAzyme is the most exploited one, which was first reported by Santoro and Joyce (1997). As the highest activity of 8–17 DNAzyme reaches its maximum in the presence of Pb^{2+}, it was mostly explored for the detection of Pb^{2+} (Peracchi et al., 2005). According to Xiang and Lu (2014), various metal ions including Zn^{2+}, Cu^{2+}, Mg^{2+}, Pb^{2+}, Hg^{2+}, UO$_2^{2+}$, Mn^{2+}, and Co^{2+} have obtained their respective specific DNAzymes.

8.3.5.2 Amino Acids and Peptide-Based Fluorescent Probes

Formation of peptide occurs by linkage of amino acids via peptide bonds (Li et al., 2019a). The application of peptides in fluorescence-based detection is limited due to their weak or no fluorescence properties. This inherent limitation may be due to the interaction of proteins with other compounds or analytes (Xu et al., 2009). Many researchers have recognized that this inherent defect of weak luminescence can be eliminated by coupling peptides with other non-biological compounds viz. metal chelates, hydrogels, polymers, or small molecular weight compounds (Yang et al., 2020; Li et al., 2019b). Moreover, fluorescent properties of peptides can also be enhanced by combining an organic fluorophore.

In nature, there are three aromatic amino acids, i.e., tryptophan (Trp), tyrosine (Tyr), and phenylalanine (Phe) that possess natural fluorescence and thus peptide comprising any of these amino acids too. However, fluorescence properties of these three amino acids are naturally influenced by their confirmation, solvent polarity, and temperature (Gopika et al., 2020; Knox et al., 2020). The fluorescent properties of Trp are because of an indole group present as the luminescent chromophore, which can absorb ultraviolet light and cause a $\pi \rightarrow \pi^*$ transition. The ground state structural heterogeneity of Trp can explain that its fluorescence lifetime is 3.1 ns with a maximum wavelength of emission, 348 nm (Grigoryan et al., 2017; Ghisaidoobe and Chung, 2014). The fluorescence properties of Tyr are due to the presence of phenol as chromophore, a derivative of benzene that can absorb UV light and cause a $\pi \rightarrow \pi^*$ transition to the excited state. As the Tyr molecules are unstable in excited state, they emit photon energy and in turn fluorescence while returning to the ground state, with a maximum wavelength of emission, 303 nm (Lakowicz, 2006; Zhang et al., 2006). The fluorescence properties of Phe are because of its aromatic ring. Due to the lowest fluorescence quantum efficiency, the fluorescence intensity of Phe is the lowest among three amino acids which exhibit natural fluorescence. The transition in the aromatic ring of Phe generates a strong peak, whereas, and electron-donating amides or electron-accepting carboxyl group $n \rightarrow \pi^*$ or $\pi \rightarrow \pi^*$ produces weaker characteristics peaks (Khan et al., 2020; Ando et al., 2019).

Peptides having the three natural fluorescent amino acids (FlAAs) possess most of the fluorescence properties similar to that of FlAAs. Moreover, fluorescent peptides can also be constructed by combining fluorophores to the side chains of amino acids including thiols, phenols, carboxylic acids, amines, and imidazols. However, this can alter the natural properties of peptides; therefore, the selection of suitable fluorophores is very important. A few examples of common fluorophores are Tide Fluor 3, Quasar 570, and 5(6)-carboxyfluorescein (Fernandez et al., 2019; Joshi and Rai, 2019).

Such fluorescence-based peptides exhibit a potential role in detection of metal ions. A fluorescent probe with a tripeptide (Lys-His-Gly-NH$_2$) and dansyl chromophore with high selectivity in the presence of other ions can be employed for detection of Zn^{2+}. A novel fluorescent sensor, H2L (Dansyl-Gly-Pro-Trp-Gly-NH$_2$), could not only detect Zn^{2+} with high accuracy and sensitivity, but also mark the concentration difference (Gopika et al., 2020). Detection of Fe^{3+} can be accomplished by a gold nanocluster (GHRP6-Au NCS), which produces the green fluorescence with high stability (Li et al., 2017). A novel fluorescent peptide chemical sensor (FITC-Ahx-GCA-NH$_2$) has been recently reported by Xu et al. (2019), which has an ability of simultaneous detection of S^{2-}, Ag^+, and Cu^{2+} in aqueous solutions. Such biomolecule-based fluorescent probes play a great role in the detection of hazardous metal ions.

8.4   APPLICATIONS OF FLUORESCENT PROBES IN DETECTION OF HAZARDOUS METAL IONS

Certain heavy metal ions, for example, copper, iron, aluminum, and chromium (III), serve as essential nutrients; however, their higher concentration can lead to toxicity. Heavy metal ions such as chromium (VI), lead (II), arsenic (III), cadmium (II), and mercury (II) are the most commonly present toxic pollutants in industrial effluents. Heavy metal ions interact with the proteins (enzymes) present in the human body leading to chronic poisoning (He and Lu, 2001). Hence, it is extremely important to decrease these hazardous heavy metal ions and prevent pollution of water from them through environment-friendly sensors. Studies are therefore being carried out to develop real-time sensors to detect hazardous ion pollutants in the environment. Various fluorescent probes are now being designed to detect several free fluorescent FNA biosensors that have also been developed for the detection of various metal ions including Ti^+, Hg^+, Ir^{3+}, K^+, Na^+, and Tb^{3+} (Ma et al., 2019a, 2019b; Zhu et al., 2018; Sun et al., 2017; Xu et al., 2017; Wang et al., 2016; Hoang et al., 2016; Chen et al., 2015).

8.4.1 DETECTION OF LEAD (Pb) AND ITS RESPECTIVE IONS

Sensors are now being designed to easily detect hazardous environmental pollutants like mercury and lead. Nucleic acid fragments are potential candidates for recognition of these elements. These nucleic acid-based biosensors can be categorized into two groups based on the target–probe interactions. The first group consists of hybridization probes wherein the target and the probe molecules are complementary to each other (Tyagi and Kramer, 1996), whereas the second group constitutes aptamers which can bind as three-dimensional bioreceptors capable of binding non-nucleic acid analytes. These aptamers are crucial for detecting proteins, small bioactive molecules, and metal cations (Liu et al., 2009; Chen et al., 2009; Juskowiak et al., 2006). One or more fluorescent groups can be attached to the oligonucleotide chain that works as an analytical signal for the binding of the analyte to the bioreceptor.

Similarly, nucleic acid enzymes can also be conjoined to aptamers to generate a new kind of probes known as aptazymes (Liu et al., 2009). Several DNAzyme-based aptazymes using various design principles are now being generated for more efficient metal-ion detection using these DNAzyme-based sensors. Several metal ions including Cu^{2+}, Ca^{2+}, Zn^{2+}, Pb^{2+}, UO_2^{2+}, and Hg^{2+} can be detected efficiently using such these sensors (Wang et al., 2009; Liu and Lu, 2007; Liu et al., 2007; Chiuman and Li, 2006; Ono and Tagashi, 2004; Thomas et al., 2004; Liu et al., 2003; Liu and Lu, 2003; Carmi and Breaker, 2001; Li and Lu, 2000; Peracchi, 2000). The oligopeptide Cysteinyl-Aspartyl-Arginyl-Valyl-Tyrosyl-Isoleucyl-Histidyl-Prolyl-Pheneylalanyl-Histidyl-Leucine (CDRVYIHPFHL) was used to detect Pb^{2+}. The possible binding of Pb^{2+} to histidine residues and two adjacent carbonyl groups explains the sensitivity of the oligopeptide to Pb^{2+}.

Bi et al. (2009) in their study developed a metal-ion sensor based on oligopeptide-modified single crystal silicon nanowire (SiNW) arrays. Various oligopeptides can used to modify the surface of SiNW cluster and peptides are immobilized onto these SiNW clusters.

Li and Lu (2000) were the first to report DNAzyme-based lead sensor (Figure 8.5). A variant of 8–17 deoxyribozyme known as 17E was used to make the FNA probe for lead-ion detection. Low fluorescence is detected due to the quenching effect of the quencher in the natural state. However, the presence of lead ions activates the 8–17 deoxyribozyme which in turn rapidly cleaves the substrate chain that is labeled with fluorescent groups leading to an enhanced fluorescent signal. The increased fluorescence signal intensity indicates the concentration of Pb^{2+}. However, in this method, two groups need to be labeled on the enzyme chain consequently leading

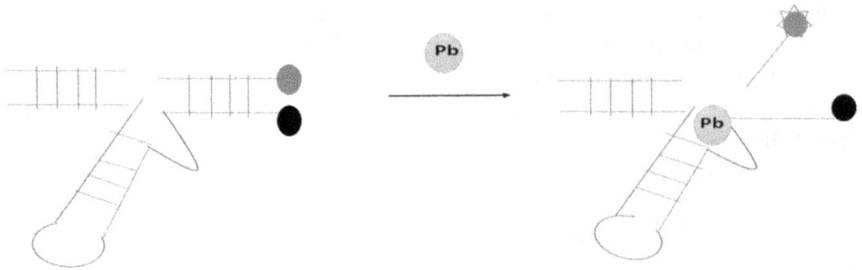

FIGURE 8.5   Schematic representation aptamer-based generation of fluorescence for Pb^{2+} ions.

to increased experimental cost and time delay. With advancement in research, sensors have been further developed for hazardous metal-ion detection. In these sensors, the substrate chain is labeled with the fluorescent group, fluorescein amidite (FAM) and subsequently DNAzyme is hybridized with it leading to formation of a bulged structure (Guo et al., 2015). In this bulged structure, ethidium bromide is embedded as a quencher which quenches the fluorescence signal. In the presence of Pb^{+2}, DNAzyme is activated cleaving ethidium bromide resulting in an enhanced fluorescence signal. Since this method does not require the modification quencher on the DNAzyme, the experimental cost is drastically decreased and less time-consuming.

8.4.2   DETECTION OF URANIUM OXIDE (Uo₂) AND ITS RESPECTIVE IONS

FNA sensors based on the contrasting principle of fluorescence turn-off also worked efficiently. Fluorescent FNA sensor based on the fluorescence turn-off principle has also been designed to detect H_2S in the air. Several sensors are being developed using labeled fluorescent FNA sensors to detect a variety of metal ions (Ji et al., 2020; Saran and Liu, 2016; Zhou et al., 2016; Torabi et al., 2015; Huang and Liu, 2015; Huang et al., 2014a and 2014b). Intercalating fluorescent dyes are now being used to intercalate FNA and develop label-free fluorescent FNA. Following the interaction of target with FNA, the fluorescence response is then changed, the most common being the G-quadruplex structure of FNA (Chen et al., 2015). G-quadruplex structure is formed by stacking multiple tetrad structures using guanine nucleotides through hydrogen bonds (Lipps et al., 2009). Such sensors are now being used in various fields as they exhibit accuracy like the labeled type without affecting the functional activity of FNA. In another study, Zhu et al. developed a sensor using UO_2^{2+}-sensitive DNAzyme and dsDNA containing G-rich substrate along with SYBR Green I (SGI) for fluorescent labeling that intercalates the nucleic acid.

High fluorescence intensity is observed when DNAzyme is not activated and the fluorescence dye is embedded in dsDNA. However, in the presence of UO_2^{2+}, the DNAzyme is activated cutting the substrate chain to form a G-quadruplex causing the release of fluorescent dye and decreased fluorescence signal. Such sensors can be easily developed as compared to the labeled type and yet exhibit similar sensitivity (Zhu et al., 2019b). Similarly, studies have been carried out to detect Pb^{2+} from the environmental samples using the same principle (Zhu, 2019b).

8.4.3   Detection of Mercury (Hg) and Its Respective Ions

Metal-ion pollutants are a major threat to our health and safety. The G-quadruplex structure is formed from the G-quadruplex sequence and the metal cation. The K^+ ions can be detected by screening of aptamer along with G-quadruplex sequence (oligo-3) (Sun et al., 2017). The fluorophore selected should be such that it binds to the generated G-quadruplex. The aptamer is induced in the presence of K^+ leading to formation of a G-quadruplex, which in turn binds the fluorophore leading to high fluorescence signal. G-quadruplex can also be used to detect Hg^{2+}. The presence of Hg^{2+} leads to formation of T-Hg^{2+}-T mismatch within the DNA sequence (Zhu et al., 2018). This causes the fluorescence to be turned off as it destroys the G-quadruplex structure.

Amino acid-based biosensors can also be used for the detection of Hg^{2+} ions. A group of proteins known as metallothioneins (MTs) are rich in cysteine residues. These MTs have a strong affinity toward divalent metal ions strongly including mercury, cadmium, zinc, and copper. A MT that is self-assembled on gold containing 6.7% cadmium and 0.5% zinc takes up Hg^{2+} rapidly from solution. Each MT molecule binds to four mercury ions (Ju and Leech, 2000).

8.4.4   Detection of Cadmium (Cd) and Its Respective Ions

Food and water containing cadmium as a contaminant is highly toxic as it slows down the metabolism and can be detrimental to the human body (Bhardiya et al., 2021). The accumulation of cadmium in the human body can not only decrease lung efficiency but can also cause renal failure. Hence research is being carried out to develop simple yet sensitive probes for rapid detection of cadmium ions (Cd^{2+}) (Hasan et al., 2021). Several Cd^{2+} aptasensors have recently been reported for the detection of Cd^{2+}. Xue et al. (2020) in their study reported a Cd^{2+} concentration-dependent interaction mechanism and subsequent conformation of aptamer using the method of dual polarization interferometry. When the concentration of Cd^{2+} is low it interacts with phosphate group of DNA leading to the formation of ssDNA; however, in high Cd^{2+} concentration, a tight and short hairpin structure is formed through coordination interaction.

Zhu et al. in their study reported a highly sensitive fluorescence sensor having a detection limit of 2.15 nM for Cd^{2+} by using Cd-4 aptamers (Zhu et al., 2017). Both the ends of Cd-4 aptamer contain modified Te 6-carboxyfuorescein (6-FAM) and GGGG sequences. The fluorescence quenching is obtained by G4 that is close to 6-FAM and PET. Zeng et al. (2019) in their study reported a PGM using Cd-4 aptamer for Cd^{2+} detection.

8.5   CONCLUSION AND FUTURE PROSPECTS

Pollution by hazardous metal ions is one of the threatening environmental issues. As hazardous metal-ion toxicity imposes serious life-threatening health issues to living beings, researchers have been motivated to develop various detection strategies. Many methods and technologies have exhibited good performance, but a few drawbacks were the limitation for extended usage. To improve performance and overcome the limitations, such methodologies have also been modified time by time. Moreover, recent studies have contributed to both theoretical and practical pools of knowledge which has accelerated the development synthesis and application of these biomolecular fluorescent probes. Still, novel technologies with advantages of being cheaper, quicker, eco-friendly, highly accurate, and sensitive with a broader detection range are lacking and in the near future need to be worked on.

REFERENCES

Aglan, R.F., H.M. Saleh, and G.G. Mohamed. 2018. Potentiometric determination of mercury (II) ion in various real samples using novel modified screen-printed electrode. *Appl Water Sci.* 8: 1–11.

Aiestaran, P., V. Dominguez, J. Arrue, and J. Zubia. 2009. A fluorescent linear optical fiber position sensor. *Opt Mater.* 31(7): 1101–1104.

Aksuner, N., B. Basaran, E. Henden, I. Yilmaz, and A. Cukurovali. 2011. A sensitive and selective fluorescent sensor for the determination of mercury (II) based on a novel triazine-thione derivative. *Dyes Pigm.* 88(2): 143–148.

Alreja, P., and N. Kaur. 2016. Recent advances in 1, 10-phenanthroline ligands for chemosensing of cations and anions. *RSC Adv.* 6(28): 23169–23217.

Amaya-González, S., De-los-Santos-Álvarez, N., Miranda-Ordieres, A. J., & Lobo-Castañón, M. J.. 2013. Aptamer-based analysis: A promising alternative for food safety control. Sensors, 13(12), 16292–16311.

Ando, D., J. Ijichi, T. Uno, T. Itoh, and M. Kubo. 2019. Preparation of donor-acceptor polyfluorenes with pendant carboxyl or amine functionalities and their photoluminescence properties. *Polym Bullet.* 76: 6137–6151.

Aragay, G., and A. Merkoçi. 2012. Nanomaterials application in electrochemical detection of heavy metals. *Electrochim Acta.* 84: 49–61.

Arora, M., B. Kiran, S. Rani, A. Rani, B. Kaur, and N. Mittal. 2008. Heavy metal accumulation in vegetables irrigated with water from different sources. *Food Chem.* 111(4): 811–815.

Ayangbenro, A.S., and O.O. Babalola. 2017. A new strategy for heavy metal polluted environments: A review of microbial biosorbents. *Int. J. Environ. Res. Public Health.* 14(1): 94.

Bagchi, D., S.J. Stohs, B.W. Downs, M. Bagchi, and H.G. Preuss. 2002. Cytotoxicity and oxidative mechanisms of different forms of chromium. *Toxicology.* 180(1): 5–22.

Bakker, E., and E. Pretsch. 2008. Nanoscale potentiometry. *Trends Ana Chem.* 27(7): 612–618.

Bansod, B., T. Kumar, R. Thakur, S. Rana, and I. Singh. 2017. A review on various electrochemical techniques for heavy metal ions detection with different sensing platforms. *Biosens Bioelectron.* 94: 443–455.

Barceloux, D.G. and Barceloux, D., 1999. Cobalt. *J Toxicol Clin Toxicol.* 37(2): 201–216.

Beltrán, B., L.O. Leal, L. Ferrer, and V. Cerdà. 2015. Determination of lead by atomic fluorescence spectrometry using an automated extraction/pre-concentration flow system. *J Anal At Spectrom.* 30(5): 1072–1079.

Bencs, L., K. Ravindra, and R. Van Grieken. 2003. Methods for the determination of platinum group elements originating from the abrasion of automotive catalytic converters. *Spectrochim Acta B At Spectrosc.* 58(10): 1723–1755.

Bhardiya, S.R., A. Asati, H. Sheshma, A. Rai, V.K. Rai, and M. Singh. 2021. A novel bioconjugated reduced graphene oxide-based nanocomposite for sensitive electrochemical detection of cadmium in water. *Sens Actuat B Chem.* 328: 129019.

Bhatti, P., P.A. Stewart, A. Hutchinson, N. Rothman, M. S. Linet, P. D. Inskip, and P. Rajaraman. 2009. Lead exposure, polymorphisms in genes related to oxidative stress, and risk of adult brain tumors. *Cancer Epidemiol Biomark Prev.* 18(6): 1841–1848.

Bi, X., A. Agarwal, and K.L. Yang. 2009. Oligopeptide-modified silicon nanowire arrays as multichannel metal ion sensors. *Biosens Bioelectron.* 24(11): 3248–3251.

Breaker, R.R., and G.F. Joyce. 1994. A DNA enzyme that cleaves RNA. *Chem Biol.* 1(4): 223–229.

Brown, D.R. 2001. Copper and prion disease. *Brain Res Bull.* 55(2): 65–173.

Bryan, A.J., A.P. de Silva, S.A. De Silva, R.D. Rupasinghe, and K.S. Sandanayake. 1989. Photo-induced electron transfer as a general design logic for fluorescent molecular sensors for cations. *Biosensors.* 4(3): 169–179.

Cai, S., X. Tian, L. Sun, H. Hu, S. Zheng, H. Jiang, L. Yu, and S. Zeng. 2015. Platinum (II)-oligonucleotide coordination based aptasensor for simple and selective detection of platinum compounds. *Anal Chem.* 87(20): 10542–10546.

Camaschella, C. 2017. New insights into iron deficiency and iron deficiency anemia. *Blood Rev.* 31(4): 225–233.

Carmi, N., and R.R. Breaker. 2001. Characterization of a DNA-cleaving deoxyribozyme. *Bioorgan Medic Chem.* 9(10): 2589–2600.

Carter, K.P., A. M. Young, and A. E. Palmer. 2014. Fluorescent sensors for measuring metal ions in living systems. *Chem Rev.* 114(8): 4564–4601.

Çelebi, M.S., H. Özyörük, A. Yıldız, and S. Abacı. 2009. Determination of Hg^{2+} on poly (vinylferrocenium) (PVF+)-modified platinum electrode. *Talanta.* 78(2): 405–409.

Chen, G.H., W.Y. Chen, Y.C. Yen, C.W. Wang, H.T. Chang, and C.F. Chen. 2014a. Detection of mercury (II) ions using colorimetric gold nanoparticles on paper-based analytical devices. *Anal Chem.* 86(14): 6843–6849.

Chen, H.W., Y. Kim, L. Meng, P. Mallikaratchy, J. Martin, Z. Tang, D. Shangguan, M. O'Donoghue, and W. Tan. 2009. Fluorescent aptamer sensors. In: Yingfu, L., Yi, L. (eds) *Functional Nucleic Acids for Analytical Applications. Integrated Analytical Systems.* Springer, New York, NY. 111–130.

Chen, J., S. Zhou, and J. Wen. 2014b. Disposable strip biosensor for visual detection of Hg^{2+} based on Hg^{2+}-triggered toehold binding and exonuclease III-assisted signal amplification. *Anal Chem.* 86(6): 3108–3114.

Chen, J., X. Zhou, and L. Zeng. 2013. Enzyme-free strip biosensor for amplified detection of Pb^{2+} based on a catalytic DNA circuit. *Chem Comm.* 49(10): 984–986.

Chen, Q., J. Zuo, J. Chen, P. Tong, X. Mo, L. Zhang, and J. Li. 2015. A label-free fluorescent biosensor for ultratrace detection of terbium (III) based on structural conversion of G-quadruplex DNA mediated by ThT and terbium (III). *Biosens Bioelectron.* 72: 326–331.

Chiuman, W., and Y. Li. 2006. Revitalization of six abandoned catalytic DNA species reveals a common three-way junction framework and diverse catalytic cores. *J Mol Biol.* 357(3): 748–754.

Chowdhury, S., B. Rooj, A. Dutta, and U. Mandal. 2018. Review on recent advances in metal ions sensing using different fluorescent probes. *J Fluoresc.* 28: 999–1021.

Conway, D. and M.A. Henderson. 2019. Iron metabolism. *Anaesth Intensive Care Med.* 20(3): 175–177.

Cui, L., J. Wu, and H. Ju. 2015. Electrochemical sensing of heavy metal ions with inorganic, organic and bio-materials. *Biosens Bioelectron.* 63: 276–286.

Dalavoy, T.S., D.P. Wernette, M. Gong, J.V. Sweedler, Y. Lu, B.R. Flachsbart, M.A. Shannon, P.W. Bohn, and D.M. Cropek. 2008. Immobilization of DNAzyme catalytic beacons on PMMA for Pb^{2+} detection. *Lab Chip.* 8(5): 786–793.

De Silva, A.P., H.N. Gunaratne, T. Gunnlaugsson, A.J. Huxley, C.P. McCoy, J.T. Rademacher, and T.E. Rice. 1997. Signaling recognition events with fluorescent sensors and switches. *Chem Rev.* 97(5): 1515–1566.

Devi, N.R., M. Sasidharan, and A.K. Sundramoorthy. 2018. Gold nanoparticles-thiol-functionalized reduced graphene oxide coated electrochemical sensor system for selective detection of mercury ion. *J Electrochem Soc.* 165(8): B3046–B3053.

Ding, Y., Y. Tang, W. Zhu, and Y. Xie. 2015. Fluorescent and colorimetric ion probes based on conjugated oligopyrroles. *Chem Soc Rev.* 44(5): 1101–1112.

Du, J., M. Liu, X. Lou, T. Zhao, Z. Wang, Y. Xue, J. Zhao, and Y. Xu. 2012. Highly sensitive and selective chip-based fluorescent sensor for mercuric ion: development and comparison of turn-on and turn-off systems. *Anal Chem.* 84(18): 8060–8066.

Duan, J., and Z.Y. Guo. 2012. Development of a test strip based on DNA-functionalized gold nanoparticles for rapid detection of mercury (II) ions. *Chinese Chem Lett.* 23(2): 225–228.

Durmuşkahya, C., H. Alp, Z.S. Hortooğlu, Ü. Toktas, and H. Kayalar. 2016. X-ray fluorescence spectroscopic determination of heavy metals and trace elements in aerial parts of Origanum sipyleum L from Turkey. *Trop J Pharm Res.* 15(5): 1013–1015.

Düzgün, A., G.A. Zelada-Guillén, G.A. Crespo, S. Macho, J. Riu, and F.X. Rius. 2011. Nanostructured materials in potentiometry. *Anal Bioanal Chem.* 399: 171–181.

El-Bahi, S.M., A.T. Sroor, N.F. Arhoma, and S.M. Darwish. 2013. XRF analysis of heavy metals for surface soil of Qarun Lake and Wadi El Rayan in Faiyum, Egypt. *Open J Metal.* 3: 21–25.

Elmas Karuk, Ş.N., Z.E. Dinçer, A.S. Ertürk, A. Bostancı, A. Karagöz, M. Koca, M, and G. Sadi. 2020. A novel fluorescent probe based on isocoumarin for Hg^{2+} and Fe^{3+} ions and its application in live-cell imaging. *Spectrochim Acta A Mol Biomol Spectrosc.* 224: 117402.

Ene, A., A. Bosneaga, and L. Georgescu. 2010. Determination of heavy metals in soils using XRF technique. *Rom J Phys.* 55(7–8): 815–820.

Engwa, G.A., P.U. Ferdinand, F.N. Nwalo, and M.N. Unachukwu. 2019. Mechanism and health effects of heavy metal toxicity in humans. *Poisoning in the Modern World-New Tricks for an Old Dog.* 10: pp. 70–90. doi: 10.5772/intechopen.82511.

Evans, N.H., and P.D. Beer. 2014. Advances in anion supramolecular chemistry: From recognition to chemical applications. *Angewandte Chemie Int Ed.* 53(44), 11716–11754.

Fang, Z., J. Huang, P. Lie, Z. Xiao, C. Ouyang, Q. Wu, Y. Wu, G. Liu, and L. Zeng. 2010. Lateral flow nucleic acid biosensor for Cu^{2+} detection in aqueous solution with high sensitivity and selectivity. *Chem Comm.* 46(47): 9043–9045.

Fernandez, A., E.J. Thompson, J.W. Pollard, T. Kitamura, and M. Vendrell. 2019. A fluorescent activatable AND-Gate Chemokine CCL2 enables in vivo detection of metastasis-associated macrophages. *Angewandte Chemie Int Ed.* 58(47): 16894–16898.

Formica, M., V. Fusi, L. Giorgi, and M. Micheloni. 2012. New fluorescent chemosensors for metal ions in solution. *Coordination Chem Rev.* 256(1–2): 170–192.

Frederickson, C.J., and A.I. Bush. 2001 Synaptically released zinc: Physiological functions and pathological effects. *Biometals.* 14: 353–366.

Fu, F. and Q. Wang. 2011. Removal of heavy metal ions from wastewaters: A review. *J Environ Manage.* 92(3): 407–418.

Fu, L., J. Zhuang, W. Lai, X. Que, M. Lu, and D. Tang. 2013. Portable and quantitative monitoring of heavy metal ions using DNAzyme-capped mesoporous silica nanoparticles with a glucometer readout. *J Mater Chem B.* 1(44): 6123–6128.

Gaggelli, E., H. Kozlowski, D. Valensin, and G. Valensin. 2006. Copper homeostasis and neurodegenerative disorders (Alzheimer's, prion, and Parkinson's diseases and amyotrophic lateral sclerosis). *Chem Rev.* 106(6): 1995–2044.

Ghisaidoobe, A.B., and S.J. Chung. 2014. Intrinsic tryptophan fluorescence in the detection and analysis of proteins: A focus on Förster resonance energy transfer techniques. *Int J Mol Sci.* 15(12): 22518–22538.

Goldhaber, S.B. 2003. Trace element risk assessment: Essentiality vs. toxicity. *Regul Toxicol Pharmacol.* 38(2): 232–242.

Gong, T., J. Liu, X. Liu, J. Liu, J. Xiang, and Y. Wu. 2016. A sensitive and selective sensing platform based on CdTe QDs in the presence of L-cysteine for detection of silver, mercury and copper ions in water and various drinks. *Food Chem.* 213: 306–312.

Gopika, G., S. Selvam, P. Kumaresan, and E. Kandasamy. 2020. Hydrophobic association of sodium cholate with human serum albumin evades protein denaturation induced by urea. *Mater Today Proc.* 33: 2167–2169.

Gouda, A.A., R. El Sheikh, A.O. Youssef, N. Gouda, W. Gamil, and H.A. Khadrajy. 2023. Preconcentration and separation of Cd (II), Co (II), Cu (II), Ni (II), and Pb (II) in environmental samples on cellulose nitrate membrane filter prior to their flame atomic absorption spectroscopy determinations. *Int J Environ Anal Chem.* 103(2): 364–377.

Grabowski, Z.R. 1979. Twisted intramolecular charge transfer states (TICT): A new class of excited states with a full charge separation. *Nouv J Chim.* 3: 443–454

Grigoryan, K.R., and H.A. Shilajyan. 2017. Fluorescence 2D and 3D spectra analysis of tryptophan, tyrosine and phenylalanine. *Proc YSU B Chem Biol Sci.* 51(1): 3–7.

Guliani, E., Taneja, A., Ranjan, K. R., and Mishra, V., 2023. Luminous insights: Exploring organic fluorescent "Turn-On" chemosensors for metal-ion (Cu^{+2}, Al^{+3}, Zn^{+2}, Fe^{+3}) detection. *J Fluoresc.* 1–37. doi: 10.1007/s10895-023-03419-5.

Gumpu, M.B., U.M. Krishnan, and J.B.B. Rayappan. 2017. Design and development of amperometric biosensor for the detection of lead and mercury ions in water matrix-a permeability approach. *Anal Bioanal Chem.* 409: 4257–4266.

Guo, Y., J. Li, X. Zhang, and Y. Tang. 2015. A sensitive biosensor with a DNAzyme for lead (II) detection based on fluorescence turn-on. *Analyst.* 140(13): 4642–4647.

Hafuka, A., A. Takitani, H. Suzuki, T. Iwabuchi, M. Takahashi, S. Okabe, and H. Satoh. 2017. Determination of cadmium in brown rice samples by fluorescence spectroscopy using a fluoroionophore after purification of cadmium by anion exchange resin. *Sensors.* 17(10): 2291.

Hahn, S.H., M.S. Tanner, D.M. Danke, and W.A. Gahl. 1995. Normal metallothionein synthesis in fibroblasts obtained from children with Indian childhood cirrhosis or copper-associated childhood cirrhosis. *Biochem Mol Med.* 54(2): 142–145.

Hao, C., X. Guo, Q. Lai, Y. Li, B. Fan, G. Zeng, Z. He, and J. Wu. 2020. Peptide-based fluorescent chemical sensors for the specific detection of Cu^{2+} and S^{2-}. *Inorganica Chim Acta.* 513: 119943.

Hasan, M.N., M.S. Salman, A. Islam, H. Znad, and M.M. Hasan. 2021. Sustainable composite sensor material for optical cadmium (II) monitoring and capturing from wastewater. *Microchem J.* 161: 105800.

He, W., and J. Lu. 2001. Distribution of Cd and Pb in a wetland ecosystem. *Sc. China Ser. B Chem.* 44(1): 178–184.

Hoang, M., P.J.J. Huang, and J. Liu. 2016. G-quadruplex DNA for fluorescent and colorimetric detection of thallium (I). *Acs Sens.* 1(2): 137–143.

Hong, Y., J.W. Lam, and B.Z. Tang. 2009. Aggregation-induced emission: Phenomenon, mechanism and applications. *Chem Comm.* 29: 4332–4353.

Hu, R., N.L. Leung, and B.Z. Tang. 2014. AIE macromolecules: Syntheses, structures and functionalities. *Chem Soc Rev.* 43(13): 4494–4562.

Huang, P.J.J., and J. Liu. 2015. Rational evolution of Cd^{2+}-specific DNAzymes with phosphorothioate modified cleavage junction and Cd^{2+} sensing. *Nucleic Acid Res.* 43(12): 6125–6133.

Huang, P.J.J., J. Lin, J. Cao, M. Vazin, and J. Liu. 2014a. Ultrasensitive DNAzyme beacon for lanthanides and metal speciation. *Anal Chem.* 86(3): 1816–1821.

Huang, P.J.J., M. Vazin, and J. Liu. 2014b. In vitro selection of a new lanthanide-dependent DNAzyme for ratiometric sensing lanthanides. *Anal Chem.* 86(19): 9993–9999.

Huff, J., R.M. Lunn, M.P. Waalkes, L. Tomatis, and P.F. Infante. 2007. Cadmium-induced cancers in animals and in humans. *Int J Occup Environ Health.* 13(2): 202–212.

Hush, N.S., and J.R. Reimers. 1998. Solvent effects on metal to ligand charge transfer excitations. *Coord Chem Rev.* 177(1): 37–60.

Hutton, L.A., G.D. O'Neil, T.L. Read, Z.J. Ayres, M.E. Newton, and J.V. Macpherson. 2014. Electrochemical X-ray fluorescence spectroscopy for trace heavy metal analysis: Enhancing X-ray fluorescence detection capabilities by four orders of magnitude. *Anal Chem.* 86(9): 4566–4572.

Inamuddin, M. N., T.A. Rangreez, and Z.A. ALOthman. 2015. Ion-selective potentiometric determination of Pb (II) ions using PVC-based carboxymethyl cellulose Sn (IV) phosphate composite membrane electrode. *Desalin Water Treat.* 56(3): 806–813.

Innuphat, C., and P. Chooto. 2017. Determination of trace levels of Cd (II) in tap water samples by anodic stripping voltammetry with an electrografted boron-doped diamond electrode. *Sci Asia.* 43(1): 33–41.

Ismaiel, A.A., M.K. Aroua, and R. Yusoff. 2012. Potentiometric determination of trace amounts of mercury (II) in water sample using a new modified palm shell activated carbon paste electrode based on kryptofix 5. *Am J Anal Chem.* 3(12): 25702.

Javanbakht, M., F. Divsar, A. Badiei, M.R. Ganjali, P. Norouzi, G.M. Ziarani, M. Chaloosi, and A.A. Jahangir. 2009. Potentiometric detection of mercury (II) ions using a carbon paste electrode modified with substituted thiourea-functionalized highly ordered nanoporous silica. *Anal Sci.* 25(6): 789–794.

Ji, X., Z. Wang, S. Niu, and C. Ding. 2020. DNAzyme-functionalized porous carbon nanospheres serve as a fluorescent nanoprobe for imaging detection of microRNA-21 and zinc ion in living cells. *Microchim Acta.* 187: 1–9.

John De Acha, N., Elosúa, C., Corres, J.M. and Arregui, F.J., 2019. Fluorescent sensors for the detection of heavy metal ions in aqueous media. *Sensors.* 19(3): 599.

Joshi, P.N., and V. Rai. 2019. Single-site labeling of histidine in proteins, on-demand reversibility, and traceless metal-free protein purification. *Chem Comm.* 55(8): 1100–1103.

Ju, H., and D. Leech. 2000. Electrochemical study of a metallothionein modified gold disk electrode and its action on Hg^{2+} cations. *J Electroanal Chem.* 484(2): 150–156.

Jung, H.S., P.S. Kwon, J.W. Lee, J.I. Kim, C.S. Hong, J.W. Kim, S. Yan, J. Y. Lee, J.H. Lee, T. Joo, and J.S. Kim. 2009. Coumarin-derived Cu^{2+}-selective fluorescence sensor: Synthesis, mechanisms, and applications in living cells. *J Am Chem Soc.* 131(5): 2008–2012.

Juskowiak, B. 2011. Nucleic acid-based fluorescent probes and their analytical potential. *Anal Bioanal Chem.* 399: 3157–3176.

Juskowiak, B., and S. Takenaka. 2006. Fluorescence Resonance Energy Transfer in the Studies of Guanine Quadruplexes. In: Didenko V.V. (ed.) *Fluorescent Energy Transfer Nucleic Acid Probes . Methods in Molecular Biology*™ vol. 335. Humana Press Inc., Totowa.

Kahlon, S.K., G. Sharma, J.M. Julka, A. Kumar, S. Sharma, and F.J. Stadler. 2018. Impact of heavy metals and nanoparticles on aquatic biota. *Environ Chem Lett.* 16: 919–946.

Kanchana, P., N. Sudhan, S. Anandhakumar, J. Mathiyarasu, P. Manisankar, and C. Sekar. 2015. Electrochemical detection of mercury using biosynthesized hydroxyapatite nanoparticles modified glassy carbon electrodes without preconcentration. *RSC Adv.* 5(84): 68587–68594.

Karimi, M., F. Aboufazeli, R. Hamid, O. Sadeghi, and E. Najafi. 2012. Determination of cadmium (II) ions in environmental samples: A potentiometric sensor. *Curr World Environ.* 7(2): 201.

Kaur, A., J.L. Kolanowski, and E.J. New. 2016. Reversible fluorescent probes for biological redox states. *Angew Chem Int Ed.* 55(5): 1602–1613.

Kempegowda, R.G., and P. Malingappa. 2012. A binderless, covalently bulk modified electrochemical sensor: Application to simultaneous determination of lead and cadmium at trace level. *Anal Chim Acta.* 728: 9–17.

Khan, M.F.S., J. Wu, C. Cheng, M. Akbar, B. Liu, C. Liu, J. Shen, and Y. Xin. 2020. Insight into fluorescence properties of 14 selected toxic single-ring aromatic compounds in water: Experimental and DFT study. *Front Environ Sci Eng.* 14: 1–16.

Kim, H.N., Ren, W.X., Kim, J.S. and Yoon, J., 2012. Fluorescent and colorimetric sensors for detection of lead, cadmium, and mercury ions. *Chem Soc Rev.* 41(8): 3210–3244.

Knox, P.P., V.V. Gorokhov, B.N. Korvatovsky, N.P. Grishanova, S.N. Goryachev, and V.Z. Paschenko. 2020. Specific features of the temperature dependence of tryptophan fluorescence lifetime in the temperature range of −170–20°C. *J Photochem Photobiol A Chem.* 393: 112435.

Kobayashi, M. and S. Shimizu. 1999. Cobalt proteins. *Eur J Biochem.* 261(1): 1–9.

Koh, J.Y., S.W. Suh, B.J. Gwag, Y.Y. He, C.Y. Hsu, and D.W. Choi. 1996. The role of zinc in selective neuronal death after transient global cerebral ischemia. *Science.* 272: 1013–1016.

Kuswandi, B., J. Huskens, and W. Verboom. 2007. Optical sensing systems for microfluidic devices: A review. *Anal chim Acta.* 601(2): 141–155.

Lakowicz, J.R. Ed., 2006. *Principles of Fluorescence Spectroscopy.* Boston, MA: Springer US.

Lee, J.S., and C.A. Mirkin. 2008. Chip-based scanometric detection of mercuric ion using DNA-functionalized gold nanoparticles. *Anal Chem.* 80(17): 6805–6808.

Li, G., M. Liu, C. Song, and Z. Yuan. 2019b. Printable and conductive supramolecular hydrogels facilitated by peptides and group 1B metal ions. *Appl Surf Sci.* 493: 94–104.

Li, H., H. Huang, J.J. Feng, X. Luo, K.M. Fang, Z.G. Wang, and A.J. Wang. 2017. A polypeptide-mediated synthesis of green fluorescent gold nanoclusters for Fe^{3+} sensing and bioimaging. *J Colloid Interface Sci.* 506: 386–392.

Li, H., J. Fan, and X. Peng. 2013. Colourimetric and fluorescent probes for the optical detection of palladium ions. *Chem Soc Rev.* 42(19): 7943–7962.

Li, J. and Y. Lu. 2000. A highly sensitive and selective catalytic DNA biosensor for lead ions. *J Am Chem Soc.* 122(42): 10466–10467.

Li, L., J. Wang, S. Xu, C. Li, and B. Dong. 2022. Recent progress in fluorescent probes for metal ion detection. *Front Chem.* 10.

Li, S., C. Zhang, S. Wang, Q. Liu, H. Feng, X. Ma, and J. Guo. 2018. Electrochemical microfluidics techniques for heavy metal ion detection. *Analyst.* 143(18): 4230–4246.

Li, X.A., D.M. Zhou, J.J. Xu, and H.Y. Chen. 2007. In-channel indirect amperometric detection of heavy metal ions for electrophoresis on a poly (dimethylsiloxane) microchip. *Talanta.* 71(3): 1130–1135.

Li, Y., and Y. Lu. 2009. *Functional Nucleic Acids for Analytical Applications.* New York: Springer.

Li, Z., C. Cui, Z. Zhang, X. Meng, Q. Yan, J. Ouyang, W. Xu, Y. Niu, and S. Zhang. 2019a. The investigation of a multi-functional peptide as gelator, dyes separation agent and metal ions adsorbent. *ChemistrySelect.* 4(27): 7838–7843.

Li, Z., W. Zhao, Y. Zhang, L. Zhang, M. Yu, J. Liu, and H. Zhang. 2011. An 'off-on' fluorescent chemosensor of selectivity to Cr^{3+} and its application to MCF-7 cells. *Tetrahedron.* 67(37): 7096–7100.

Liang, R.P., Z.X. Wang, L. Zhang, and J.D. Qiu. 2013. Label-free colorimetric detection of arsenite utilizing G-/T-rich oligonucleotides and unmodified Au nanoparticles. *Chem-A Euroean J.* 19(16): 5029–5033.

Lipps, H.J., and D. Rhodes. 2009. G-quadruplex structures: In vivo evidence and function. *Trends Cell Biol.* 19(8): 414–422.

Liu, C., Y. Li, J. Liu, L. Liao, R. Zhou, W. Yu, Q. Li, L. He, Q. Li, and X. Xiao. 2022. Recent advances in the construction of functional nucleic acids with isothermal amplification for heavy metal ions sensor. *Microchem J.* 175: 107077.

Liu, J., A.K. Brown, X. Meng, D.M. Cropek, J.D. Istok, D.B. Watson, and Y. Lu. 2007. A catalytic beacon sensor for uranium with parts-per-trillion sensitivity and millionfold selectivity. *Proc Natl Acad Sci.* 104(7): 2056–2061.

Liu, J., and Y. Lu. 2003. Improving fluorescent DNAzyme biosensors by combining inter-and intramolecular quenchers. *Anal Chem.* 75(23): 6666–6672.

Liu, J., and Y. Lu. 2007. Rational design of "turn-on" allosteric DNAzyme catalytic beacons for aqueous mercury ions with ultrahigh sensitivity and selectivity. *Angew Chem Int Ed.* 46(40): 7587–7590.

Liu, J., Z. Cao, and Y. Lu. 2009. Functional nucleic acid sensors. *Chem Rev.* 109(5): 1948–1998.

Liu, Z., S.H. Mei, J.D. Brennan, and Y. Li. 2003. Assemblage of signaling DNA enzymes with intriguing metal-ion specificities and pH dependences. *J Am Chem Soc.* 125(25): 7539–7545.

Losev, V.N., O.V. Buyko, A.K. Trofimchuk, and O.N. Zuy. 2015. Silica sequentially modified with polyhexamethylene guanidine and Arsenazo I for preconcentration and ICP-OES determination of metals in natural waters. *Microchem J.* 123: 84–89.

Luo, J., Z. Xie, J.W. Lam, L. Cheng, H. Chen, C. Qiu, H.S. Kwok, X. Zhan, Y. Liu, D. Zhu, and B.Z. Tang. 2001. Aggregation-induced emission of 1-methyl-1, 2, 3, 4, 5–pentaphenylsilole. *Chem Comm.* 18: 1740–1741.

Ma, G., Z. Yu, W. Zhou, Y. Li, L. Fan, and X. Li. 2019a. Investigation of Na^+ and K^+ competitively binding with a G-quadruplex and discovery of a stable K^+-Na^+-quadruplex. *J Physic Chem B.* 123(26): 5405–5411.

Ma, Q., P. Li, Z. Gao, and S.F.Y. Li. 2019b. Rapid, sensitive and highly specific label-free fluorescence biosensor for microRNA by branched rolling circle amplification. *Sens Actuat B: Chem.* 281: 424–431.

Malik, L.A., A. Bashir, A. Qureashi, and A.H. Pandith. 2019. Detection and removal of heavy metal ions: A review. *Environ Chem Lett.* 17: 1495–1521.

Mallampati, S.R., Y. Mitoma, T. Okuda, S. Sakita, and M. Kakeda. 2013. Total immobilization of soil heavy metals with nano-Fe/Ca/CaO dispersion mixtures. *Environ Chem Lett.* 11: 119–125.

Mao, J., L. Wang, W. Dou, X. Tang, Y. Yan, and W. Liu. 2007. Tuning the selectivity of two chemosensors to Fe (III) and Cr (III). *Org Lett.* 9(22): 4567–4570.

Marti, A.A., S. Jockusch, N. Stevens, J. Ju, and N.J. Turro. 2007. Fluorescent hybridization probes for sensitive and selective DNA and RNA detection. *Acc Chem Res.* 40(6): 402–409.

Mazloum-Ardakani, M., M.K. Amini, M. Dehghan, E. Kordi, and M.A. Sheikh-Mohseni. 2012. Nanomolar determination of Pb (II) ions by selective templated electrode. *J Serb Chem Soc.* 77(7): 899–910.

Merget, R. and G. Rosner. 2001. Evaluation of the health risk of platinum group metals emitted from automotive catalytic converters. *Sci Total Environ.* 270(1–3): 165–173.

Moraes, P.M., F.A. Santos, B. Cavecci, C.C. Padilha, J.C. Vieira, P.S. Roldan, and P.D.M. Padilha. 2013. GFAAS determination of mercury in muscle samples of fish from Amazon, Brazil. *Food Chem.* 141(3): 2614–2617.

Mugheri, A.Q., A. Tahira, S.T.H. Sherazi, M.I. Abro, M. Willander, and Z.H. Ibupoto. 2016. An amperometric indirect determination of heavy metal ions through inhibition of glucose oxidase immobilized on cobalt oxide nanostructures. *Sensor Lett.* 14(12): 1178–1186.

Nagajyothi, P.C., M. Pandurangan, M. Veerappan, D.H. Kim, T.V.M. Sreekanth, and J. Shim. 2018. Green synthesis, characterization and anticancer activity of yttrium oxide nanoparticles. *Mater Lett.* 216: 58–62.

Naghdi, S., K.Y. Rhee, D. Hui, and S.J. Park. 2018. A review of conductive metal nanomaterials as conductive, transparent, and flexible coatings, thin films, and conductive fillers: Different deposition methods and applications. *Coatings.* 8(8): 278.

Nakamura, T., I. Naguro, and H. Ichijo. 2019. Iron homeostasis and iron-regulated ROS in cell death, senescence and human diseases. *Biochim Biophys Acta Gen Subj.* 1863(9): 1398–1409.

Nan, X., Y. Huyan, H. Li, S. Sun, and Y. Xu. 2021. Reaction-based fluorescent probes for Hg^{2+}, Cu^{2+} and Fe^{3+}/Fe^{2+}. *Coord Chem Rev.* 426: 213580.

Naushad, M., Inamuddin, and T.A. Rangreez. 2015. Potentiometric determination of Cd (II) ions using PVC-based polyaniline Sn (IV) silicate composite cation-exchanger ion-selective membrane electrode. *Desalin Water Treat.* 55(2): 463–470.

Nayak, P. 2002. Aluminum: Impacts and disease. *Environ Res.* 89(2): 101–115.

Nie, L.X., H.Y. Jin, G.L. Wang, J.G. Tian, and R.C. Lin. 2008. Study of determination method for heavy metals and harmful elements residues in four traditional Chinese medicine injections. *China J Chinese Materia Medica, Zhongguo Zhong Yao Za Zhi* 33(23): 2764–2767.

Odonchimeg, S., J. Oyun, and N. Javkhlantugs. 2016. Determination of plantinum in rocks by graphite furnace atomic absorption spectrometry after separation on sorbent. *Int Res J Eng Technol*. 3: 753–757.

Okamoto, S. and L.D. Eltis. 2011. The biological occurrence and trafficking of cobalt. *Metallomics*. 3(10): 963–970.

Ono, A., and H. Togashi. 2004. Highly selective oligonucleotide-based sensor for mercury (II) in aqueous solutions. *Angew Chemie Int Ed*. 116(33): 4400–4402.

Palmer, P. T., Jacobs, R., Baker, P. E., Ferguson, K., & Webber, S.. 2009. Use of field-portable XRF analyzers for rapid screening of toxic elements in FDA-regulated products. Journal of agricultural and food chemistry, 57(7), 2605–2613.

Park, S.H., N. Kwon, J. H. Lee, J. Yoon, and I. Shin. 2020. Synthetic ratiometric fluorescent probes for detection of ions. *Chem Soc Rev*. 49 (1): 143–179.

Peracchi, A. 2000. Preferential activation of the 8-17 deoxyribozyme by Ca^{2+} ions: Evidence for the identity of 8-17 with the catalytic domain of the Mg5 deoxyribozyme. *J Biol Chem*. 275(16): 11693–11697.

Peracchi, A., M. Bonaccio, and M. Clerici. 2005. A mutational analysis of the 8-17 deoxyribozyme core. *J Molecul Biol*. 352(4): 783–794.

Peng, Y.-Z., Y.M. Huang, D.-X. Yuan, Y. Li, and Z.-B. Gong. 2012. Rapid analysis of heavy metals in coastal seawater using preconcentration with precipitation/co-precipitation on membrane and detection with X-ray fluorescence. *Chinese J Anal Chem*. 40(6): 877–882.

Powers, J.M. and G.R. Buchanan. 2019. Disorders of iron metabolism: New diagnostic and treatment approaches to iron deficiency. *Hematol Oncol Clin*. 33(3): 393–408.

Pöykiö, R., and P. Perämäki. 2003. Acid dissolution methods for heavy metals determination in pine needles. *Environ Chemistry Lett*. 1: 191–195.

Radulescu, C., I.D. Dulama, C. Stihi, I. Ionita, A. Chilian, C. Necula, and E.D. Chelarescu. 2014. Determination of heavy metal levels in water and therapeutic mud by atomic absorption spectrometry. *Rom J Phys*. 59(9–10): 1057–1066.

Rashid, M., and N. Rahman. 2012. Potentiometric sensor for the determination of lead (II) ion based on zirconium (IV) iodosulphosalicylate. *Sci Adv Mater*. 4(12): 1232–1237.

Ratner, N., and D. Mandler. 2015. Electrochemical detection of low concentrations of mercury in water using gold nanoparticles. *Anal Chem*. 87(10): 5148–5155.

Raux, E., H.L. Schubert, and M.J. Warren. 2000. Biosynthesis of cobalamin (vitamin B12): A bacterial conundrum. *Cell Mol Life Sci*. 57: 1880–1893.

Rice, K.M., E.M. Walker Jr, M. Wu, C. Gillette, and E.R. Blough. 2014. Environmental mercury and its toxic effects. *J Prev Med Public Health*. 47(2): 74.

Saran, R., and J. Liu. 2016. A silver DNAzyme. *Anal Chem*. 88(7): 4014–4020.

Santoro, S. W., & Joyce, G. F. 1997. A general purpose RNA-cleaving DNA enzyme. Proceedings of the national academy of sciences, 94(9), 4262–4266.

Schäferling, M. 2012. The art of fluorescence imaging with chemical sensors. *Angew Chem Int Ed*. 51(15): 3532–3554.

Shaikh, K.A., K.S. Ryu, E.D. Goluch, J.M. Nam, J. Liu, C.S. Thaxton, T.N. Chiesl, A.E. Barron, Y. Lu, C.A. Mirkin, and C. Liu. 2005. A modular microfluidic architecture for integrated biochemical analysis. *Proc Natl Acad Sci*. 102(28): 9745–9750.

Shirkhanloo, H., Z.H. Mousavi, and A. Rouhollahi. 2011. Preconcentration and determination of heavy metals in water, sediment and biological samples. *J Serb Chem Soc*. 76(11): 1583–1595.

Shuai, H., C. Xiang, L. Qian, F. Bin, L. Xiaohui, D. Jipeng, Z. Chang, L. Jiahui, and Z. Wenbin. 2021. Fluorescent sensors for detection of mercury: From small molecules to nano-probes. *Dyes Pigms*. 187: 109–125.

Silva, A.P., H.Q. NimaláGunaratne, P.L. MarkáLynch, E.M. Glenn, and K.R.A. Samankumaraá-Sandanayake. 1992. Molecular fluorescent signalling with 'fluor-spacer-receptor' systems: Approaches to sensing and switching devices via supramolecular photophysics. *Chem Soc Rev*. 21(3): 187–195.

Silva, D.H., D.A. Costa, R.M. Takeuchi, and A.L. Santos. 2011. Fast and simultaneous deter-mination of Pb^{2+} and Cu^{2+} in water samples using a solid paraffin-based carbon paste electrode chemically modified with 2-aminothiazole-silica-gel. *J Brazil Chem Soc.* 22: 1727–1735.

Singh, H., A. Bamrah, S. K. Bhardwaj, A. Deep, M. Khatri, K. H. Kim, and N. Bhardwaj. 2021. Nanomaterial-based fluorescent sensors for the detection of lead ions. *J Hazard Mater.* 407: 124379.

Sitko, R., P. Janik, B. Zawisza, E. Talik, E. Margui, and I. Queralt. 2015. Green approach for ultratrace determination of divalent metal ions and arsenic species using total-reflection X-ray fluorescence spectrometry and mercapto-modified graphene oxide nanosheets as a novel adsorbent. *Anal Chem.* 87(6): 3535–3542.

Soni, M.G., S.M. White, W.G. Flamm, and G.A. Burdock. 2001. Safety evaluation of dietary aluminum. *Regul Toxicol Pharmacol.* 33(1): 66–79.

Sore, H.F., W.R. Galloway, and D.R. Spring. 2012. Palladium-catalyzed cross-coupling of organosilicon reagents. *Chem Soc Rev.* 41(5): 1845–1866.

Stafilov, T. 2000. Determination of trace elements in minerals by electrothermal atomic absorp-tion spectrometry. *Spectrochim Acta Part B At Spectrosc.* 55(7): 893–906.

Sun, C., Y. Zhang, J. Ying, L. Jin, A. Tian, and X. Wang. 2022. A series of POM compounds constructed using a flexible ligand containing three coordination groups: Electrocatalytic and photocatalytic reduction and amperometric detection of Cr (VI). *New J Chem.* 46(6): 2798–2807.

Sun, X., Q. Li, J. Xiang, L. Wang, X. Zhang, L. Lan, S. Xu, F. Yang, and Y. Tang. 2017. Novel fluorescent cationic benzothiazole dye that responds to G-quadruplex aptamer as a novel K^+ sensor. *Analyst.* 142(18): 3352–3355.

Taha, K. 2017. Heavy elements analyses in the soil using X-ray fluorescence and inductively coupled plasma-atomic emission spectroscopy. *Int J Adv Sci Eng Technol.* 5: 118–120.

Talat, M., P. Tripathi, and O.N. Srivastava. 2018. Highly sensitive electrochemical detection of mercury present in the beauty creams using graphene modified glassy carbon electrode. *Innovat Corros Mater Sci.* 8(1): 24–31.

Tareen, A.K., I.N. Sultan, P. Parakulsuksatid, M. Shafi, A. Khan, M.W. Khan, and S. Hussain, 2014. Detection of heavy metals (Pb, Sb, Al, As) through atomic absorption spectros-copy from drinking water of District Pishin, Balochistan, Pakistan. *Int J Curr Microbiol Appl Sci.* 3(1): 299–308.

Thomas, J.M., R. Ting, and D.M. Perrin. 2004. High affinity DNAzyme-based ligands for transition metal cations-a prototype sensor for Hg^{2+}. *Org Biomol Chem.* 2(3): 307–312.

Thompson, R.B. 2005. *Fluorescence Sensors and Biosensors.* CRC Press, Boca Raton, FL.

Torabi, S.F., P. Wu, C.E. McGhee, L. Chen, K. Hwang, N. Zheng, J. Cheng, and Y. Lu. 2015. In vitro selection of a sodium-specific DNAzyme and its application in intracellular sens-ing. *Proc Natl Acad Sci.* 112(19): 5903–5908.

Torigoe, H., Ono, A., & Kozasa, T. 2010. HgII ion specifically binds with T: T mismatched base pair in duplex DNA. Chemistry–A European Journal, 16(44), 13218–13225.

Tsade, H.K. 2016. Atomic absorption spectroscopic determination of heavy metal concentra-tions in Kulufo River, Arbaminch, Gamo Gofa, Ethiopia. *J Environ Anal Chem.* 3(177): 2.

Tyagi, S., and F.R. Kramer. 1996. Molecular beacons: Probes that fluoresce upon hybridiza-tion. *Nat Biotechnol.* 14(3): 303–308.

Valeur, B., and I. Leray. 2000. Design principles of fluorescent molecular sensors for cation recognition. *Coord Chem Rev.* 205(1): 3–40.

Vardhan, K.H., P.S. Kumar, and R.C. Panda. 2019. A review on heavy metal pollution, toxicity and remedial measures: Current trends and future perspectives. *J Mol Liq.* 290: 111197.

Voegelin, A., S. Pfister, A.C. Scheinost, M.A. Marcus, and R. Kretzschmar. 2005. Changes in zinc speciation in field soil after contamination with zinc oxide. *Environ Sci Technol.* 39(17): 6616–6623.

Waggoner, D.J., T.B. Bartnikas, and J.D. Gitlin. 1999. The role of copper in neurodegenerative disease. *Neurobiol Dis.* 6(4): 221–230.

Wallace, D.R. and Djordjevic, A.B., 2020. Heavy metal and pesticide exposure: A mixture of potential toxicity and carcinogenicity. *Curr Opin Toxicol.* 19: 72–79.

Wang, H., E. Zhao, J.W. Lam, and B.Z. Tang. 2015. AIE luminogens: Emission brightened by aggregation. *Mater Today.* 18(7): 365–377.

Wang, H., Y. Kim, H. Liu, Z. Zhu, S. Bamrungsap, and W. Tan. 2009. Engineering a unimolecular DNA-catalytic probe for single lead ion monitoring. *J American Chem Soc.* 131(23): 8221–8226.

Wang, H., Y. Wang, J. Jin, and R. Yang. 2008. Gold nanoparticle-based colorimetric and "turn-on" fluorescent probe for mercury (II) ions in aqueous solution. *Anal chem.* 80(23): 9021–9028.

Wang, H., Z. Wu, B. Chen, M. He, and B. Hu. 2015. Chip-based array magnetic solid phase microextraction on-line coupled with inductively coupled plasma mass spectrometry for the determination of trace heavy metals in cells. *Analyst.* 140(16): 5619–5626.

Wang, M., W. Wang, T.S. Kang, C.H. Leung, and D.L. Ma. 2016. Development of an Iridium (III) complex as a G-quadruplex probe and its application for the G-quadruplex-based luminescent detection of picomolar insulin. *Anal Chem.* 88(1): 981–987.

Wang, S. and X. Shi. 2001. Molecular mechanisms of metal toxicity and carcinogenesis. *Mol Cell Biochem.* 222: 3–9.

Wang, X., A.K. Mandal, H. Saito, J.F. Pulliam, E.Y. Lee, Z.J. Ke, J. Lu, S. Ding, L. Li, B.J. Shelton, and T. Tucker. 2012. Arsenic and chromium in drinking water promote tumorigenesis in a mouse colitis-associated colorectal cancer model and the potential mechanism is ROS-mediated Wnt/β-catenin signaling pathway. *Toxicol Appl Pharmacol.* 262(1): 11–21.

Wang, Y., Y. Wu, W. Liu, L. Chu, Z. Liao, W. Guo, G.Q. Liu, X. He, and K. Wang. 2018. Electrochemical strategy for pyrophosphatase detection based on the peroxidase-like activity of G-quadruplex-Cu^{2+} DNAzyme. *Talanta.* 178: 491–497.

Wasim, A.A., S. Naz, M.N. Khan, and S. Fazalurrehman. 2019. Assessment of heavy metals in rice using atomic absorption spectrophotometry-a study of different rice varieties in Pakistan. *Pak J Anal Environ Chem.* 20(1): 67–74.

Wu, J., W. Liu, J. Ge, H. Zhang, and P. Wang. 2011a. New sensing mechanisms for design of fluorescent chemosensors emerging in recent years. *Chem Soc Rev.* 40(7): 3483–3495.

Wu, X.F., H. Neumann, and M. Beller. 2011b. Palladium-catalyzed carbonylative coupling reactions between Ar-X and carbon nucleophiles. *Chem Soc Rev.* 40(10): 4986–5009.

Wu, Y., S. Zhan, L. Xu, W. Shi, T. Xi, X. Zhan, and P. Zhou. 2011c. A simple and label-free sensor for mercury (II) detection in aqueous solution by malachite green based on a resonance scattering spectral assay. *Chem Comm.* 47(21): 6027–6029.

Xia, F., X. Zhang, C. Zhou, D. Sun, Y. Dong, and Z. Liu. 2010. Simultaneous determination of copper, lead, and cadmium at hexagonal mesoporous silica immobilized quercetin modified carbon paste electrode. *J Anal Methods Chem.* 2010.

Xiang, Y., and Y. Lu. 2011. Using personal glucose meters and functional DNA sensors to quantify a variety of analytical targets. *Nat Chem.* 3(9): 697–703.

Xiang, Y., and Y. Lu. 2013. An invasive DNA approach toward a general method for portable quantification of metal ions using a personal glucose meter. *Chem Comm.* 49(6): 585–587.

Xiang, Y., and Y. Lu. 2014. DNA as sensors and imaging agents for metal ions. *Inorgan Chem.* 53(4): 1925–1942.

Xie, W., C. Peng, H. Wang, and W. Chen. 2017. Health risk assessment of trace metals in various environmental media, crops and human hair from a mining affected area. *Int J Eviron.* 14(12): 1595.

Xu, H., S.L. Gao, J.B. Lv, Q.W. Liu, Y. Zuo, and X. Wang. 2009. Spectroscopic investigations on the mechanism of interaction of crystal violet with bovine serum albumin. *J Mol Struct.* 919(1–3), 334–338.

Xu, J., N. Liu, C. Hao, Q. Han, Y. Duan, and J. Wu. 2019. A novel fluorescence "on-off-on" peptide-based chemosensor for simultaneous detection of Cu^{2+}, Ag^+ and S^{2-}. *Sens Actuators B Chem.* 280: 129–137.

Xu, L., Y. Chen, R. Zhang, T. Gao, Y. Zhang, X. Shen, and R. Pei. 2017. A highly Sensitive Turn-on Fluorescent Sensor for Ba^{2+} Based on G-Quadruplexes. *J Fluores.* 27: 569–574.

Xu, W., Z. Zeng, J.H. Jiang, Y.T. Chang, and L. Yuan. 2016. Discerning the chemistry in individual organelles with small-molecule fluorescent probes. *Angew Chem Int Ed.* 55(44): 13658–13699.

Xu, Z. and L. Xu. 2016. Fluorescent probes for the selective detection of chemical species inside mitochondria. *Chem Commun.* 52(6): 1094–1119. '

Xue, Y., Y. Wang, S., Wang, M. Yan, J. Huang, and X. Yang. 2020. Label-free and regenerable aptasensor for real-time detection of cadmium (II) by dual polarization interferometry. *Anal Chem.* 92(14): 10007–10015.

Yadav, N., R. P. Gaikwad, V. Mishra, and M. B. Gawande. 2022. Synthesis and photocatalytic applications of functionalized carbon quantum dots. *Bull Chem Soc Jpn.* 95(11), 1638–1679.

Yadav, N., D. Mudgal, and V. Mishra. 2023. In-situ synthesis of ionic liquid-based-carbon quantum dots as fluorescence probe for hemoglobin detection. *Anal Chim Acta.* 1272: 341502.

Yang, F., J. Duan, M. Li, Z. Wang, and Z. Guo. 2012. Visual and on-site detection of mercury (II) ions on lateral flow strips using DNA-functionalized gold nanoparticles. *Anal Sci.* 28(4): 333–333.

Yang, S., G. Li, C. Song, M. Liu, and Z. Yuan. 2020. Ultrashort peptide-stabilized copper nanoclusters with aggregation-induced emission. *Colloids Surf A Physicochem Eng Aspects.* 606: 125514.

Yao, X.Z., Z. Guo, Q.H. Yuan, Z.G. Liu, J.H. Liu, and X.J. Huang. 2014. Exploiting differential electrochemical stripping behaviors of Fe_3O_4 nanocrystals toward heavy metal ions by crystal cutting. *ACS Appl Mater Inter.* 6(15): 12203–12213.

Yin, B.C., B.C. Ye, W. Tan, H. Wang, and C.C. Xie. 2009. An allosteric dual-DNAzyme unimolecular probe for colorimetric detection of copper (II). *J Am Chem Soc.* 131(41): 14624–14625.

Yousef, M.I., A. M. El-Morsy, and M.S. Hassan. 2005. Aluminium-induced deterioration in reproductive performance and seminal plasma biochemistry of male rabbits: Protective role of ascorbic acid. *Toxicology.* 215(1–2): 97–107.

Yuan, L., W. Lin, K. Zheng, and S. Zhu. 2013. FRET-based small-molecule fluorescent probes: Rational design and bioimaging applications. *Acc Chem Res.* 46(7): 1462–1473.

Zarazúa, G., K. Girón-Romero, S. Tejeda, C. Carreño-De León, and P. Ávila-Pérez. 2014. Total reflection X-ray fluorescence analysis of toxic metals in fish tissues. *Am J Anal Chem.* 2014.

Zayed, A.M. and N. Terry. 2003. Chromium in the environment: Factors affecting biological remediation. *Plant and soil.* 249: 139–156.

Zeng, L., J. Gong, P. Rong, C. Liu, and J. Chen. 2019. A portable and quantitative biosensor for cadmium detection using glucometer as the point-of-use device. *Talanta.* 198:.412–416.

Zhan, S., Y. Wu, L. He, F. Wang, X. Zhan, P. Zhou, and S. Qiu. 2012. A silver-specific DNA-based bio-assay for Ag (I) detection via the aggregation of unmodified gold nanoparticles in aqueous solution coupled with resonance Rayleigh scattering. *Anal Methods.* 4(12): 3997–4002.

Zhan, S., Y. Wu, L. Wang, X. Zhan, and P. Zhou. 2016. A mini-review on functional nucleic acids-based heavy metal ion detection. *Biosens Bioelectron*. 86: 353–368.

Zhan, S., H. Xu, D. Zhang, B. Xia, X. Zhan, L. Wang, J. Lv, and P. Zhou. 2015. Fluorescent detection of Hg^{2+} and Pb^{2+} using GeneFinder(tm) and an integrated functional nucleic acid. *Biosens Bioelectron*. 72: 95–99.

Zhang, M., L. Ge, S. Ge, M. Yan, J. Yu, J. Huang, and S. Liu. 2013. Three-dimensional paper-based electrochemiluminescence device for simultaneous detection of Pb^{2+} and Hg^{2+} based on potential-control technique. *Biosens Bioelectron*. 41: 544–550.

Zhang, L., G.H. Peslherbe, and H.M. Muchall. 2006. Ultraviolet absorption spectra of substituted phenols: A computational study. *Photochem Photobiol*. 82(1): 324–331.

Zhang, H., Y. Ruan, L. Lin, M. Lin, X. Zeng, Z. Xi, and F. Fu. 2015a. A turn-off fluorescent biosensor for the rapid and sensitive detection of uranyl ion based on molybdenum disulfide nanosheets and specific DNAzyme. *Spectrochim Acta Part A Mol Biomol Spectrosc*. 146: 1–6.

Zhang, J., Y. Tang, L. Teng, M. Lu, and D. Tang. 2015b. Low-cost and highly efficient DNA biosensor for heavy metal ion using specific DNAzyme-modified microplate and portable glucometer-based detection mode. *Biosens Bioelectron*. 68: 232–238.

Zhang, L., J.X. Wong, Y. Li, Y. Li, and H.Z. Yu. 2015c. Detection and quantitation of heavy metal ions on bona fide DVDs using DNA molecular beacon probes. *Anal Chem*. 87(10): 5062–5067.

Zhao, B., H. Jiang, Z. Lin, S. Xu, J. Xie, and A. Zhang. 2019. Preparation of acrylamide/acrylic acid cellulose hydrogels for the adsorption of heavy metal ions. *Carbohydr Polym*. 224: 115022.

Zhao, G., H. Wang, and G. Liu. 2017. Direct quantification of Cd^{2+} in the presence of Cu^{2+} by a combination of anodic stripping voltammetry using a Bi-film-modified glassy carbon electrode and an artificial neural network. *Sensors*. 17(7): 1558.

Zhao, L., Z. Zhang, Y. Liu, J. Wei, Q. Liu, P. Ran, and X. Li. 2020. Fibrous strips decorated with cleavable aggregation-induced emission probes for visual detection of Hg^{2+}. *J hazard Mater*. 385: 121556.

Zheng, X., W. Cheng, C. Ji, J. Zhang, and M. Yin. 2020. Detection of metal ions in biological systems: A review. *Rev Anal Chem*. 39 (1): 231–246.

Zhong, W.S., T. Ren, and L.J. Zhao. 2016. Determination of Pb (Lead), Cd (Cadmium), Cr (Chromium), Cu (Copper), and Ni (Nickel) in Chinese tea with high-resolution continuum source graphite furnace atomic absorption spectrometry. *J Food Drug Anal*. 24(1): 46–55.

Zhou, W., M. Vazin, T. Yu, J. Ding, and J. Liu. 2016. In vitro selection of chromium-dependent DNAzymes for sensing chromium (III) and chromium (VI). *Chem-A Eur J*. 22(28): 9835–9840.

Zhou, Y., F. Wang, Y. Kim, S.J. Kim, and J. Yoon. 2009. Cu^{2+}-selective ratiometric and "off-on" sensor based on the rhodamine derivative bearing pyrene group. *Org Lett*. 11(19): 4442–4445.

Zhou, Z., M. Yu, H. Yang, K. Huang, F. Li, T. Yi, and C. Huang. 2008. FRET-based sensor for imaging chromium (III) in living cells. *Chem commun*. 29: 3387–3389.

Zhu, J., Q. Fu, G. Qiu, Y. Liu, H. Hu, Q. Huang, and A. Violante. 2019a. Influence of low molecular weight anionic ligands on the sorption of heavy metals by soil constituents: A review. *Environ Chem Lett*. 17: 1271–1280.

Zhu, Q., L. Liu, Y. Xing, and X. Zhou. 2018. Duplex functional G-quadruplex/NMM fluorescent probe for label-free detection of lead (II) and mercury (II) ions. *J Haz Mat*. 355: 50–55.

Zhu, Y.F., Y.S. Wang, B. Zhou, J.H. Yu, L.L. Peng, Y.Q. Huang, X.J. Li, S.H. Chen, X. Tang, and X.F. Wang. 2017. A multifunctional fluorescent aptamer probe for highly sensitive and selective detection of cadmium (II). *Anal Bioanal Chem*. 409: 4951–4958.

Zhu, P., Y. Zhang, S. Xu, and X. Zhang. 2019b. G-quadruplex-assisted enzyme strand recycling for amplified label-free fluorescent detection of UO_2^{2+}. *Chinese Chem Lett.* 30(1): 58–62.

Zuo, P., B.C. Yin, and B.C. Ye. 2009. DNAzyme-based microarray for highly sensitive determination of metal ions. *Biosens Bioelectron.* 25(4): 935–939.

9 Protein Labelling for Molecular Recognition and Application

Ruchi Sharma, Preeti Kasana,
Ashima Sharma, and Vinod Kumar

9.1 PROTEIN AS CRUCIAL BIOMOLECULES AND ITS LABELLING

Proteins are crucial biomolecules responsible for carrying out the majority of the reactions in a living system. Proteins are composed of polypeptide chains and their biological functions are determined in part by their correct folding, size, and the number of reactive functional groups present throughout the polypeptide chain [1]. With the recent advancements in the field of biotechnology and protein biochemistry, researchers can carry out site-specific protein modifications, which enable them to explore broadly the properties and their overall biological function. Labelling of a protein is a method to tag the protein with a label to facilitate identification, monitoring, characterization, and even purification of the protein of interest (POI). Moreover, the identification of the protein opens up the whole paradigm of its associated moieties interacting directly or indirectly to carry out function in the cell at a given point of time. This aids in further deciphering and provides a comprehensive understanding of various complex even unexpected activities carried out by the target protein in the body [2].

Protein labelling empowers the researcher to monitor target protein in the living cell or in vitro and to recognize the related behaviour, role, and characteristics of the POI. It gives the opportunity to highlight the protein in its milieu so that the tracking of its activity and related observations can be carried out in real time [3].

The most commonly adopted protein-labelling and modification technologies include adding fluorophores, biotin, and other small molecules to examine protein–protein interactions, protein folding, to investigate the overall protein structure, and their biological functions. Genetically encoded tags have revolutionized cell biology during the past two decades. These are either fluorophores that are bound to tiny molecules or protein sequences that fold to form the protein sequence. During cell development, proteins can be marked by the inclusion of amino acids with various isotopes or by attaching certain molecules to samples of biological fluids, cells, or tissues. Bio-conjugation and chemical labelling are the two majorly exploited methods for protein labelling. Bio-conjugation involves the introduction of a reactive group by modification of a specific amino acid residue in a protein. The predominantly exploited sites in the protein are N-terminus, cysteine residues, and the amino group of lysine. The method cannot

DOI: 10.1201/9781003352372-9

be employed for selective labelling of a protein in a living cell. To overcome this disadvantage, the chemical labelling method of protein has been adopted. For the chemical labelling of POI in living cells, there are two different approaches: Bio-orthogonal chemical reactions and molecular recognition-derived chemical reactions. The former method entirely depends on the chemical reaction's selectivity. In the beginning, this technique adds a non-canonical functional group to a POI (as a bio-orthogonal reaction handle). A similar reactant with high bio-orthogonality is then used to modify the initially introduced reaction handle (bio-orthogonal reaction). In the case of latter approach involving molecular recognition-based methods, the interaction between reactive groups of labelling reagent and the POI is mainly driven by the molecular recognition between the two moieties leading to proximity-driven pair formation in a selective and efficient manner. This molecular recognition-based labelling can be carried out in endogenously living cells. There are different strategies by which we can carry out chemical-based labelling. In this chapter, we however focus on discussing in detail the molecular recognition methods employed in protein labelling.

9.2   PROTEIN-LABELLING REAGENTS

A number of fluorophores are available for covalent and non-covalent labelling of proteins. The covalent probes can have a variety of reactive groups, for coupling with amines and sulfhydryl or histidine side chains in proteins. Due to early introduction in the literature and favourable lifetime (10 ns), dansyl chloride (Figure 9.1) can be used for labelling proteins especially where polarization measurements are expected. One can excite dansyl groups at 350 nm, where proteins do not absorb and emission maxima are obtained typically near 520 nm. Since dansyl groups absorb near 350 nm, these groups can serve as acceptors of protein fluorescence. The emission spectrum of the dansyl moiety is also highly sensitive to solvent polarity. Fluoresceins and rhodamines are also widely used as extrinsic labels for various proteins. These dyes play a significant role in protein labelling due to favourably long absorption maxima near 480 and 600 nm and emission wavelengths from 510 to 615 nm, respectively. Furthermore, it has been observed that unlike the dansyl group, rhodamines, and fluoresceins are not sensitive to solvent polarity. More importantly, their high molar extinction coefficients near 80,000 $M^{-1}cm^{-1}$ are also an additional reason for their widespread use as protein labels.

Fluorescein and rhodamine are commonly used for labelling of antibodies. A wide variety of fluorescein- and rhodamine-labelled immunoglobulins are commercially available, and these proteins are frequently used in fluorescence microscopy and immunoassays. The reasons for selecting these probes include high quantum yields and the long wavelengths of absorption and emission, which minimize the problems of background fluorescence from biological samples and eliminate the need for quartz optics. The lifetimes of these dyes are near 4 ns and their emission spectra are not significantly sensitive to solvent polarity. These dyes are suitable for quantifying the associations of small labelled molecules with proteins via changes in fluorescence polarization. The boron-containing fluorophore-based BODIPY dyes have been introduced as replacements for fluorescein and rhodamines. One can obtain a wide range of emission wavelengths from 510 to 675 nm depending on the

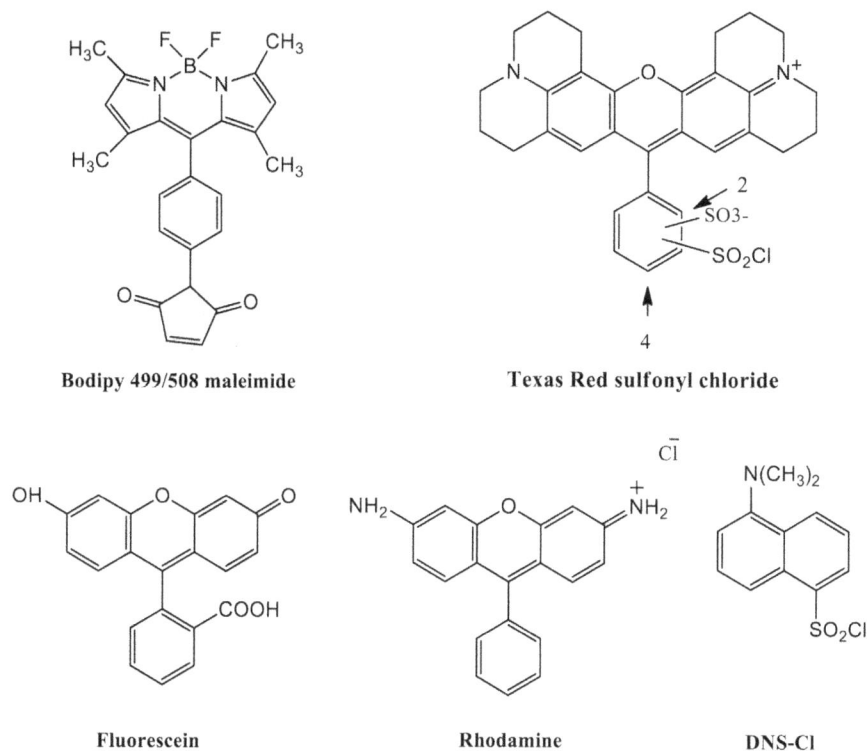

Bodipy 499/508 maleimide   Texas Red sulfonyl chloride

Fluorescein   Rhodamine   DNS-Cl

FIGURE 9.1 Structures of some important protein-labelling reagents.

precise structure. The BODIPY dyes have the additional advantage of displaying high quantum yields approaching unity, extinction coefficients near 80,000 $M^{-1}cm^{-1}$, and insensitivity to solvent polarity and pH. The emission spectra are narrower than those of fluorescein and rhodamines, so that more of the light is emitted at the peak wavelength, possibly allowing more individual dyes to be resolved. A disadvantage of the BODIPY dyes is a very small Stokes shift. As a result, the dyes transfer to each other with a Förster distance near 57 Å.

9.2.1 ROLE OF THE STOKES SHIFT IN PROTEIN LABELLING

Fluoresceins and rhodamines have a tendency to self-quench. Furthermore, it is well reported that the brightness of fluorescein-labelled proteins does not increase linearly with the extent of labelling. Sometimes, the intensity can decrease as the extent of labelling increases. Fluorescein displays a small Stokes shift. When more than a single fluorescein group is bound to the protein within 40 Å of each other which is within the Förster distance for fluorescein-to-fluorescein transfer, there can be a possibility for energy transfer between these groups. Fluorescein and Texas-Red both show substantial self-quenching. The two Alexa Fluor dyes show much less self-quenching, which allows the individually labelled antibodies to be more highly fluorescent. It is not clear why the Alexa Fluor dyes showed less self-quenching since their Stokes shift is similar to that of fluorescein and rhodamine. The BODIPY dyes have a small Stokes shift and usually display self-quenching. Development of new dyes such as Cascade Yellow with large Stokes shift and good water solubility, which displays excitation and emission maximum near 409 and 558 nm, respectively, have contributed a lot to labelling technology. The large Stokes shift minimizes the tendency for homo transfer, and the charges on the aromatic rings aid solubility.

9.2.2 NON-COVALENT PROTEIN-LABELLING PROBES

Naphthylamine sulfonic acids, of which 1-anilinonaphthalene-6-sulfonic acid (ANS) and 2-(p-toluidinyl)naphthalene-6-sulfonic acid (TNS), are most commonly used to non-covalently label proteins. These dyes are generally weak or non-fluorescent in water, but fluoresce strongly when bound to proteins or membranes. ANS-type dyes are amphiphatic, and hence nonpolar region prefers to adsorb onto nonpolar regions of macromolecules. Since the water-phase dye does not contribute to the emission, the observed signal is due to the area of interest and the probe binding site on the macromolecule.

9.3 RECOGNITION IN PROTEIN LABELLING

A significant role of recognized protein labelling is to monitor biological processes, accurate chemical quantification, targeted recognition of protein structural changes and its iso-forms, simplicity of detection procedure, and upgraded detection sensitivity. In biological fluids, cells, or tissue samples, POI can be identified by attaching specific groups to the lysine, cysteine, or N-terminal amino acid residues. Proteins can also be identified during cell division by inserting amino acids with different isotopes [4]. In biological research, molecular labels that have been covalently attached to a POI are routinely employed to help identify or purify the labelled protein and/ or its binding partners. Using labelling techniques, various compounds, including the target protein or nucleotide sequence, are covalently linked to biotin, reporter enzymes, fluorophores, and radioactive isotopes [4]. Different types of protein-labelling strategies are discussed below.

9.3.1 BIOTIN-BASED LABELLING

Biotin is a very small 244.3 Da molecule. Due to its extremely strong affinity for binding to avidin and streptavidin, biotin is a useful label for protein detection, purification, and isolation [5]. Biotin-mediated protein interaction is one of the strongest non-covalent interactions between protein and ligand. The smaller size of biotin does not hamper the native biological activity of the target protein in the bound state. The proteins or nucleotides can be tagged with biotin molecules using the enzymatic and chemical processes of biotinylation.

9.3.2 ACTIVE SITE PROBE-BASED TARGET ENZYME LABELLING

Active site probes (ASPs) are one of the types of chemical labelling agents comprising engineered reactive clusters intended to bind and tag specific active sites in the enzyme. The primary structural elements of this probe consist of a spacer moiety, detectable tags, and a reactive cluster liable for attaching the target class of enzyme's active site. Reactive groups of active sites are mostly electrophilic compounds forming covalent bonds with nucleophilic residues in the site [6]. These probes can be used to evaluate the specificity and affinities of enzyme inhibitors or to profile, identify, and selectively enhance target enzyme classes throughout the samples. Active site probes have been developed to mark a variety of different enzyme types, including phosphatases, cytochrome P450 enzymes cysteine proteases, metalloproteases, kinases, and serine hydrolases. It is possible to utilize all active site probes to identify whether small compounds have inhibited an enzyme, and certain probes selectively react with specific functional proteins only, enabling activity-based proteome profiling (ABPP). ABPP is a potent tool to evaluate protein activity also along with the quantization of the target protein and is therefore considered superior as compared to the conventional RNA or protein expression-based analysis approach which can only assess quantity. Figure 9.2 illustrates the chemical structure of some of the active site probes, Azido-FP, Desthiobiotin-FP, and TAMARA-FP, used to detect active serine hydrolase enzymes [4].

9.3.3 ENZYME CONJUGATES: ENZYME REPORTERS AS DETECTION TAGS

In comparison to biotin, enzyme labels are significantly bigger and need a substrate for creating a fluorescent, chromogenic, or chemi-luminescent signal that may be sensed in a variety of ways. Some enzymes can function as very sensitive and stable probes with versatility for the detection of proteins in tissues, whole cells, or lysates [6]. The great variety of signal outputs, signal amplification, and enzyme-labelled

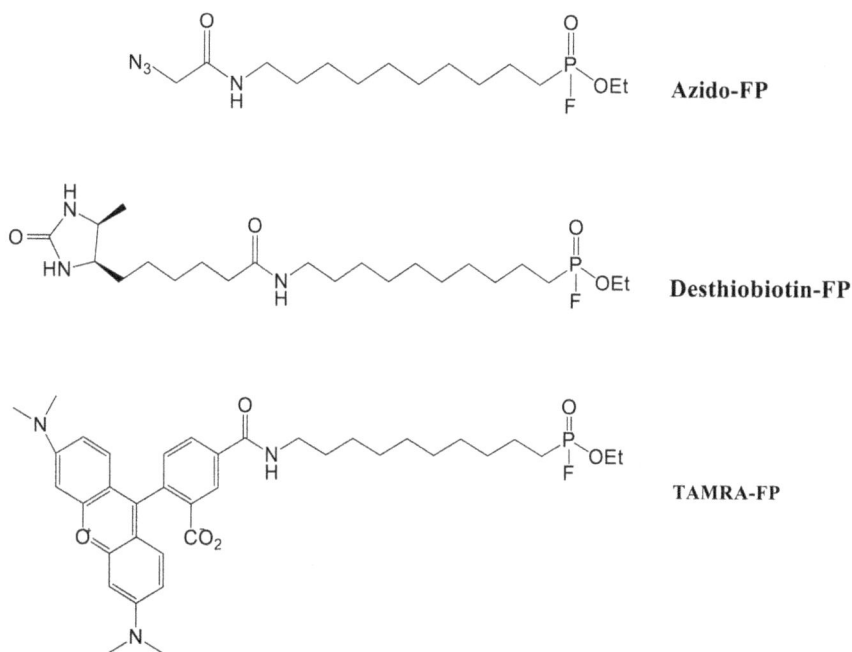

FIGURE 9.2   Chemical structure of active site probes. Azido, Desthiobiotin, and fluorescently-tagged fluorophosphonate probe structures.

products, particularly antibodies, make enzyme labels extremely popular. Alkaline phosphatase, glucose oxidase, β-galactosidase, and horseradish peroxidase are a few enzymes that are frequently employed as labels, and each enzyme has a particular set of substrates. Using a primary antibody that binds to the particular epitope of the target protein, the target protein or antigen is identified for immunoassays such as western blots, ELISA, and immune histochemistry. Either direct detection or indirect detection can be used to find out the POI. The main antibody used in direct detection is conjugated to a detectable enzyme, fluorophore, or other substances. A labelled secondary antibody is eventually employed to bind to the original antibody, thereby detecting anything in an indirect manner. In contrast to the earlier method, signal amplification occurs when several secondary antibodies interact with primary antibodies, thereby enhancing the signal and increasing the sensitivity of detection.

9.3.4 FLUORESCENT PROBES

Fluorescent probes are molecules that examine biological material by absorbing light of a certain wavelength and emitting light of another, usually longer, wavelength (Figure 9.3) [7]. Proteins and other biomolecules can all be stained or chemically marked with different fluorescent reagents.

Target proteins or antigens and the interacting partners can be identified by using biomolecules or antibodies tagged chemically with the fluorescent probes. These probes are used in applications such as flow cytometry, western blotting, high content analysis, cell imaging, and ELISA. Fluorescent labels do not need any extra reagents for detection, unlike enzymes or biotin. Fluorescent molecules, commonly known as fluorophores or just fluors, react to light immediately and noticeably and provide a discernible signal [8]. Fluorescence microscopes, flow cytometers, and cell sorters

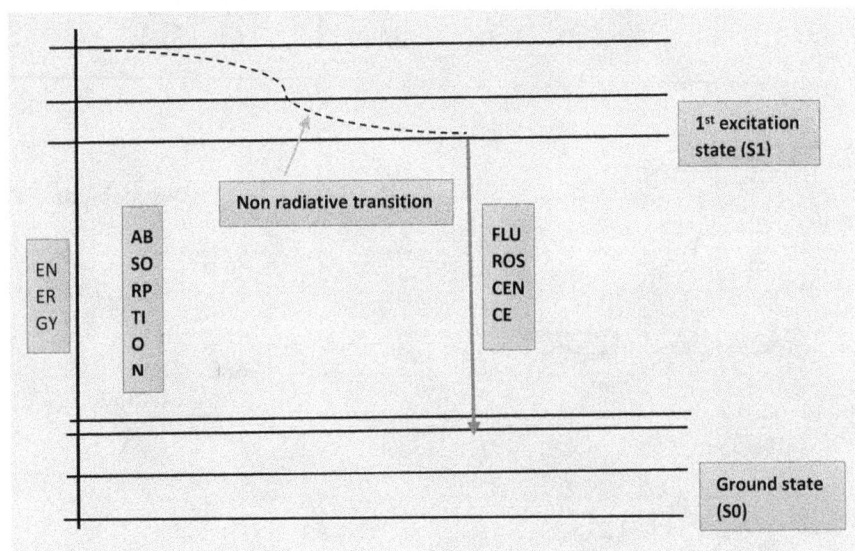

FIGURE 9.3 Fluorescence: Energy level diagram.

are just a few examples of the specialized equipment needed to detect fluorescent probes [9]. It also has a detector, filter set, and excitation light source. This apparatus enables the fluorescence-based absolute measurement of proteins, which is a key advantage of fluorescent probes over other kinds of probes.

9.4 STRATEGIES INVOLVED IN RECOGNITION IN PROTEIN-LABELLING METHODS

9.4.1 RECOGNITION-DRIVEN PROTEIN LABELLING USING A GENETICALLY ENCODED TAG

Many desired biological processes, such as enzymatic activities, immunological responses, signal transduction, and protein synthesis, depend on molecular recognition. Such molecular identification enables a molecule to appropriately identify the relevant target binding partner in question, which also enables quickening of biochemical processes. In enzymatic processes, for example, the related enzymes' reaction substrates are efficiently forced into close proximity by molecular recognition, enabling highly selective and swift reactions with the proximity effect. The CLIP-tag, mutant AGT-based tag, and CLIP-tag, which only reacts with O_2-benzylcytosin derivatives, were used by Johnsson's team to create an orthogonal covalent labelling technique [10]. The orthogonality of the SNAP-tag and CLIP-tag systems allowed for the creation of FRET (fluorescence resonance energy transfer) biosensors and the detection of protein–protein interactions, among other biological applications. For protein tag-based covalent labelling, similar techniques utilizing an enzyme-suicide inhibitor pair have now been developed. BL-tag, eDHFR-tag, Halo-tag, Cutinase-tag, and PYP-tag are some of these methods. Amaike et al. (2017) and Tsien and colleagues have shown that a tetra-Cys synthetic peptide (CCXXCC) could bind to a dinuclear fluorophore (FlAsH tag) with a strong and specific affinity even in living cells [K_d (dissociation constant) = 4–70 pM] [11]. Due to the reduced size of the peptides used in these approaches compared to protein tag techniques, appropriate fusion proteins must still be produced. As a result, none of the techniques mentioned above can be used to mark endogenous proteins.

9.4.2 RECOGNITION-DRIVEN ENDOGENOUS PROTEIN LABELLING

One of the most useful techniques for examining the real activities of proteins in living cells is endogenous protein labelling. Yet, it is challenging to chemically identify an essential protein in a highly selective manner due to multi-molecular crowding events, like the ones occurring in live cells [12]. Various types of strategies used in protein-labelling approaches for recognition include affinity labelling and activity-based labelling which are discussed below.

9.4.2.1 Affinity Labelling

These are reversible substrates that permit entry into the enzyme's active site. Once inside, they often covalently alter the catalytic residue, preventing the enzyme from doing its job. One of the examples can be found in glycolysis when bromo acetone phosphate reacts with triose phosphate isomerase it forms a covalent bond in which carbon of bromo reacts with an oxygen molecule and covalent bond or modification and kicks off the Br from triose phosphate isomerase and essentially inactivates the active site of triose phosphate isomerase. As a result, triose phosphate isomerase does not get converted to dihydroacetone phosphate into its isomeric form (Figure 9.4).

Protein affinity labelling has long been employed for the selective chemical labelling of endogenous POIs [14]. A synthetic reagent including an affinity ligand for the POI, a reactive moiety, and a reporter tag, is used in this approach.

9.4.2.2 Activity-Based Labelling

Activity-based probes (ABPs) consist of a receptor tag and suicide enzyme inhibitors. Suicide inhibitors provide the most specific means of active site binding. Penicillin and aspirin are some of the examples of suicide inhibitors. These compounds, also known as mechanical-based inhibitors, get attached to the active site and start the catalysis process as if they were the natural substrate. Yet along the chemical process, formation of a reactive intermediate results in changing the active site residue covalently thereby irreversibly blocking the enzyme [15].

9.5 LIMITATIONS OF PROTEIN LABELLING IN VITRO

While being widely used, the approach yet comprises a number of drawbacks summarized in Table 9.1.

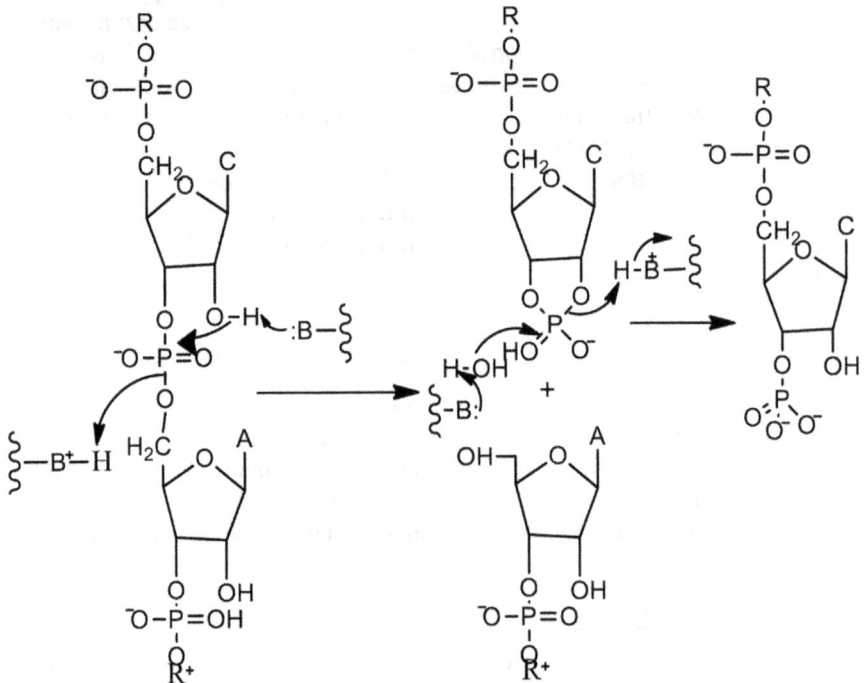

FIGURE 9.4 General acid/base-catalysed mechanism for ribonuclease A [13].

TABLE 9.1

Limitations Associated with Respective Employed Applications for Recognized Protein Labelling

Application	Methodology	Labelling Requirements	Example Fluorophore	Problems/Recommendations
Conformational change	Intenseness change	Specific location cysteine	MDCC	Requires a fluorophore that is sensitive to the environment.
	FRET/FLIM	Specific location cysteine fluorescent protein	Cy3/Cy5 GFP/RFP	Luminescent proteins are ideal for massive rearrangements due to the necessary spectrum overlap.
Protein–protein interaction	smFRET	Specific location cysteine	Cy3/ Cy5	Fluorophores must be bright and photostable.
	Intenseness change	Specific location cysteine	MDCC	Requires a fluorophore that is sensitive to the environment.
	Anisotropy	Specific location cysteine	Fluorescein	LABEL must be tiny and the fluorescence lifespan must be close to 5 ns.
Single-molecule/ particle tracking	smFRET	Specific location cysteine	Cy3/Cy5	Proteins can be separated apart and fluorophores can be directed to inactive regions.
	Single-molecule imaging	Specific location cysteine Peptide tag Biotinylation	Cy3B Quantum dot	Oversized label could obstruct activity.
Protein counting	Photobleaching	Fluorescent protein	eGFP	Utilize fluorescent proteins that are monomeric. Correct stoichiometry can only be obtained from fluorescent protein.
Live-cell localization	Live-cell imaging	Fluorescent protein	eGFP	Monomeric fluorescent protein should be used.
		Protein tag	FLAsH	Organic fluorophores are less precise in their labelling but are brighter and more stable than fluorescent proteins.
		Protein tag	SNAP-Tag	

9.6 BIO-MOLECULAR LABELLING STRATEGIES

To meet the demand for all varieties of bio-molecular probes, techniques for labelling proteins and nucleic acids have been developed both in vitro and in vivo and in vitro labelling uses a tag conjugated to chemical groups which reacts with particular amino acids, chemical techniques of protein labelling require the covalent attachment of the label to amino acids. The appropriate polymerases, ATP, and tagged amino acids or nucleotides are needed for in vitro procedures. While in vitro DNA transcription is rather simple, in vitro translation of labelled proteins can be challenging due to the need for the correct protein folding and post-translational amendments, all of which are beyond the capabilities of commercially available kits. By cultivating them with tagged nucleotides or amino acids, respectively, metabolic labelling is a technique for marking the entire population of proteins or nucleic acids present in a cell.

Molecular probes are biomolecules that are used to track a target protein inside the cell. A molecular probe consists of a binding domain, involved in binding with the specific target, and a reporter unit for detection purposes (Figure 9.5).

If you want to know or track a gene, especially the diseased gene, so you will expect mutation to be present in DNA/gene normal/disease cells. Peptide nucleic acid (PNA) (Figure 9.6) and locked nucleic acid (LNA) (Figure 9.7) are artificially synthesized and are used as a molecular probe. Their activity is better than DNA as a molecular probe for protein labelling and detection.

Many different types of cancer, diabetes, and epilepsy are due to single-nucleotide polymorphism. So, using DNA as a molecular probe is not the best idea because the

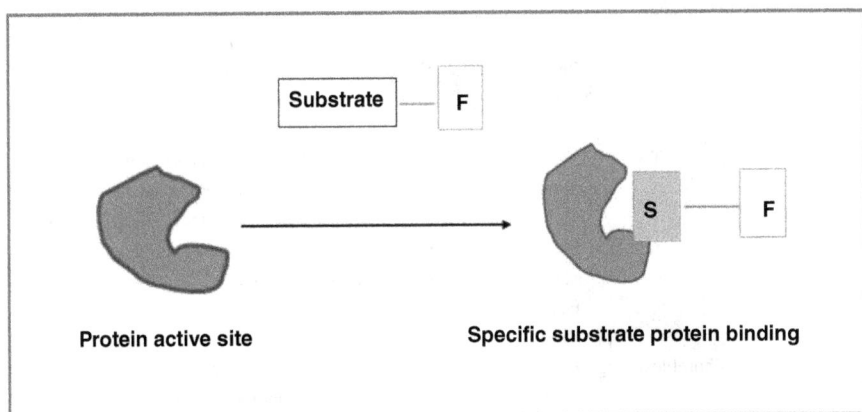

FIGURE 9.5 Fluorescence (F) tagged substrate (S) to observe the activity of proteins in the cell.

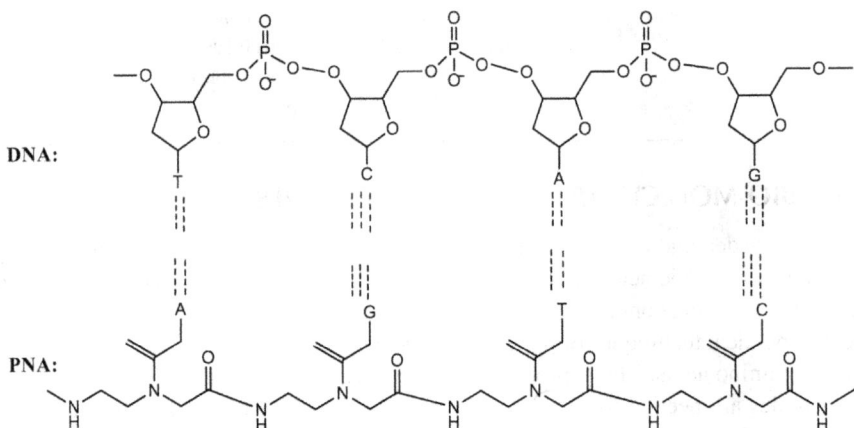

FIGURE 9.6 Structure and interaction of PNA and DNA [16].

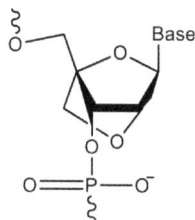

FIGURE 9.7 Structure of locked nucleic acid [17].

differences in the healthy and mutated gene properties, like melting temperature (T_m), will not be much high or distinctive. Therefore, the need is to develop molecules that will have a stronger interaction with the target. If the interaction is stronger, then a mismatch pair will be equally weaker and the difference between strong and weak binding would be large enough. The T_m of DNA and PNA is DNA-DNA < DNA-PNA < PNA-PNA. The melting temperature of DNA-DNA is significant because if two DNA molecules are brought closer to each other, repulsion between the phosphates acts as a destabilizing factor for DNA double helix formation. DNA backbone, if neutral, does not have any charge; therefore, if you make a hybridization of DNA with PNA, then this repulsion will not be observed. PNA also has a more favourable conformation as compared to DNA. It allows better hybridization and better stalking between the nucleobases. PNA can easily enter inside the cell whereas for DNA the entry is limited by the presence of nuclease enzymes. If DNA is used as a molecular probe, then most of the time cell will not be allowed to enter inside the cell.

PNA, on the other hand, has more peptide backbone which facilitates its swift entry inside the cell. PNA is a better molecular probe when it comes to living cell studies and has stronger binding.

9.7 APPLICATIONS OF MOLECULAR RECOGNITION IN PROTEIN LABELLING

Monitoring biological processes, accurate chemical quantification, targeted detection of protein changes and isoforms in multiplexed samples, improvement of detection sensitivity, and simplicity of detection procedures are the main goals of protein labelling. Biotin is a valuable label for protein detection, purification, and immobilization, because of its incredibly high binding to avidin and streptavidin. For the detection of proteins in tissues, entire cells, or lysates, certain enzymes can be used as highly sensitive probes with a long shelf life and a variety of uses [3]. Fluorophores have a wide range of applications and are the new standard for locating and activating proteins, detecting protein complex formation and conformational changes, and tracking biological processes. Nucleic acids and proteins can both be labelled using enzymatic techniques. The appropriate polymerases, ATP, and tagged amino acids or nucleotides are needed for these in vitro procedures. While in vitro DNA transcription is rather simple, in vitro translation of tagged proteins can be challenging since these processes need for several commercial kits that are unable to provide the appropriate post-translational modifications, folding, or protein length [5].

REFERENCES

1. The Shape and Structure of Proteins - Molecular Biology of the Cell - NCBI Bookshelf [Internet]. [cited 2023 Mar 24]. Available from: https://www.ncbi.nlm.nih.gov/books/NBK26830/.
2. Protein Labeling with Fluorescent Probes - Theory and Methods [Internet]. [cited 2023 Mar 24]. Available from: https://scigine.com/blog/protein-labeling-fluorescent-probe/.
3. Obermaier C., Griebel A., Westermeier R. Principles of protein labeling techniques. *Methods in Molecular Biology*. 2015; 1295:153–65. Available from: https://pubmed.ncbi.nlm.nih.gov/25820721/.
4. Overview of Protein Labeling | Thermo Fisher Scientific - IN [Internet]. [cited 2023 Mar 25]. Available from: https://www.thermofisher.com/in/en/home/life-science/protein-biology/protein-biology-learning-center/protein-biology-resource-library/pierce-protein-methods/overview-protein-labeling.html.
5. Obermaier C., Griebel A., Westermeier R. Principles of protein labeling techniques. *Methods in Molecular Biology*. 2021; 2261:549–62.
6. Protein Labelling 101: Understanding the Basics [Internet]. [cited 2023 Mar 25]. Available from: https://info.gbiosciences.com/blog/bid/182082/protein-labelling-101-understanding-the-basics.
7. Holmes K.L., Lantz L.M. Protein labeling with fluorescent probes. *Methods in Cell Biology*. 2001; 63:185–204.
8. Mao L., Han Y., Zhang Q.W., Tian Y. Two-photon fluorescence imaging and specifically biosensing of norepinephrine on a 100-ms timescale. *Nature Communications*. 2023; 14:1419. Available from: https://www.nature.com/articles/s41467-023-36869-3.
9. Fluorescence guide | Abcam [Internet]. [cited 2023 Mar 25]. Available from: https://www.abcam.com/secondary-antibodies/fluorescence-guide.
10. Amaike K., Tamura T., Hamachi I. Recognition-driven chemical labeling of endogenous proteins in multi-molecular crowding in live cells. *Chemical Communications*. 2017; 53(88):11972–83.
11. Griffin B.A., Adams S.R., Tsien R.Y. Specific covalent labeling of recombinant protein molecules inside live cells. *Science*. 1998; 281:269.
12. Amaike K., Tamura T., Hamachi I. Recognition-driven chemical labeling of endogenous proteins in multi-molecular crowding in live cells. *Chemical Communications*. 2017; 53(88):11972–83. Available from: https://pubs.rsc.org/en/content/articlehtml/2017/cc/c7cc07177a.
13. Silverman R.B. *The Organic Chemistry of Enzyme-Catalyzed Reactions*. Academic Press; 2002 [cited 2022 Oct 17]. 717 p. Available from: https://www.sciencedirect.com:5070/book/9780080513362/organic-chemistry-of-enzyme-catalyzed-reactions.
14. Wofsy L., Metzger H., Singer S.J. Affinity labeling-a general method for labeling the active sites of antibody and enzyme molecules. *Biochemistry*. 1962; 1:1031.
15. Geurink P.P., Prely L.M., Van Der Marel G.A., Bischoff R., Overkleeft H.S. Photoaffinitylabeling in activity-based protein profiling. *Topics in Current Chemistry*. 2012; 324: 85–113. Available from: https://pubmed.ncbi.nlm.nih.gov/22028098/.
16. Wu J.C., Meng Q.C., Ren H.M., Wang H.T., Wu J., Wang Q. Recent advances in peptide nucleic acid for cancer bionanotechnology. *Acta Pharmacologica Sinica*. 2017; 38(6):798–805. Available from: https://www.nature.com/articles/aps201733.
17. Structure of locked nucleic acid (LNA). | Download Scientific Diagram [Internet]. [cited 2022 Oct 18]. Available from: https://www.researchgate.net/figure/Structure-of-locked-nucleic-acid-LNA_fig1_11004043.

10 Organic–Inorganic Nanohybrids for Sensing and Optoelectronics Applications

Deepak Kumar, Ashish Kumar Rajayan,
Ishpal Rawal, and Rishi Pal

10.1 INTRODUCTION

In view of environmental safety, the concern of scientific community is continuously gaining attention towards the real-time fabrication of novel sensors and optoelectronic devices. In this direction, a plethora of sensing devices has been widely established in the fields of medicine, agriculture, industries, etc. [1]. To fabricate these sensors, both organic [2] and inorganic [3] materials are used for the detection of various hazardous gases and chemicals such as NH_3, NO, NO_2, acetylene, and CO_2. Usually, the inorganic material-based thin film sensors have high response, selectivity, and stability [4]. Further, these sensors operate in high-temperature ranges [5], which weaken their impact on targeted applications in materials science. Besides these inorganic materials, organic materials also contribute significantly to the field of sensors [6], which can be employed rapidly at room temperature [7]. But, these organic sensors are comparatively lesser stable as compared to inorganic material-based sensors.

Moreover, conducting polymers including polythiophene (PTP), polypyrrole (PPy), and polyaniline (PANI) exhibit reactivity to specimen analytes at room temperature in these sensors [8]. Therefore, the constraints of operating temperature for inorganic sensors and stability issues of organic sensors can be solved out with rational integration of organic/inorganic materials. Such synthesized hybrids have been shown to increase mechanical strength and chemical stability [9]. The most often approach has been to use inorganic materials in the organic matrix to serve as an improved sensing platform [10]. It is probable to modify the morphology of the sensing materials to increase the response towards a specimen analyte because organic compounds exhibit synthetic plasticity and reactivity [11]. In the last decade, various inorganic/organic nanohybrids [12,13] have been synthesized and their sensing performances towards various toxic gases are investigated. At ambient temperature or low temperatures, gas sensing experiments revealed that the hybrids performed better than the single component indicating a potential use for the former ones.

DOI: 10.1201/9781003352372-10

Moving ahead, engineering the molecular interactions of these nanohybrids (like metal oxide nanoparticles, carbon dots, nanotubes, and graphene) can pave the way for better biocompatibility and lesser toxicity to be useful for imaging in clinical sectors, catalysis, light harvesting, energy storage, and memory applications [14–16]. Also, these possess high photosensitivity and good solution processibility to be exploited for commercialization of state-of-the-art opto-electrical components [17]. Recently, fascinating non-linear optical effects are being realized utilizing hybrid perovskites [18]. Further, incorporating organic templates [19] in designing of hybrid nanomaterials with exotic properties can provide an additional degree of freedom in tailoring their morphologies. Therefore, it is anticipated that a journey through the current status of this field will be advantageous to potential researchers for developing functional materials [20,21] with synergistic features. Figure 10.1 depicts the outline of the proposed chapter with inclinations on the sensing and optoelectronics aspects of organic–inorganic nanohybrids.

10.1.1 Sensing Employing Organic–Inorganic Nanohybrids

Here, we elaborate numerous platforms adopted for sensing applications based on organic–inorganic nanohybrid systems.

10.1.1.1 PANI/V₂O₅ Hybrid Gas Sensor

The PANI/V_2O_5 nanohybrid samples were synthesized by Pal et al. [22] using an in-situ polymerization synthesis route (Figure 10.2a) to detect methanol vapours at room temperature. When the methanol ppm level is increased from 40 to 60 ppm, pure PANI-based sensors reveal the change in response from 16.56% to 18.27%. Nevertheless, when methanol ppm level is enhanced from 40 to 60 ppm, the response of the PANI/V_2O_5 nanohybrid (30 wt% doped) increases from 32.62% to 36.41%.

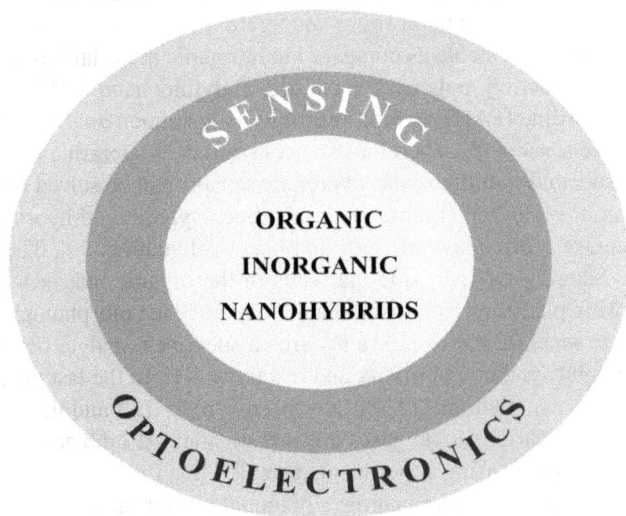

FIGURE 10.1 Outline of sensing and optoelectronics aspects of organic–inorganic nanohybrids.

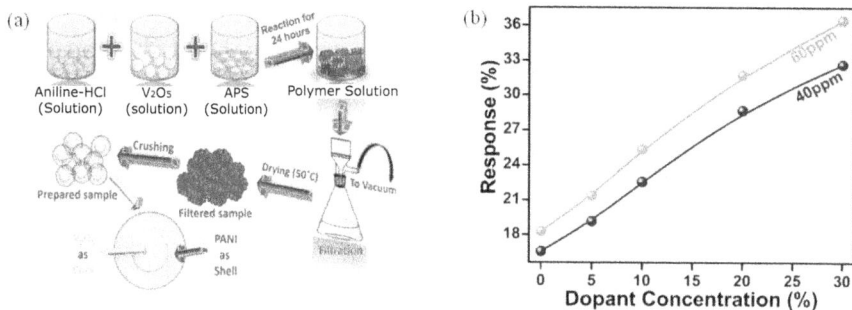

FIGURE 10.2   (a) Synthesis route followed for the PANI/V_2O_5 nanohybrids and (b) enhancement in response (%) as a function of doping concentration [22].

When V_2O_5 nanoparticles are introduced in the PANI matrix up to 30 wt%, the response gets enhanced from 18.27% to 36.41% at 60 ppm level of methanol (as shown in Figure. 10.2b). Moreover, the response time decreases from 230 to 170 s, whereas recovery time increases from 190 to 290 s with an increase in V_2O_5 nanoparticles loading concentration.

10.1.1.2   Organic/SnO$_2$ Hybrid Gas Sensor

Recently, a number of research groups have reported the development of SnO$_2$ nanoparticles embedded PANI hybrid gas sensors (Figure 10.3a,b). The SnO$_2$ or TiO$_2$ nanoparticle-embedded PANI matrix was used as ultrathin films for CO gas detection [23]. The repeatability, stability, and dependability of these ultrathin nanocomposite films for CO gas sensing suggested that the fabricated sensors would be appropriate for real-time detection. Moreover, the inorganic/organic hybrids such as PPy/SnO$_2$, PTP/SnO$_2$, and PANI/SnO$_2$ are being developed with various types of morphologies. By using the hydrothermal process, the PANI/SnO$_2$ hybrid sensor was fabricated. At various temperatures (30°C, 60°C, and 90°C), these fabricated sensors are utilized to detect ethanol and acetone vapours. It was discovered that the hybrid material of PANI and SnO$_2$ (3 wt%) only responded to vapours when operated at 60°C and 90°C. At 90°C, the responses directly deal with the ppm level of ethanol and acetone.

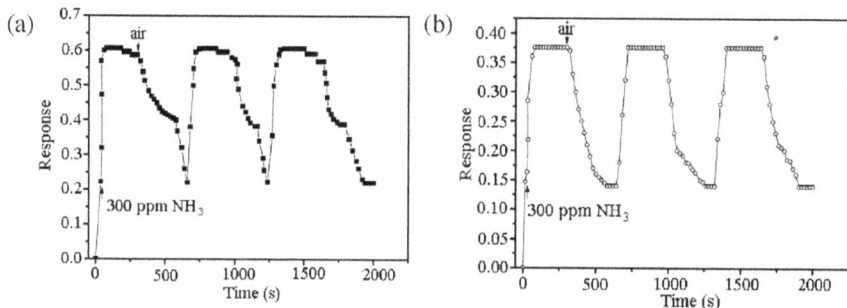

FIGURE 10.3   Variation in resistance for (a) PANI/SnO$_2$ nanohybrid and (b) PANI sample [23].

By following in-situ polymerization synthesis route, PTP/SnO$_2$ hybrid-based sensors were fabricated for the detection of NO$_x$, ethanol, methanol, and acetone at various temperatures. It was discovered that PTP/SnO$_2$ materials could detect NO$_x$ molecules with considerably efficient response and have selectivity even at low temperatures.

10.1.1.3 Strain Sensor

Gu et al. designed an organic/inorganic hybrid hydrogel strain sensor for application in highly flexible soft electronic devices [24]. Polyanionpolyacrylic acid (PAA) hydrogel acts as the organic part and barium ferrite (BaFe$_{12}$O$_{19}$) nanoparticles (NPs) act as the inorganic part of the hybrid hydrogel. To fabricate the hybrid hydrogel, crystallized BaFe$_{12}$O$_{19}$ nanoparticles are synthesized by the sol gel and self-propagating combustion method. Further, in-situ polymerization of acrylic acid monomer in aqueous suspension of BaFe$_{12}$O$_{19}$ results in the formation of BaFe$_{12}$O$_{19}$/PAA hydrogel hybrid. BaFe$_{12}$O$_{19}$/PAA hydrogel hybrid shows the optimum values of mechanical strength and flexibility (Figure 10.4a). Scanning electron microscopy (SEM) images of the synthesized hydrogel depict the smooth surface of BaFe$_{12}$O$_{19}$ with an average size of

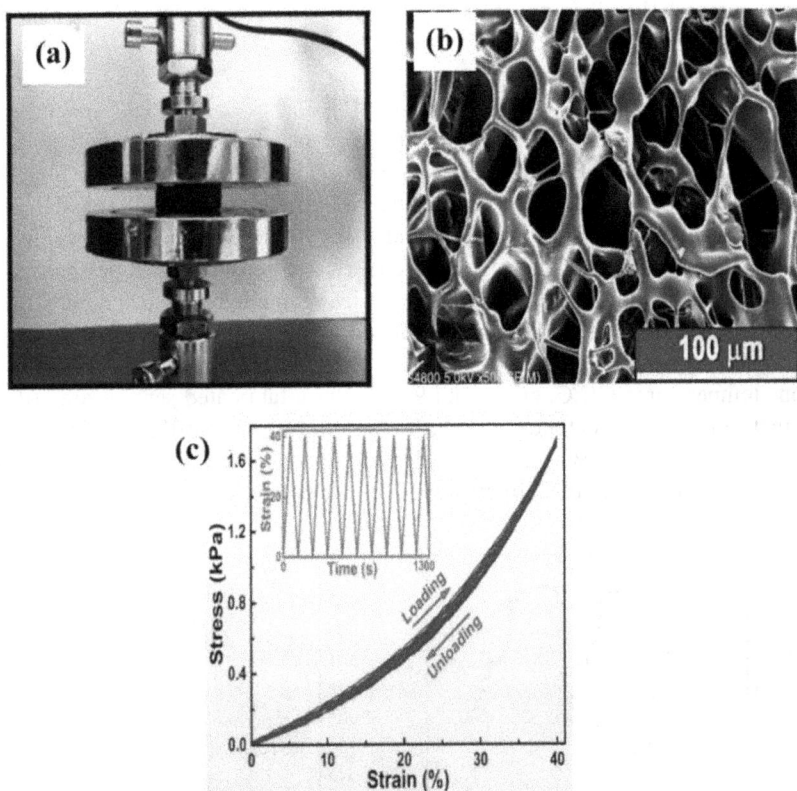

FIGURE 10.4   (a) BaFe$_{12}$O$_{19}$/PAA compression process under 133N strain, (b) SEM micrographs with BaFe$_{12}$O$_{19}$ NPs loading of 0.5 wt%, and (c) compressive strain curves with BaFe$_{12}$O$_{19}$ NPs loading of 0.3 wt% up to 40% strain rate at 5 mm/min. The inset shows the corresponding strain as a function of time [24].

60–80 nm (Figure 10.4b). The inset of Figure 10.4c shows the response of sensor as a function of strain up to 40% at 5 mm/min compression rate under 10 cyclic compressions for $BaFe_{12}O_{19}$/PAA hydrogels and the corresponding stress–strain curve is plotted in Figure 10.4c. It is also observed that the electrical conductivity of the hybrid hydrogel increases with increasing strain as a result of the formation of conductive networks during compression loading. Once the stress is released, the resistance of hydrogel returns to the initial high value due to its elasticity. In conclusion, $BaFe_{12}O_{19}$/PAA hydrogel has proven to be an outstanding material for strain sensing and hence can be utilized in designing hi-tech flexible wearable soft electronic devices.

10.1.1.4   ODBA–ZnO Sensor for CO Sensing

Mandal et al. synthesized an organic/inorganic nanohybrid for the fabrication of a sensor detecting CO at a relatively low-temperature environment (125°C). Organo-di-benzoic-acid (ODBA) forms the organic part and zinc oxide (ZnO) forms the inorganic part of the nanohybrid system [25]. The sensor fabricated from this hybrid, on exposure to CO (5–500 ppm) shows a noticeable increase in its resistance as a result of the formation of complex ions. In addition, the time taken by the sensor to reach saturated response is recorded as 91 s for 100 ppm of CO and 175 s is recorded as the recovery time. Four resultant hybrid powders having different concentrations of ODBA are synthesized, named ZnO NPs, OZ10, OZ50, and OZ100. The synthesized ODBA–ZnO hybrid material is highly porous and comprises a net-like hierarchical structure. To fabricate the sensor, the synthesized ODBA–ZnO powder is dispersed in DI water to form a suspension. This suspension is deposited on the inter-digital electrodes realized on corning glass substrates by drop-casting technique followed by calcination at 120°C for overnight in air (Figure 10.5a–d). All four types of fabricated hybrid sensors are investigated for CO sensing properties. The response of these sensors when exposed to 100 ppm of CO is recorded over a varying range of temperature up to 350°C and the results are shown in Figure 10.5e. The sensing response shown by OZ10, OZ50, and OZ100 is 28%, 35%, and 18%, respectively. It is concluded that the OZ50 sensor shows the best response among all four sensors. In parallel to the temperature dependency, the response of the sensor also depends on the concentration of CO gas.

Further, OZ50 sensor is exposed to varying concentrations of CO for a specific period of time and the recorded data with varying CO concentration v/s sensor response is shown in Figure 10.5f. It is observed that as the concentration of CO increases, the response also increases and shows saturation at higher concentrations. In conclusion, the fabricated ODBA–ZnO hybrid sensor shows remarkable CO sensing properties at a low optimal temperature of 125°C.

10.1.2   ORGANIC–INORGANIC NANOHYBRIDS FOR OPTOELECTRONIC DEVICES

Here, we discuss the potential nano-composites used for bringing out their applications in optoelectronics.

10.1.2.1   ZnO-Graphene Quantum Dots (GQDs) Nanocomposite

Aaryashree et al. reported a comparative analysis of the optical and structural properties of ZnO, GQDs, and ZnO/GQDs hybrid nanocomposite films [26].

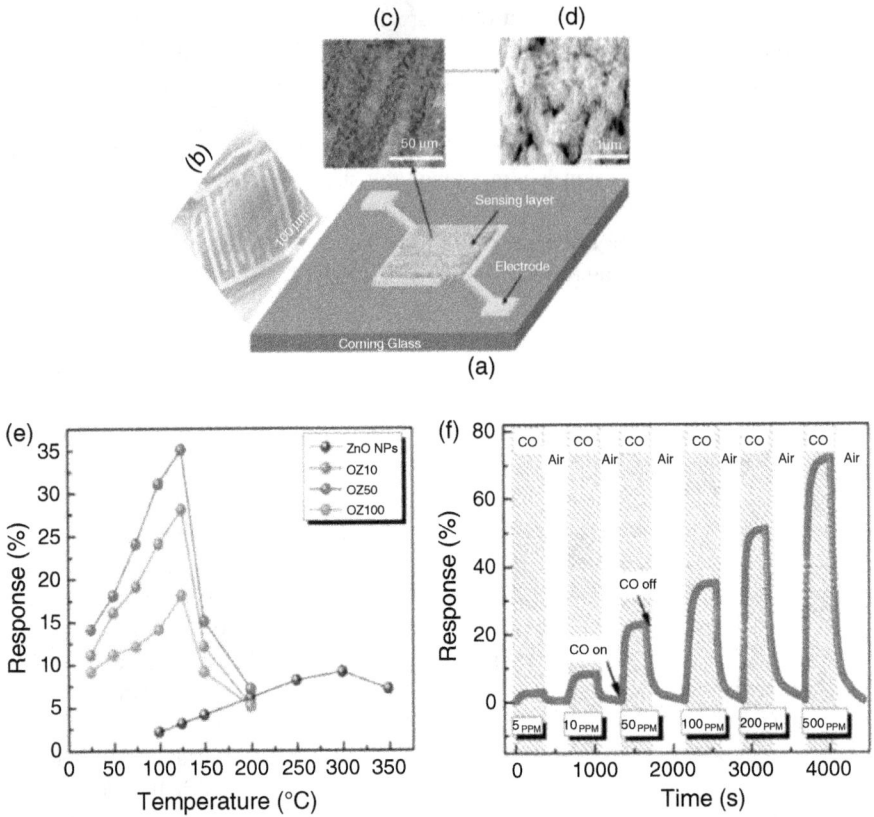

FIGURE 10.5 (a) The fabricated sensor. FESEM (field emission scanning electron microscope) images of (b) electrodes and (c) sensing layer along with its (d) magnified view. (e) Sensors response as a function of temperature. (f) Dynamic response of OZ50 sensor at different CO concentrations [25].

This hybrid nanocomposite of ZnO/GQDs shows vast optoelectronic applications. GQDs form the organic part and ZnO forms the inorganic part of the hybrid. Further studies are conducted to understand the properties of electrodeposited ZnO/GQD hybrid and electrodeposited ZnO and GQDs' film separately as well. The addition of GQDs does not show any significant change in the properties of ZnO; however, a shift in the photoluminescence (PL) peak of ZnO can be observed with some changes in optical and electrical properties. The process of electrochemical deposition is used to deposit thin films of ZnO/GQDs hybrid on the desired substrate. Further, the morphology of the films grown is studied by FESEM. The highly magnified images of ZnO, GQD, and ZnO/GQD are shown in Figure 10.6a–c. The photosensitivity of ZnO/GQD hybrid is reported to be 14.67 times of photosensitivity of ZnO and 1.9 times of GQDs, respectively. Afterwards, the optical properties of the hybrid are also studied by the PL spectra. It is observed that at room temperature the PL peak of ZnO is at 380 nm, GQD at 423 nm, and that of ZnO/GQD hybrid is at 370 nm (Figure 10.6d).

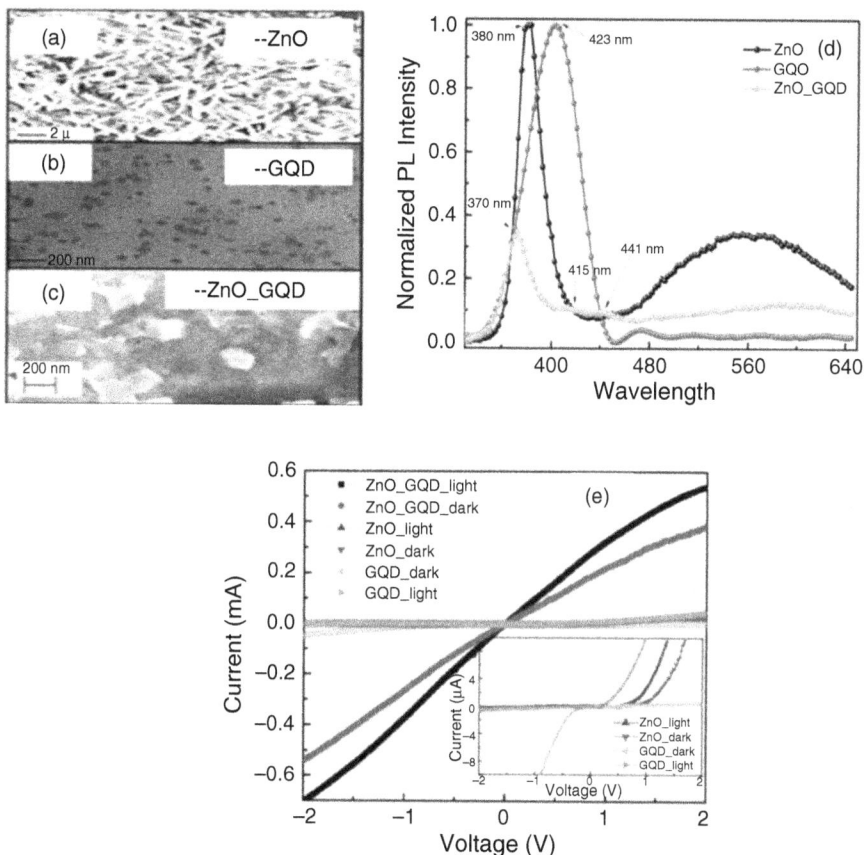

FIGURE 10.6 (a–c) FESEM images of ZnO, GQD, and ZnO/GQD nanocomposite, respectively. (d) PL spectra of ZnO, GQD, and ZnO/GQD nanocomposite at room temperature and (e) I–V characteristics of samples under light and dark conditions at room temperature, I–V feature of ZnO and GQD is shown in the inset [26].

Further, the current–voltage characteristics of the hybrid are also studied under dark and irradiated conditions. The Zn/GQD nanocomposite shows a larger magnitude of current variation with a change in bias voltage in comparison to ZnO and GQD (Figure 10.6e). In conclusion, the ease and cost-effectiveness of growing this hybrid on different substrates along with the different properties of ZnO/GQD hybrid nanocomposite such as high optical band gap improved current sensitivity, enhanced photosensitivity (in comparison to individual films of ZnO and GQDs), and eco-friendly nature make it an ideal material for applications in UV (ultraviolet)–Visible optoelectronic devices.

10.1.2.2 Polyaniline/ZnSe Nanocomposite

Shokr et al. synthesized a novel organic/inorganic hybrid compound of PANI/ZnSe nanocomposite for the potential application in optoelectronic devices [27]. PANI having a semicrystalline structure forms the organic part and ZnSe having a cubic crystal structure forms the inorganic part of the hybrid. The addition of ZnSe into the PANI matrix results in an increase in the particle size. It is observed that the optical band gap of PANI decreases with an increase in ZnSe concentration in the hybrid, which results in an enhanced DC electrical conductivity with a decrease in activation energy. In addition, an increase in the ZnSe concentration increases the mobility and number of charge carriers but decreases the potential barrier of the hybrid. Further, it is observed that holes are the majority charge carriers in the hybrid. Initially, each component of the hybrid (ZnSe and PANI) is synthesized separately. ZnSe nanoparticles with a particle size of 8 nm are prepared by colloidal technique and PANI is prepared by carrying specific reactions. Further, PANI and ZnSe (10–50 wt% of ZnSe nanoparticles (NPs)) are mixed in appropriate amounts in ethanol and sonicated for 5 h to form a homogenous suspension. Afterwards, to obtain uniform films of the hybrid, the suspension is spin coated on quartz substrates at 3000 rpm for 1 min, followed by drying the films at 90°C in an electric oven for 2 h under vacuum conditions. Further, the synthesized hybrid is characterized to study its structure, morphology, and optical behaviour. SEM (scanning electron microscope) images (Figure 10.7a) show the micrographs of hybrid at 50 wt% concentration of ZnSe. The optical band

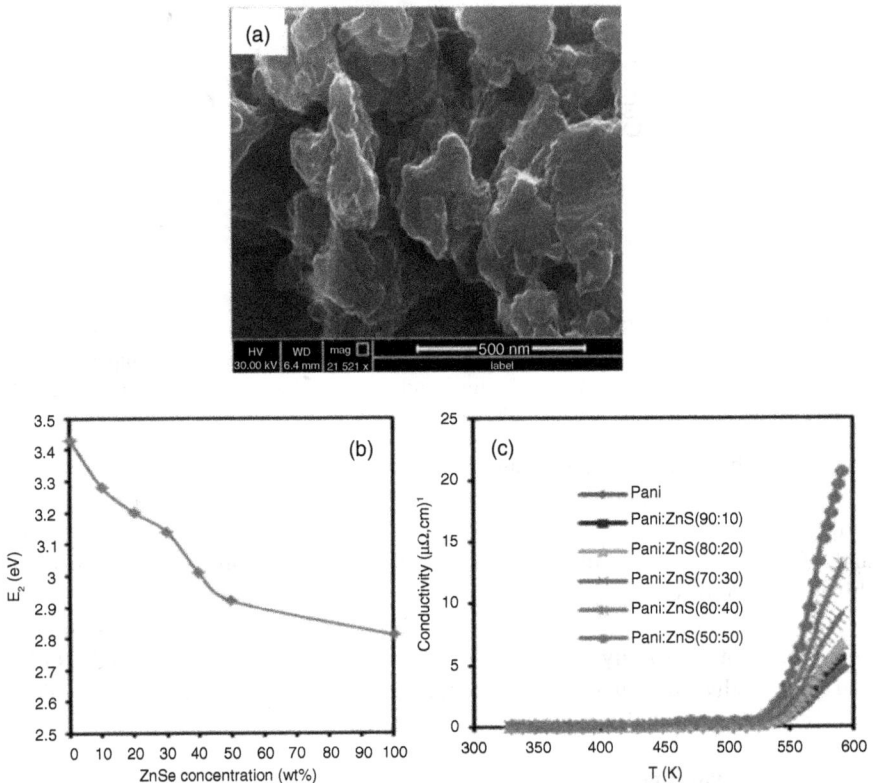

FIGURE 10.7 (a) SEM micrograph polyaniline:ZnSe (50:50). (b) Bandgap variation with ZnSe concentration and (c) temperature-conductivity relation for different ZnSe concentrations [27].

gap of PANI is influenced by the doping of ZnSe NPs. It is observed that the doping of PANI with 50 wt% of ZnSe results in a decrease in its band gap from 3.43 to 2.92 eV.

Hence, high optical absorption and low band gap collectively enhance the properties of hybrid and make it efficient for solar cell application (Figure 10.7b). At a fixed temperature, the conductivity of the hybrid is increased with an increasing concentration of ZnSe NPs, embedded in polymer matrix. Hence, Figure 10.7c shows the effect of temperature on DC conductivity for undoped and PANI-doped ZnSe at various concentrations and the best results among all is shown by PANI:ZnSe (50:50) hybrid. In conclusion, the synthesized hybrid shows optimum properties and is a suitable material to be utilized in the solar cell and other optoelectronic applications.

10.1.2.3 TiO$_2$ Nanotubes/Porphyrin Nanoparticles Hybrid

Chen et al. fabricated an inorganic/organic nanohybrid with enhanced photoelectrochemical (PEC) properties for high-performance optoelectronic applications [28]. The inorganic part is an array of one-dimensional TiO$_2$ nanotubes (TiNTs) with rapid charge transport and high charge carrier mobility. On the other hand, the organic part is formed by porphyrin nanoparticles (TPPs) having good solution processability and extensive visible light absorption properties. The weak electron hole binding of the inorganic part results in low light absorption efficiency, whereas the organic part shows enhanced UV-Visible absorption. Therefore, in comparison to individual constituents, the hybrid shows better solar absorption and enhanced photocurrent generation and is suitable for optoelectronics applications. The fabrication of hybrid starts with the preparation of TiNTs and TPP NPs separately. Further, a thin foil of Ti is rinsed ultrasonically with acetone, ethanol, and DI water in a sequence. Afterwards, anodization is performed under a constant applied voltage of 60 V for 30 min followed by drying, rinsing, and annealing processes. In parallel to this, TPP NPs colloid is prepared by the solvent exchange method. Further, to fabricate the inorganic/organic hybrid, the prepared anodic TiNTs are dropped in dispersion of TPP NPs, followed by drying in vacuum oven at 100°C for 1 h. Repeating the dipping and drying process results in the formation of different hybrids denoted by I, II, III, IV, and V. The surface morphology and structure of the hybrid fabricated by coating TiNTs with different amounts of TPP NPs are studied by the SEM images (Figure 10.8a–e). The hybridized TPP NPs are dispersed evenly on the TiNTs surface, as observed in Figure 10.8a. Also, the surface pores become narrow and finally are covered due to the growing loading of TPP NPs, which results in an increase in the size of NPs coated on TiNTs. Figure 10.8f represents the photocurrent density of the hybrid as a result of applied potential in order to investigate the PEC performance. Different hybrids (I, II, III, IV, and V) show a variation in performance in comparison to each other. As observed from the graph, upon illumination, hybrid III shows the largest anodic current generation across applied potentials followed by hybrid II. In addition, hybrids IV and V generate lower photocurrent. It is also observed that the current density of hybrid I is much lower than that of TiNTs. In comparison to all the samples, TPP NPs show the lowest current density of about 0.01 mA/cm^2 due to the low charge carrier mobility of organic material. Hence, the overall PEC performance of TiNTs is improved by moderate coverage of TPP NPs over TiNTs. Further, the absorption spectra of the sample are studied by UV-Vis spectroscopy. TPP shows wide absorption in the visible light region as well as a common porphyrin S-band is observed at 430 nm.

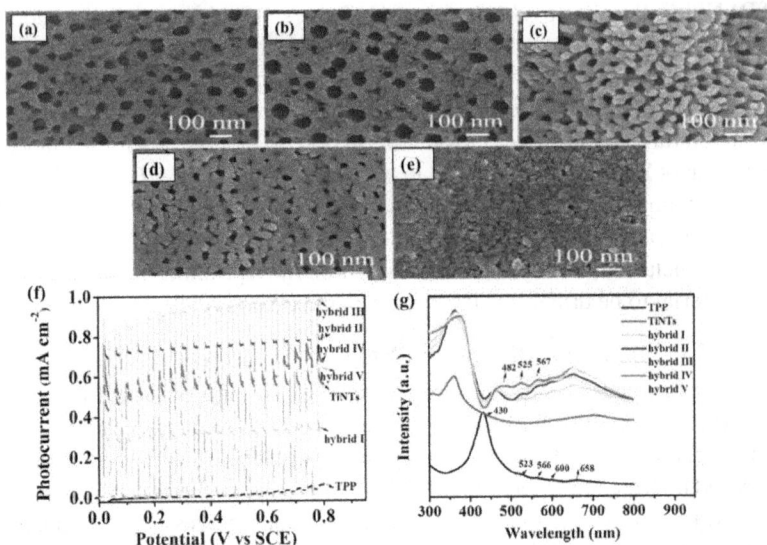

FIGURE 10.8 SEM images of hybrids (a) I, (b) II, (c) III, (d) IV, and (e) V. (f) Photocurrent density v/s applied potential for synthesized samples in 1 mmol/L NaOH solution and (g) UV-Visible absorption spectra of TiNTs, TPP, and TiO$_2$/TPP hybrids [28].

In addition, four weak absorption peaks at 523, 566, 600, and 652 nm are also observed due to Q-band. The hybrid shows characteristic absorption peaks of TiO$_2$ and TPP. It is observed that TiNTs show an absorption edge around 400 nm as well as its absorption spectrum is extended to the visible region (Figure 10.8g). As a result, the hybrid shows greater absorption intensity in comparison to TiNTs and TPP NPs.

Hence, in the visible light region, the hybrid is capable of providing a large number of photocharges that are required for PEC applications. Thus, the fabricated inorganic/organic hybrid's excellent response to the visible light as well as high photocurrent generation make it an ideal material to be utilized as photoanode in PEC water splitting. In addition to this, the hybrid opens new pathways towards the fabrication of devices for high-performance optoelectronic applications.

10.1.3 Conclusions

Here, we have shown a substantial progress in the emerging fields of sensing and optoelectronic devices based on organic–inorganic nanohybrids. This is being accelerated by facilitating potential research outcomes in this direction. We are pretty hopeful that these endeavours will be helpful to flourish advancements to develop multifunctional material templates.

REFERENCES

1. Parvatikar, Narsimha, Shilpa Jain, Syed Khasim, M. Revansiddappa, S. V. Bhoraskar, and M.V.N. Ambika Prasad. "Electrical and humidity sensing properties of polyaniline/ WO3 composites." *Sensors and Actuators B: Chemical* 114, no. 2 (2006): 599–603, https://doi.org/10.1016/j.snb.2005.06.057.

2. Navale, Sachin T., Sanjit Manohar Majhi, Ali Mirzaei, Hyoun Woo Kim, and Sang Sub Kim. "Metal oxide ceramic gas sensors." (2023). https://doi.org/10.1016/B978-0-12-819728-8.00083-8.

3. Pal, Rishi, Sneh Lata Goyal, Ishpal Rawal, and Smriti Sharma. "Efficient room temperature methanol sensors based on polyaniline/graphene micro/nanocomposites." *Iranian Polymer Journal* 29 (2020): 591–603, https://doi.org/10.1007/s13726-020-00822-8.

4. Zhang, Chao, Kaichun Xu, Kewei Liu, Jinyong Xu, and Zichen Zheng. "Metal oxide resistive sensors for carbon dioxide detection." *Coordination Chemistry Reviews* 472 (2022): 214758, https://doi.org/10.1016/j.ccr.2022.214758.

5. Sharma, Asha, Anoop Singh, Vinay Gupta, Ashok K. Sundramoorthy, and Sandeep Arya. "Involvement of metal organic frameworks in wearable electrochemical sensor for efficient performance." *Trends in Environmental Analytical Chemistry* (2023): e00200, https://doi.org/10.1016/j.teac.2023.e00200.

6. Yang, Chia-Ming, Yu-Cheng Yang, Bing-Huang Jiang, Jiun-Han Yen, Xuan-Ming Su, and Chih-Ping Chen. "An organic semiconductor obtained with a low-temperature process for light-addressable potentiometric sensors." *Sensors and Actuators B: Chemical* 381 (2023): 133449, https://doi.org/10.1016/j.snb.2023.133449.

7. Zhu, Yangyang, Yiqun Zhang, Jiajia Yu, Chengren Zhou, Chaojie Yang, Lu Wang, Li Wang, Libo Ma, and Li Juan Wang. "Highly-sensitive organic field effect transistor sensors for dual detection of humidity and NO2." *Sensors and Actuators B: Chemical* 374 (2023): 132815, https://doi.org/10.1016/j.snb.2022.132815.

8. Ma, Xingfa, Mang Wang, Guang Li, Hongzheng Chen, and Ru Bai. "Preparation of polyaniline-TiO2 composite film with in situ polymerization approach and its gas-sensitivity at room temperature." *Materials Chemistry and Physics* 98, no. 2–3 (2006): 241–247, https://doi.org/10.1016/j.matchemphys.2005.09.027.

9. Chuang, Feng-Yi, and Sze-Ming Yang. "Titanium oxide and polyaniline core-shell nanocomposites." *Synthetic Metals* 152, no. 1–3 (2005): 361–364, https://doi.org/10.1016/j.synthmet.2005.07.299.

10. Ulman, Abraham. *An Introduction to Ultrathin Organic Films: From Langmuir–Blodgett to Self–Assembly.* Academic Press, Cambridge, MA, 2013.

11. Nardis, Sara, Donato Monti, Corrado Di Natale, Arnaldo D'Amico, Pietro Siciliano, Angiola Forleo, Mauro Epifani, Antonella Taurino, Roberto Rella, and Roberto Paolesse. "Preparation and characterization of cobalt porphyrin modified tin dioxide films for sensor applications." *Sensors and Actuators B: Chemical* 103, no. 1–2 (2004): 339–343, https://doi.org/10.1016/j.snb.2004.04.063.

12. Kong, Fanhong, Yan Wang, Jun Zhang, Huijuan Xia, Baolin Zhu, Yanmei Wang, Shurong Wang, and Shihua Wu. "The preparation and gas sensitivity study of polythiophene/SnO2 composites." *Materials Science and Engineering: B* 150, no. 1 (2008): 6–11, https://doi.org/10.1016/j.mseb.2008.01.003.

13. Xu, Mijuan, Jun Zhang, Shurong Wang, Xianzhi Guo, Huijuan Xia, Yan Wang, Shoumin Zhang, Weiping Huang, and Shihua Wu. "Gas sensing properties of SnO2 hollow spheres/polythiophene inorganic-organic hybrids." *Sensors and Actuators B: Chemical* 146, no. 1 (2010): 8–13, https://doi.org/10.1016/j.snb.2010.01.053.

14. Taylor, Arthur, Katie M. Wilson, Patricia Murray, David G. Fernig, and Raphael Levy. "Long-term tracking of cells using inorganic nanoparticles as contrast agents: Are we there yet?" *Chemical Society Reviews* 41, no. 7 (2012): 2707–2717, https://doi.org/10.1039/C2CS35031A.

15. Naoi, Katsuhiko, Wako Naoi, Shintaro Aoyagi, Jun-ichi Miyamoto, and Takeo Kamino. "New generation "Nanohybrid Supercapacitor"." *Accounts of Chemical Research* 46, no. 5 (2013): 1075–1083, https://doi.org/10.1021/ar200308h.

16. Gogoi, K. K., and A. Chowdhury. "Organic-inorganic nanohybrids for low-powered resistive memory applications." *Journal of Physics: Conference Series* 1706, no. 1 (2020): 012010, https://doi.org/10.1088/1742-6596/1706/1/012010.

17. Kang, Dong Jun, Jong-Pil Jeong, and Byeong-Soo Bae. "Direct photofabrication of focal-length-controlled microlens array using photoinduced migration mechanisms of photosensitive sol-gel hybrid materials." *Optics Express* 14, no. 18 (2006): 8347–8353, https://doi.org/10.1364/OE.14.008347.

18. Han, Xiao, Yongshen Zheng, Siqian Chai, Songhua Chen, and Jialiang Xu. "2D organic-inorganic hybrid perovskite materials for nonlinear optics." *Nanophotonics* 9, no. 7 (2020): 1787–1810, https://doi.org/10.1515/nanoph-2020-0038.

19. Yao, Hong-Bin, Min-RuiGao, and Shu-Hong Yu. "Small organic molecule templating synthesis of organic-inorganic hybrid materials: their nanostructures and properties." *Nanoscale* 2, no. 3 (2010): 322–334, https://doi.org/10.1039/B9NR00192A.

20. Kumar, Deepak, Koijam Monika Devi, Ranjan Kumar, and Dibakar Roy Chowdhury. "Dynamically tunable slow light characteristics in graphene based terahertz metasurfaces." *Optics Communications* 491 (2021): 126949, https://doi.org/10.1016/j.optcom.2021.126949.

21. Han, Song-De, Ji-Xiang Hu, Jin-Hua Li, and Guo-Ming Wang. "Anchoring polydentate N/O-ligands in metal phosphite/phosphate/phosphonate (MPO) for functional hybrid materials." *Coordination Chemistry Reviews* 475 (2023): 214892, https://doi.org/10.1016/j.ccr.2022.214892.

22. Pal, Rishi, Sneh Lata Goyal, and Ishpal Rawal. "Selective methanol sensors based on polyaniline/V2O5 nanocomposites." *Iranian Polymer Journal* 31, no. 4 (2022): 519–532, https://doi.org/10.1007/s13726-022-01024-0.

23. Wang, Shurong, Yanfei Kang, Liwei Wang, Hongxin Zhang, Yanshuang Wang, and Yao Wang. "Organic/inorganic hybrid sensors: A review." *Sensors and Actuators B: Chemical* 182 (2013): 467–481, https://doi.org/10.1016/j.snb.2013.03.042.

24. Gu, Hongbo, Hongyuan Zhang, Chao Ma, Hongling Sun, Chuntai Liu, Kun Dai, Jiaoxia Zhang, Renbo Wei, Tao Ding, and Zhanhu Guo. "Smart strain sensing organic-inorganic hybrid hydrogels with nano barium ferrite as the cross-linker." *Journal of Materials Chemistry C* 7, no. 8 (2019): 2353–2360, https://doi.org/10.1039/C8TC05448G.

25. Mandal, Biswajit, Sayan Maiti, Aaryashree, Gaurav Siddharth, Mangal Das, Ajay Agarwal, Apurba K. Das, and Shaibal Mukherjee. "Organo-di-benzoic-acidified ZnO nanohybrids for highly selective detection of CO at low temperature." *The Journal of Physical Chemistry C* 124, no. 13 (2020): 7307–7316, https://doi.org/10.1021/acs.jpcc.0c01044.

26. Aaryashree, Biswas, Sagar, Pankaj Sharma, Vishnu Awasthi, Brajendra S. Sengar, Apurba K. Das, and Shaibal Mukherjee. "Photosensitive ZnO-graphene quantum dot hybrid nanocomposite for optoelectronic applications." *ChemistrySelect* 1, no. 7 (2016): 1503–1509, https://doi.org/10.1002/slct.201600149.

27. Shokr, F. S., and S. A. Al-Gahtany. "Synthesis, characterization, and charge transport mechanism of polyaniline/ZnSenanocomposites for promising optoelectronic applications." *Polymer Composites* 39, no. 5 (2018): 1724–1730, https://doi.org/10.1002/pc.24123.

28. Chen, Yingzhi, Aoxiang Li, Ming Jin, Lu-Ning Wang, and Zheng-Hong Huang. "Inorganic nanotube/organic nanoparticle hybrids for enhanced photoelectrochemical properties." *Journal of Materials Science & Technology* 33, no. 7 (2017): 728–733, https://doi.org/10.1016/j.jmst.2016.08.030.

11 Role of Triplet–Triplet Annihilation Mechanism in Molecular Recognition

Sumit Kumar Panja

11.1 INTRODUCTION

The triplet state is considered an excited quantum state wherein the same spin state of unpaired electrons can co-exist within molecules (Figure 11.1a). The triplet excited states in organic molecules show interesting photophysics and a wide range of applications in various research fields.[1-4] The triplet states of organic chromophore have two unpaired electrons at different energy states and show the paramagnetic character in the excited state.[5] Triplet excited states are generally formed from singlet excited states through the non-radiative intersystem crossing (ISC) process (Figure 11.1a). Organic chromophores with triplet excited states character show special applications for photovoltaics,[6,7] photocatalysis,[7,8] and photodynamic therapy.[9,10]

Photon upconversion via triplet–triplet annihilation (TTA-UC) plays an important role in emerging artificial light-harvesting systems. A typical TTA-UC system is composed of chromophores (annihilator) and sensitizers having the triplet state. After excitation of sensitizer at low energy, the energy of the sensitizer is

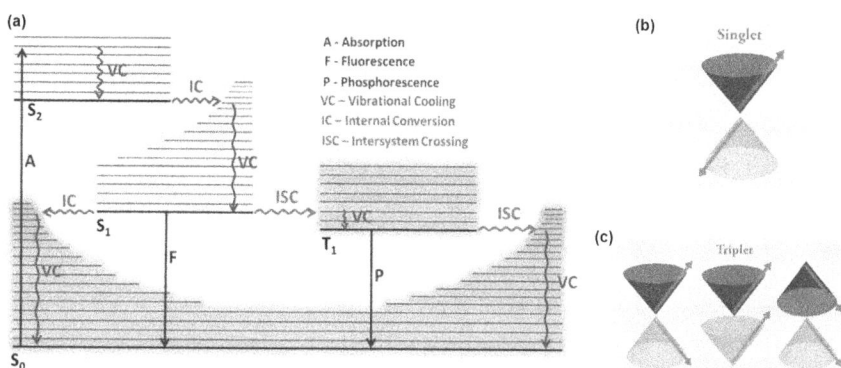

FIGURE 11.1 (a) Jablonski diagram indicating the various radiative (fluorescence, F; and phosphorescence, P) and non-radiative transitions (ISC, IC, VC). Electron spin diagrams of (b) singlet (out-of-phase) and (c) triplet (in-phase) states represented with respect to the direction of the spin vectors.[11] (Reproduced from ref. [11] with permission from Royal Society Chemistry, copyright@ 2020.)

DOI: 10.1201/9781003352372-11

transferred via the triplet state to the annihilator (chromophore). After annihilation of two sensitized annihilators (chromophore having the triplet state), anti-Stokes delayed fluorescence is observed at higher energy (Figures 11.2–11.4).[6] In the TTA-UC process, the triplet energy (Sensitizer and annihilator) transfer occurs via the Dexter energy transfer (DET) mechanism. DET mechanism is mainly based on non-radiative electron exchange between the overlapping wave functions of molecules within 10 Å. In the TTA-UC photochemical process, low-energy photons (two photons) are converted to high-energy photons (one photon) via triplet–triplet DET processes between annihilated molecules (Figure 11.3).[12] TTA-UC has opened up the various research fronts including, inorganic–organic hybrid systems,[13,14] oxygen sensitivity,[15] and applications beyond photovoltaics.[6] However, near-infrared (NIR) to Vis TTA-UC is very appealing for several practical photonic and bioimaging applications. Body tissues are unable to absorb NIR radiation; therefore, NIR probe is useful in probing biological systems for imaging purposes. Furthermore, to increase the efficiency and operation of both semiconductor-based and dye-sensitized solar cells in the visible region, the broad NIR region is highly desirable.[16] Due to a deep penetration depth of NIR light, high-energy photons via TTA-UC can be generated from NIR probe for efficient photocatalysis with higher yields compared to direct sensitization.[8] Therefore, development and applications of NIR-to-Vis TTA-UC systems are reported and explained on the basis of chromophores, sensitization, energy transfer mechanisms, hybrid systems, etc. Existing challenges and future directions of recent research are discussed as a guiding pathway for further advancement of research.

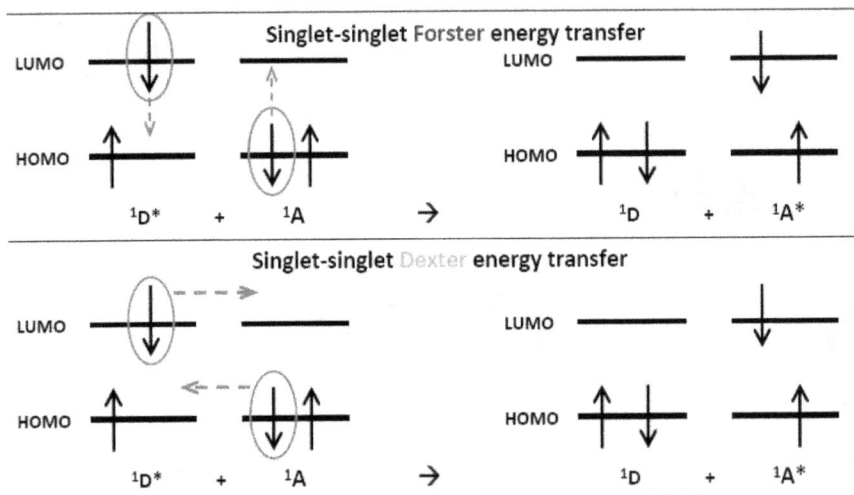

FIGURE 11.2 Singlet–singlet Foster and Dexter energy transfer mechanism.

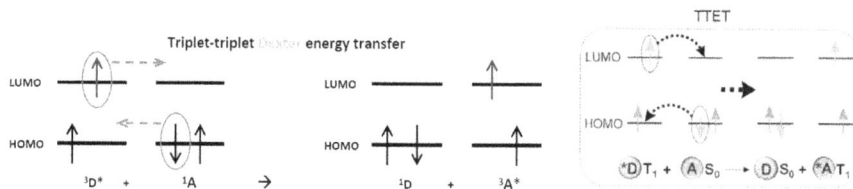

FIGURE 11.3 Triplet–triplet Dexter energy transfer mechanism. (Reproduced from ref. [1] with permission from Royal Society Chemistry, copyright@ 2020.)

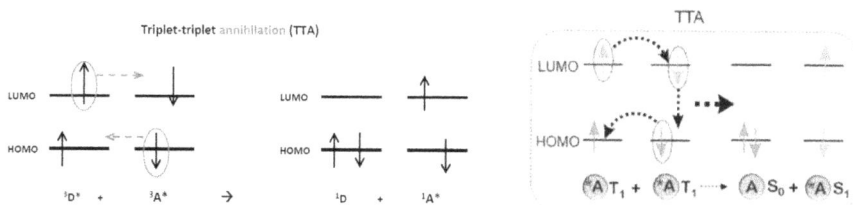

FIGURE 11.4 Triplet–triplet anhilation (TTA) energy transfer mechanism. (Reproduced from ref. [1] with permission from Royal Society Chemistry, copyright@ 2020.)

11.2 BACKGROUND OF FOSTER AND DEXTER ENERGY TRANSFER PROCESS

The intermolecular energy transfer process is observed without emitting a photon when the interaction between excited molecule (D*) and ground-state molecule (A). From the Förster energy transfer model, it could be stated that the energy released from an excited molecule (excited donor molecule: D*) could simultaneously excite the ground-state molecule (ground-state acceptor: A) via Coulombic interaction. Independently, David L. Dexter provided another mechanism that an excited molecule (excited donor molecule: D*) and a ground-state molecule (ground-state acceptor: A) might indeed exchange electrons to accomplish the non-radiative process associated with fluorescence quenching. Bilaterally exchange of electrons is observed in the DET mechanism. Unlike the sixth-power dependence of Förster energy transfer, the reaction rate constant of DET exponentially decays as the distance between these two parties increases. Dexter energy exchange mechanism typically follows the exponential relationship to the distance within 10 Å. Hence, the Dexter energy exchange mechanism is also considered a short-range energy transfer mechanism. Based on the Wigner spin conservation rule, the spin-allow process could be represented as follows:

Singlet–singlet energy transfer: $^1D^* + {}^1A \rightarrow {}^1D + {}^1A^*$

It could be better understood by considering that a singlet group could produce another singlet group.

Triplet–triplet energy transfer: $^3D^* + {}^1A \rightarrow {}^1D + {}^3A^*$

It is very clear that the singlet–singlet energy transfer process is feasible via Coulombic interaction but triplet–triplet exchange is not possible because of violation of the Wigner spin conservation law. The TTA is an example of an energy exchange process between two excited triplet molecules. Two triplet molecules (excited triplet state donor molecule: D* and excited triplet state acceptor molecule: A*) react with each other to produce two singlet states (ground single state molecule and excited singlet state molecule). Generally, the energy difference between S_0 and T_1 is larger than the energy gap between T_1 and S_1. When two triplet excited-state molecules are combined, two singlet state molecules are generated. These two singlet state molecules have different energy. One singlet state molecule remains at the ground state and another molecule remains at the excited state. This higher singlet excited-state molecule can show the fluorescence property. The energy level that the electron occupies could be twice the lowest triplet energy gap after the annihilation process. It is called triplet–triplet annihilation upconversion (TTA-UC) process.

In the typical TTA-UC, the singlet excited donor (sensitizer: $^DS_1^*$) could cross the spin barrier and move to the triplet state ($^DT_1^*$) via ISC pathway in the presence of strong spin–orbit coupling. The singlet excited donor (sensitizer) transfers its triplet energy to the organic molecule (annihilator) by triplet–triplet energy transfer (TTET) pathway, followed by molecular diffusion of a sensitized triplet organic molecule (annihilated molecule).

When two such sensitized triplet organic molecules (annihilated molecule) collide in the space-time ($^AT_1^* \leftrightarrow {}^AT_1^*$), the TTA may result in a higher energy singlet excited state ($^AS_1^* = 2 \, ^AT_1^*$) and shows the anti-Stokes delayed fluorescence (Figure11.3). In TTA-UC process, the emission energy band appeared at a higher energy level compared to the lower excitation energy level, resulting in unique practical application research fields (Figure 11.5).

TTA-UC is a classic example of nonlinear optical process, where triplet excited states of organic chromophores produce higher energy emissive singlet state molecule via triplet–triplet annihilated energy transfer process. Since the triplet energy

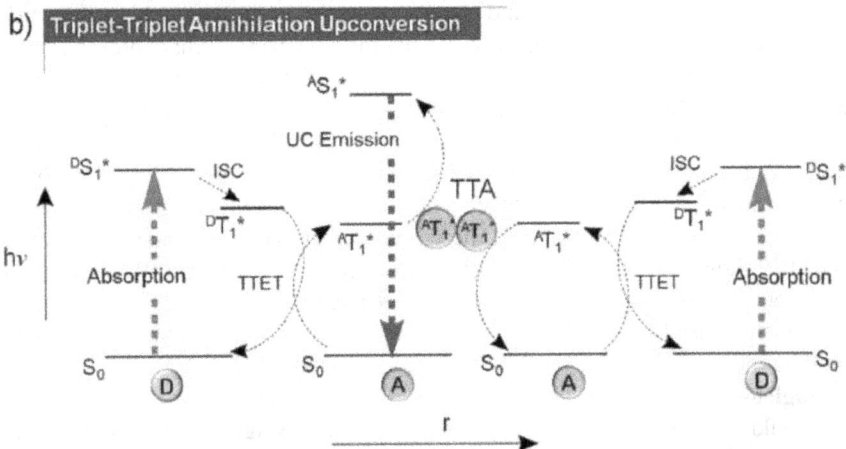

FIGURE 11.5 Triplet–triplet anhilation upconversion (TTA-UC) and anti-stoke fluorescence. (Reproduced from ref. [1] with permission from Royal Society Chemistry, copyright@ 2020.)

transfer (TET) among excited molecules (excited-state donor and excited-state acceptor molecules) occurs through DET mechanism,[17] the chromophores must be present within the distance of 10 Å with respect to each other (Figure 11.3). The rate of DET (k_{ET}) can be expressed with respect to the distance as shown in Equation (1).[18]

$$k_{ET} \propto J_{exp}\left[\frac{-2r}{L}\right] \qquad (1)$$

where r represents the distance between donor and acceptor, L is the sum of van der Waals' radii of donor and acceptor, and J is the spectral overlap integral defined by Equation (2):

$$J = \int f_D(\lambda)\varepsilon_A(\lambda)\lambda^4\,d\lambda \qquad (2)$$

where λ is the wavelength and ε is the molar absorption coefficient. The distance between the donor and the acceptor is the key for effective overlap of wavefunctions for an electron exchange to occur.

11.3 REQUIREMENTS FOR THE SENSITIZER AND ACCEPTOR/ANNIHILATOR MOLECULE

The sensitized TTA involves the energy transfer from triplet sensitizer (donor) molecule to triplet annihilator (acceptor) at the excited state. Later on, two triplet annihilator molecules take part in TTA process and TTA upconversion is observed. Hence, several factors are to be considered for creating an efficient TTA-UC scheme via a proper combination of chromophores and sensitizer. Sensitizers must have the ability to absorb light in the visible-to-near-IR region with low-energy excitation and a long triplet excited-state lifetime, typically on the order of several microseconds and beyond. Further, enabling efficient diffusional-based quenching is also a necessary condition to make an efficient TTA-UC system.

Heavy metal-containing organic molecules having metal-to-ligand charge transfer excited states as sensitizers are generally studied due to the presence of absorption and emission maxima toward the near-IR region. Mainly, heavy metal (Pd or Pt) present in the organic chromophore like porphyrins and phthalocyanines strongly enhances singlet–triplet ISC efficiencies near unity due to creation of strong spin–orbit coupling. Generally, bimolecular quenching of the triplet excited state of the sensitizer happens when the triplet acceptor energy must be lower than the triplet energy of the sensitizer. More favorably, the TET process occurs when the energy difference between the triplet sensitizer and triplet acceptor is high and it becomes the driving force for the triplet (sensitizer) to triplet (annihilator) energy transfer process.

Fluorescence quantum yields of acceptors/annihilators influence the overall upconversion (UC) quantum efficiency. The singlet excited state of sensitizer (donor) molecule lies below that of the acceptor's singlet, but the sensitizer's triplet state should lie above that of the acceptor at the excited state. An important parameter is that the singlet and triplet excited states of the sensitizer should be positioned between the singlet and triplet excited states of the acceptor/annihilator. If these specific energy criteria are met, then only combined triplet energy from two acceptor molecules after TTA is greater than or equal to the acceptor's singlet state energy, resulting in the observation of upconverted fluorescence.

The primary evidence supporting TTA involves the observation of anti-Stokes fluorescence from excited singlet state of the molecule after TTA process with respect to excitation.[19] Interesting factor is that spectral profile of TT-UC fluorescence is identical to the annihilator (acceptor) molecule. The intensity of TT-UC fluorescence displays a quadratic (x2) incident light power dependence of TTA process.

11.4　DESIGN CRITERIA FOR EFFICIENT TRIPLET–TRIPLET ANNIHILATION UPCONVERSION (TTA-UC)

The following factors should be considered for getting efficient chromophores in a viable homogeneous solution based on the TTA-UC mechanism.

I. A large absorption coefficient (ε_{max}) of triplet sensitizer should have the excitation wavelength (in the visible-to-near-IR region of the spectrum), thus allowing the low-energy and low-power photoexcitation of sensitizer.
II. Necessity of a long-lived triplet state of sensitizer with lifetime (>μs).
III. Efficient diffusional-based quenching of sensitizer is highly desirable.
IV. Large spin–orbit coupling coefficients (χ_{so}) of transition metal-based sensitizers are favorable due to the efficient ISC to enhance the triplet state population manifold.
V. To observe bimolecular quenching of the triplet excited state of the sensitizer, the triplet acceptor energy must be lower than the triplet energy of the sensitizer. The greater the energy difference between the triplet sensitizer and triplet acceptor, the greater the driving force for this reaction, and the more favorable the TET process.
VI. The singlet excited state of the sensitizer generally lies below the singlet manifold of the acceptor while the triplet state of the sensitizer should lie above that of the acceptor. In a general sense, the singlet and triplet excited states of the sensitizer should be tactically nuzzled between the singlet and triple excited states of the annihilator.
VII. If the above specific energy criteria are fulfilled, then only combined triplet energy from two annihilator molecules should be greater than or equal to the singlet state energy of annihilator, resulting in the upconverted fluorescence (TTA-UC).

11.5　TTA-UC THEORETICAL LIMIT

After considering the absolute quenching efficiency (Φ_{TTET}, 100%) and the fluorescence quantum yield (Φ_{Fl}, 100%), the theoretical limit for TTA-UC efficiency should be expected to be about 11.1%.[21,22] Triplet–triplet UC fluorescence efficiency can be expressed as

$$\Phi_{UC} = \Phi_{TTET}, \Phi_{TTA}, \Phi_{Fl}$$

From spin statistics, spin-statistical factors play an important role in triplet–triplet UC fluorescence efficiency and arise from the spin state of the triplet acceptors. When interaction of two excited acceptor triplets ($^3A*$) takes place, there can be up to nine excited-state dimers having spin states $^n(AA)*$ with equal probability. Spin statistics predicts that the $^1A*$ state represents just $1/9^{th}$ (11.1%) of the annihilation products. Sometimes, it is reported that TTA-UC may exceed the proposed 11% limit because of indirect triplet state formation from the quintet excimer state[5] $(AA)*$.

TTA-UC shows advantages over traditional UC techniques for a number of reasons:

1. Lower excitation power density (<100 mW/cm^2) is required for TTA-UC and need not be coherent.
2. Solar light is a renewable energy source with lower energy density (few mW/cm^2) and can be used as a sufficient excitation source to sensitize the TTA-UC process.
3. The excitation wavelength and emission wavelength of TTA-UC are remained far ways with respect to energy in the UC scheme (Figure 11.6). There must have energy level matching between the triplet sensitizer and the triplet acceptor (annihilator/emitter) (Figures 11.7 and 11.8).

11.5.1 HISTORY OF MOLECULAR TRIPLET–TRIPLET ANNIHILATION UPCONVERSION (TTA-UC)

Historically, Parker and Hatchard first reported the anti-Stokes delayed fluorescence using sensitizer in 1962.[4] The anti-Stokes delayed fluorescence was observed for phenanthrene/naphthalene systems from UV to UV region and proflavine hydrochloride/anthracene system in ethanol at low temperature from Vis to Vis region (Figure 11.6).[4]

The TTA-UC system had not been developed effectively due to less availability of a long-lived triplet state molecule at room temperature. After development of heavy metal–organic complexes with a long-lived triplet state at room temperature, it has been eventually created huge opportunities for creating TTA-UC system as a vibrant research

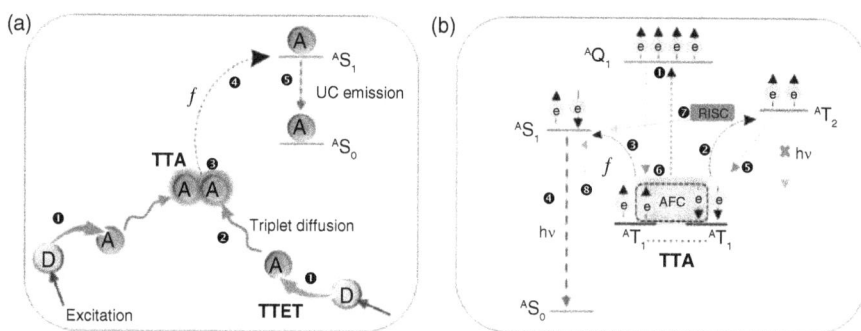

FIGURE 11.6 (a) Events of TTA-UC emission and (b) post TTA events showing formation of different energy states of anhilation with different spin multiplicitirs includeing singlet ($^AS_1$), triplet ($^AT_2$), and quintet ($^AQ_1$). (Reproduced from ref. [1] with permission from Royal Society Chemistry, copyright@ 2020.)

(a)

(b)

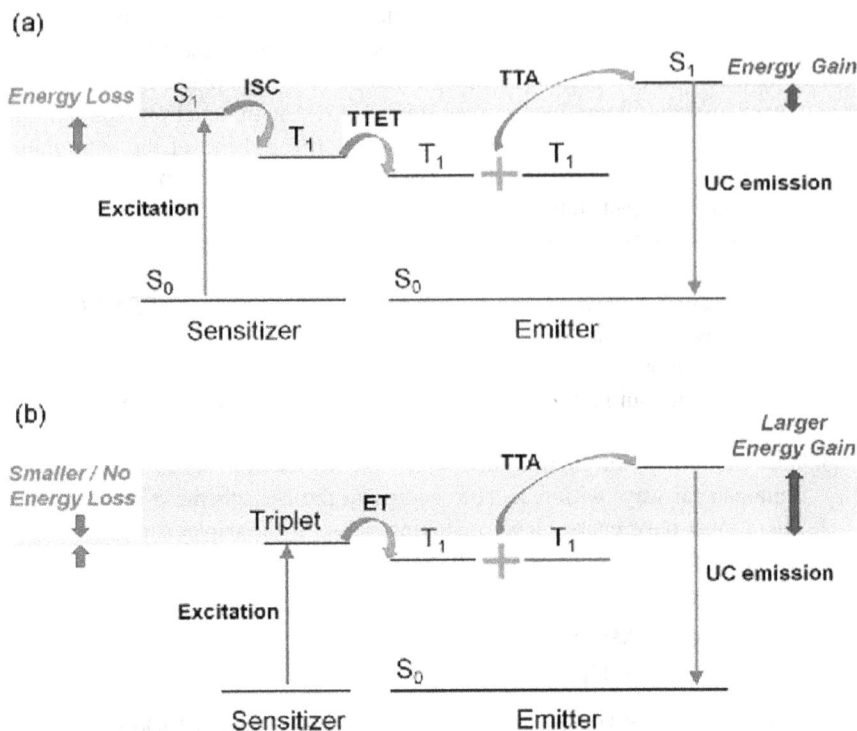

FIGURE 11.7 (a) Conventional triplet sensitization route for TTA-UC emission and (b) newly post-triplet sensitization route for TTA-UC emission. (Reproduced from ref. [33] with permission from American Chemical Society Chemistry, copyright@ 2017.)

field. First time, the sensitized delayed TTA-UC (Green to blue) at room temperature was demonstrated by Baluschev and co-workers for increasing the efficiency of solar cells.[23] Further, NIR-to-Vis TTA-UC was first reported by Baluschev and co-workers.[24,25] The TTA-UC process is a nonlinear process and TTA-UC emission intensity showed quadratic dependence on the excitation intensity of laser (Figure 11.9).[20]

The TTA-UC is becoming an attractive research field in present day and eventually leads to new research areas like.

11.5.1.1 Liquid Crystal for Triplet–Triplet Annihilation Upconversion (TTA-UC)

The liquid crystalline has shown interesting orientational control emission properties by chromophores, which can be used for minimizing the energy loss in a device like solar cells using UC emission photon.[24] A system of dissolving sensitizer/annihilator couples of palladium(II) octaethylporphyrin/anthracene has shown the directed photon UC emission (Visible to Visible) derivatives in orientationally ordered nematic liquid crystalline matrix.[26] Further, controlled switching of directional TTA-UC emission (axial and longitudinal emission) is also achieved for 9-(4-cyanophenyl)and 10-phenylanthracene annihilator (Figure 11.10).[26]

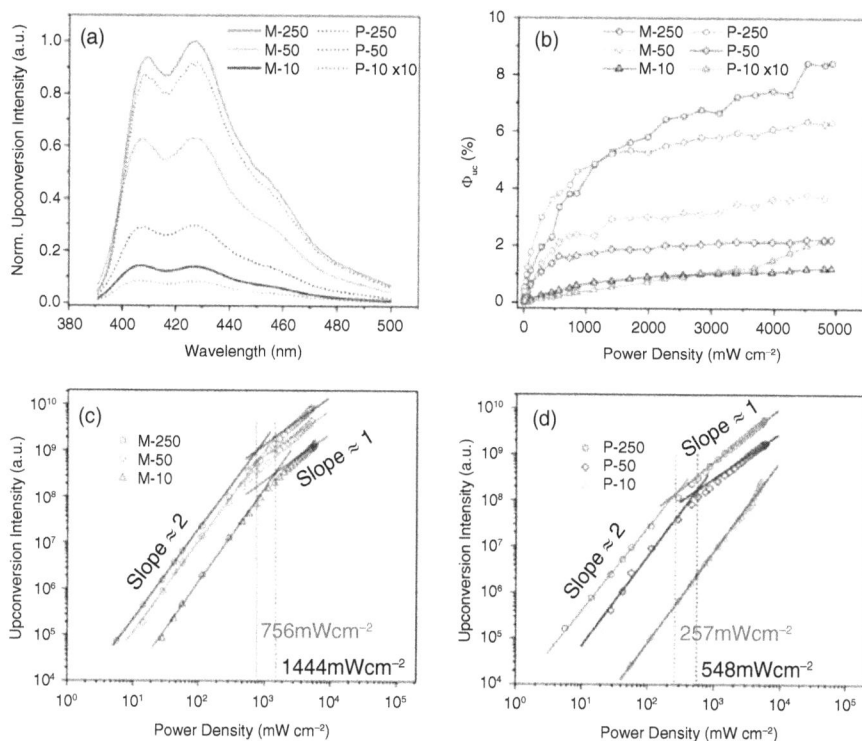

FIGURE 11.8 (a) Upconversion emission of M-10, M-50, M-250, P-10, P-50, and P-250 with an excitation density of 850 mW/cm^2 and are normalized at the emission maximum of M-250 at 427 nm. (b) Power density vs upconversion quantum yield; (c) and (d) log (power density) vs log (integrated upconversion emission intensity) for P-250, the low annihilation regime slope is 1.8 and the high-annihilation regime slope is 1.2, which are considered as 2 and 1, respectively).[20] (Reproduced from ref. [20] with permission from American Chemical Society Chemistry, copyright@ 2022.)

11.5.1.2 Metal–Organic Frameworks (MOFs) for Triplet–Triplet Annihilation Upconversion (TTA-UC)

In the solid state, TTA-UC of NIR light to visible light is a great challenge due to the aggregation of sensitizer that hampers the efficient TET. A molecular sensitizer exhibiting direct singlet-to-triplet (S to T) absorption into a new emitter-based MOF is designed to achieve an efficient triplet sensitizer assisted NIR-to-visible TTA-UC in solid state. The new Zr-based MOF is useful to create an effective S–T absorption-based sensitizer to make a highly efficient donor-to-acceptor TET system (Figure 11.11).[27]

Nanocrystal–MOF hybrids are used to design TTA_UC system in solid state. Benchmark work for TTA-UC using nanocrystal–MOF hybrids is confirmed by CdSe/CdS with the anthracene-containing MOF system and PbS with the tetracene-containing MOF system. These two systems show a green-to-blue TTA-UC but also interesting NIR-to-visible TTA-UC. Further investigation is required to enhance

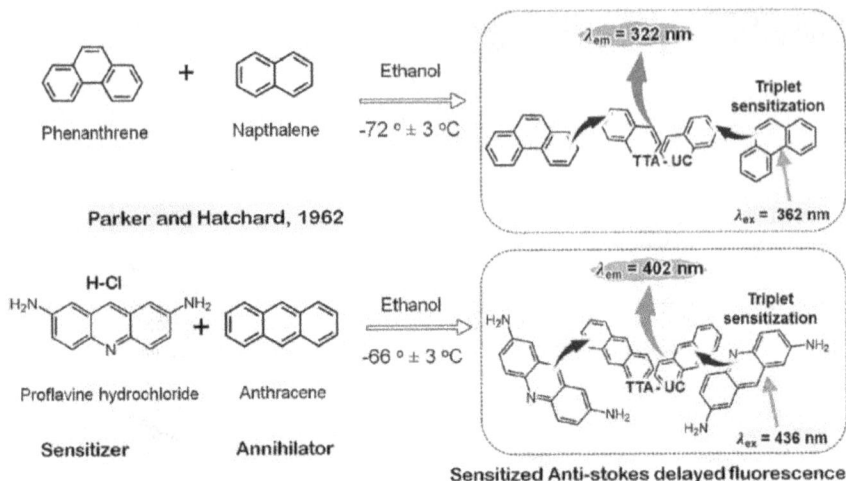

FIGURE 11.9 Anti-Stokes fluorescence in the phenanthrene/naphthalene system (362–322 nm) and the proflavine hydrochloride/anthracene system (436 –402 nm).[4]

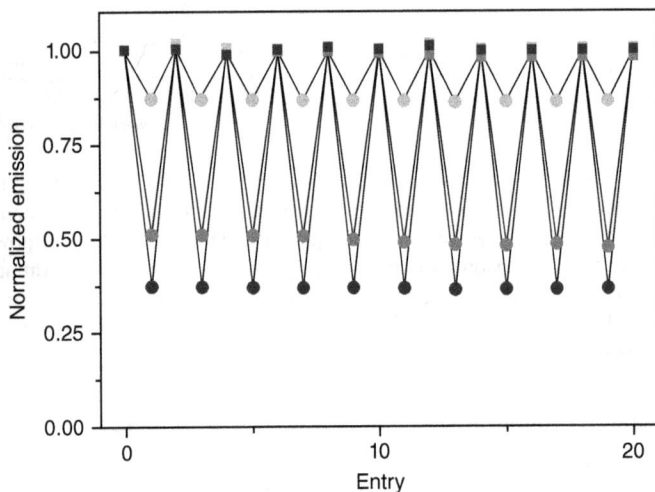

FIGURE 11.10 Normalized switching of upconverted emission of 1 (black), 2 (blue), and 3 (red) in the presence (circles) and absence (squares) of an applied electric field of 14 Vrms (excitation wavelength was 547 nm and the emission was monitored at 430 nm).[26] (Reproduced from ref. [26] with permission from Springer Nature, copyright@ 2016.)

the UC efficiency with respect to other MOFs and related materials such as covalent organic frameworks. Next-generation triplet sensitizers and ordered emitter arrays would be useful for solid upconverters with high efficiency at low excitation intensity.[28] The TTA-UC quantum yield of the solid samples is very low as expected, more efforts are required to improve the UC efficiency.[29–31]

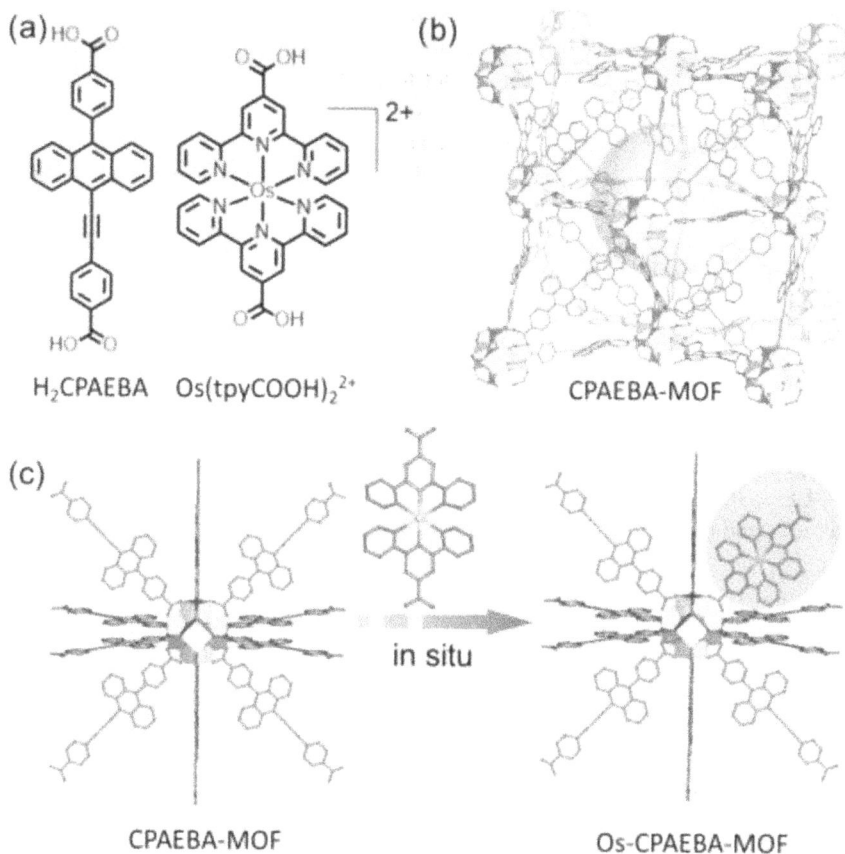

FIGURE 11.11 (a) Molecular structures of H2CPAEBA and Os(tpyCOOH)$_2$$^{2+}$, (b) structure of CPAEBA-MOF with UiO-69 type topology, and (c) schematic diagram of in-situ incorporation of Os-donor into CPAEBA-MOF framework.[27] (Reproduced from ref. [20] with permission from Wiley-VCH, copyright@ 2022.)

11.6 SEMICONDUCTOR QUANTUM DOTS, TADF, AND PEROVSKITES NANOCRYSTALS FOR NEW SENSITIZERS

For minimizing the energy loss during ISC, new triplet sensitization routes are developed to significantly enlarge the range of conversion wavelength of triplet sensitizers. The ISC inevitably is associated with the energy loss of hundreds of millielectronvolts and plays a significant role in limiting the TTA-UC with large anti-Stokes shifts. The small S_1–T_1 gap of molecules showing thermally activated delayed fluorescence (TADF) allows the sensitization of emitters with the highest T_1 and S_1 energy levels ever employed in TTA-UC, which results in efficient vis-to-UV UC (Figure 11.4). Inorganic nanocrystals with broad NIR absorption or bypassing the ISC process bands are used as effective sensitizers for NIR-to-Vis

TTA-UC. The modification of nanocrystal surfaces with organic acceptors also plays a significant role in efficient energy transfer between the components and succeeding TTA processes. Direct singlet-to-triplet (S–T) excitation is employed by using nanocrystal as triplet sensitizers to remove restrictions on the energy loss during the ISC process. Although the S-T absorption is spin forbidden, large spin–orbital coupling occurs for appropriately designed metal complexes, which allow S–T absorption in the NIR region with large absorption coefficients.[32] While the triplet lifetime of such S–T absorption sensitizers is often short (less than microseconds), the integration of the molecular sensitizers with emitter assemblies allows facile DET to the surrounding emitter molecules, leading to efficient NIR-to-Vis UC emission through triplet energy migration in the condensed state (Figure 11.12).

The molecular assembly's triplet sensitization route is highly effective to populate emitter triplets but the issue of back energy transfer process is also to be considered.[34] The high efficiency of sensitized (S_0-to-T_1) is useful to achieve the NIR-to-Vis TTA-UC systems and is highly suitable for photocatalytic synthesis and integrated solar cells and biological systems.[35,36]

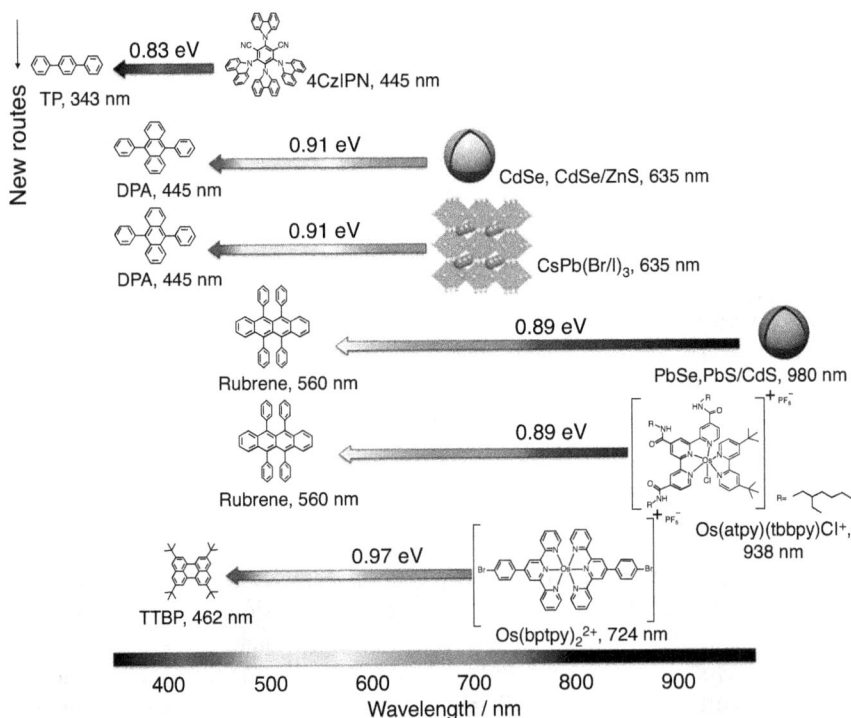

FIGURE 11.12 Representative excitation and emission wavelengths of sensitizer (right) and emitter (left) with anti-Stokes shift.[33] (Reproduced from ref. [33] with permission from American Chemical Society Chemistry, copyright@ 2017.)

11.7   CHALLENGES AND FUTURE DIRECTIONS

Over the last several decades, TTA pathway via triplet excited states of chromophore has been an exciting research area due to a wide range of photonic applications, photocatalysis, bioimaging, and solar cell. It is considered that the molecular TTA-UC is advantageous over two-photon absorption processes. The rare earth metal-doped UC nanocrystals are known as a useful and efficient TTA-UC system for getting high UC quantum yield at low-threshold excitation intensities. Molecular engineering of annihilators is important and required to avoid the loss of triplet photons because competition with quintet and triplet channels is sought.

However, a big challenge is to achieve high efficiencies and low-intensity radiation in aerated environments. The efficiency of TTA-UC depends upon photo-degradation in aerated conditions, and aggregation-induced fluorescence quenching in solid state by secondary quenching pathways like singlet fission. In future, molecular TTA-UC should be exploited for night-time vision by developing NIR-to-Vis TTA-UC contact lenses and many more.

REFERENCES

1. Bharmoria, P.; Bildirir, H.; Moth-Poulsen, K., Triplet-triplet annihilation based near infrared to visible molecular photon upconversion. *Chemical Society Reviews* 2020, *49* (18), 6529–6554.
2. Sasaki, Y.; Oshikawa, M.; Bharmoria, P.; Kouno, H.; Hayashi-Takagi, A.; Sato, M.; Ajioka, I.; Yanai, N.; Kimizuka, N., Near-infrared optogenetic genome engineering based on photon-upconversion hydrogels. *Angewandte Chemie International Edition* 2019, *58* (49), 17827–17833.
3. Gray, V.; Dzebo, D.; Abrahamsson, M.; Albinsson, B.; Moth-Poulsen, K., Triplet-triplet annihilation photon-upconversion: Towards solar energy applications. *Physical Chemistry Chemical Physics* 2014, *16* (22), 10345–10352.
4. Parker, C. A.; Hatchard, C. G., Delayed fluorescence from solutions of anthracene and phenanthrene. *Proceedings of the Royal Society of London. Series A, Mathematical and Physical Sciences* 1962, *269*, 574–584.
5. Turro, N. J.; Ramamurthy, V.; Scaiano, J. C., *Principles of Molecular Photochemistry: An Introduction*, University science books, Sausalito, California (USA), 2009.
6. Beery, D.; Schmidt, T. W.; Hanson, K., Harnessing sunlight via molecular photon upconversion. *ACS Applied Materials & Interfaces* 2021, *13* (28), 32601–32605.
7. Richards, B. S.; Hudry, D.; Busko, D.; Turshatov, A.; Howard, I. A., Photon upconversion for photovoltaics and photocatalysis: A critical review. *Chemical Reviews* 2021, *121* (15), 9165–9195.
8. Ravetz, B. D.; Pun, A. B.; Churchill, E. M.; Congreve, D. N.; Rovis, T.; Campos, L. M., Photoredox catalysis using infrared light via triplet fusion upconversion. *Nature* 2019, *565* (7739), 343–346.
9. Liu, Q.; Xu, M.; Yang, T.; Tian, B.; Zhang, X.; Li, F., Highly photostable near-IR-excitation upconversion nanocapsules based on triplet-triplet annihilation for in vivo bioimaging application. *ACS Applied Materials & Interfaces* 2018, *10* (12), 9883–9888.
10. Lu, C.; Joulin, E.; Tang, H.; Pouri, H.; Zhang, J. Upconversion nanostructures applied in theranostic systems. *International Journal of Molecular Sciences* 2022, *23*, 9003.
11. Sasikumar, D.; John, A. T.; Sunny, J.; Hariharan, M., Access to the triplet excited states of organic chromophores. *Chemical Society Reviews* 2020, *49* (17), 6122–6140.

12. Bai, S.; Zhang, P.; Beratan, D. N., Predicting dexter energy transfer interactions from molecular orbital overlaps. *The Journal of Physical Chemistry C* 2020, *124* (35), 18956–18960.

13. Huang, Z.; Li, X.; Mahboub, M.; Hanson, K. M.; Nichols, V. M.; Le, H.; Tang, M. L.; Bardeen, C. J., Hybrid molecule-nanocrystal photon upconversion across the visible and near-infrared. *Nano Letters* 2015, *15* (8), 5552–5557.

14. Huang, Z.; Simpson, D. E.; Mahboub, M.; Li, X.; Tang, M. L., Ligand enhanced upconversion of near-infrared photons with nanocrystal light absorbers. *Chemical Science* 2016, *7* (7), 4101–4104.

15. Filatov, M. A.; Baluschev, S.; Landfester, K., Protection of densely populated excited triplet state ensembles against deactivation by molecular oxygen. *Chemical Society Reviews* 2016, *45* (17), 4668–4689.

16. Schulze, T. F.; Czolk, J.; Cheng, Y.-Y.; Fückel, B.; MacQueen, R. W.; Khoury, T.; Crossley, M. J.; Stannowski, B.; Lips, K.; Lemmer, U.; Colsmann, A.; Schmidt, T. W., Efficiency enhancement of organic and thin-film silicon solar cells with photochemical upconversion. *The Journal of Physical Chemistry C* 2012, *116* (43), 22794–22801.

17. Kimizuka, N.; Yanai, N.; Morikawa, M.-A., Photon upconversion and molecular solar energy storage by maximizing the potential of molecular self-assembly. *Langmuir* 2016, *32* (47), 12304–12322.

18. Zhou, Y.; Castellano, F. N.; Schmidt, T. W.; Hanson, K., On the quantum yield of photon upconversion via triplet-triplet annihilation. *ACS Energy Letters* 2020, *5* (7), 2322–2326.

19. Fan, C.; Wei, L.; Niu, T.; Rao, M.; Cheng, G.; Chruma, J. J.; Wu, W.; Yang, C., Efficient triplet-triplet annihilation upconversion with an anti-stokes shift of 1.08 eV achieved by chemically tuning sensitizers. *Journal of the American Chemical Society* 2019, *141* (38), 15070–15077.

20. Jha, K. K.; Prabhakaran, A.; Burke, C. S.; Schulze, M.; Schubert, U. S.; Keyes, T. E.; Jäger, M.; Ivanšić, B. D., Triplet-triplet annihilation upconversion by polymeric sensitizers. *The Journal of Physical Chemistry C* 2022, *126* (8), 4057–4066.

21. Olesund, A.; Johnsson, J.; Edhborg, F.; Ghasemi, S.; Moth-Poulsen, K.; Albinsson, B., Approaching the spin-statistical limit in visible-to-ultraviolet photon upconversion. *Journal of the American Chemical Society* 2022, *144* (8), 3706–3716.

22. Singh-Rachford, T. N.; Castellano, F. N., Photon upconversion based on sensitized triplet-triplet annihilation. *Coordination Chemistry Reviews* 2010, *254* (21–22), 2560–2573.

23. Keivanidis, P. E.; Baluschev, S.; Miteva, T.; Nelles, G.; Scherf, U.; Yasuda, A.; Wegner, G., Up-conversion photoluminescence in polyfluorene doped with Metal(II)-octaethyl porphyrins. *Advanced Materials* 2003, *15* (24), 2095–2098.

24. Baluschev, S.; Yakutkin, V.; Miteva, T.; Avlasevich, Y.; Chernov, S.; Aleshchenkov, S.; Nelles, G.; Cheprakov, A.; Yasuda, A.; Müllen, K.; Wegner, G., Blue-green upconversion: Noncoherent excitation by NIR light. *Angewandte Chemie International Edition* 2007, *46* (40), 7693–7696.

25. Baluschev, S.; Yakutkin, V.; Miteva, T.; Wegner, G.; Roberts, T.; Nelles, G.; Yasuda, A.; Chernov, S.; Aleshchenkov, S.; Cheprakov, A., A general approach for non-coherently excited annihilation up-conversion: Transforming the solar-spectrum. *New Journal of Physics* 2008, *10* (1), 013007.

26. Börjesson, K.; Rudquist, P.; Gray, V.; Moth-Poulsen, K., Photon upconversion with directed emission. *Nature Communications* 2016, *7* (1), 12689.

27. Joarder, B.; Mallick, A.; Sasaki, Y.; Kinoshita, M.; Haruki, R.; Kawashima, Y.; Yanai, N.; Kimizuka, N., Near-infrared-to-visible photon upconversion by introducing an S–T absorption sensitizer into a metal-organic framework. *ChemNanoMat* 2020, *6* (6), 916–919.

28. Amemori, S.; Gupta, R. K.; Böhm, M. L.; Xiao, J.; Huynh, U.; Oyama, T.; Kaneko, K.; Rao, A.; Yanai, N.; Kimizuka, N., Hybridizing semiconductor nanocrystals with metal-organic frameworks for visible and near-infrared photon upconversion. *Dalton Transactions* 2018, *47* (26), 8590–8594.

29. Gharaati, S.; Wang, C.; Förster, C.; Weigert, F.; Resch-Genger, U.; Heinze, K., Triplet-triplet annihilation upconversion in a MOF with acceptor-filled channels. *Chemistry - A European Journal* 2020, *26* (5), 1003–1007.

30. Roy, I.; Goswami, S.; Young, R. M.; Schlesinger, I.; Mian, M. R.; Enciso, A. E.; Zhang, X.; Hornick, J. E.; Farha, O. K.; Wasielewski, M. R.; Hupp, J. T.; Stoddart, J. F., Photon upconversion in a glowing metal-organic framework. *Journal of the American Chemical Society* 2021, *143* (13), 5053–5059.

31. Li, Y.; Jiang, C.; Chen, X.; Jiang, Y.; Yao, C., Yb^{3+}-doped two-dimensional upconverting Tb-MOF nanosheets with luminescence sensing properties. *ACS Applied Materials & Interfaces* 2022, *14* (6), 8343–8352.

32. Amemori, S.; Sasaki, Y.; Yanai, N.; Kimizuka, N., Near-infrared-to-visible photon upconversion sensitized by a metal complex with spin-forbidden yet strong S0-T1 absorption. *Journal of the American Chemical Society* 2016, *138* (28), 8702–8705.

33. Yanai, N.; Kimizuka, N., New triplet sensitization routes for photon upconversion: Thermally activated delayed fluorescence molecules, inorganic nanocrystals, and singlet-to-triplet absorption. *Accounts of Chemical Research* 2017, *50* (10), 2487–2495.

34. Nienhaus, L.; Wu, M.; Geva, N.; Shepherd, J. J.; Wilson, M. W. B.; Bulović, V.; Van Voorhis, T.; Baldo, M. A.; Bawendi, M. G., Speed limit for triplet-exciton transfer in solid-state PbS nanocrystal-sensitized photon upconversion. *ACS Nano* 2017, *11* (8), 7848–7857.

35. Sasaki, Y.; Amemori, S.; Kouno, H.; Yanai, N.; Kimizuka, N., Near infrared-to-blue photon upconversion by exploiting direct S-T absorption of a molecular sensitizer. *Journal of Materials Chemistry C* 2017, *5* (21), 5063–5067.

36. Haruki, R.; Sasaki, Y.; Masutani, K.; Yanai, N.; Kimizuka, N., Leaping across the visible range: Near-infrared-to-violet photon upconversion employing a silyl-substituted anthracene. *Chemical Communications* 2020, *56* (51), 7017–7020.

12 Fluorescent Sensors/ Materials to Detect Analytes and Their Applications

Abhinav Sharma, Hendrik Faber, and Thomas D. Anthopoulos

12.1 INTRODUCTION

12.1.1 FLUORESCENCE-BASED BIOSENSORS

Fluorescence biosensors have a broad range of sensing applications, including biomedical diagnostics, environmental monitoring, and food safety (Gaviria-Arroyave et al. 2020; Qu et al. 2021; Kakkar et al. 2023). Fluorescence biosensors provide a diverse range of opportunities to investigate various factors related to fluorescence, such as fluorescence intensity, fluorescence anisotropy, energy transfer, decay time, quantum yield, and quenching efficiency (Lee and Kang 2023; Sultangaziyev and Bukasov 2020; Zhang et al. 2022). These factors can offer valuable insights into the characteristics of the analyte detection and can be utilized to optimize the performance of the biosensor. In fluorescence detection, synthetic organic dyes based on cyanine and xanthene dyes are frequently used as tags for biomolecules due to being easily available, cost-effective, fluorescence reporters (as a donor and acceptor), and suitable for chemical reactivity and spectroscopic properties (Wycisk et al. 2017; Sargazi et al. 2022; Keller et al. 2020). First-generation organic dyes for fluorescent labeling include fluorescein and rhodamine (Yuan et al. 2012). These dyes have some limitations, like pH sensitivity, photo-bleaching, and hydrophobicity. These dyes used in fluorescence biosensors have some limitations, such as pH sensitivity, photo-bleaching, and hydrophobicity. The pH sensitivity of the dyes can affect their fluorescence properties and lead to inaccurate measurements. Photo-bleaching can occur when the dye is exposed to light for an extended period, resulting in a decrease in fluorescence intensity over time. Hydrophobicity can also be an issue as it can affect the solubility of the dye in aqueous solutions and lead to aggregation or precipitation. These limitations must be considered when designing and using fluorescence biosensors. Nanomaterials, such as semiconductor quantum dots (Anfossi et al. 2018) and metal nanoparticles (MNPs) (Wu et al. 2012; Rhouati et al. 2016), have been investigated as excellent alternatives to organic dyes to overcome technical challenges. Moreover, several natural fluorescence molecules have been used

 DOI: 10.1201/9781003352372-12

for fluorescence-based detection. The NAD/NADH enzymatic reactions are widely used for the fluorescence-based detection of different analytes. Numerous other biomolecules, such as DNA, reduced pyridine nucleotides (NADH), flavin nucleotides, and green fluorescent proteins, have inherent fluorescence properties (Sharma et al. 2018; Müllerová et al. 2022; Heikal 2010). When these molecules bind with the target analyte, fluorescent behavior changes such as polarization and photoluminescence emission intensity take place. In contrast, the majority of the target analytes are labeled with fluorescence probes to enable fluorescence spectroscopy to identify them. The fluorescence probes are attached to the target analyte (nucleic acids, proteins, and other biomolecules) via covalent bonding through -OH, -COOH, -NH$_2$, and -SH functional groups. The binding between fluorescent probe and biomolecules is highly dependent on the type of functional groups, pH, temperature, and ionic strength of the buffer solution. The bioreceptor elements, including antibodies, enzymes, peptides, and aptamers are tremendous candidates because of their high specificity, straightforward chemical modification, and selectivity to bind fluorescent probes to develop the high sensitive and stable fluorescence biosensors (Yamamoto and Kumar 2000). For example, Yamamoto and Kumar (2000) developed a Förster resonance energy transfer detection method using a hairpin-like aptamer structure modified with the fluorescent probe. Pyrene dye is also used as an effective alternative for fluorescence-based sensing techniques. However, pyrene monomer has very little fluorescence; when two monomeric units are in close proximity to one another, they create pyrene excimers, which have a long stokes shift and a long fluorescence lifetime (~40 ns), while <10 ns is the average fluorescence lifespan for chromophores.

12.1.2 FLUORESCENCE DETECTION TECHNIQUES

The fluorescence detection technique is a method used to detect and measure the amount of fluorescent light emitted by a sample. This technique involves exciting the sample with a specific wavelength of light, which causes the molecules in the sample to absorb energy and become excited. As these molecules return to their ground state, they emit light at a longer wavelength, which can be detected and measured using a fluorescence detector. Fluorescence detection is commonly used in various fields, such as cell imaging, protein analysis, DNA sequencing, drug discovery, and environmental monitoring (Zhang et al. 2023; Yan et al. 2022; Shin et al. 2021). Fluorescence is a specific type of luminescence that occurs when a fluorophore molecule emits light quanta upon absorption of light energy. The excited state is transient, and the molecule returns to its ground state by releasing light quanta, which can be detected as fluorescence. This phenomenon is referred to as fluorescence emission and is distinguished by a characteristic wavelength or color of light emitted by the fluorophore. The duration of fluorescence emission is determined by the characteristics of the fluorophore and its interactions with bioreceptors. In the case of organic dyes, fluorescence lifetime typically ranges from picoseconds (ps) to nanoseconds (ns), usually between 10^{-8} and 10^{-11} s (Drummen 2012). For semiconductor nanocrystals, this timescale can be extended to tens of nanoseconds (ns), while for organometallic compounds and lanthanide complexes, it can range from hundreds of nanoseconds up to milliseconds (ms) (Pu et al. 2016). These variations in fluorescence lifetime are attributed to differences in excited-state relaxation pathways and interactions with other biomolecules.

While luminescence is a broader term that encompasses various phenomena involving the emission of light by excited species. These can include chemiluminescence, which involves the emission of light during chemical reactions; bioluminescence, which involves the emission of light by living organisms as a result of biochemical reactions (Cinquanta et al. 2017); and electrochemiluminescence, which involves the emission of light upon oxidation or reduction on an electrode surface (Hong et al. 2021). Each form of luminescence exhibits unique characteristics and can be applied in various types of biosensing applications. It is important to note that the duration of luminescence types, such as chemiluminescence, bioluminescence, and electrochemiluminescence, can be much longer than fluorescence. The fluorescence biosensors using quantum dots operate on the principle of size-dependent fluorescence properties of quantum dots. Upon excitation by a light source, quantum dots emit fluorescent light at a specific wavelength that is determined by their size. By attaching a biomolecule to the surface of the quantum dot, the presence of a specific analyte can be detected through changes in the fluorescence signal (Cardoso Dos Santos et al. 2020). Various methods have been employed to detect different biomolecules, including proteins, viruses, and bacteria. These methods include the culture method, polymerase chain reaction (PCR), and ELISA. However, these techniques have several limitations such as the requirement for sophisticated instruments, multiple-step processing, long processing time, cross-contamination risks, expensive chemicals and reagents, and the need for skilled personnel (Liu and Lei 2021; Sciuto et al. 2021). To overcome these limitations, various biosensing techniques have been developed that can detect multiple biomolecules using mechanical, optical, electrical, and electrochemical techniques. The optical methods include surface plasmon resonance (SPR), surface-enhanced Raman scattering (SERS), fluorescence resonance energy transfer (FRET), fluorescence-based biosensors, and colorimetric assays (García-Hernández et al. 2023; Das et al. 2023). Among these methods, fluorescence detection methods have grown the interest of researchers due to their numerous advantages, such as simple operation, fast response, and high sensitivity (Xu et al. 2023; Kim et al. 2016; Qu et al. 2022). When compared to optical techniques that rely on absorbance, fluorescence sensing technology is superior due to its higher sensitivity (approximately 100 times better than absorbance), selectivity (due to the fluorescent probe), and faster detection times (Sargazi et al. 2022). The higher sensitivity can be attributed to the effective interaction between fluorescent probes and the surface of metallic nanostructures.

12.1.3   FLUORESCENT CARBON NANOMATERIALS

Carbon nanomaterials have been used as a potential material for the development of numerous types of biosensors, including electrical biosensors (Singh et al. 2014; Sharma et al. 2015; Sharma, et al. 2016; Sharma and Jang 2019), electrochemical (Joshi et al. 2020; Sharma et al. 2020; Sharma et al. 2022), colorimetric (Zhao et al. 2020), and fluorescence biosensors (Qu et al. 2021) due to high electrical, good mechanical, thermal, and optical properties; chemical stability; biocompatibility;

and good transduction element (Rajakumar et al. 2020; Hwang et al. 2020). Carbon nanotubes (CNTs) and graphene nanoribbons are categorized as one-dimensional carbon nanomaterials, graphene and its derivatives such as reduced graphene oxide (rGO) and carbon nitride nanosheets are categorized as 2-D, carbon and graphene quantum dots (GQDs) are categorized as 0-D (dimensionless), and graphite and covalent organic frameworks are categorized as 3-D and have been used for biosensing applications to detect a variety of analytes (Pandey and Chusuei 2021; Burdanova et al. 2021; Lu et al. 2023; Hwang et al. 2020). In 1991, Iijima made the initial discovery of CNTs, six-membered bands made of sp^2-hybridized carbon atoms (Iijima 1991). Single-walled carbon nanotubes (SWCNTs) and multi-walled carbon nanotubes (MWCNTs) are the two most commonly studied types of CNTs. SWCNTs are composed of a single layer of graphene sheet that is rolled up into a cylindrical tube. The diameter of the tube can range from less than 1 nm to several nanometers, while the length can extend up to several micrometers (Saifuddin et al. 2013). The unique structure and properties of SWCNTs make them useful for different biosensing applications (Anzar et al. 2020). MWCNTs are composed of several layers of graphene sheets that are arranged in a concentric manner. The layers are held together by van der Waals forces and can vary in thickness from a few to tens of layers, depending on the synthesis method employed. The electronic properties of SWCNTs are determined by their chirality, which refers to the direction in which the graphene sheet is rolled to form the nanotube. Depending on the chirality, SWCNTs can exhibit either metallic or semiconducting behavior. Metallic SWCNTs have a high electrical conductivity, while semiconducting SWCNTs have a bandgap and exhibit properties similar to those of conventional semiconductors. The ability to control the electronic properties of SWCNTs based on their chirality makes them attractive for various applications in electronic applications (Saifuddin et al. 2013). MWCNTs can exhibit both metallic and semiconducting behavior depending on their diameter, the number of walls, and the arrangement of the graphene layers. While some MWCNTs may exhibit metallic behavior due to the presence of a large number of graphene layers, others may exhibit semiconducting behavior due to the confinement of electrons in the radial direction.

12.2 CARBON NANOTUBES (CNTS)-BASED FLUORESCENCE BIOSENSOR

CNTs possess distinctive electronic and optical properties that make them suitable for integration into optical biosensors (Farrera et al. 2017; Kruss et al. 2013). CNTs can enhance electron transfer between biomolecules and electrodes, which is useful for developing electrochemical biosensors to detect biomarkers (Yang et al. 2022; Zhang and Du 2020; Hu and Hu 2009). Additionally, CNTs exhibit unique optical properties such as strong light absorption and fluorescence quenching, which can be exploited for developing optical biosensors for biomarker detection and molecular imaging (Gong et al. 2013). The integration of CNTs into biosensors has the potential to improve their sensitivity, selectivity, and speed of detection, making them valuable tools for disease diagnosis and monitoring (Hong et al. 2009; Kruss et al. 2013; Farrera et al. 2017; Barone et al. 2004). Optical biosensors operate on the principle of measuring light emission, such as UV, visible, infrared, or fluorescence, to detect the interaction between target biomolecules and bioreceptors (Strianese et al. 2012, Borisov and Wolfbeis 2008). Among the types of optical biosensors, fluorescence biosensors rely on changes in the fluorescence properties of the sensing element to detect analytes (Borisov and Wolfbeis 2008; Leopold, Shcherbakova, and Verkhusha 2019). The binding of a target biomolecule to a bioreceptor on the sensing surface causes a change in the fluorescence properties of the sensing element that can be detected and quantified using various fluorescence techniques. Fluorescence biosensors offer several advantages over electrochemical biosensors, including easy operation, fast response, high sensitivity, and selectivity (Senutovitch et al. 2015). They have been widely used for detecting biomolecules such as proteins, nucleic acids, and small molecules in various biological samples such as blood, urine, and saliva (Kocheril et al. 2022; Nawrot et al. 2018; Gaviria-Arroyave et al. 2020). SWCNTs have been investigated for their potential use in optical imaging and biosensing applications due to their distinctive near-infrared (NIR) photoluminescence properties. The fluorescence of SWCNTs in the 900–1600 nm range is highly penetrant to living tissue and fluids and is photostable, which makes them a promising tool for detecting biomarkers (Hendler-Neumark and Bisker 2019). The NIR region of the electromagnetic spectrum offers several advantages over visible light, including deeper tissue penetration, reduced autofluorescence, and lower phototoxicity (Ackermann et al. 2022). SWCNTs can be functionalized with various biomolecules such as antibodies or aptamers to selectively bind to target biomarkers, enabling their detection with high sensitivity and specificity. The unique optical properties of SWCNTs make them a valuable tool for developing noninvasive diagnostic tools for disease detection and monitoring. Various fluorescence-based CNT biosensors have been developed to detect ions, metabolites, and protein biomarkers (J. Lee 2023; Pasinszki et al. 2017; Ferrier and Honeychurch 2021; Hendler-Neumark and Bisker 2019).

The early detection and monitoring of cancer are crucial for improving patient outcomes. The development of biomarker-based sensing technologies has revolutionized cancer diagnosis and monitoring (Kumar et al. 2023; Bohunicky and Mousa 2010; Roberts and Gandhi 2022) Biomarkers, such as proteins and nucleic acids found in blood, are indicators of disease states and can be used to detect cancer at an early stage or monitor disease progression (Alaimo et al. 2022; Erkocyigit et al. 2023). Prostate cancer is the most common type of cancer that develops in the prostate gland of males (Chan et al. 2022). The risk of developing prostate cancer increases with age and family history, causing a high number of deaths globally. The standard methods for detecting prostate cancer rely on a combination of serum-specific biomarker prostate-specific antigen (PSA) and digital rectal exam, which can give contradictory results in up to 85% of cases (Andersson et al. 2022; Thompson and Ankerst 2007). PSA-based detection has a high false-positive ratio and fails to detect early-stage disease. In addition, the diagnostic challenge with prostate cancer is the inability to differentiate between indolent disease and metastatic disease. Urokinase plasminogen activator (UPA) is a potential biomarker that may help to differentiate metastatic from indolent prostate cancer. The serum levels of UPA have also been found a direct correlation with the presence of metastatic disease. Williams et al. (2018) developed a highly sensitive fluorescence biosensor for detection of UPA (metastatic prostate cancer biomarker) using CNTs as the transducer element. In order to fabricate the CNT-based biosensor, HiPCO single-walled CNTs were functionalized with amine-modified ssDNA and then conjugated with a monoclonal antibody using carbodiimide chemistry against UPA. The resulting CNT-antibody complex was characterized using dynamic light scattering, UV/VIS/NIR spectroscopy, and electrophoretic light scattering. Analyte interaction with the antibody resulted in a modulation of the CNTs' optical bandgap, leading to a shift in fluorescence emission wavelength that was proportional to the concentration of UPA. The proposed CNT sensor showed high sensitivity and selectivity for UPA detection. The sensor was able to detect UPA in plasma, serum, and whole blood with limits of detection of 0.5, 1, and 2 ng/mL, respectively. The sensor also showed negligible interference from other proteins commonly found in the blood. The CNT-based sensor demonstrated the potential for biomarker detection in cancer diagnosis and monitoring and holds promise as a clinical diagnostic tool for prostate cancer.

12.3 GRAPHENE-BASED FLUORESCENCE BIOSENSOR

Graphene is a two-dimensional, crystalline allotrope of the carbon atom made of a single sheet of graphite layer (Xu et al. 2013). It consists of layers of sp^2-hybridized carbon atoms arranged in a hexagonal lattice-like honeycomb structure, in which graphene layers are bonded together via weak van der Waals forces (Armano and Agnello 2019). After its first discovery in 2004 by Novoselov and Geim, graphene has been used for biosensing applications due to its unique properties like large surface area, physicochemical properties, mechanical strength, biocompatibility, and good thermal and electronic conductivity (Bai et al. 2020; Geim and Novoselov 2007; Justino et al. 2017). Several methods, including chemical and mechanical exfoliation, thermal decomposition, and chemical vapor deposition, have been used to synthesize high-quality and bulk production of graphene (Mbayachi et al. 2021; Bhuyan et al. 2016; Santhiran et al. 2021; Moosa et al. 2021). Carbon nanomaterials are a diverse group of materials that include CNTs, graphene oxide (GO), rGO, GQDs, and carbon quantum dots (CQDs) (Pandey and Chusuei 2021; Guan et al. 2023; Lin, Chen, and Huang 2016). Among these carbon nanomaterials, GO exhibits mild fluorescence, whereas GQDs, carbon dots, and CNTs exhibit strong fluorescence (Li et al. 2019; Zheng and Wu 2017). In recent years, GO has been utilized to develop biosensors with different signal transduction methods, including optical, electrical, and electrochemical. Among these strategies, the fluorescence detection method has been widely used in biomedical applications due to its ease of operation, rapidity, and high sensitivity. Moreover, the FRET-based fluorescent technique is widely used for optical biosensing applications. GO is useful in developing highly sensitive fluorescence biosensors. GO is considered the best fluorescence quencher through FRET, although the mechanism behind this is not fully understood. Additionally, GO can interact with various biomolecules through π–π interactions and hydrogen bonding and has a strong affinity for single-stranded DNA and a relatively lower affinity for double-stranded DNA. These properties make GO an efficient material for developing highly sensitive fluorescence biosensors. A recent study has shown a relationship between the introduction of oxygen-containing groups into graphene and fluorescence emission (Xiao et al. 2022). The presence of C-O, C=O, and O=C-OH groups in the graphene structure can cause GO to act as an energy donor rather than an acceptor, depending on its structure (Zheng and Wu 2017; Feng et al. 2017). In many cases of fluorescent GO-based biosensors, a relatively small size of GO is used, such as GQDs, which are zero-dimensional nanomaterials consisting of a single- or few-layer graphene (Fan et al. 2015; Kalkal et al. 2020; Yim et al. 2021). Additionally, distance control between the FRET donor and acceptor is often achieved by using linker molecules or probe–protein bioconjugates via covalent or electrostatic bonding.

Ebola virus (EBOV) is a highly infectious and deadly virus that has caused several outbreaks in Africa (Ilkhani and Farhad 2018). Early detection of EBOV is crucial for the effective management of the disease and prevention of its spread. In this study, a fluorescence biosensor was developed for the detection of EBOV using GO-aided rolling circle amplification (RCA) platform (Wen et al. 2016). The GO was synthesized via a modified Hummers method. The GO-assisted RCA platform works by using a specific detection probe labeled with fluorescein amidate (FAM) that is adsorbed on the GO surface. The GO-assisted RCA platform shows excellent sensitivity toward EBOV gene (5′-CTACCAGCAGCGCCAGACGG-3′). The GO-aided RCA platform does not generate any RCA products in the absence of EBOV gene, and the FAM-labeled probe (5′-FAM-GGGCTGCCAGATACTCTTCGCAATTTT-3′, detection probe) remains adsorbed on the GO surface, causing fluorescence quenching. However, upon addition of the EBOV gene, RCA occurs and leads to the formation of dsDNA between FAM-labeled detection probe and RCA products. The platform was able to detect EBOV in both serum sample and aqueous solution. The GO-assisted RCA platform developed in this study offers several advantages over other methods for detection of EBOV gene. First, it is highly sensitive due to its ability to amplify small amounts of target DNA through RCA. Second, it is simple and easy to use, making it suitable for use in resource-limited settings. Third, it is highly specific for EBOV due to the use of a specific detection probe. Finally, it is cost-effective compared to other methods such as PCR. The specificity of the GO-assisted RCA platform was confirmed by utilizing four distinct DNA sequences, including complementary EBOV genes and three types of mismatched sequences at the same concentration. The inclusion of the mismatched sequences resulted in only a slight increase in fluorescence intensity, indicating that this method can efficiently differentiate between base mismatched sequences and perfect complementary target. This result shows the high specificity of the platform for the detection of EBOV.

12.4 CARBON QUANTUM DOTS-BASED FLUORESCENCE BIOSENSOR

CQDs are a novel class of zero-dimensional carbon nanostructures with sizes smaller than 10 nm that has gained interest in optical biosensing applications owing to the biocompatibility, good photostability, physical–chemical properties, and strong tunable photoluminescence (Raveendran and Kizhakayil 2021; Azam et al. 2021; Ji et al. 2020). CQDs have a sp^2-conjugated structure with various functional groups oxygen, -OH, and -COOH groups. These functional groups can be further efficiently attached to bioreceptors for selective detection of biomolecules, imaging, and other biological applications. In 2004, CQDs were unintentionally discovered during the purification of single-wall nanotubes (Xu et al. 2004). There are commonly two methods to produce CQDs: top-down and bottom-up. In a top-down approach, carbon bulk materials (i.e., carbon black CNTs, carbon fibers, graphite, graphene oxide, etc.) cleave into smaller pieces using physical forces including laser ablation, arc discharge, and electrochemical exfoliation, thermal and microwave irradiation, hydrothermal method, and oxidation by strong acids (Khayal et al. 2021; Yadav et al. 2023; Cui et al. 2021). The carbon dots were found as a byproduct during the production of CNTs made using the arc-discharge technique. Then, the additional surface modification is used to enhance their fluorescence characteristics. In the bottom-up approach, various small carbon precursors were used, including citrates, carbohydrates, and other natural materials, to synthesize the controlled size of CQDs. The surface modification can be applied after or during synthesis via surface passivation, doping, or functionalization. Pacquiao et al. (2018) reported the synthesis of CQDs from mushrooms using hydrothermal synthesis. The mushroom pieces were first boiled in a solution containing H_2SO_4 (5% v/v) in DI water, followed by filtration, and transferred into a Teflon tube kept in an autoclave at 250°C for 4 h. The brown color solution was obtained and centrifuged at fixed rpm for 25 min to isolate the quantum dots. The CQDs produced from mushrooms were used to detect the different bioanalytes and toxic metal ions (Boobalan et al. 2020). The surface modification of CQDs provides large surface areas and better chemical and optical properties, such as increased fluorescence intensity, solubility, and chemical stability, as well as the ability to bind bioreceptors for the recognition of target biomolecules. Numerous organic compounds, including PEI, ethylenediamine, and oleylamine, can also passivate CQDs. The soft shell of carbon dots can be strengthened by cross-linking the branches in order to maintain surface function and enhance fluorescence emission. In contrast to heavy metal and semiconductors quantum dots, CQDs have gained a lot of attention in biosensing applications due to their tunable luminescence property, photostability, water solubility, biocompatibility, and sustainability (Khan et al. 2023; Farshbaf et al. 2017). Several methods have been used for the production of carbon dots, including arc discharge, laser ablation, pyrolysis, hydrothermal, solvothermal, and electrochemical. (Chauhan et al. 2022; Anwar et al. 2019). Among these approaches, the hydrothermal method is one-pot and bottom-up process, an effective and widely used process for producing the CQDs using a choice of natural precursors, including fruits, juices, milk, and eggs (Lou et al. 2021). Raveendran and Kizhakayil (2021) reported the synthesis of CQDs using an extract of mint leaf via a green approach applied for fluorometric detection of folic acid. In order to synthesize CQDs via green method, first, well-crushed mint leaves were dissolved in DI water under constant stirring for 30 min, followed by a hydrothermal process for 5 h at 200°C. The resultant brown color CQDs mixture was filtered out with centrifugation at constant rpm (2500) for 1 h. The average size of as-synthesized CQDs was <10 nm. Then, AgNPs were synthesized using mint-based CQDs as a stabilizer and reducing agent added to silver nitrate solution under magnetic stirring for 10 min. The obtained yellow-brown color mixture showed the reduction of Ag (from Ag(I) to Ag(0)). The average size of AgNPs is 12 nm, characterized by HRTEM. Afterward, AuNPs were prepared using $HAuCl_4$ solution added to DI water and heated at 60°C, followed by adding the solution of mint-based CQDs and starch solution (1%) under constant stirring.

The resultant solution with the purple color indicated that reduction from Au(III) to Au(0) was obtained, indicating the production of AuNPs with an average diameter of 60 nm. The various concentrations of the folic acid were mixed with the CQDs solution and incubated only for a few seconds, followed by measuring the fluorescence intensity at 360 nm (excitation wavelength). The optical performance shows photoluminescence intensity at 441 nm decreased progressively with the addition of the folic acid concentration due to the fluorescence quenching. The results show high sensitivity toward folic acid and established good linearity between luminescence intensities of mint-based CQDs with folic acid concentrations (0.5–5.1 µM) and achieved a limit of detection of 280 nM. For the selectivity test, no significant change was observed after adding a series of various interference biomolecules (i.e., ascorbic acid, alanine cholesterol, cysteine, dopamine hydrochloride, glycine, phenylalanine, glucose, urea, glutathione, glutamic acid, methionine, nicotinic acid, tryptophan, and valine), indicating the high selectivity toward folic acid (Liu et al. 2017). The strong and stable photoluminescence of the multi-colored carbon dots makes them a promising candidate for fluorescent live-cell imaging. The fluorescence images of HeLa cells modified with multi-colored carbon dots were recorded for cell imaging. The hydrophilic nature and functional groups on multi-colored carbon dots surface allow for easy internalization into cells via endocytosis. Long-wavelength UV light (between 360 and 380 nm), blue light (between 460 and 480 nm), and green light (between 510 and 590 nm) were used to stimulate the cell images, and the results demonstrate the successful application of multi-colored carbon dots as an effective probe for live-cell imaging.

According to the World Health Organization (WHO), cancer is one of the life-threatening diseases and the foremost cause of death globally (Bray et al. 2018). Cancer is caused by abnormal cells that expand uncontrollably, cross their usual boundaries, and spread to neighboring tissues or other organs (called metastasis). It resulted in a variety of diseases that can develop in practically any organ or bodily tissue. In 2018, an estimated 8.2 million projected deaths and ~9.6 million new cancer diagnosed cases were reported from the various forms of cancers (i.e., lung, breast, prostate, intestine, liver, etc.) (Bray et al. 2018). The cancer blood biomarker can be an indicator of the prognosis and diagnosis of cancer in the early stage for patients' survival. To detect cancer in its earliest stages, a number of biomarkers are used for the prognosis and diagnosis of different types of cancer (Tang et al. 2017; Wu and Qu 2015). For example, PSA is a specific biomarker for prostate cancer, while human epidermal growth factor receptor 2 (HER2) and estrogen and progesterone receptors are biomarkers for breast cancer. Epidermal growth factor is a biomarker for colorectal cancer, while CD20 and CD30 cytokines, platelet-derived growth factor receptor, and promyelocytic leukemia protein are biomarkers for leukemia and lymphoma. High levels of leucine, isoleucine, and valine are biomarkers for pancreatic cancer. Additionally, cancer antigen 125 is a specific biomarker for a certain type of ovarian cancer. Alarfaj et al. (2018) proposed a fluorescence immunoassay using Au-modified CQDs for carbohydrate antigen 19-9 (CA 19-9) detection in human serum (Alarfaj et al. 2018). The CA 19-9 antigen is a specific blood biomarker for the prognosis and diagnosis of pancreatic cancer. First, CQDs were prepared via green synthesis, glucose solution in water heated for 15 min at 120°C in a microwave oven (at 270 W). The yellow solution was obtained, and CQDs (d: <10 nm) were collected after dialysis with DI water for 3 h. The CQDs solution was added to $HAuCl_4$, NH_3, and DI water solutions and incubated under controlled conditions. The purple color resultant solution was obtained, indicating the formation of Au-modified CQDs. The surface of Au-modified CQDs was incubated with HRP-labeled CA 19-9 antibody for 60 min via EDC/NHS chemistry. The different concentration of CA 19-9 antigen from 0.01 to 350 U/mL in human serum was incubated with Au-modified CQDs/HRP-anti-CA 19-9 antibody. The fluorescence intensity was measured with increasing CA 19-9 antigen and calculated LOD of 0.007 U/mL. The immunoassay shows good selectivity against various cations, sugars, proteins, and other substances. Vascular endothelial growth factor (VEGF) biomarker (also known as a hypoxia-inducible protein) is a disease-cause biomarker responsible for the growth of endothelial cells, physiological vascular development during embryogenesis (called vasculogenesis), and formation of blood vessels (called angiogenesis) (Kim et al. 2019). The physiological levels of the VEGF biomarker are elevated beyond their normal range in the blood, which can be a sign of a severe disease like cancer. Therefore, VEGF is an effective biomarker for accurate and early diagnosis of cancer angiogenesis. In this research, Deb et al. (2023) developed a fluorescence sandwich immunoassay for disease biomarker VEGF detection using CQDs. To prepare fluorescent CQD-modified IgG antibody bioconjugate, first carbon dots were prepared using orange juice via a microwave-assisted process a 160°C for 5 min under pressure (10–12 bar) and power (20–23 W). The carbon dots were produced after centrifuging at constant rpm for 30 min, followed by filtration using a filter (0.22 μm syringe). The carbon dots solution was incubated with IgG detection antibody solution vis EDC/NHS chemistry, and IgG detection antibody conjugate solution was centrifuged to remove the unbound site of quantum dots (Figure12.1a). The average size of quantum dots is ~3–5.5 nm, characterized by HRTEM. The different concentrations of VEGF were incubated to the capture antibody solution to form the antigen–antibody immunocomplex. The resultant bioconjugate was further incubated with the fluorescent carbon dots modified with secondary antibody. The fluorescence signals were recorded with various concentration ranges of VEGF from 0.1 fg/mL to 1 ng/mL at an excitation wavelength of 390 nm. The result shows the fluorescent intensity was increasing with the increase in VEGF concentration owing to the formation of a passivation layer on the carbon dots surface, which produced extrinsic surface states, causing electron holes to recombine at the surface vacancies and decreasing the trap on the surface at the same time (Idowu et al. 2008; Zhu et al. 2014). The fluorescence immunosensor demonstrated that photoluminescence intensity increased with increasing a range of VEGF concentration from 0.1 fg/mL to 10 pg/mL (Figure 12.1b), achieved good linearity in logarithmic scale and LOD of 0.1 fg/mL (Figure 12.1c). The immunosensor shows good selectivity for sandwich immunocomplex (c-IgG/VEGF/d-IgG-CDs), against other interference substances such as ions, proteins, and other substances.

12.5   GQD-BASED FLUORESCENCE BIOSENSOR

GQD is a new class of fluorescence nanomaterial produced from carbon sources and has drawn a lot of interest in various applications such as biosensing, bioimaging, and drug delivery (Dong et al. 2018; Nair et al. 2017; Hua Liu et al. 2017; N. Cai et al.

(a)

OCDs   IgG Antibody   IgG-OCD

(b)

(c)

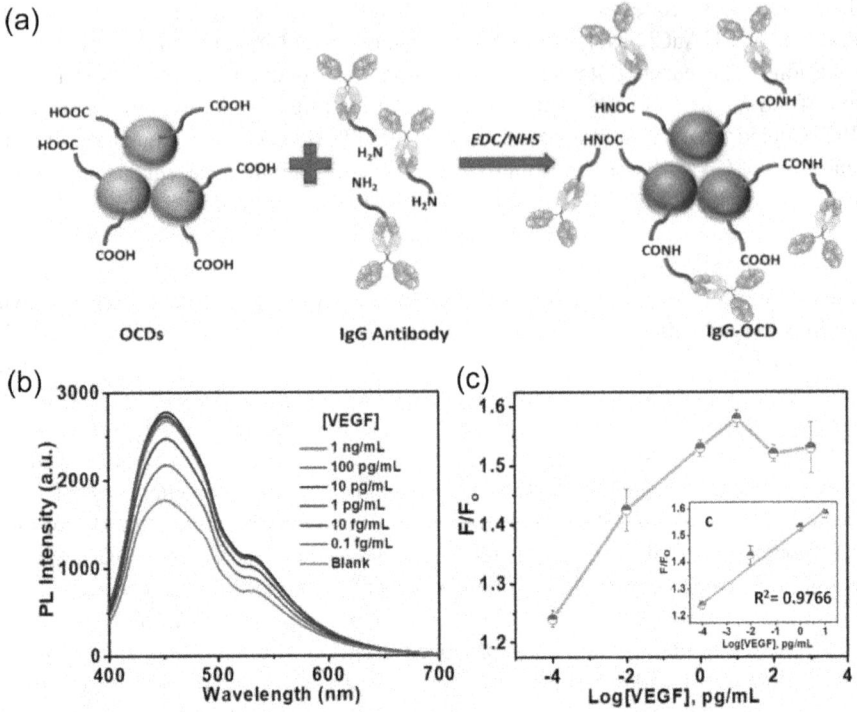

FIGURE 12.1 (a) Schematic illustration of the formation of a CQD-IgG antibody bioconjugate using EDC/NHS coupling chemistry for detection of VEGF biomarker. (b) PL emission spectrum was measured for various concentrations of VEGF biomarker at 390 nm (excitation wavelength), and (c) calibration plot was generated based on relative fluorescent intensity changes vs. logarithmic concentration of VEGF. The inset figure in the calibration plot shows a linear relation between VEGF concentrations ranging from 0.1 fg/mL to 10 pg/mL. (Reprinted from Deb et al. 2023. Analytica Chimica Acta 1242: 340808, with permission from Elsevier.)

2017; Huang et al. 2013). GQDs can be classified as the minimum size of graphene sheets (dimension <100 nM within one to 10 layers) (Elvati et al. 2017). Single-sheet carbon dots, which are frequently used in optical sensing applications, can be classified as GQD with a typical size of 1–10 nm. Since its discovery by Ponomarenko et al. in 2008, GQD consists of π conjugated structure and highly crystalline sp^2 carbon atom and has been extensively used in optical due to multi-color emission, high photostability, electrochemiluminescence, photoluminescence, biocompatibility, chemical stability, good solubility in water (due to -COOH groups), and tiny size of GQDs results in edge effects and quantum confinement. Moreover, GQD has a high surface area that acts as an electron donor and acceptor that also improves the optical properties and a large number of active sites to bind the selective biorecognition element. However, the optical sensor achieved low detection sensitivity due to the low quantum yield of GQDs. Therefore, the surface modification via organic layer and heteroatoms (i.e., boron, phosphorous, nitrogen, and sulfur) doped GQDs efficiently used in optical biosensors (Anh et al. 2017). The GQDs surface modification improves quantum yield, biofunctionality, chemical stability, and optical properties, which improve the sensitivity and selectivity of biomolecules (Tam et al. 2014; Qu et al. 2015). The "bottom-up" and "top-down" are the two primary methods that can be used to synthesize GQDs. Chemical approaches such as physical stripping, hydrothermal/solvothermal, electrochemical exfoliation, and microwave-assisted methods are widely used to produce a distinct size of GQDs. Hu et al. (2012) used GO in an ammonia solution and employed a one-pot hydrothermal method to synthesize the nitrogen-doped GQDs (N-GQDs). Santiago et al. (2017) used the pulse laser ablation technique with GO as a carbon source and diethylenetriamine (DETA) as a nitrogen source to synthesize the N-GQDs. Zheng et al. (2018) produced N-GQDs via the CVD method using C_{60} molecules under N_2 gas. These methods are time-consuming and complicated and need sophisticated instruments like CVD. Therefore, the one-step, "bottom-up" technique is the best method for creating N-GQDs owing to its simplicity and fast. Various methods including hydrothermal method, microwave irradiation, and pyrolysis have been used to prepare N-GQDs using different carbon and nitrogen sources like citric acid, glucose, and glycine and nitrogen sources like ammonia, glycine, dicyandiamide, tris(hydroxymethyl) aminomethane, or 3,4-dihydroxy-l-phenylalanine have also been utilized in the synthesis of N-GQDs (Zheng et al. 2017). The aforementioned techniques for synthesizing N-GQDs involve a series of processes, including polymerization, carbonization, and nitrogen heteroatom doping, carried out in a single reaction to develop N-GQDs. Safardoust-Hojaghan et al. (2017) produced N-GQDs by using citric acid and ethylenediamine via the hydrothermal method. Zheng et al. (2017) developed fast, high-quality N-GQDs using one-step microwave irradiations using glucose (carbon source) and ammonia (nitrogen source and act as a catalyst). The rapid glucose polymerization and carbonization occur under microwave heating, speed up the reaction rate, and cause the rapid production of N-GQDs within 1 min. Yan et al. (2016) reported the synthesis of water-soluble N-GQDs via one-step one-pot pyrolysis (high-temperature process) using citric acid (carbon source) and glycine (nitrogen source) at 200°C. Kalkal et al. (2020) reported highly sensitive fluorescence biosensors for lung cancer biomarker (neuron-specific enolase) detection using AuNP-modified GQDs. First GQDs were prepared by one-pot hydrothermal synthesis (bottom-up) at low temperatures. Briefly, the mixture of citric acid and DETA was dissolved in DI water under magnetic stirring. The resultant mixture was moved to a Teflon-lined vessel and heated for 5 h at 170°C. The obtained dispersion was filtered (using 0.2 μm syringe filter), followed by dialysis for 2 days to collect amine-N-GQDs. The antibody (anti-NSE Ab) was activated via EDC/NHS chemistry. In order to activate the anti-NSE antibody, the antibody was incubated with the mixed solution of EDC/NHS under optimized conditions. The activated antibody solution was added to the amine-N-GQDs solution (1:1 v/v) and incubated under optimized conditions, resulting in bioconjugate (anti-NSE Ab/amine-N-GQDs) produced via stable amid bond between the -COOH groups of antibody and -NH_2 groups of N-GQDs. The bioconjugate (anti-NSE Ab/amine-N-GQDs) was modified with AuNPs. For the synthesis of AuNPs, trisodium citrate dihydrate solution was dissolved in DI water under constant stirring for 2 h at 95°C, followed by slowly added gold (III) chloride trihydrate for further stirring for 5–10 min. The red-colored solution was obtained, indicating the formation of AuNPs.

The well-distributed, spherical AuNPs have an average dimension of about 20 nm. The different concentrations of antigen were added to AuNP-modified bioconjugate solution and incubated under optimized conditions. The fluorescence intensity was recorded with increasing concentrations of NSE antigen from 0.1 pg/mL to 1000 ng/mL, achieved good linear behavior, and the LOD is about 0.09 pg/mL. The highly sensitive response is attributed to the formation of immunocomplex between AuNP-modified anti-NSE Ab/amine-N-GQDs and antigen, causing the hindrance of the energy transfer process while restoring fluorescence for adding NSE antigen concentration ("ON" state). Figure 12.2 schematically depicts the sensing mechanism of the fluorescence immunoassay that utilizes anti-NSE Ab/amine-N-GQDs@ AuNPs. The mechanism behind this process is the bioconjugate (anti-NSE Ab/ amine-N-GQDs) releases strong blue fluorescence under the exposure of excitation wavelength (at 360 nm, "ON" state), which is quenched by AuNPs (as an acceptor), resulting in accepting the donor energy. The AuNPs provide a large surface area, which allows an efficient transfer of energy through a multiple-donor-single-acceptor configuration, which enhances the sensitivity and selectivity of biosensors.

12.6   METAL NANOPARTICLE (MNP)-BASED FLUORESCENCE BIOSENSORS

MNPs, including silver nanoparticles (AgNPs), gold nanoparticles (AuNPs), and platinum nanoparticles, are introduced in optical biosensors for the detection of a variety of biomolecules (Yaraki and Tan 2020; Choi et al. 2021). In particular, AuNPs are widely used as excellent transducers for optical biosensors due to their high surface-to-volume ratio, biocompatibility, chemical stability, and unique electrical and optical properties (Anh et al. 2022). Several Au nanostructures have been explored, including nanorods

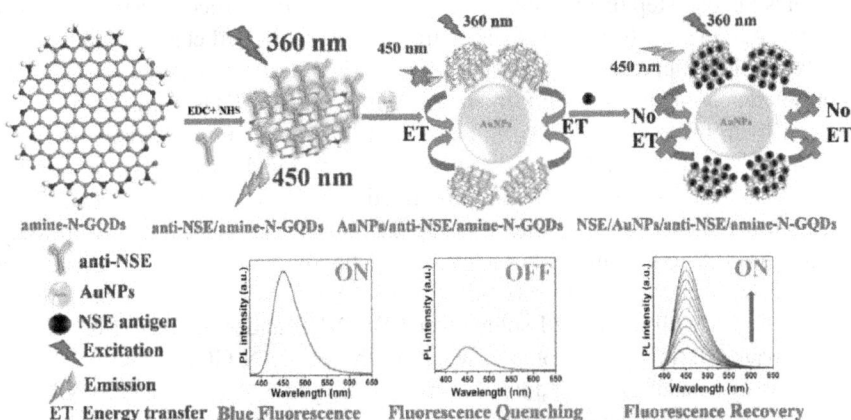

FIGURE 12.2 Schematic illustration of biofunctionalization steps to fabricate the fluorescence biosensor of AuNP-modified anti-NSE Ab/GQDs and fluorescence sensing mechanism for small cell lung cancer biomarker detection. (Reprinted from Kalkal et al. 2020. ACS Applied Bio Materials 3: 4922–493290, with permission from the American Chemical Society.)

(AuNRs), nanospheres (AuNSs), nanorings (AuNRgs), and nanodisks (AuNDs), which offer unique optical properties. When compared to Au electrodes, AuNPs with diameters between 1 and 100 nm have distinctive catalytic and good optical characteristics. In addition, the surface of AuNPs is especially amenable to conjugation with thiol groups (-SH), which allows surface functionalization for efficient binding of various bioreceptors. AuNPs have been easily synthesized via chemical and physical approaches. The optical characteristics indicating wavelength shift, color change, and change in Raman scattering are highly dependent on size, shape, and aggregation. AuNPs have been employed to develop different types of optical sensors, including fluorescence, colorimetric, FRET, SERS, and local surface plasmon resonance.

Food safety is a critical global issue that has a significant impact on human health, particularly during the processing and storage of foods. Foodborne illnesses are a major public health concern worldwide, affecting both developing and developed countries. Foodborne microbes are responsible for millions of infections, poisonings, and fatalities. Bacterial diarrhea is caused by pathogens such as *Salmonella*, *Shigella*, and diarrheagenic (Scallan et al. 2011; Kotloff et al. 2013). *E. coli* is prevalent in children, with *Shigella* being a significant cause of dysentery and inflammatory diarrhea. Moreover, there are common foodborne microbes, including *Salmonella*, *Staphylococcus aureus* (*S. aureus*), *Listeria monocytogenes*, and *Escherichia coli* (O157:H7), which are responsible for the majority of foodborne illnesses, hospitalizations, and deaths (Majowicz et al. 2010; "Foodborne Germs and Illnesses CDC" 2023; Doyle and Erickson 2012). The primary cause of foodborne disease is the consumption of contaminated food or water. Standard culture methods are typically used to isolate and enumerate pathogens in food samples. These methods are considered the "gold standard" because they are sensitive, inexpensive, and provide both qualitative and quantitative detection of microorganisms present in food samples. However, these methods have some limitations such as being time-consuming and requiring multistep processing that includes pre-enrichment steps. In recent years, fluorescent detection technology has been widely used to detect food pathogens. Nanomaterial-based fluorescent sensors have gained significant attention due to their rapid response, high sensitivity, and wide detection range. Therefore, the development of ultrasensitive, rapid, and reliable methods for early detection and analysis of foodborne pathogens to prevent outbreaks of foodborne diseases. Elahi et al. (2019) reported a highly sensitive fluorescence-based sandwich immunoassay for the detection of *Shigella* species. For AuNP synthesis, $HAuCl_4$ was heated under continuous stirring, then slowly added sodium citrate solution. The dark purple color resultant solution was obtained after 20 min magnetic stirring at 100 °C, confirming the formation of AuNPs. Afterward, MNPs were synthesized by the following process; $FeCl_3 \cdot 6H_2O$ and ethylene glycol were mixed together, followed by addition of CH_3COONa, NaOH, and ethylene diamine (to generate $-NH_2$ groups) under continuous stirring for 30 min. The resultant mixture was autoclaved for 6 h at 121°C and 105 KPa. The obtained $-NH_2$-MNPs were separated out using a magnet, followed by washing with ethanol and DI water. To activate the DNA probes, both the probes DNA probe1 (5′AAGCTGGTTCGAAAGTATTthiol-3′) and fluorescence-DNA probe2 (5′TEX613TTATTCGTAGCTAAAAAAAAAAthiol3′) were added to dithiothreitol solution (1N DTT: 0.001M sodium acetate) for 15 min at 25°C. Then, DTT was extracted from the reaction mixture after the breakdown of the disulfide bond between the DNA probes by adding ethyl acetate. After centrifugation, the supernatant was discarded, and an activated DNA probe was collected. The activated DNA probe and fluorescent DNA probe were added to the AuNPs solution and incubated under a dark environment for 14 h at room temperature. The AuNPs modified with two probes (DNA probe and fluorescent DNA probe) were obtained and analyzed by fluorescence spectrophotometry. For the functionalization of MNPs, an optimized amount of MNPs was mixed with the coupling buffer (composed of NaCl, PBS buffer, and sulfosuccinimidyl 4-nmaleimidomethyl cyclohexane-1-carboxylate), kept the reaction mixture for 2–3 h. The modified MNPs were filtered out and washed with buffer, then added to the DNA probe3 (5′thiolTTAAGAGTGGGGTTTGATG3′) solution, followed by incubation for 8–10 h at room temperature. The resultant bioconjugate (MNP@SMCC/DNA probe3) was separated by a magnetic field, followed by washing with buffer solution. The different concentrations of single-strand target DNA were added to the bioconjugate (MNP@SMCC/DNA probe3) and incubated for 45 min at an optimized temperature. The binding of target ssDNA to MNPs modified DNA probe3 was separated by the magnetic field, followed by washing with buffer. Afterward, AuNP-modified DNA probe1 and fluorescence-DNA probe2 conjugate were added to MNP-modified DNA probe3 and incubated under optimized conditions. The resulting sandwich immunoassay (MNPs/DNA probe3/target DNA/ fluorescence-DNA probe2/AuNPs/DNA probe1) was collected after several washing steps with buffer using a magnetic field. Figure 12.3 illustrates the step-by-step process for the fabrication of a fluorescence-based sandwich immunoassay, which includes the synthesis of both nanoparticles (MNPs and AuNPs) modified with DNA probes. The different concentrations of extracted target DNA of bacteria ranging from 2.3×10^1 to 2.3×10^9 CFU/mL were mixed with the bioconjugate.

The biosensor formed a biocomplex that released fluorescence-DNA probe2 upon detecting the target DNA under optimal conditions. The fluorescence-DNA probe2 was then isolated using magnetic nanoparticles and a magnetic field, and its fluorescence intensity was measured via fluorescence spectrophotometry. The results show the fluorescence intensity increased with increasing target DNA concentration ranging from 10^2 to 10^9 CFU/mL, achieving good linearity and LOD of 90 CFU / mL. The study evaluated the specificity of a fluorescent DNA biosensor designed to detect *Shigella* species by testing it with target DNA from various pathogens (i.e., *S. aureus*, *E. coli*, *P. aeruginosa*, and *V. cholera*). The result shows that the biosensor was specific to detecting *Shigella* species, with no cross-reactivity observed with other pathogens tested in this study.

12.7 QUANTUM DOT-BASED FLUORESCENCE BIOSENSOR

Quantum dots (QDs) are semiconductor colloidal nanoparticles (average size 2–10 nm) that have been utilized for developing sensitive fluorescence biosensors. QDs have unique optical and electronic properties, such as high quantum yield, broad absorption spectra, narrow and tunable emission spectrum, good photostability, and chemical stability (Petryayeva et al. 2013; Shao et al. 2011; Wegner and Hildebrandt 2015). These properties make QDs an attractive alternative to organic fluorophores for developing

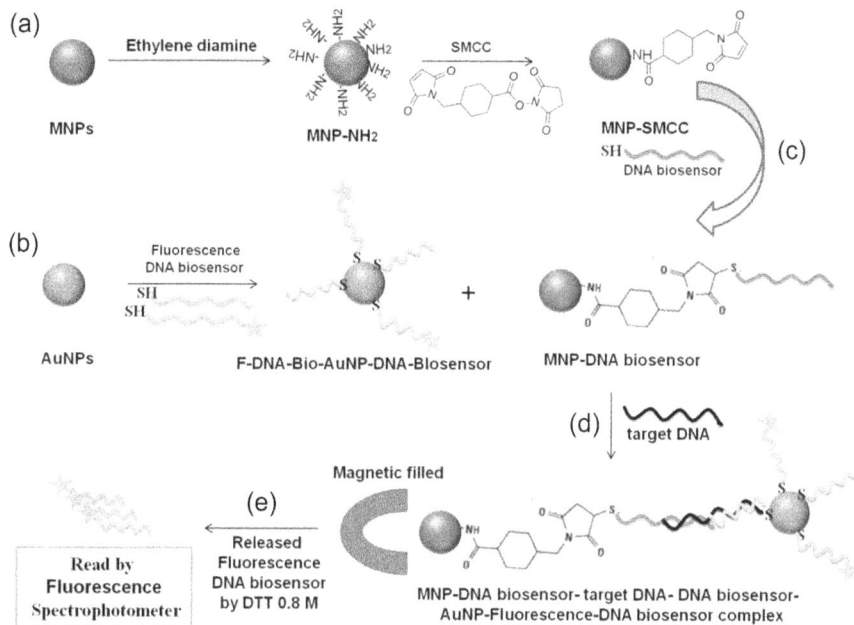

FIGURE 12.3 A schematic illustration of fabrication and measurement fluorescence-based sandwich immunoassay includes (a) modification of MNPs using ethylene diamine and DNA probe3, (b) modification of AuNPs with DNA probe1 and probe2, (c) formation of MNP/AuNP-modified probe DNA conjugate, (d) addition of target DNA to the formation of MNP/AuNP-modified probe DNA conjugate/target DNA, separated via a magnetic field, (e) measurement of the fluorescence intensity using a spectrophotometer. (Reprinted from Elahi et al. 2019. Materials Science and Engineering: C 105: 110113, with permission from Elsevier.)

highly sensitive fluorescence biosensors. The high quantum yield of QDs allows for efficient light emission, while their broad absorption spectra enable excitation with a wide range of wavelengths. Additionally, the narrow and tunable emission spectrum of QDs allows for multiplexed detection of multiple analytes simultaneously. Finally, the good chemical stability and photostability of QDs make them suitable for long-term use in biosensing applications. The composition of colloidal nanocrystals is typically made up of elements from groups III–V, II–VI, and IV–VI of the periodic table. To improve the photoluminescence quantum yield and stability of these core nanocrystals, a higher-band gap semiconductor layer such as ZnS, ZnSe, or CdS is often used to encapsulate the core nanocrystals such as CdS or CdSe (Tajarrod et al. 2016; Yu and Peng 2002). This results in the formation of core-shell nanocrystals such as CdS/ZnSe or CdS/ZnS. The use of a higher-band gap semiconductor layer helps to enhance the optical properties and stability of the core nanocrystals by reducing surface defects and preventing oxidation. The resulting core-shell nanocrystals have improved photoluminescence quantum yield and stability, making them useful for various applications in fields such as optoelectronics and biomedicine. The toxicity of quantum dots is a crucial factor in the study of cellular imaging and in vivo applications. For example, cadmium and lead are both toxic heavy metals in cadmium-based and lead-based QDs, respectively. Therefore, the newly developed QDs with better biocompatibility support their biological uses even more. Cadmium-free quantum dots such as CQDs, GQDs, and silicon quantum dots (Si QDs) are less toxic and biocompatible than group II–VI QDs (Morozova et al. 2020; Lu et al. 2019; Sun et al. 2013). The CQDs and GQDs are simply produced from carbon sources that are abundant in nature using a cost-effective chemical process. The surface functionalization and passivation using the organic layer enable simple tuning of the optical characteristics and solubility carbon dots. Therefore, CQDs and GQDs have been widely used in fluorescence-based biosensors. CDs have several advantages, including cost-effectiveness, photostability, and easy to produce compared to other fluorescent organic dyes. Moreover, semiconductor quantum dots (ZnS, ZnSe, etc.) are also alternate options for biosensing applications due to their photostability, quantum efficiency, and tunable fluorescence properties. Di-n-butyl phthalate (DBP) is a plasticizer that is commonly used in the food packing, chemical, and pipe industries (Ostrovský et al. 2011). Moreover, in order to increase the flexibility and plasticity of polyvinyl chloride polymers, DBP was frequently used (Silva et al. 2007). Earlier research demonstrated that DBP is an endocrine-disrupting substance which can cause cancer and lower fertility in humans (Yue et al. 2020). The normal concentration of DBP is 0.01 μM which should not be exceeded in drinking water (Zhu et al. 2019). Therefore, researchers are focused on the accurate detection of DBP in food samples and drinking water (Xu et al. 2017; Liang et al. 2017). Chen et al. (2022) reported a fluorescence imprinted biosensor for rapid and selective detection of di-n-butyl phthalate in food samples using the CdTe/ZIF-67 nanostructure. Figure 12.4a shows the complete demonstration of the synthesis of nanostructure complex (CdTe/ZIF-67/MIP). In order to synthesize CdTe QDs, the appropriate amount of tellurium powder and $NaBH_4$ was mixed in DI water, followed by ultrasonication for 1 h to produce a colorless NaHTe solution. Then, cadmium chloride hemipentahydrate ($CdCl_2 \cdot 2.5H_2O$) and 3-mercaptopropionic acid were mixed together, and pH was maintained at 12. The resulting mixture was added slowly into NaHTe solution under anaerobic conditions. The reaction was kept for 50 min in a water bath at an optimized temperature. The fluorescence intensity of the CdTe QDs solution was measured at 695 nm when excited with a wavelength of 400 nm. Afterward, zeolite imidazolate framework-67 (ZIF-67) nanoparticles were prepared according to the reported study (Yang et al. 2018). First, cobalt nitrate hexahydrate $Co(NO_3)_2 \cdot 6H_2O$ was dispersed in methanol, followed by adding the 2-methylimidazole (HmIM) in methanol at room temperature under constant stirring for 60 min. The resultant solution was centrifuged, followed by washing with ethanol and dried under controlled conditions, followed by mixing with CdTe QDs solution. Thus, the -COOH groups on the CdTe QDs covalently attached with $-NH_2$ groups on ZIF-67 surface. The CdTe/ZIF-67/MIP was synthesized using the sol-gel method (Cai et al. 2019). The optimized amount of DBP, ZIF-67, APTES, and CdTe QDs solution was mixed into ethanol under magnetic stirring for 15 min at room temperature, followed by addition of the tetraethoxysilane and ammonium hydroxide ($NH_3 \cdot H_2O$), kept the reaction for 12 h. The resultant solution (CdTe/ZIF-67/MIP) was centrifuged, followed by washing with ethanol, and dried under controlled conditions. The structure and surface characterization of the prepared nanostructure complex (CdTe/ZIF-67/MIP) was characterized using XRD, SEM, and TEM techniques. The average diameter of CdTe nanoparticles was ~4.8 nm, and SEM images show a rhombic dodecahedron structure of ZIF-67 microcrystals with a uniform average size of 650 nm (Figure 12.4b). According to TEM images of complex (CdTe/ZIF-67/MIP), the average thickness of the shell of the imprinted layer coated on the ZIF-67-modified CdTe surface was ~90 nm. The optimum concentration of DBP and ions was incubated with a complex (CdTe/ZIF-67/MIP) solution under optimized conditions. The fluorescence sensor was tested with real food samples such as tap and bottled water, juice, milk, and fish muscle tissue. The complex (CdTe/ZIF-67/MIP) solution was incubated with a fixed concentration of real sample solution, followed by measuring the fluorescence intensity at the wavelength (695 nm). The results show the fluorescence quenching intensities decreased with increasing the DBP concentration ranging from 0.05 to 18 μM (Figure 12.4c). These results attributed to ZIF-67 with a high specific surface area attached MIP layer to improve the rate of mass transfer and binding capacity of CdTe QDs due to the specific recognition cavities formed on the complex (CdTe/ZIF-67/MIP) surface for selective binding of DBP. The fluorescence imprinted sensor demonstrated the linear behaviors for the increasing concentration range of DBP in drinking water (Figure 12.4d), achieved with a LOD of 1.6 nM;

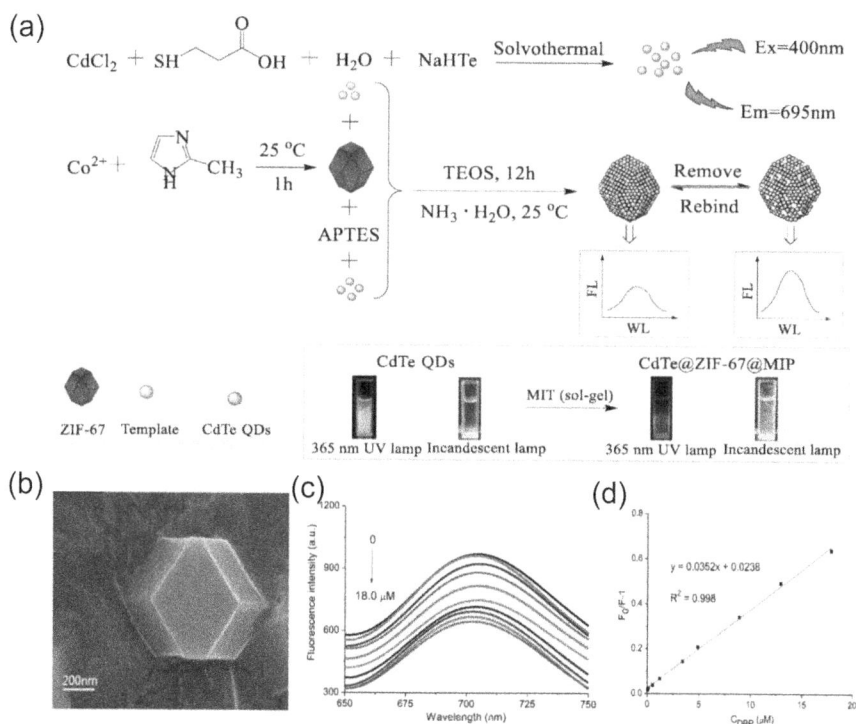

FIGURE 12.4 (a). Schematic illustrations of CdTe QDs synthesis were prepared by the one-pot solvothermal method, ZIF-67 microparticle and CdTe/ZIF-67/MIP were produced by the facile chemical method. (b) SEM images of ZIF-67, (c) fluorescence emission spectrum of CdTe/ZIF-67/MIP for increasing concentration of di-n-butyl phthalate solution, and (d) corresponding calibration plot. (Reprinted from Chen et al. 2022. Food Chemistry 367: 130505, with permission from Elsevier.)

this value is less than the normal concentration of DBP in drinking water. For the selectivity of the fluorescence sensor, the sensor was tested against interference ions. The sensor shows a higher fluorescence quenching effect for the concentration of DBP compared to other interference ions.

12.8   ADVANTAGES AND LIMITATIONS

As biosensors become increasingly important for clinical analysis, the use of fluorescent biosensors for detecting bioanalytes has become a promising area of research. In this book chapter, we will also discuss the advantages and limitations of using fluorescent biosensors. Fluorescent biosensors offer several advantages over other types of biosensing techniques for detecting various bioanalytes. One major advantage is their high sensitivity and selectivity. Fluorescent dyes can be designed to specifically bind to certain analytes, allowing for accurate detection even in complex biological samples. Another advantage of fluorescent biosensors is their versatility in terms of analyte detection. In addition to proteins, viruses, neurotransmitters, hormones, and metabolites like glucose and lactate, fluorescent biosensors can also be used to detect pathogens and environmental toxins. This makes them useful not only in clinical settings but also in environmental monitoring and food safety. Compared to other types of biosensors such as electrochemical or optical biosensors, fluorescent biosensors offer several advantages. For example, electrochemical biosensors require a conductive surface for detection and may be affected by interference from other electroactive species in the sample. Optical biosensors such as SPR require specialized equipment and may be limited by the availability of specific antibodies or bioreceptors for target analytes. In contrast, fluorescent biosensors are relatively easy to construct and can be designed to detect a wide range of substances with high sensitivity and selectivity.

There are some limitations and challenges to using fluorescent biosensors for continuous and time-resolved measurements. One challenge is the potential interference from autofluorescence or other background signals in biological samples. This can lead to false positives or inaccurate measurements. Additionally, the cost and complexity of constructing fluorescence-based detection systems can be a barrier to widespread adoption. To address these challenges, researchers have developed various strategies. For example, one approach is to use advanced signal processing techniques to filter out background signals and improve the accuracy of measurements. Another approach is to use alternative detection methods such as SPR or electrochemical detection, which can offer improved accuracy and selectivity. In terms of cost and complexity, researchers have developed microfluidic devices based on fluorescence that offer several advantages over traditional detection methods. These devices are small-scale systems that allow for precise control over fluid flow and reaction conditions. They can be constructed using a variety of materials, such as glass or polymers, and can incorporate multiple channels for different reactions. Microfluidic devices based on fluorescence have shown promising applications in several areas, including immunoassays and nucleic acid amplification tests. Overall, ongoing research in this area will continue to explore new strategies for improving the accuracy, sensitivity, selectivity, and affordability of fluorescent biosensors for continuous and real-time measurements.

12.9 CONCLUSIONS

Fluorescence biosensors have been used for detecting a variety of biomolecules for early-stage disease diagnosis and food safety analyses. The nanostructured materials such as carbon nanomaterials, quantum dots, and metallic nanoparticles have unique fluorescence properties that can enhance the quantum yield, quenching efficiency, and photostability of fluorescence biosensors. Additionally, these nanomaterials can be used as signal amplifiers to increase fluorescence polarization. This chapter discusses different types of nanomaterials used in fluorescence-based biosensors to detect the various bioanalytes (proteins, viruses, pathogens, and other substances) and their roles as fluorescent elements and/or signal amplifiers. For example, quantum dots are semiconductor nanocrystals that exhibit unique optical properties such as high fluorescence, narrow emission spectra, and long-term photostability. These properties make them ideal for use in fluorescence-based biosensors. Metallic nanoparticles, such as AuNPs or AgNPs, can also be used as signal amplifiers in fluorescence-based biosensors. These nanoparticles exhibit strong surface plasmon resonance effects that can enhance the fluorescence polarization signal by several orders of magnitude. Moreover, nanostructured materials used in fluorescence-based biosensors have high photostability, which allows for the detection of different analytes with high sensitivity. This is an advantage over conventional fluorescence biosensors made of organic dyes. However, there are also limitations to using nanomaterials in fluorescence-based biosensors. For example, the synthesis and functionalization of these materials can be complex and expensive. Additionally, there may be concerns about the toxicity or environmental impact of some types of nanomaterials. Overall, the use of nanomaterials in fluorescence-based biosensors offers several advantages such as improved sensitivity and selectivity.

REFERENCES

Ackermann, Julia, Justus T. Metternich, Svenja Herbertz, and Sebastian Kruss. 2022. "Biosensing with fluorescent carbon nanotubes." *Angewandte Chemie International Edition* 61 (18). John Wiley & Sons, Ltd: e202112372.

Alaimo, Alessandro, Valentina Vaira, Virinder Kaur Sarhadi, and Gemma Armengol. 2022. "Molecular biomarkers in cancer." *Biomolecules* 12 (8). Multidisciplinary Digital Publishing Institute: 1021.

Alarfaj, Nawal Ahmad, Maha Farouk El-Tohamy, and Hesham Farouk Oraby. 2018. "CA 19-9 pancreatic tumor marker fluorescence immunosensing detection via immobilized carbon quantum dots conjugated gold nanocomposite." *International Journal of Molecular Sciences* 19 (4). Multidisciplinary Digital Publishing Institute: 1162.

Andersson, Joel, Thorgerdur Palsdottir, Anna Lantz, Markus Aly, Henrik Grönberg, Lars Egevad, Martin Eklund, and Tobias Nordström. 2022. "Digital rectal examination in Stockholm3 biomarker-based prostate cancer screening." *European Urology Open Science* 44. Elsevier: 69–75.

Anfossi, Laura, Fabio Di Nardo, Simone Cavalera, Cristina Giovannoli, Giulia Spano, Elena S. Speranskaya, Irina Y. Goryacheva, and Claudio Baggiani. 2018. "A lateral flow immunoassay for straightforward determination of fumonisin mycotoxins based on the quenching of the fluorescence of CdSe/ZnS quantum dots by gold and silver nanoparticles." *Microchimica Acta* 185 (2). Springer-Verlag Wien: 1–10.

Anh, Nguyen Ha, Mai Quan Doan, Ngo Xuan Dinh, Tran Quang Huy, Doan Quang Tri, Le Thi Ngoc Loan, Bui Van Hao, and Anh Tuan Le. 2022. "Gold nanoparticle-based optical nanosensors for food and health safety monitoring: Recent advances and future perspectives." *RSC Advances* 12 (18). The Royal Society of Chemistry: 10950–10988.

Anh, Nguyen Thi Ngoc, Ankan Dutta Chowdhury, and Ruey an Doong. 2017. "Highly sensitive and selective detection of mercury ions using N, S-codoped graphene quantum dots and its paper strip based sensing application in wastewater." *Sensors and Actuators B: Chemical* 252 (November). Elsevier: 1169–1178.

Anwar, Sadat, Haizhen Ding, Mingsheng Xu, Xiaolong Hu, Zhenzhen Li, Jingmin Wang, Li Liu, et al. 2019. "Recent advances in synthesis, optical properties, and biomedical applications of carbon dots." *ACS Applied Bio Materials* 2 (6). American Chemical Society: 2317–2338.

Anzar, Nigar, Rahil Hasan, Manshi Tyagi, Neelam Yadav, and Jagriti Narang. 2020. "Carbon nanotube - a review on synthesis, properties and plethora of applications in the field of biomedical science." *Sensors International* 1 (January). Elsevier: 100003.

Armano, Angelo, and Simonpietro Agnello. 2019. "Two-dimensional carbon: A review of synthesis methods, and electronic, optical, and vibrational properties of single-layer graphene." *C – Journal of Carbon Research* 5 (4). Multidisciplinary Digital Publishing Institute: 67.

Azam, Nayab, Murtaza Najabat Ali, and Tooba Javaid Khan. 2021. "Carbon quantum dots for biomedical applications: Review and analysis." *Frontiers in Materials* 8 (August). Frontiers Media S.A.: 272.

Bai, Yunlong, Tailin Xu, and Xueji Zhang. 2020. "Graphene-based biosensors for detection of biomarkers." *Micromachines* 11 (1). Multidisciplinary Digital Publishing Institute: 60.

Barone, Paul W., Seunghyun Baik, Daniel A. Heller, and Michael S. Strano. 2004. "Near-infrared optical sensors based on single-walled carbon nanotubes." *Nature Materials* 4 (1). Nature Publishing Group: 86–92.

Bhuyan, Md Sajibul Alam, Md Nizam Uddin, Md Maksudul Islam, Ferdaushi Alam Bipasha, and Sayed Shafayat Hossain. 2016. "Synthesis of graphene." *International Nano Letters* 6 (2). Springer: 65–83.

Bohunicky, Brian, and Shaker A. Mousa. 2010. "Biosensors: The new wave in cancer diagnosis." *Nanotechnology, Science and Applications* 4 (1). Dove Press: 1–10.

Boobalan, T., M. Sethupathi, N. Sengottuvelan, Ponnuchamy Kumar, P. Balaji, Balázs Gulyás, Parasuraman Padmanabhan, Subramanian Tamil Selvan, and A. Arun. 2020. "Mushroom-derived carbon dots for toxic metal ion detection and as antibacterial and anticancer agents." *ACS Applied Nano Materials* 3 (6). American Chemical Society: 5910–5919.

Borisov, Sergey M., and Otto S. Wolfbeis. 2008. "Optical biosensors." *Chemical Reviews* 108 (2). American Chemical Society: 423–461.

Bray, Freddie, Jacques Ferlay, Isabelle Soerjomataram, Rebecca L. Siegel, Lindsey A. Torre, and Ahmedin Jemal. 2018. "Global cancer statistics 2018: GLOBOCAN estimates of incidence and mortality worldwide for 36 cancers in 185 countries." *CA: A Cancer Journal for Clinicians* 68 (6). American Cancer Society: 394–424.

Burdanova, Maria G., Marianna V. Kharlamova, Christian Kramberger, and Maxim P. Nikitin. 2021. "Applications of pristine and functionalized carbon nanotubes, graphene, and graphene nanoribbons in biomedicine." *Nanomaterials* 11 (11). Multidisciplinary Digital Publishing Institute: 3020.

Cai, Lei, Zhaohui Zhang, Haimei Xiao, Shan Chen, and Jinli Fu. 2019. "An eco-friendly imprinted polymer based on graphene quantum dots for fluorescent detection of p-nitroaniline." *RSC Advances* 9 (71). The Royal Society of Chemistry: 41383–41391.

Cai, Nan, Lu Tan, Yan Li, Tingting Xia, Tianyu Hu, and Xingguang Su. 2017. "Biosensing platform for the detection of uric acid based on graphene quantum dots and G-quadruplex/hemin DNAzyme." *Analytica Chimica Acta* 965. Elsevier: 96–102.

Cardoso Dos Santos, Marcelina, W. Russ Algar, Igor L. Medintz, and Niko Hildebrandt. 2020. "Quantum dots for Förster resonance energy transfer (FRET)." *TrAC Trends in Analytical Chemistry* 125 (April). Elsevier: 115819.

Chan, Kit Man, Jonathan M. Gleadle, Michael O'Callaghan, Krasimir Vasilev, and Melanie MacGregor. 2022. "Prostate cancer detection: A systematic review of urinary biosensors." *Prostate Cancer and Prostatic Diseases* 25 (1). Nature Publishing Group: 39–46.

Chauhan, Dheeraj Singh, M. A. Quraishi, and Chandrabhan Verma. 2022. "Carbon nanodots: Recent advances in synthesis and applications." *Carbon Letters* 32 (7). Springer: 1603–1629.

Chen, Shan, Jinli Fu, Shu Zhou, Pengfei Zhao, Xiaodan Wu, Sisi Tang, and Zhaohui Zhang. 2022. "Rapid recognition of Di-n-butyl phthalate in food samples with a near infrared fluorescence imprinted sensor based on zeolite imidazolate framework-67." *Food Chemistry* 367 (January). Elsevier: 130505.

Choi, Hye Kyu, Myeong Jun Lee, Sang Nam Lee, Tae Hyung Kim, and Byung Keun Oh. 2021. "Noble metal nanomaterial-based biosensors for electrochemical and optical detection of viruses causing respiratory illnesses." *Frontiers in Chemistry* 9 (May). Frontiers Media S.A.: 298.

Cinquanta, Luigi, Desré Ethel Fontana, and Nicola Bizzaro. 2017. "Chemiluminescent immunoassay technology: What does it change in autoantibody detection?" *Autoimmunity Highlights* 8 (1). Springer-Verlag Italia s.r.l.: 1–8.

Cui, Lin, Xin Ren, Mengtao Sun, Haiyan Liu, and Lixin Xia. 2021. "Carbon dots: Synthesis, properties and applications." *Nanomaterials* 11 (12). Multidisciplinary Digital Publishing Institute: 3419.

Das, Sreyashi, Ram Devireddy, and Manas Ranjan Gartia. 2023. "Surface plasmon resonance (SPR) sensor for cancer biomarker detection." *Biosensors* 13 (3). Multidisciplinary Digital Publishing Institute: 396.

Deb, Ankita, Gaurav Raghunath Nalkar, and Devasish Chowdhury. 2023. "Biogenic carbon dot-based fluorescence-mediated immunosensor for the detection of disease biomarker." *Analytica Chimica Acta* 1242 (February). Elsevier: 340808.

Dong, Jian, Kaiqi Wang, Liping Sun, Baoliang Sun, Mingfeng Yang, Hongyu Chen, Yi Wang, Jingyi Sun, and Lifeng Dong. 2018. "Application of graphene quantum dots for simultaneous fluorescence imaging and tumor-targeted drug delivery." *Sensors and Actuators B: Chemical* 256 (March). Elsevier: 616–623.

Doyle, Michael P., and Marilyn C. Erickson. 2012. "Opportunities for mitigating pathogen contamination during on-farm food production." *International Journal of Food Microbiology* 152 (3). Elsevier: 54–74.

Drummen, Gregor P.C. 2012. "Fluorescent probes and fluorescence (Microscopy) techniques - illuminating biological and biomedical research." *Molecules* 17 (12). Multidisciplinary Digital Publishing Institute: 14067–14090.

Elahi, Narges, Mehdi Kamali, Mohammad Hadi Baghersad, and Bahram Amini. 2019. "A fluorescence nano-biosensors immobilization on iron (MNPs) and Gold (AuNPs) nanoparticles for detection of shigella Spp." *Materials Science and Engineering: C* 105 (December). Elsevier: 110113.

Elvati, Paolo, Elizabeth Baumeister, and Angela Violi. 2017. "Graphene quantum dots: Effect of size, composition and curvature on their assembly." *RSC Advances* 7 (29). The Royal Society of Chemistry: 17704–17710.

Erkocyigit, Bilge Asci, Ozge Ozufuklar, Aysenur Yardim, Emine Guler Celik, and Suna Timur. 2023. "Biomarker detection in early diagnosis of cancer: Recent achievements in point-of-care devices based on paper microfluidics." *Biosensors* 13 (3). Multidisciplinary Digital Publishing Institute: 387.

Fan, Zetan, Shuhua Li, Fanglong Yuan, and Louzhen Fan. 2015. "Fluorescent graphene quantum dots for biosensing and bioimaging." *RSC Advances* 5 (25). The Royal Society of Chemistry: 19773–19789.

Farrera, Consol, Fernando Torres Andón, and Neus Feliu. 2017. "Carbon nanotubes as optical sensors in biomedicine." *ACS Nano* 11 (11). American Chemical Society: 10637–10643.

Farshbaf, Masoud, Soodabeh Davaran, Fariborz Rahimi, Nasim Annabi, Roya Salehi, and Abolfazl Akbarzadeh. 2017. "Carbon quantum dots: Recent progresses on synthesis, surface modification and applications." *Artificial Cells, Nanomedicine, and Biotechnology* 46 (7). Taylor & Francis: 1331–1348.

Feng, Jianguang, Hongzhou Dong, Liyan Yu, and Lifeng Dong. 2017. "The optical and electronic properties of graphene quantum dots with oxygen-containing groups: A density functional theory study." *Journal of Materials Chemistry C* 5 (24). The Royal Society of Chemistry: 5984–5993.

Ferrier, David C., and Kevin C. Honeychurch. 2021. "Carbon nanotube (CNT)-based biosensors." *Biosensors* 11 (12). Multidisciplinary Digital Publishing Institute: 486.

"Foodborne Germs and Illnesses | CDC." 2023. Accessed April 9. https://www.cdc.gov/foodsafety/foodborne-germs.html.

García-Hernández, Luis Abraham, Eduardo Martínez-Martínez, Denni Pazos-Solís, Javier Aguado-Preciado, Ateet Dutt, Abraham Ulises Chávez-Ramírez, Brian Korgel, Ashutosh Sharma, and Goldie Oza. 2023. "Optical detection of cancer cells using lab-on-a-chip." *Biosensors* 13 (4). Multidisciplinary Digital Publishing Institute: 439.

Gaviria-Arroyave, María Isabel, Juan B. Cano, and Gustavo A. Peñuela. 2020. "Nanomaterial-based fluorescent biosensors for monitoring environmental pollutants: A critical review." *Talanta Open* 2 (December). Elsevier: 100006.

Geim, A. K., and K. S. Novoselov. 2007. "The rise of graphene." *Nature Materials* 6 (3). Nature Publishing Group: 183–191.

Gong, Hua, Rui Peng, and Zhuang Liu. 2013. "Carbon nanotubes for biomedical imaging: The recent advances." *Advanced Drug Delivery Reviews* 65 (15). Elsevier: 1951–1963.

Guan, Xinwei, Zhixuan Li, Xun Geng, Zhihao Lei, Ajay Karakoti, Tom Wu, Prashant Kumar, Jiabao Yi, and Ajayan Vinu. 2023. "Emerging trends of carbon-based quantum dots: Nanoarchitectonics and applications." *Small* 19. John Wiley & Sons, Ltd, 2207181.

Heikal, Ahmed A. 2010. "Intracellular coenzymes as natural biomarkers for metabolic activities and mitochondrial anomalies." *Biomarkers in Medicine* 4 (2). Future Medicine Ltd London, UK: 241–263. doi: 10.2217/Bmm.10.1.

Hendler-Neumark, Adi, and Gili Bisker. 2019. "Fluorescent single-walled carbon nanotubes for protein detection." *Sensors* 19 (24). Multidisciplinary Digital Publishing Institute: 5403.

Hong, Guolin, Canping Su, Zhongnan Huang, Quanquan Zhuang, Chaoguo Wei, Haohua Deng, Wei Chen, and Huaping Peng. 2021. "Electrochemiluminescence immunoassay platform with immunoglobulin G-encapsulated gold nanoclusters as a 'Two-In-One' Probe." *Analytical Chemistry* 93 (38). American Chemical Society: 13022–13028.

Hong, Hao, Ting Gao, and Weibo Cai. 2009. "Molecular imaging with single-walled carbon nanotubes." *Nano Today* 4 (3). Elsevier: 252–261.

Hu, Chaofan, Yingliang Liu, Yunhua Yang, Jianghu Cui, Zirong Huang, Yaling Wang, Lufeng Yang, Haibo Wang, Yong Xiao, and Jianhua Rong. 2012. "One-step preparation of nitrogen-doped graphene quantum dots from oxidized debris of graphene oxide." *Journal of Materials Chemistry B* 1 (1). The Royal Society of Chemistry: 39–42.

Hu, Shengshui, and Chengguo Hu. 2009. "Carbon nanotube-based electrochemical sensors: Principles and applications in biomedical systems." *Journal of Sensors* 2009. doi: 10.1155/2009/187615.

Huang, Hongduan, Lei Liao, Xiao Xu, Mingjian Zou, Feng Liu, and Na Li. 2013. "The electron-transfer based interaction between transition metal ions and photolumines-cent graphene quantum dots (GQDs): A platform for metal ion sensing." *Talanta* 117 (December). Elsevier: 152–157.

Hwang, Hye Suk, Jae Won Jeong, Yoong Ahm Kim, and Mincheol Chang. 2020. "Carbon nanomaterials as versatile platforms for biosensing applications." *Micromachines* 11 (9). Multidisciplinary Digital Publishing Institute: 814.

Idowu, Mopelola, Emmanuel Lamprecht, and Tebello Nyokong. 2008. "Interaction of water-soluble thiol capped CdTe quantum dots and bovine serum albumin." *Journal of Photochemistry and Photobiology A: Chemistry* 198 (1). Elsevier: 7–12.

Iijima, Sumio. 1991. "Helical microtubules of graphitic carbon." *Nature* 354. Nature Publishing Group: 56–58.

Ilkhani, Hoda, and Siamak Farhad. 2018. "A novel electrochemical DNA biosensor for ebola virus detection." *Analytical Biochemistry* 557 (September). Academic Press: 151–155.

Ji, Chunyu, Yiqun Zhou, Roger M. Leblanc, and Zhili Peng. 2020. "Recent developments of carbon dots in biosensing: A review." *ACS Sensors* 5 (9). American Chemical Society: 2724–2741.

Joshi, Shalik R., Abhinav Sharma, Gun Ho Kim, and Jaesung Jang. 2020. "Low cost synthe-sis of reduced graphene oxide using biopolymer for influenza virus sensor." *Materials Science and Engineering: C* 108 (March). Elsevier: 110465.

Justino, Celine I.L., Ana R. Gomes, Ana C. Freitas, Armando C. Duarte, and Teresa A.P. Rocha-Santos. 2017. "Graphene based sensors and biosensors." *TrAC Trends in Analytical Chemistry* 91 (June). Elsevier: 53–66.

Kakkar, Saloni, Payal Gupta, Navin Kumar, and Krishna Kant. 2023. "Progress in fluorescence biosensing and food safety towards point-of-detection (PoD) system." *Biosensors* 13 (2). Multidisciplinary Digital Publishing Institute: 249.

Kalkal, Ashish, Rangadhar Pradhan, Sachin Kadian, Gaurav Manik, and Gopinath Packirisamy. 2020. "Biofunctionalized graphene quantum dots based fluorescent biosensor toward efficient detection of small cell lung cancer." *ACS Applied Bio Materials* 3 (8). American Chemical Society: 4922–4932.

Keller, Sascha G., Mako Kamiya, and Yasuteru Urano. 2020. "Recent progress in small spiro-cyclic, xanthene-based fluorescent probes." *Molecules* 25 (24). Multidisciplinary Digital Publishing Institute: 5964.

Khan, Ajahar, Parya Ezati, Jun Tae Kim, and Jong-Whan Rhim. 2023. "Biocompatible carbon quantum dots for intelligent sensing in food safety applications: Opportunities and sus-tainability." *Materials Today Sustainability* 21 (March). Elsevier: 100306.

Khayal, Areeba, Vinars Dawane, Mohammed A. Amin, Vineet Tirth, Virendra Kumar Yadav, Ali Algahtani, Samreen Heena Khan, Saiful Islam, Krishna Kumar Yadav, and Byong Hun Jeon. 2021. "Advances in the methods for the synthesis of carbon dots and their emerging applications." *Polymers* 13 (18). Multidisciplinary Digital Publishing Institute: 3190.

Kim, Haseong, Gui Hwan Han, Yaoyao Fu, Jongsik Gam, and Seung Goo Lee. 2016. "Highly sensitive and rapid fluorescence detection with a portable FRET analyzer." *Journal of Visualized Experiments* 2016 (116): 54144.

Kim, Minsoo, Raymond Iezzi, Bong Sup Shim, and David C. Martin. 2019. "Impedimetric biosensors for detecting vascular endothelial growth factor (VEGF) based on poly (3,4-Ethylene Dioxythiophene) (PEDOT)/gold nanoparticle (Au NP) composites." *Frontiers in Chemistry* 7 (MAR). Frontiers Media S.A.: 234.

Kocheril, Philip A., Kiersten D. Lenz, Daniel E. Jacobsen, and Jessica Z. Kubicek-Sutherland. 2022. "Amplification-free nucleic acid detection with a fluorescence-based waveguide biosensor." *Frontiers in Sensors* 3 (October). Frontiers: 28.

Kotloff, Karen L., James P. Nataro,hhhhhhhh William C. Blackwelder, Dilruba Nasrin, Tamer H. Farag, Sandra Panchalingam, Yukun Wu, et al. 2013. "Burden and aetiology of diarrhoeal disease in infants and young children in developing countries (the Global Enteric Multicenter Study, GEMS): A prospective, case-control study." *The Lancet* 382 (9888). Elsevier: 209–222.

Kruss, Sebastian, Andrew J. Hilmer, Jingqing Zhang, Nigel F. Reuel, Bin Mu, and Michael S. Strano. 2013. "Carbon nanotubes as optical biomedical sensors." *Advanced Drug Delivery Reviews* 65 (15). Elsevier: 1933–1950.

Kumar, Rajkumar Rakesh, Amit Kumar, Cheng-Hsin Chuang, and Muhammad Omar Shaikh. 2023. "Recent advances and emerging trends in cancer biomarker detection technologies." *Industrial & Engineering Chemistry Research* 62. American Chemical Society: 5691.

Lee, Jinyoung. 2023. "Carbon nanotube-based biosensors using fusion technologies with biologicals & chemicals for food assessment." *Biosensors* 13 (2). Multidisciplinary Digital Publishing Institute: 183.

Lee, Seungah, and Seong Ho Kang. 2023. "Wavelength-dependent metal-enhanced fluorescence biosensors via resonance energy transfer modulation." *Biosensors* 13 (3). Multidisciplinary Digital Publishing Institute: 376.

Leopold, Anna V., Daria M. Shcherbakova, and Vladislav V. Verkhusha. 2019. "Fluorescent biosensors for neurotransmission and neuromodulation: Engineering and applications." *Frontiers in Cellular Neuroscience* 13 (October). Frontiers Media S.A.: 474.

Li, Meixiu, Tao Chen, J. Justin Gooding, and Jingquan Liu. 2019. "Review of carbon and graphene quantum dots for sensing." *ACS Sensors* 4 (7). American Chemical Society: 1732–1748.

Liang, Ya Ru, Zong Mian Zhang, Zhen Jiang Liu, Kun Wang, Xiang Yang Wu, Kun Zeng, Hui Meng, and Zhen Zhang. 2017. "A highly sensitive signal-amplified gold nanoparticle-based electrochemical immunosensor for dibutyl phthalate detection." *Biosensors and Bioel ectronics* 91 (May). Elsevier: 199–202.

Lin, Jing, Xiaoyuan Chen, and Peng Huang. 2016. "Graphene-based nanomaterials for bioimaging." *Advanced Drug Delivery Reviews* 105 (October). Elsevier: 242–254.

Liu, Haomin, and Yu Lei. 2021. "A critical review: Recent advances in 'Digital' biomolecule detection with single copy sensitivity." *Biosensors and Bioelectronics* 177 (April). Elsevier: 112901.

Liu, Hua, Weidan Na, Ziping Liu, Xueqian Chen, and Xingguang Su. 2017. "A novel turn-on fluorescent strategy for sensing ascorbic acid using graphene quantum dots as fluorescent probe." *Biosensors and Bioelectronics* 92 (June). Elsevier: 229–233.

Lou, Ying, Xinyu Hao, Lei Liao, Kaiyou Zhang, Shuoping Chen, Ziyuan Li, Jun Ou, Aimiao Qin, and Zhou Li. 2021. "Recent advances of biomass carbon dots on syntheses, characterization, luminescence mechanism, and sensing applications." *Nano Select* 2 (6). John Wiley & Sons, Ltd: 1117–1145.

Lu, Fei, Yi Hua Zhou, Li Hui Wu, Jun Qian, Sheng Cao, Ya Feng Deng, and Yuan Chen. 2019. "Highly fluorescent nitrogen-doped graphene quantum dots' synthesis and their applications as Fe(III) ions sensor." *International Journal of Optics* 2019. Hindawi Limited. doi: 10.1155/2019/8724320.

Lu, Zhenyu, Yingying Wang, and Gongke Li. 2023. "Covalent organic frameworks-based electrochemical sensors for food safety analysis." *Biosensors* 13 (2). Multidisciplinary Digital Publishing Institute: 291.

Majowicz, Shannon E., Jennie Musto, Elaine Scallan, Frederick J. Angulo, Martyn Kirk, Sarah J. O'Brien, Timothy F. Jones, Aamir Fazil, and Robert M. Hoekstra. 2010. "The global burden of nontyphoidal salmonella gastroenteritis." *Clinical Infectious Diseases* 50 (6). Oxford Academic: 882–889.

Mbayachi, Vestince B., Euphrem Ndayiragije, Thirasara Sammani, Sunaina Taj, Elice R. Mbuta, and Atta ullah khan. 2021. "Graphene synthesis, characterization and its applications: A review." *Results in Chemistry* 3 (January). Elsevier: 100163.

Moosa Mayyadah Abed, Ahmed, Turk J. Chem, Ahmed A. Moosa, and Mayyadah S Abed. 2021. "Graphene preparation and graphite exfoliation." *Turkish Journal of Chemistry* 45 (3). TUBITAK: 493–519.

Morozova, Sofia, Mariya Alikina, Aleksandr Vinogradov, and Mario Pagliaro. 2020. "Silicon quantum dots: Synthesis, encapsulation, and application in light-emitting diodes." *Frontiers in Chemistry* 8 (April). Frontiers Media S.A.: 191.

Müllerová, Lucie, Kateřina Marková, Stanislav Obruča, and Filip Mravec. 2022. "Use of flavin-related cellular autofluorescence to monitor processes in microbial biotechnology." *Microorganisms* 10 (6). Multidisciplinary Digital Publishing Institute: 1179.

Nair, Raji V., Reny Thankam Thomas, Vandana Sankar, Hanif Muhammad, Mingdong Dong, and Saju Pillai. 2017. "Rapid, acid-free synthesis of high-quality graphene quantum dots for aggregation induced sensing of metal ions and bioimaging." *ACS Omega* 2 (11). American Chemical Society: 8051–8061.

Nawrot, Witold, Kamila Drzozga, Sylwia Baluta, Joanna Cabaj, and Karol Malecha. 2018. "A fluorescent biosensors for detection vital body fluids' agents." *Sensors* 18 (8). Multidisciplinary Digital Publishing Institute: 2357.

Ostrovský, Ivan, Radomír Čabala, Róbert Kubinec, Renáta Górová, Jaroslav Blaško, Janka Kubincová, Lucie Řimnáčová, and Wilhelm Lorenz. 2011. "Determination of phthalate sum in fatty food by gas chromatography." *Food Chemistry* 124 (1). Elsevier: 392–395.

Pacquiao, Melvin R., Mark Daniel G. de Luna, Nichaphat Thongsai, Sumana Kladsomboon, and Peerasak Paoprasert. 2018. "Highly fluorescent carbon dots from enokitake mushroom as multi-faceted optical nanomaterials for Cr6+ and VOC detection and imaging applications." *Applied Surface Science* 453 (September). North-Holland: 192–203.

Pandey, Raja Ram, and Charles C. Chusuei. 2021. "Carbon nanotubes, graphene, and carbon dots as electrochemical biosensing composites." *Molecules* 26 (21). Multidisciplinary Digital Publishing Institute: 6674.

Pasinszki, Tibor, Melinda Krebsz, Thanh Tran Tung, and Dusan Losic. 2017. "Carbon nanomaterial based biosensors for non-invasive detection of cancer and disease biomarkers for clinical diagnosis." *Sensors* 17 (8). Multidisciplinary Digital Publishing Institute: 1919.

Petryayeva, Eleonora, W. Russ Algar, and Igor L. Medintz. 2013. "Quantum dots in bioanalysis: A review of applications across various platforms for fluorescence spectroscopy and imaging." *Appl Spectrosc* 67 (3). SAGE PublicationsSage UK: London, England: 215–252. doi: 10.1366/12-06948.

Pu, Chaodan, Junliang Ma, Haiyan Qin, Ming Yan, Tao Fu, Yuan Niu, Xiaoli Yang, Yifan Huang, Fei Zhao, and Xiaogang Peng. 2016. "Doped semiconductor-nanocrystal emitters with optimal photoluminescence decay dynamics in microsecond to millisecond range: Synthesis and applications." *ACS Central Science* 2 (1). American Chemical Society: 32–39.

Qu, Binhong, Jianhui Sun, Peng Li, and Liqiang Jing. 2022. "Current advances on g-C3N4-based fluorescence detection for environmental contaminants." *Journal of Hazardous Materials* 425 (March). Elsevier: 127990.

Qu, Dan, Zaicheng Sun, Min Zheng, Jing Li, Yongqiang Zhang, Guoqiang Zhang, Haifeng Zhao, et al. 2015. "Three colors emission from s, n co-doped graphene quantum dots for visible light H2 production and bioimaging." *Advanced Optical Materials* 3 (3). John Wiley & Sons, Ltd: 360–367.

Qu, Hongke, Chunmei Fan, Mingjian Chen, Xiangyan Zhang, Qijia Yan, Yumin Wang, Shanshan Zhang, et al. 2021. "Recent advances of fluorescent biosensors based on cyclic signal amplification technology in biomedical detection." *Journal of Nanobiotechnology* 19 (1). BioMed Central: 1–28.

Rajakumar, Govindasamy, Xiu Hua Zhang, Thandapani Gomathi, Sheng Fu Wang, Mohammad Azam Ansari, Govindarasu Mydhili, Gnanasundaram Nirmala, Mohammad A. Alzohairy, and Ill Min Chung. 2020. "Current use of carbon-based materials for biomedical applications-A prospective and review." *Processes* 8 (3). Multidisciplinary Digital Publishing Institute: 355.

Raveendran, Varsha, and Renuka Neeroli Kizhakayil. 2021. "Fluorescent carbon dots as biosensor, green reductant, and biomarker." *ACS Omega* 6 (36). American Chemical Society: 23475–23484.

Rhouati, Amina, Akhtar Hayat, Rupesh K. Mishra, Diana Bueno, Shakir Ahmad Shahid, Roberto Muñoz, and Jean Louis Marty. 2016. "Ligand assisted stabilization of fluorescence nanoparticles; an insight on the fluorescence characteristics, dispersion stability and DNA loading efficiency of nanoparticles." *Journal of Fluorescence* 26 (4). Springer, New York LLC: 1407–1414.

Roberts, Akanksha, and Sonu Gandhi. 2022. "A concise review on potential cancer biomarkers and advanced manufacturing of smart platform-based biosensors for early-stage cancer diagnostics." *Biosensors and Bioelectronics: X* 11 (September). Elsevier: 100178.

Safardoust-Hojaghan, Hossein, Masoud Salavati-Niasari, Omid Amiri, and Mohammad Hassanpour. 2017. "Preparation of highly luminescent nitrogen doped graphene quantum dots and their application as a probe for detection of *Staphylococcus aureus* and *E. Coli.*" *Journal of Molecular Liquids* 241 (September). Elsevier: 1114–1119.

Saifuddin, N., A. Z. Raziah, and A. R. Junizah. 2013. "Carbon nanotubes: A review on structure and their interaction with proteins." *Journal of Chemistry.* doi: 10.1155/2013/676815.

Santhiran, Anuluxan, Poobalasuntharam Iyngaran, Poobalasingam Abiman, Navaratnarajah Kuganathan, and Stefano Bellucci. 2021. "Graphene synthesis and its recent advances in applications-A review." *C* 7 (4). Multidisciplinary Digital Publishing Institute: 76. doi: 10.3390/c7040076.

Santiago, Svette Reina Merden Solante, Toe Naing Lin, Chiao-Hsin Chang, Yee-Ann Wong, Cheng-An J. Lin, Chi-Tsu Yuan, and J L Shen. 2017. "Synthesis of N-doped graphene quantum dots by pulsed laser ablation with diethylenetriamine (DETA) and their photoluminescence." *Physical Chemistry Chemical Physics* 19 (33). The Royal Society of Chemistry: 22395–22400.

Sargazi, Saman, Iqra Fatima, Maria Hassan Kiani, Vahideh Mohammadzadeh, Rabia Arshad, Muhammad Bilal, Abbas Rahdar, Ana M. Díez-Pascual, and Razieh Behzadmehr. 2022. "Fluorescent-based nanosensors for selective detection of a wide range of biological macromolecules: A comprehensive review." *International Journal of Biological Macromolecules* 206 (May). Elsevier: 115–147.

Scallan, Elaine, Robert M. Hoekstra, Frederick J. Angulo, Robert V. Tauxe, Marc Alain Widdowson, Sharon L. Roy, Jeffery L. Jones, and Patricia M. Griffin. 2011. "Foodborne illness acquired in the United States-major pathogens." *Emerging Infectious Diseases* 17 (1): 7–15.

Sciuto, Emanuele Luigi, Antonio Alessio Leonardi, Giovanna Calabrese, Giovanna De Luca, Maria Anna Coniglio, Alessia Irrera, and Sabrina Conoci. 2021. "Nucleic acids analytical methods for viral infection diagnosis: state-of-the-art and future perspectives." *Biomolecules* 11 (11). Multidisciplinary Digital Publishing Institute: 1585.

Senutovitch, Nina, Lawrence Vernetti, Robert Boltz, Richard DeBiasio, Albert Gough, and D. Lansing Taylor. 2015. "Fluorescent protein biosensors applied to microphysiological systems." 240 (6). SAGE PublicationsSage UK: London, England: 795–808. doi: 10.1177/1535370215584934.

Shao, Lijia, Yanfang Gao, and Feng Yan. 2011. "Semiconductor quantum dots for biomedicial applications." *Sensors* 11 (12). Molecular Diversity Preservation International: 11736–11751.

Sharma, Abhinav, Wejdan S. AlGhamdi, Hendrik Faber, and Thomas D. Anthopoulos. 2022. "Paper-based microfluidics devices with integrated nanostructured materials for glucose detection." *Advanced Microfluidics-Based Point-of-Care Diagnostics.* CRC Press, 191–228. ISBN: 9781003033479.

Sharma, Abhinav, Jyoti Bhardwaj, and Jaesung Jang. 2020. "Label-free, highly sensitive electrochemical aptasensors using polymer-modified reduced graphene oxide for cardiac biomarker detection." *ACS Omega* 5 (8). American Chemical Society: 3924–3931.

Sharma, Abhinav, Chang Ho Han, and Jaesung Jang. 2016. "Rapid electrical immunoassay of the cardiac biomarker troponin i through dielectrophoretic concentration using imbedded electrodes." *Biosensors and Bioelectronics* 82 (August). Elsevier: 78–84.

Sharma, Abhinav, Seongkyeol Hong, Renu Singh, and Jaesung Jang. 2015. "Single-walled carbon nanotube based transparent immunosensor for detection of a prostate cancer biomarker osteopontin." *Analytica Chimica Acta* 869. Elsevier: 68–73.

Sharma, Abhinav, and Jaesung Jang. 2019. "Flexible electrical aptasensor using dielectrophoretic assembly of graphene oxide and its subsequent reduction for cardiac biomarker detection." *Scientific Reports* 9 (1): 5970.

Sharma, Atul, Reem Khan, Gaelle Catanante, Tauqir A. Sherazi, Sunil Bhand, Akhtar Hayat, and Jean Louis Marty. 2018. "Designed strategies for fluorescence-based biosensors for the detection of mycotoxins." *Toxins* 10 (5). Multidisciplinary Digital Publishing Institute: 197.

Shin, Young-Ho, M. Teresa Gutierrez-Wing, and Jin-Woo Choi. 2021. "Review-recent progress in portable fluorescence sensors." *Journal of The Electrochemical Society* 168 (1). IOP Publishing: 017502.

Silva, Manori J., Ella Samandar, John A. Reidy, Russ Hauser, Larry L. Needham, and Antonia M. Calafat. 2007. "Metabolite profiles of Di-n-butyl phthalate in humans and rats." *Environmental Science and Technology* 41 (21). American Chemical Society: 7576–7580.

Singh, Renu, Abhinav Sharma, Seongkyeol Hong, and Jaesung Jang. 2014. "Electrical immunosensor based on dielectrophoretically-deposited carbon nanotubes for detection of influenza virus H1N1." *Analyst* 139 (21). Royal Society of Chemistry: 5415–5421.

Strianese, Maria, Maria Staiano, Giuseppe Ruggiero, Tullio Labella, Claudio Pellecchia, and Sabato D'Auria. 2012. "Fluorescence-based biosensors." *Methods in Molecular Biology* 875. Humana Press Inc.: 193–216.

Sultangaziyev, Alisher, and Rostislav Bukasov. 2020. "Review: Applications of surface-enhanced fluorescence (SEF) spectroscopy in bio-detection and biosensing." *Sensing and Bio-Sensing Research* 30 (December). Elsevier: 100382.

Sun, Hanjun, Li Wu, Weili Wei, and Xiaogang Qu. 2013. "Recent advances in graphene quantum dots for sensing." *Materials Today* 16 (11). Elsevier: 433–442.

Tajarrod, Narjes, Mohammad Kazem Rofouei, Majid Masteri-Farahani, and Reza Zadmard. 2016. "A quantum dot-based fluorescence sensor for sensitive and enzymeless detection of creatinine." *Analytical Methods* 8 (30). The Royal Society of Chemistry: 5911–5920.

Tam, Tran Van, Nguyen Bao Trung, Hye Ryeon Kim, Jin Suk Chung, and Won Mook Choi. 2014. "One-pot synthesis of N-doped graphene quantum dots as a fluorescent sensing platform for Fe3+ ions detection." *Sensors and Actuators B: Chemical* 202 (October). Elsevier: 568–573.

Tang, Yong, Guibin Qiao, Enwu Xu, Yiwen Xuan, Ming Liao, and Guilin Yin. 2017. "Biomarkers for early diagnosis, prognosis, prediction, and recurrence monitoring of non-small cell lung cancer." *OncoTargets and Therapy* 10 (September). Dove Press: 4527–4534.

Thompson, Ian M., and Donna P. Ankerst. 2007. "Prostate-specific antigen in the early detection of prostate cancer." *CMAJ* 176 (13) 1853–1858.

Wegner, K. David, and Niko Hildebrandt. 2015. "Quantum dots: Bright and versatile in vitro and in vivo fluorescence imaging biosensors." *Chemical Society Reviews* 44 (14). The Royal Society of Chemistry: 4792–4834.

Wen, Jia, Weisi Li, Jiaqi Li, Binbin Tao, Yongqian Xu, Hongjuan Li, Aiping Lu, and Shiguo Sun. 2016. "Study on rolling circle amplification of ebola virus and fluorescence detection based on graphene oxide." *Sensors and Actuators B: Chemical* 227 (May). Elsevier: 655–659.

William Yu, W., and Xiaogang Peng. 2002. "Formation of high-quality CdS and other II ± VI semiconductor nanocrystals in noncoordinating solvents: Tunable reactivity of monomers**." *Angewandte Chemie International Edition* 41 (13): 2368.

Williams, Ryan M., Christopher Lee, and Daniel A. Heller. 2018. "A fluorescent carbon nanotube sensor detects the metastatic prostate cancer biomarker UPA." *ACS Sensors* 3 (9). American Chemical Society: 1838–1845.

Wu, Li, and Xiaogang Qu. 2015. "Cancer biomarker detection: Recent achievements and challenges." *Chemical Society Reviews* 44 (10). The Royal Society of Chemistry: 2963–2997.

Wu, Shijia, Nuo Duan, Xiaoyuan Ma, Yu Xia, Hongxin Wang, Zhouping Wang, and Qian Zhang. 2012. "Multiplexed fluorescence resonance energy transfer aptasensor between upconversion nanoparticles and graphene oxide for the simultaneous determination of mycotoxins." *Analytical Chemistry* 84 (14). American Chemical Society: 6263–6270.

Wycisk, Virginia, Katharina Achazi, Ole Hirsch, Christian Kuehne, Jens Dernedde, Rainer Haag, and Kai Licha. 2017. "Heterobifunctional dyes: highly fluorescent linkers based on cyanine dyes." *ChemistryOpen* 6 (3). John Wiley & Sons, Ltd: 437–446.

Xiao, Xinzhe, Yumin Zhang, Lei Zhou, Bin Li, and Lin Gu. 2022. "Photoluminescence and fluorescence quenching of graphene oxide: A review." *Nanomaterials* 12 (14). Multidisciplinary Digital Publishing Institute: 2444.

Xu, Mingsheng, Tao Liang, Minmin Shi, and Hongzheng Chen. 2013. "Graphene-like two-dimensional materials." *Chemical Reviews* 113 (5). American Chemical Society: 3766–3798.

Xu, Qian, Fangbin Xiao, and Hengyi Xu. 2023. "Fluorescent detection of emerging virus based on nanoparticles: From synthesis to application." *TrAC Trends in Analytical Chemistry* 161 (April). Elsevier: 116999.

Xu, Wanzhen, Tao Li, Weihong Huang, Yu Luan, Yanfei Yang, Songjun Li, and Wenming Yang. 2017. "A magnetic fluorescence molecularly imprinted polymer sensor with selectivity for dibutyl phthalate via Mn doped ZnS quantum dots." *RSC Advances* 7 (81). The Royal Society of Chemistry: 51632–51639.

Xu, Xiaoyou, Robert Ray, Yunlong Gu, Harry J. Ploehn, Latha Gearheart, Kyle Raker, and Walter A. Scrivens. 2004. "Electrophoretic analysis and purification of fluorescent single-walled carbon nanotube fragments." *Journal of the American Chemical Society* 126 (40). American Chemical Society: 12736–12737.

Yadav, Pradeep Kumar, Subhash Chandra, Vivek Kumar, Deepak Kumar, and Syed Hadi Hasan. 2023. "Carbon quantum dots: synthesis, structure, properties, and catalytic applications for organic synthesis." *Catalysts* 13 (2). Multidisciplinary Digital Publishing Institute: 422.

Yamamoto, Rika, and Penmetcha K.R. Kumar. 2000. "Molecular beacon aptamer fluoresces in the presence of tat protein of HIV-1." *Genes to Cells* 5 (5). John Wiley & Sons, Ltd: 389–396.

Yan, Ze, Yi Cai, Jing Zhang, and Yong Zhao. 2022. "Fluorescent sensor arrays for metal ions detection: A review." *Measurement* 187 (January). Elsevier: 110355.

Yan, Zhengyu, Xincheng Qu, Qianqian Niu, Chunqing Tian, Chuanjian Fan, and Baofen Ye. 2016. "A green synthesis of highly fluorescent nitrogen-doped graphene quantum dots for the highly sensitive and selective detection of Mercury(II) ions and biothiols." *Analytical Methods* 8 (7). The Royal Society of Chemistry: 1565–1571.

Yang, Qingxiang, Shuang Shuang Ren, Qianqian Zhao, Ran Lu, Cheng Hang, Zhijun Chen, and Hegen Zheng. 2018. "Selective separation of methyl orange from water using magnetic ZIF-67 composites." *Chemical Engineering Journal* 333 (February). Elsevier: 49–57.

Yang, Zhongjie, Xiaofei Zhang, and Jun Guo. 2022. "Functionalized carbon-based electrochemical sensors for food and alcoholic beverage safety." *Applied Sciences* 12 (18). Multidisciplinary Digital Publishing Institute: 9082.

Yaraki, Mohammad Tavakkoli, and Yen Nee Tan. 2020. "Metal nanoparticles-enhancedbi-osensors: synthesis, design and applications in fluorescence enhancement and sur-face-enhanced Raman scattering." *Chemistry - An Asian Journal* 15 (20). John Wiley & Sons, Ltd: 3180–3208.

Yim, Yeajee, Hojeong Shin, Seong Min Ahn, and Dal Hee Min. 2021. "Graphene oxide-based fluorescent biosensors and their biomedical applications in diagnosis and drug discov-ery." *Chemical Communications* 57 (77). The Royal Society of Chemistry: 9820–9833.

Yuan, Lin, Weiying Lin, Yueting Yang, and Hua Chen. 2012. "A unique class of near-infra-red functional fluorescent dyes with carboxylic-acid-modulated fluorescence ON/OFF switching: Rational design, synthesis, optical properties, theoretical calculations, and applications for fluorescence imaging in living animals." *Journal of the American Chemical Society* 134 (2). American Chemical Society: 1200–1211.

Yue, Qi, Yu Ying Huang, Xiao Fang Shen, Cheng Yang, and Yue Hong Pang. 2020. "In situ growth of covalent organic framework on titanium fiber for headspace solid-phase microextraction of 11 phthalate esters in vegetables." *Food Chemistry* 318 (July). Elsevier: 126507.

Zhang, Congcong, and Xin Du. 2020. "Electrochemical sensors based on carbon nanomate-rial used in diagnosing metabolic disease." *Frontiers in Chemistry* 8 (August). Frontiers Media S.A.: 651.

Zhang, Jialin, Ming Zhou, Xin Li, Yaqi Fan, Jinhui Li, Kangqiang Lu, Herui Wen, and Jiali Ren. 2023. "Recent advances of fluorescent sensors for bacteria detection-A review." *Talanta* 254 (March). Elsevier: 124133.

Zhang, Yingqi, Howyn Tang, Wei Chen, and Jin Zhang. 2022. "Nanomaterials used in fluores-cence polarization based biosensors." *International Journal of Molecular Sciences* 23 (15). Multidisciplinary Digital Publishing Institute: 8625.

Zhao, Victoria Xin Ting, Ten It Wong, Xin Ting Zheng, Yen Nee Tan, and Xiaodong Zhou. 2020. "Colorimetric biosensors for point-of-care virus detections." *Materials Science for Energy Technologies* 3 (January). Elsevier: 237–249.

Zheng, Binjie, Yuanfu Chen, Pingjian Li, Zegao Wang, Bingqiang Cao, Fei Qi, Jinbo Liu, Zhiwen Qiu, and Wanli Zhang. 2017. "Ultrafast ammonia-driven, microwave-assisted synthesis of nitrogen-doped graphene quantum dots and their optical properties." *Nanophotonics* 6 (1). Walter de Gruyter GmbH: 259–267.

Zheng, Hui, Peng Zheng, Liang Zheng, Yuan Jiang, Zhangting Wu, Feimei Wu, Lihuan Shao, Yan Liu, and Yang Zhang. 2018. "Nitrogen-doped graphene quantum dots synthesized by C60/Nitrogen plasma with excitation-independent blue photoluminescence emis-sion for sensing of ferric ions." *Journal of Physical Chemistry C* 122 (51). American Chemical Society: 29613–29619.

Zheng, Peng, and Nianqiang Wu. 2017. "Fluorescence and sensing applications of graphene oxide and graphene quantum dots: A review." *Chemistry - An Asian Journal* 12 (18). John Wiley & Sons, Ltd: 2343–2353.

Zhu, Fang, Hu Zhang, Min Qiu, Nan Wu, Kun Zeng, and Daolin Du. 2019. "Dual-label time-resolved fluoroimmunoassay as an advantageous approach for investigation of diethyl phthalate & dibutyl phthalate in surface water." *Science of The Total Environment* 695 (December). Elsevier: 133793.

Zhu, Lei, Xin Cui, Jing Wu, Zhenni Wang, Peiyao Wang, Yu Hou, and Mei Yang. 2014. "Fluorescence immunoassay based on carbon dots as labels for the detection of human immunoglobulin G." *Analytical Methods* 6 (12). The Royal Society of Chemistry: 4430–4436.

13 Fluorescent Nanomaterials and Their Optoelectronic Properties

Weijing Yao, Xiaoman Bi, Youfusheng Wu,
Xuying Liu, and Qingyong Tian

13.1 INTRODUCTION

Nowadays, the design of nanomaterials supplies an additional dimension to regulate the physical and chemical relationships of materials at the nanoscale, extending the exploitation of multifunctional nanomaterials. Recent explorations of various nanomaterials have established a yardstick of application promise that can be used to enable application across multiple disciplines. The continuous improvements of commercial electronics and cutting-edge computing technologies have resulted in the exponential growth of data traffic, posing challenges for the optoelectronics industry. Nanomaterials are at the forefront of nanotechnology and show great promise for applications in optoelectronics.

Particularly, rapidly developed fluorescent nanomaterials, such as semiconductor quantum dots (QDs), rare-earth (RE)-doped nanoparticles (NPs), perovskite nanomaterials, and carbon dots (CDs), exhibit many fascinating advantages of tunable luminescence properties, multiple luminous categories, low toxicity, easy preparation, and excellent physicochemical properties.[1-4] These merits provide new and promising ideas for fluorescent nanomaterials in photoelectronic applications such as light-emitting diodes (LEDs), photovoltaics (PVs), and photodetectors (PDs). Fluorescent nanomaterials can be used not only as a stand-alone component of the photoluminescence but also as an accessory ingredient in the fabrication of optoelectronic devices. Nevertheless, the exploration and development of new fluorescent nanomaterials are still important for expanding the application in the field of optoelectronics. Moreover, with the increasing attention to environmental safety and human health, it is urgent to develop fluorescent materials with non-toxic and eminent performances for further applying them in optoelectronics.

In this chapter, we present a comprehensive review that generalizes developments in fluorescent nanomaterials and their corresponding optoelectronic properties. The fundamentals of fluorescent nanomaterials including semiconductor QDs, perovskite nanomaterials, and RE-doped NPs and CDs are briefly introduced in turn.

DOI: 10.1201/9781003352372-13

Further, the utility of corresponding fluorescent nanomaterials for optoelectronic applications (LEDs, PVs, and PDs) are discussed in detail. To this end, we hope that this chapter can play a beneficial role in guiding and paving the way for the wider application of fluorescent nanomaterials in optoelectronics fields.

13.2 SEMICONDUCTOR QDS AND THEIR OPTOELECTRONIC APPLICATIONS

Semiconductor QDs refer to a collection of nanocrystals (NCs) and molecules with quantum-confined boundaries in three-dimensional space. The size distribution of QDs is commonly ranges from 2 to 20 nm and usually constitutes hundreds to thousands of atoms.[1,5] Quantum and dielectric confinement effects occur when the NC boundaries are confined in zero dimension, which means electrons are confined to a range equivalent to their de Broglie wavelength. Therefore, the physicochemical properties of semiconductor QDs changed dramatically different with homologous bulk materials.[6] Diverse with bulk counterparts, the continuous valence and conduction bands (VB and CB) energetically split into discrete energy-level structures with molecular characteristics, as the carriers of QDs are confined to quantum spaces. Simultaneously, the discrete energy and configuration of QDs are highly rest with its size and shape, and the bandgap broadens as QDs size decreased.[7] Therefore, the luminescence emissions of QDs are able to range a wide spectrum from the ultraviolet (UV) to near-infrared (NIR) region.

It is well known that when a semiconductor is illuminated by incident light with energy higher than that of bandgap, electrons in the CB are excited and photocarriers are generated. In semiconductor QDs, the quantum and dielectric confinement effects induce intense light absorption resonance accompanied by admissible excitations between quantized energy levels. The highly exposed surface atoms of QDs generate plenty of dangling bonds that act as intermediate-gap trapping state, introducing a competitive pathway to radiative band-to-band transition. When the QDs in the ground state excite the inner electrons by photons or electrons, it creates commensurable holes in the VB of the QDs. From an energy point of view, this situation is unstable and will spontaneously transition to a lower energy state, which is namely deexcitation process. One possible nonradiative deexcitation process is the transfer of energy to another electron in the outer layer, allowing it to overcome the binding energy and emit outward, called the Auger process. The Auger process is more likely to occur in semiconductor QDs because momentum conservation is broken with respect to the bulk material. Excitons with large binding energies and complex multi-quantum states can exist in QDs due to the strong Coulomb interaction between charge carriers. The multiple exciton generation (MEG) process refers to the physical process by which semiconductor QDs absorb a photon with an energy equal to or greater than 2.7 times the band gap of the semiconductor material and generates multiple excitons (electron–hole pairs).[8,9] The nonradiative relaxation process provides competent energy to excite neighboring carrier to higher states.

Carrier concentration, mobility, and density of states are the basic electrical properties of semiconductor QDs. The undercoordinated dangling bonds and exposed surface atoms change the photoelectric property of QDs by affecting the electrical properties.[10] These unsaturated surface hanging bonds serve as effective charge traps, greatly reducing the quantum yield (QY), and easily react with oxygen and become unstable.[11] Ligand passivation is an effective strategy to tailor the electronic states, photoluminescence, and power conversion efficiency (PCE) of QDs.[12] Because the ligand passivation is capable of satisfying dangling bond and decreases the unintentional trap states and hence the nonradiative trap rate.

Additionally, the configuration engineering is another efficient strategy to adjust the electronic states of QDs by construction of inorganic core/shell or multi-shell nano-heterostructures. Three acknowledged core/shell heterostructures include type I, quasi-type II, and type II bands alignment.[5] In type I core/shell heterostructure QDs, both the CB and VB of cores are situated between the shell alignments, thus confining the carriers in the core. In type II or quasi-type II core/shell QDs, the VB edge or CB edge of the shell semiconductor material is located between the band gap of the core. After excitation, electrons and holes are separated in different regions of the core and shell, which decreases exchange interaction and radiative recombination probability. Therefore, type I QDs are often used in the display domain due to their high photonic luminescence QY, while type II QDs are commonly used in PVs or PDs due to their good carrier separation efficiency.

Due to their prominent quantum confinement effects, semiconductor QDs represent peculiar physicochemical and fluorescence properties different from macroscopic bulk materials, making them promising for many applications. After more than 40 years of development, semiconductor QD materials have realized the "green synthesis route," and process with uniform distribution, good photoluminescence QYs, narrow photoluminescence peak, and remarkable restrained blinking. Nowadays, the QDs are capable of large-scale industrial product production and commercial supply. The semiconductor QDs have good application potential in the next generation of LED display, optoelectronic devices, and biomedicine.[13-15]

13.2.1   LEDs Based on Semiconductor QDs

The organic light-emitting diode (OLED) with uniform emissions with large response and nearly 100% internal energy conversion efficiency has been widely applied to displayers, portable equipment, and flexible display devices. The color rendering index (CRI, Ra) value of OLEDs for indoor illumination can reach 90 attributes to its broad emission spectrum. However, the commercially used OLEDs always characterized with broad full widths at half maximum (FWHM) (Figure 13.1a), usually > 60 nm. The superposition of emission spectra also limited the application of OLEDs in high-quality display technology, which needs high color purity.[16]

Comparatively, the QD-based LEDs (QLEDs) have attracted great attention in academic studies and commercialization fields in the past 10 years. QLEDs mainly have the following merits: (1) narrow FWHM. Because of the highly monodisperse and atomic-like structural electronic states, the QDs always occupied with narrow FWHM around 30 nm under environmental conditions. This would be very well suited to commercial computers and high-color-purity displays. (2) Color tunability. As described before, the quantum confinement effect of semiconductor QDs causes peculiar physicochemical and fluorescence properties. Thus, the

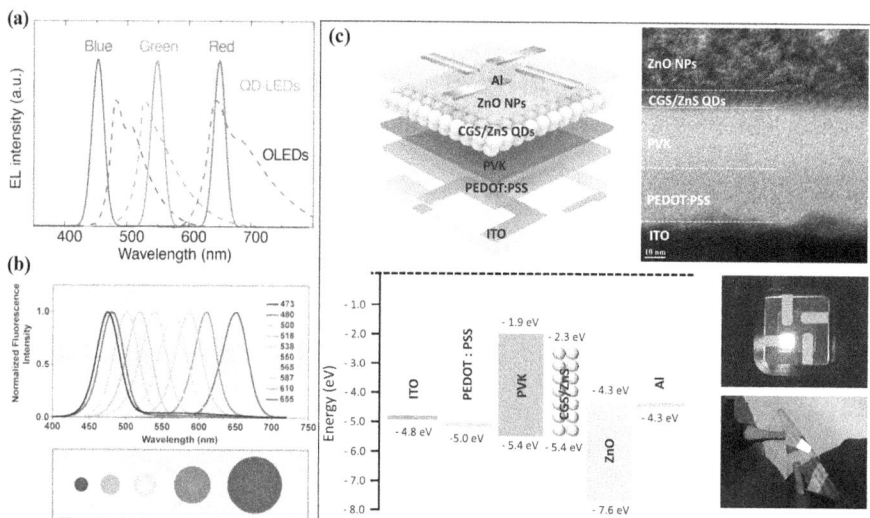

FIGURE 13.1 (a) OLED and QLED emissions. (Reproduced with permission from ref. 16. Copyright 2013 Springer Nature.) (b) The size-tunable fluorescence emissions of CdSe QDs. (Reproduced with permission from ref. 18. Copyright 2004 RSC.) (c) Multilayered white QLEDs prepared with Cu-Ga-S/ZnS core@shell QDs. (Reproduced with permission from ref. 26. Copyright 2016 Wiley.)

confinement of excitons can precisely adjust the emissions across the full spectrum and reasonably regulate the sizes and compositions of QDs.[17] Due to the quantum confinement that occurs when the QDs are in the size range of the exciton Bohr radius, the fluorescence emission is highly dependent on the size of the QDs. For example, the fluorescence emissions of CdSe QDs can be precisely regulated from the UV to NIR region by the tunable size distribution (Figure 13.1b).[18] (3) High stability. The LED devices always operate under the condition of strong light irradiation or heavy current, and *Joule* heat is a predominant factor affecting the working life.[19] Semiconductor NCs commonly indicate preferable thermostability than organic matters; therefore, semiconductor QLEDs are anticipated to possess a longer service life. The high resistance to photobleaching makes them useful for continuous monitoring of fluorescence, which is particularly important for biological imaging and biomedical applications.[20]

Over the past decade, innovative attempts have been explored to optimize the performances of the QDs or QLEDs for the purpose of satisfying the requirements of practical applications. By imputing a PMMA insulating barrier in the middle of QDs phosphor and oxide electron-transport layer, Peng realized the preparation of state-of-the-art QLEDs.[21] The immediate contact of CdSe/CdS core/shell QDs with ZnO electron-transport interlayer induces the spontaneous carriers migration because of different work functions. The insertion of 6 nm PMMA layers keeps the charge neutrality of QDs and maintains their preferable emission features. The as-obtained red QLEDs device demonstrated the highest external quantum efficiency (EQE) of 20.5%. More importantly, the insulating layer increases the photoluminescence lifetime of QDs from 10.6 to 19.5 ns, causing remarkable operational stability of QLEDs device over 100,000 h at 100 cd/m^2. Recently, Kwak got the high values of red QLEDs that show a maximum brightness of 3,300,000 cd/m^2 and an operational lifetime of 125,000,000 h at 100 cd/m^2.[22] Qian and workers reported blue, green, and red QD-based (Cd$_{1-x}$Zn$_x$Se$_{1-y}$S$_y$) LED devices with super EQE more than 10%.[23] With specially tailoring the thickness of intermediate layer and outer shell, the devices present maximum current and EQE of 63 cd/A and 14.5% for green QLEDs, 15 cd/A and 12.0% for red QLEDs, and 4.4 cd/A and 10.7% for blue QLEDs. Besides, the devices have also been extrapolated with excellent operating lifetimes over 90,000 h and 300,000 h for the green and red QLEDs, respectively, for a luminance at 100 cd/m^2. Similarly, Jin et al. introduce hole-transport polymers between QDs and hole-transport layer PEDOT:PSS, which can effectively decrease electron leakage and change the intermediate state of QDs. The EQE of green and blue QLEDs achieved 28.7% and 21.9%, respectively.[24] Remarkably, the blue LEDs demonstrate excellent stability with operating lifetimes over 4,400 h.

On the consideration of ultimate application as lighting devices, the efforts should be focused on the development of QLEDs with white emission.[25] As displayed in Figure 13.1c, a white QLED structured with single Cu-Ga-S/ZnS QDs was successfully constructed through a solution processed approach. Obviously, the advantage of preparing white LEDs (W-LEDs) with a single QDs emitter is greater than that of the three-primary color stacking method. Because the structure of three-color stacking QLEDs is complex, there is an energy transfer process between diverse QDs, which reduces the electroluminescence efficiency. The non-Cd QD-based white QLEDs possessed satisfactorily high CRI (83–88), 7494–8234 K of color temperature, the high electroluminescent intensity of 1007 cd/m^2, 3.6 cd/A in current efficiency, 1.9 lm/W in power efficiency, and 1.9% in EQE.[26]

13.2.2 SCs Based on Semiconductor QDs

Because of the superior optical and electrical properties, semiconductor QDs have demonstrated potential application prospects in solar energy conversion, especially in SCs.[19] Nowadays, the most used monocrystalline and polycrystalline silicon-based SCs have the disadvantages of lower absorption coefficients and higher energy consumption during fabrication. The merits of QDs including solution processability, low cost, high extinction coefficients, MEG, and high theoretical efficiency (Shockley–Queisser limitation) make them a potential candidate for next-generation SCs.[27] The semiconductor QD-based SCs (QDSCs) have achieved a remarkable breakthrough in both fundamental principles and device performances.

Advances in device architectures play a crucial role in accelerating the performances of SCs. A range of device architectures have been developed for QDSCs, including Schottky junction, heterojunction architecture, p-n structure, and sensitized-type (Figure 13.2a–d). The Schottky junction configuration was the first-generation structure of QDSCs with the typical structure "ITO/QDs/back electrode."[28] However, due to the insurmountable drawbacks of highly increased carrier recombination rate and Fermi-level pinning effect, the highest PCE of Schottky junction QDSCs was stalled at 5.2%.[29] Subsequently, broad-bandgap n-type semiconductors

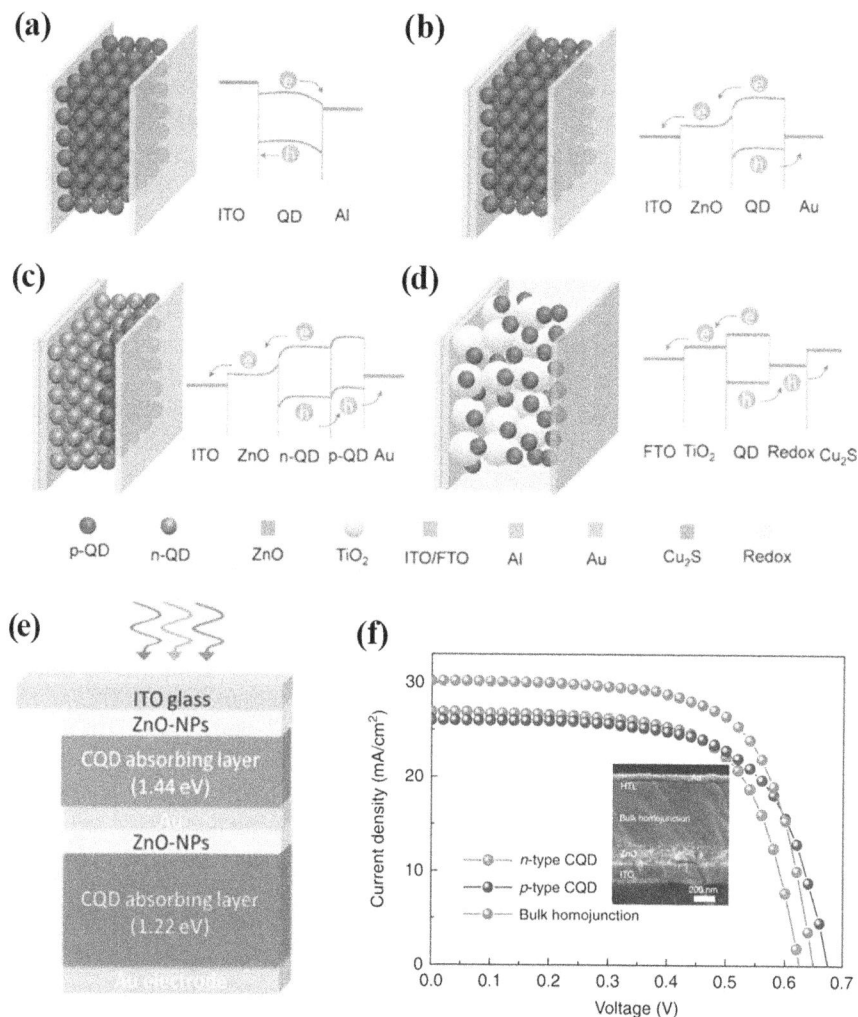

FIGURE 13.2 The device structures of QDSCs. (a) Schottky junction, (b) heterojunction architecture, (c) p-n structure, (d) sensitized-type, and (e) tandem structure. (Reproduced with permission from ref. 19. Copyright 2021 Wiley. Reproduced with permission from ref. 38. Copyright 2019 ACS.) (f) Current–voltage characteristics of p-n structural QDSCs under AM 1.5 irradiation. (Reproduced with permission from,[31] Copyright 2020 Springer Nature.)

like TiO$_2$ have been introduced between ITO and QDs interlayer to form the heterojunction architecture QDSCs. The p-n heterojunction interface and TiO$_2$ or ZnO have large hole barriers for QDs, which efficiently improve carrier extraction efficiency and reduce recombination. In addition, the formed p-n heterojunctions remove the Fermi-level pinning effect in the Schottky junction, which weakens electron–hole recombination. Accordingly, the heterojunction structure observably enhances the PCE above 9.2% with amine-stabilized PbS QDs.[30] The dot-to-dot mutual surface passivation method induced record charge carrier diffusion length also significant attributes to the current density and photovoltaic efficiency. Furthermore, appropriate high-structured QDs are developed to boost the performance of QDSCs. As displayed in Figure 13.2c, the p-type and n-type QDs as main light absorber can further increase carrier separation and current leakage, thus greatly enhancing SCs performance. For example, Sargent et al. employed n-type and p-type PbS QDs as bulk homojunction SC devices (Figure 13.2f).[31] The QDSCs achieved a AM1.5 PCE of 13.3% through the combination of open-circuit voltage (V_{oc}) of 0.65 V, short-circuit current density (J_{sc}) of 30.2 mA/cm^2, and fill factor of 68%. Similarly, Ma and workers implemented an in-situ synergistic passivation approach to remarkably enhance passivation with bringing in multiple short functional molecules.[32] The synergistic effect of coordinating ligands resulted in inferior surface trap density and favorable doping behavior. The as-prepared QDSCs realized a high V_{oc} of 0.71 V, low V_{oc} loss (0.35 eV), and PCE of 13.3%. Sensitized QDSCs have analogous device structure and working principle to dye-sensitized SCs (DSSCs).[33] As shown in Figure 13.2d, p-type QDs are embedded inside n-type semiconductor oxides with mesoporous structure. Unlike p-n junction QDSCs, there is no built-in electric field excitation in the sensitized structure. The range of energy potential between QDs and electron transport material (ETM) provides the driving force for electrons diffusion.[34–36] In addition, the close contact between QDs and ETM facilitates the effective carrier transmission and collection.[37] The performances of QDSCs with the sensitization construction have been continuously improved, and the as-obtained PCE has exceeded 15%.[33]

As we all know, the thermodynamic efficiency of single-junction QDSCs is only about 31% that is limited by the Shockley–Queisser limitation.[27] Typically, tandem SCs are constructed by stacking sub-cells consisting of multiple absorbers with distinct band gaps, as displayed in Figure 13.2e.[38] The multi-junction tandem SCs are benefited for increasing the utilization of solar irradiation and decreasing the energy loss of photons, which are expected to break the Shockley–Queisser limitation (42.5% for two-junction). For example, PbS QDs with the band gaps of 1.44 and 1.22 eV were used to construct a two-junction QDSC.[38] By precise tuning the composition of QDs and thickness of sub-cells to enhance charges extraction, a monolithic tandem QDSC with a PCE of 6.8% was carried out. Similarly, Sargent and workers successfully synthesized a high-efficiency and long-term stable organic/QDs monolithic tandem SCs, realizing a superior PCE of 13.7% with a J_{SC} value of 15.2 mA/cm.[39]

13.3 PEROVSKITE NANOMATERIALS AND THEIR OPTOELECTRONICS APPLICATIONS

Perovskite was first discovered as calcium titanate ($CaTiO_3$) compounds present in perovskite ores, named after the Russian geologist Lev Perovski. Due to its unique properties, metal halide perovskite (MHP) has stirred great concern in the field of optoelectronics in the last decade. A typical MHP presents a cubic structure under ideal conditions, which can expressed by the formula of ABX_3.[40] Thereinto, an occupied vertex could be a monovalent organic cation (e.g., methylammonium [$CH_3NH_3^+$, MA^+], formamidinium [$CH(NH_2)_2^+$, FA^+]), or inorganic RE or alkaline metal ions (e.g., Cs^+, Rb^+). B positioning at the center of the cube could be a bivalent transition metal ion (e.g., Ti^{2+}, Pb^{2+}, Sn^{2+}, Mn^{2+}, Fe^{2+}, and Co^{2+}). X sitting either the vertexes or face-centers could be halide or its analog (e.g., Cl$^-$, Br$^-$, I$^-$, or SCN$^-$).[41–43] Perovskite materials, which are composed of the above three terms through van der Waals forces and feature as large optical absorption coefficient, tunable band gap, high charge mobility, and tough defect tolerance, are substantial elements for high-performance optoelectronic devices.[44]

Different from bulk perovskites, low-dimensional perovskites have low-dimensionally layered alternating between organic amine and MHP crystals. It can be categorized according to various structures and compositions. In "material-level" category, since organic cations partially isolate the metal halide species from each other by the corner-share pattern between $[BX_6]^{4-}$ octahedrons, perovskites can be described as 0D, 1D, 2D, quasi-2D, and 3D perovskites.[45] In contrast, the "structure-level" cares about the various nano-geometric morphologies referring NCs, nanowires (NWs)/nanorods (NRs), nanoplatelets/nanosheets, and bulk materials, corresponding to 0D, 1D, 2D/quasi-2D, and 3D perovskites, respectively.[45] As one of the promising categories of perovskite, halide perovskite nanomaterials are well-studied on not only the size/surface-tunable photoelectronic properties similar to traditional nano-semiconductor materials but also attractive component-dependent performance. It is going to be narrated involving features, manipulation strategies in size/surface/component engineering, and photoelectronic applications of perovskite nanomaterials in fields of LEDs, PDs, and PV devices, which can finally guide us to take a view of the landform then look forward to the bright future of perovskite nanomaterials.

The typical features of perovskite nanomaterials are highly dependent on the quantum confinement effect. Essentially, excitons in 3D perovskites have low binding energy and are easy to be captured by defects to generate nonradiative recombination. However, relying on the quantum confinement effect, low-dimension perovskite nanomaterials exhibit large exciton binding energy leading to stable and efficient radiative recombination. Yamauchi et al. reported that mesoporous silica templated monodisperse $MAPbX_3$ NCs appear size-dependent photophysical properties for the quantum confinement effect. Particularly, the trend of the absorption and emission position toward higher energy with decreasing size of perovskite NCs provides direct evidence that the quantum-confined domain effect can adjust the photophysical properties of perovskite NCs.[46]

In addition, the defect in perovskite nanomaterials is considerable for its direct influence on photoelectronic properties. To further improve the photoelectric properties of MHP, it is necessary to eliminate the deep-level defects, including point defects, grain boundaries, surfaces, and interfaces, which can be used as nonradiative recombination centers and seriously affect the device performance. Theoretical simulations indicate that unique electron band structure configurations of perovskite NCs essentially decide their defect tolerance.[47,48] Different from the conventional defect-tolerant band structure where the energy level of most trap states sits in the band gap, the energy level of most trap in the case of perovskite NCs locates in the CB or the VB, resulting in negligible effects on PLQY. Density functional theory simulations on the energy states relating to various defects provide evidence to the above perspective by demonstrating the absence of trap states in bandgap.[49] As researches step forward, the stability of $CsPbI_3$ NCs was further explored and could be ascribed that lattice strain deriving from low-dimension structure could instructively affect the defect formation energy leading to its strong defect tolerance.[50]

13.3.1   LEDs Based on Perovskite Nanomaterials

Colloidal MHP nanomaterials have gained great interest in optoelectronic applications by virtue of their excellent optical properties involving high near-unity PLQY and adjustable emission in visible light range. LEDs research based on halogenated perovskite NCs has also aroused wide attention with extensive research.

Poly(maleic anhydride-alt-1-octadecene) was added to the β-CsPbI$_3$ nanolayer with a stable tetragonal phase (Figure 13.3a), and resulted in the PLQY of LED device was significantly increased from 34% to 89%.[51] Therefore, β-CsPbI$_3$ NC-based red LEDs exhibit an outstanding EQE of 17.8% with the operating T$_{50}$ of 300h, meaning its excellent stability and optoelectronic properties. By contrast, CsPbI$_3$ NCs were doped into a perovskite matrix as active layer to fabricate red LEDs with high luminance and stability (Figure 13.3b).[52] The as-obtained LEDs perform a high EQE of 18% with bright emission over 4700cd/m^2 and an outstanding operating T$_{50}$ of 2400h.

Competition of perovskite nanomaterials on green LEDs is also intense. In order to improve device efficiency, balanced surface passivation and carrier injection is closely related to ligand density by optimizing the number of purification cycles. Combining the solution process and device engineering, CsPbBr$_3$ NC-based green LEDs finally

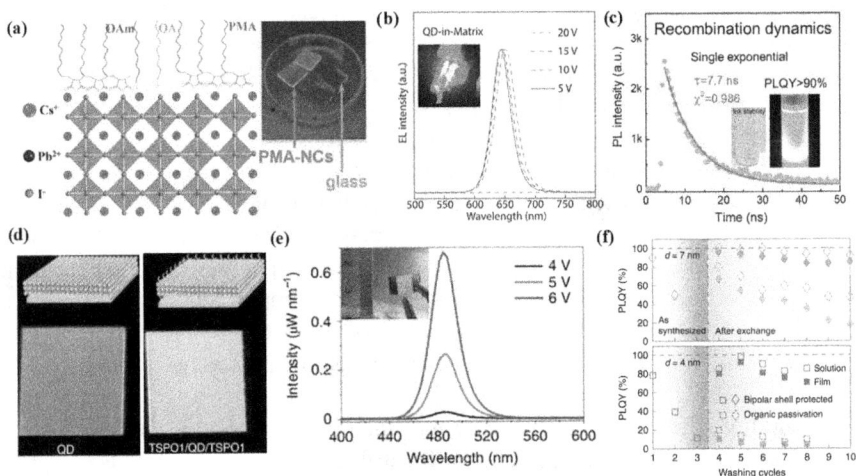

FIGURE 13.3 LEDs based on perovskite nanomaterials. (a) Schematic diagram of the PMA-incorporating β-CsPbI$_3$ NCs, and comparison of the aged colloidal β-CsPbI$_3$ NCs with and without PMA incorporation. (Reproduced with permission from ref. 51. Copyright 2021 Wiley.) (b) Electroluminescence spectrum of CsPbI$_3$-NC-in-perovskite LEDs with various driving voltages. (Reproduced with permission from ref. 52. Copyright 2021 ACS.) (c) PL decay curves of CsPbBr$_3$ QD inks exhibiting near-unity radiative channel. (Reproduced with permission from ref. 53. Copyright 2017 Wiley.) (d) Schematic diagram and photograph of QD films without passivation and with TSPO1 on both sides of QD film under UV irradiation. (Reproduced with permission from ref. 54. Copyright 2020 Springer Nature.) (e) Electroluminescence spectra under forward biases of 4, 5, and 6 V. (Reproduced with permission from ref. 55. Copyright 2019 Springer Nature.) (f) PLQYs of QDs and QD solid films made from DDAB-treated perovskite QDs and bipolar-shell-stabilized QDs of two different diameters. (Reproduced with permission from ref. 56. Copyright 2020 Springer Nature.)

achieved an EQE of 6.27%. The as-obtained EQE was further increased to 11.6% by surface treatment that can collaboratively promote charge injection, recombination dynamics, and printing ink stability (Figure 13.3c).[53] Bilateral passivation is proved to be an achievable strategy modifying the two interfaces of perovskite NCs film to process TSPO1-passivated $CsPbBr_3$ NCs. The relative LEDs show an impressive EQE of 18.7% and 20-fold enhanced operational lifetime (Figure 13.3d).[54]

There is a breakthrough for EQE of blue perovskite nanomaterial-based LEDs, which operate with EQE hovering around 2% until 2019. A device breaking this dilemma emits blue fluorescence at 483 nm with peak EQE of up to 9.5% and color coordinates of (0.094, 0.184) under a luminance of 54 cd/m^2 (Figure 13.3e).[55] Advanced developments in blue LEDs technology blossom out for the past years. Bipolar-shell-passivation strategy developed by Dong et al. refers an electrostatic interaction between the inner anionic shell and the outer cationic shell to provide enhanced NCs coupling, promoted charge transport, and decreased trap density (Figure 13.3f).[56] The as-fabricated LEDs maintain favorable PLQY of over 70% and peak EQE of 22% even under excitation density lower than 1 mW/cm^2. The research team further developed an alloyed $CsPb_{1-x}Sr_xBr_3$ NC possessing high luminescent efficiency with an unprecedented EQE of 13.8%.[57] They adopted bis(4-fluorophenyl) phenylphosphine oxide (DFPPO) to passivate the mixed Sr/Pb perovskite matrix to alleviate hygroscopic property of Sr^{2+} and thus strengthen the air- and photostability of the relative LEDs while maintaining efficient charge mobility.

Compared to red, green, and blue LEDs, more challenging W-LEDs based on perovskite nanomaterials can stay flourishing mainly benefited from the coping strategy, adopting composite structure, and the anchorage of perovskite nanomaterials on yttrium aluminum garnet (YAG)-based phosphor.[58,59] Although the efficiency and stability of W-LEDs still require further improvements, the huge and tangible application requirement for advanced lighting and display technology will propel much profound exploration in novel structure and strategy for building W-LEDs.

13.3.2 PDs Based on Perovskite Nanomaterials

As a kind of device which can convert incident photon into electrical signal, PDs have a broad application prospect in image sensing, night-vision system, and environment monitoring. PDs can generally be divided into three types, namely photodiode, phototransistor, and photoconductance.[60] The device structures are shown in Figure 13.4a–c. Photodiode with vertical structure is also called photovoltaic-type PDs with a similar operating principle of SCs. It has fast response speed and self-driven characteristics. Phototransistor is a passive device with three terminals, namely source, drain, and gate. In the presence of a thin insulating layer between gate and semiconductor layer, the photoelectric response in the semiconductor can be modulated by gate voltage applied.[61,62] Phototransistors can effectively reduce noise and amplify photocurrent signals.[63,64] Photoconductance is equivalent to photosensitive resistance where the two electrodes of device are in the same plane and form ohmic contact with semiconductor material. The conductivity of devices varies with the incident light. Photoconductance is well known for its simple structure and optical gain. However, it takes a long time for the photogenerated carrier to transport

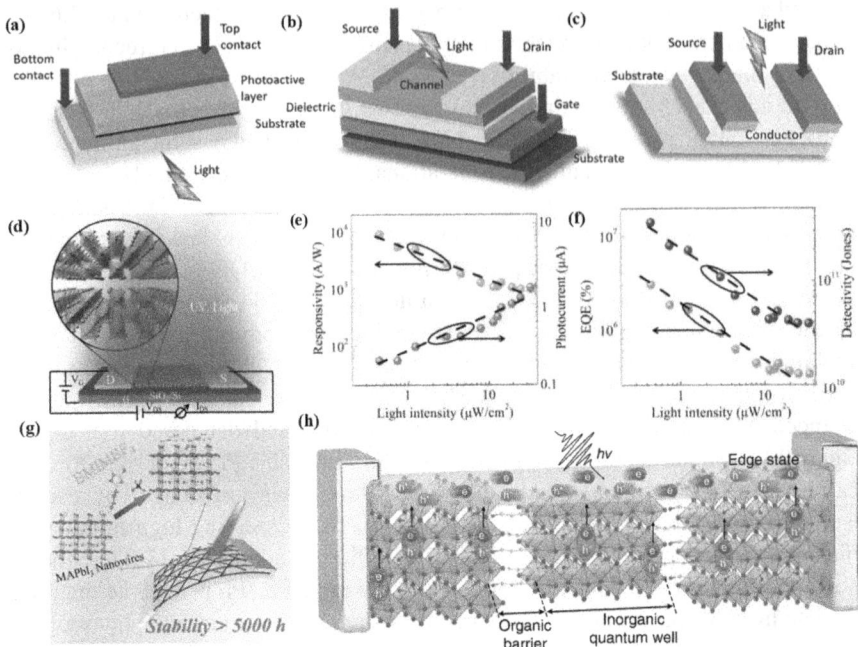

FIGURE 13.4 PDs based on perovskite nanomaterials. (a–c) Schematic diagrams of PDs with photodiode, phototransistor, and photoconductor structure, respectively. (Reproduced with permission from ref. 60. Copyright 2021 Wiley.) (d) Schematic of PDs based on $Cs_4Cd_{0.75}Mn_{0.25}Bi_2Cl_{12}$. (e–f) Light-intensity-dependent photocurrent, responsivity, EQE, and detectivity (at VG = 15 V, VD = 0 V). (Reproduced with permission from ref. 65. Copyright 2021 Wiley.) (g) The untight PDs maintain 100% of initial performance after exposing an open-air environment for over 5000 h. (Reproduced with permission from ref. 66. Copyright 2022 Wiley.) (h) Photocurrent density of $n = 4$ perovskite nanowires with different sizes. With increasing wire height, the dramatic fall of photoconductivity, together with the rise in the exciton population, suggests that free carriers localized at layer edges, instead of excitons in crystal interiors, dominate the photoconduction. (Reproduced with permission from ref. 67. Copyright 2018 Springer Nature.)

from semiconductor to electrode for the considerable distance between two electrodes.[65] As a result, this type of PD commonly exhibits a long response time and requires a high driving voltage.

In recent years, PDs based on perovskite nanomaterials have attracted wide attention and made a series of remarkable progress with great application potential. Bai et al. reported a colloidal synthesis to obtain a series of quadruple-perovskite NCs. The relative fluorescence QY of NCs could be increased by nearly 100-fold by alloying $Cs_4MnBi_2Cl_{12}$ NCs (Figure 13.4d–f). Alloying can eliminate the ultrafast trap state capture process, which competes with energy transfer, and can improve the crystallinity of NCs to significantly improve the luminescence efficiency. The PDs based on these NCs exhibit ultra-high responsiveness (0.98×104 A/W) and excellent EQE3 $\times$ 106%), relatively higher than previously reported PDs on account of non-lead perovskite nanocrystalline.[65] To enhance the long-term photostability of perovskite, Wu et al. doped the ionic liquid, 1-butyl-3-methylimidazolium tetrafluoroborate (BMIMBF$_4$), into MAPbI$_3$ NWs for significant improvement in electrical properties and long-term device stability (exposure to open air for over 5,000h, Figure 13.4g).[66] Lei et al. prepared high-quality 2D perovskite single-crystalline NW arrays and discovered the high-photoconductivity crystal edges of two-dimensional perovskite NWs.[67] The edge of the perovskite layer can effectively split excitons to generate charge carriers, thus achieving excellent photoconductance (Figure 13.4h). The as-prepared PDs based on these NWs performed high responsivities (1.5 × 10^4 A/W) and excellent detectivities (7 × 10^{15} Jones). Xia's group proposed a simple method of in-situ encapsulation by using hydrophobic molecules to further improve the stability of perovskite NWs.[68] The corresponding PDs exhibit high performance (responsivities of 20 A/W and detectivities of 4.1 × 10^{11} Jones) and stability that maintains 96% of initial photocurrent after 1-year exposure to air.

13.3.3 PSCs Based on Perovskite Nanomaterials

Perovskite nanomaterials have been demonstrated as a promising candidate as active layer source in perovskite SCs (PSCs) by virtue of both the excellent photoelectric properties inheriting from bulk perovskite materials and additional quantum-confined advantages. In exchange chemistry, by incorporating CsPbI$_3$ QDs into FAPbI$_3$ QD film, PCE of as-obtained PSCs is measured to be 15.6% which promoted the ambient stability (Figure 13.5a).[69] Similarly, surface engineering is also a common but effective method adopted in PSCs field. Modulation of Cs$^+$ and FA$^+$ cations to a suitable ratio was demonstrated as an available approach to simultaneously facilitate PCE (16.6%) and stability enhancement of Cs$_{1-x}$FA$_x$PbI$_3$ QD-based PSCs (Figure 13.5b and c).[70] Organic ligand property is another controllable factor in PSCs that hydrophobic phenethylammonium (PEA) short-chain ligands was adopted to take place of the original long-chain ones vesting stronger electronic coupling and moisture resistance to CsPbI$_3$ QDs films, resulting in a PCE of 14.1% and maintenance over 90% of this value under 30%–35% humidity after 15 days (Figure 13.5d).[71] In addition to hydrophobicity enhancement, aromatic short-chain ligand, 2-(4-fluorophenyl)ethyl ammonium, was applied in perovskite QDs arrays to facilitate interdot electronic transport, defects reduction, ion migration inhibition, and recombination suppression to finally get PSCs with enhanced efficiency and stability (Figure 13.5e and f).[72] In passivation engineering, cation passivation is utilized to accomplish effective vacancy occupation on the surface of perovskite QDs film by the corresponding metal salt solution. For instance, the cesium acetate (CsAc) passivated CsPbI$_3$ QD film was used to fabricate PSCs with a high PCE of 14.10%.[73] Moreover, a ligand passivator can help charge transport, thus improving optical properties and operational stability of devices. It is reported that the loss of surface ligand on CsPbI$_3$ QDs can be largely compensated by ligands like oleylamine, oleic acid, octylamine, and OTAc with different absorption energy as passivators contributing to upgrading a PCE of 11.87% in resulting PSCs (Figure 13.5g).[74] In device-structure engineering, the utilization of electron transfer layer (ETL) can usually enhance surface energy and lattice contraction, thus subsequently benefiting operational stability in conventional devices.[70] As reported, the incorporation of compact TiO$_2$ ETL in planar-structured perovskite QDs SCs can distinctly improve device optoelectronic performance.[75]

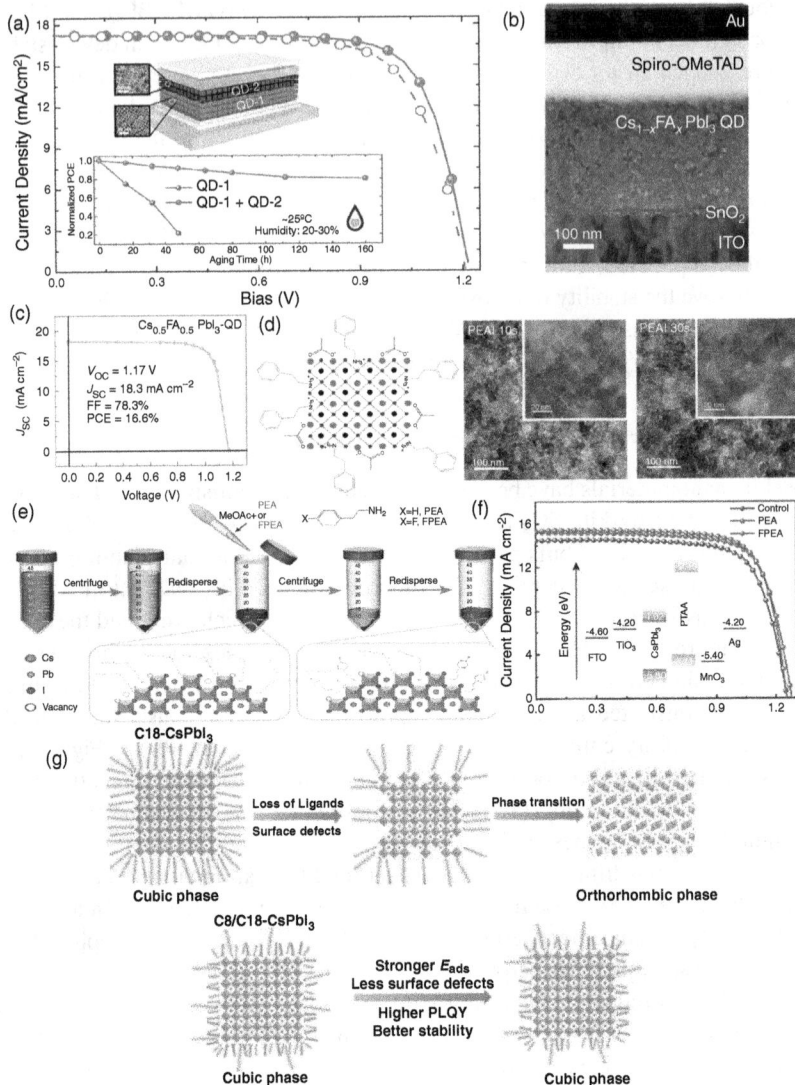

FIGURE 13.5 SCs based on perovskite nanomaterials. (a) J–V curves from reverse and forward scan of the best performance with α-CsPbI$_3$ and α-CsPbI$_3$ + FAPbI$_3$ QDs SCs, measured under AM 1.5G solar illumination. (Reproduced with permission from ref. 69. Copyright 2019 ACS.) (b–c) Cross-section TEM image and certificated J-V curve of the Cs$_{1-x}$FA$_x$PbI$_3$ QDSC-based device. (Reproduced with permission from ref. 70. Copyright 2020 Springer Nature.) (d) PEA-incorporated CsPbI$_3$-QDs as a function of each post-treatment time. (Reproduced with permission from ref. 71. Copyright 2020 Elsevier.) (e) A schematic diagram of the pseudo-solution-phase ligand exchange (p-SPLE) process. (f) J–V curves of control and p-SPLE QD SCs measured under the influence of the AM 1.5G solar spectrum at 100 mW/cm^2 (inset: the energy-level diagram of the CsPbI$_3$ QD SCs). (Reproduced with permission from ref. 72. Copyright 2021 RSC.) (g) Schematic of CsPbI$_3$ α-phase stabilization mechanism due to the presence of C8. (Reproduced with permission from ref. 74. Copyright 2019 Wiley.)

Advances in synthesis and post-processing of perovskite nanomaterials make it possible to actively adjust the size, surface, and composition to achieve excellent photoelectronic properties for a variety of applications. Although there have been significant recent achievements in perovskite nanomaterial-based photoelectronic devices, vital issues such as mechanism exploration, device performance improvement, lifetime extension, stability enhancement, and lead toxicity solution should be addressed adequately to usher a real era of perovskite nanomaterials for commercial application.

13.4 RE-DOPED FLUORESCENT NPS AND THEIR OPTOELECTRONICS APPLICATIONS

In recent years, RE-doped NPs have become a potential and promising member of fluorescent material family. RE elements are a group of 17 metal elements involving 15 lanthanide (Ln) elements (La-Lu) and 2 elements of Sc and Y. The Ln ions include two ground-state electron configurations of [Xe] $4f^n$ and [Xe] $4f^{n-1}5d^1$. The electronic configurations of [Xe] $4f$ ($n = 0$–14) generate an abundant electronic level. The Ln ions from Ce^{3+} to Yb^{3+} with abundant energy levels can induce sharp and diverse absorption or emission covering wavelengths from UV to infrared.[76,77] In general, RE-doped nanomaterials have been widely prepared by the doping techniques in which low concentrations of atoms or ions are incorporated into the host lattice to produce hybrid materials. The rigid host lattices can offer a stable micro-environment for the Ln emitters. Some host materials or other co-doping ions have superior absorption coefficients than Ln emitters, resulting in an effective energy transfer to Ln ions. Because the partially filled 4f shell of Ln ions is effectively shielded by the outer complete 5s and 5p shells, the energy level of Ln ions is not strongly influenced by the ligand ions in the host.[76,78] For obtaining high-efficiency luminescence, the choice of composition including host matrixes, activators, and sensitizers for RE-doped materials is very critical. In general, the single-doping models are consisted of inorganic matrix and activated Ln ions (activators). The inorganic matrixes not only serve on a host crystal to hold Ln ions firmly together but also play the role of sensitizing Ln ions to emission. Comparatively, the co-doping systems, in which activators co-doping with sensitizers, can produce high-effective florescence emissions by exploiting the energy transfer from sensitizers to activators.[2]

RE-based NPs exhibit unique advantages, including lower toxicity, high thermal and chemical stability, high resistance to photobleaching, low auto-fluorescence background signal, and long fluorescence decay time. Benefiting from the essential advantage and irreplaceable role, RE-doped florescence NPs have been intensively explored in the optoelectronics over the past three decades.[79-81]

13.4.1 LEDS BASED ON RE-DOPED FLUORESCENT MATERIALS

One of the landmarks in RE fluorescence is the discovery of emissive Y_2O_3:Eu(III) material, which is the beginning of phosphors for cathode-ray tubes and fluorescent lamps. Other landmarks are the findings of neodymium YAG lasers in 1964 and Er-doped optical fibers in 1987.[77] Afterwards, Eu^{2+}- or Ce^{3+}-doped fluorescent materials with tunable optical emissions are more suitable for the development of LED applications. For free

Eu²⁺ or Ce³⁺ ions, the energy gap between the lowest 5d excited state and the 4f ground state is about 34,000 and 50,000/cm, respectively. After Eu²⁺ enters the matrix, the 5d energy levels are affected by the surrounding environment (solid A), which leads to the centroid shift (ε_c) and crystal field splitting (ε_{cfs}) (Figure 13.6). The photoluminescence emission depends on the occupied 5d electronic orbit and the Stokes shift ($\Delta S(A)$). $\Delta S(A)$ is attributed to the adjacent lattice relaxation because of the coupling of 5d electrons with lattice phonons under the irradiation, which is determined by the host as well.

W-LEDs are diffusely applied in the field of lighting and display because of their higher luminous efficiency, lower power consumption, energy conservation, longer service life, and fast responsiveness. The combination of blue, green, and red LEDs can produce LED-based white light, but color stability and reproducibility are difficult to achieve and costly. Next, researchers began to consider making W-LEDs by composing LED chips with light-emitting materials that could be efficiently excited by the chips. Typically, the conventional W-LEDs are consisted of blue LED chips and yellow-emitting YAG:Ce³⁺, which have the shortcomings with high correlated color temperature (CCT > 4000 K), low CRI (<80) and narrow color gamut because of lack of red light in the emission spectra.[82] Afterwards, researchers developed a large number of commercially available blue LED-pumped phosphors for W-LEDs, for example, blue-yellow-emitting NaAlSiO₄:Eu²⁺, green-yellow-emitting (Sr,Ba)₂SiO₄:Eu²⁺, yellow-emitting Y₃Al₅O₁₂:Ce³⁺, yellow-orange-emitting (Sr,Ba)₃SiO₅:Eu²⁺, and red-emitting K₃YSi₂O₇:Eu²⁺.[83–85] Studies have shown that CRI can be increased to 90 after employing red phosphors. However, due to the absence of cyan emission in the 470–500 nm region, it is difficult to further improve the CRI values. To address this problem, Xia et al. developed full-spectrum W-LEDs by employing Na₀.₅K₀.₅Li₃SiO₄:Eu²⁺ (NKLSO:Eu²⁺) phosphors and blue chips pumped materials. The NKLSO:Eu²⁺ can provide cyan emission at 486 nm with a narrow FWHM of 20.7 nm, resulting in that the Ra of as-fabricated W-LEDs was improved from 86 to 95.2 (Figure 13.7a–c).[86]

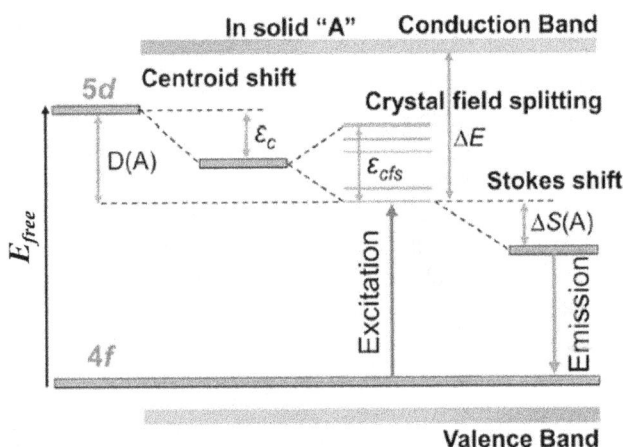

FIGURE 13.6 Schematic diagram of the effect of crystalline environment on the 4f^{n-1}5d^l levels for Eu²⁺ or Ce³⁺ doped in an inorganic solid A. (Reproduced with permission from ref. 90. Copyright 2020 ACS.)

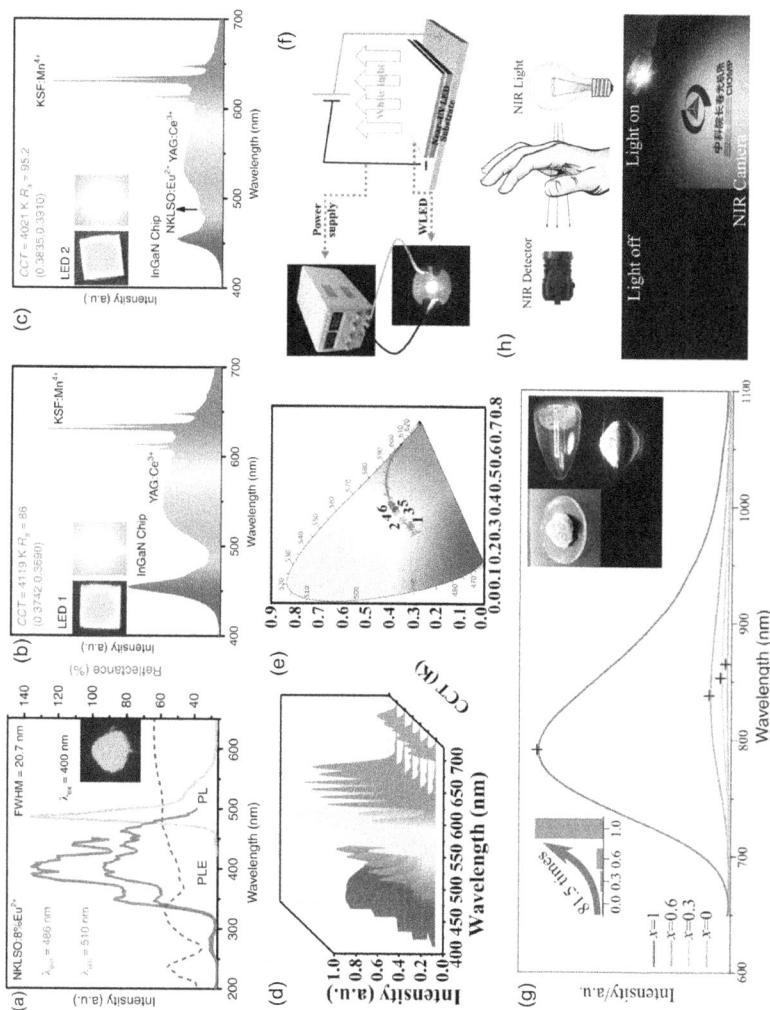

FIGURE 13.7 (a–c) $Na_{0.5}K_{0.5}Li_3SiO_4$:Eu^{2+} phosphors for W-LEDs, the reflectance, photoluminescence, and photoluminescence excitation spectra (a) and emission spectra and photographs of the W-LEDs (b and c). (Reproduced with permission from ref. 86. Copyright 2019 Springer Nature.) (d–f) Performance of W-LEDs fabricated with $NaGdF_4$:Tb^{3+}/Eu^{3+}@C:N/Eu^{3+} phosphors, luminescence spectra (d) and corresponding CIE chromaticity diagram (e) of W-LEDs, schematics and photographs of W-LEDs (f). (Reproduced with permission from ref. 87. Copyright 2022 Springer Nature.) (g and h) NIR LEDs based on $Ca_2LuHf_2Al_3O_{12}$:$0.08Cr^{3+}$ phosphors, emission spectra of Cr^{3+} in $Ca_{3-x}Lu_xHf_2Al_{2+x}Si_{1-x}O_{12}$ (x = 0,0.3, 0.6, 1.0) (g), NIR detection device (h). (Reproduced with permission from ref. 89. Copyright 2019 Wiley.)

Moreover, researchers have also turned their attention to near-ultraviolet light (NUV) chips excited phosphors for LEDs lighting. For example, Wang et al. developed $NaGdF_4$:Tb^{3+}/Eu^{3+}@C:N/Eu^{3+} single phosphor and NUV chips for constructing W-LEDs (Figure 13.7d–f). The as-obtained phosphor with tunable luminescence short-wavelength can emit a steady light range from blue to red or even white, resulting in the Ra up to 95 and controllable CCT (3568–6562 K) for the fabricated W-LEDs.[87] However, W-LED devices based on NUV chips can achieve high CRI and different CCT. There are some drawbacks, such as large energy loss during wavelength conversion and high requirements for UV light, which limit the application.

NIR phosphor-converted LEDs based on RE-doped fluorescent materials have also attracted increasing interest because of special optoelectronics.[88,89] For example, Xia et al. developed a variety of NIR-emitted phosphors by allowing Eu^{2+} to take up a selective site with low coordination number and further produce large crystal field splitting.[88,90] They reported that Eu^{2+}-activated NIR phosphor ($K_3LuSi_2O_7$:Eu^{2+}) showed a broad emission band centered at 740 nm with a high FWHM of 160 nm under 460 nm blue light excitation. Subsequently, the as-obtained NIR LEDs have been demonstrated for night-vision device applications.[88] Their researches provide a new and feasible strategy to develop Eu^{2+}-doped NIR phosphor for optoelectronic applications. Zhang et al. developed NIR LEDs by using the optimized $Ca_2LuHf_2Al_3O_{12}$:$0.08Cr^{3+}$ phosphor and commercial 460 nm LED chips for biosensing applications (Figure 13.7g and h), which had photoelectric efficiencies of 15.75%@100 mA and NIR output powers of 6.09 mW@100 mA.[89] Afterwards, Wang et al. prepared shortwave infrared (SWIR) emitting $Lu_{0.2}Sc_{0.8}BO_3$:Cr^{3+}, Yb^{3+} phosphor, which possessed high luminescence efficiency (QY of 73.6%) and good thermal stability. The as-obtained SWIR LEDs deliver good performances with an optical power of 18.4 mW with 9.3% of blue-to-SWIR PCE, endowing the applications in various fields (e.g. night-vision light, anti-counterfeiting, and nondestructive testing).[91]

13.4.2 PVs BASED ON RE-DOPED NPs

RE-doped NPs show a significant prospect in enhancing the performance of PVs, where they have been integrated as energy converters for energy loss minimization and full exploitation of the solar spectrum.[2,79] The efficiency of SCs was improved by utilizing down-shifting (DS), down-conversion (DC), and up-conversion (UC) processes to adjust the spectra. DS is a photoluminescence process with the QY of 1 or less, which can capture higher energy photons and emit lower energy photos for efficiently absorbing by the cells. However, it cannot be employed to break through the Shockley–Queisser limitation, while the DC and UC processes can improve the efficiency of SCs beyond this limitation. The DS or DC layer can be placed in front of the SCs due to the possible absorption of high-energy photons by the SCs. The UC layer should be placed on the back of the SCs so that low-energy photons can be transmitted through it.[92]

In 2014, Komarala et al. employed YVO_4:Eu^{3+} DS NPs into DSSCs to increase photocurrent and reduce UV-induced degradation.[93] In 2018, Dai incorporated YVO_4:Eu/Bi DS NPs into the mesoporous TiO_2 layer of PSCs, resulting in an enhancement of PCE from 16.3% to 17.9%.[94] These fluorescent NPs could provide short-wavelength spectral responsiveness and long-term stability for protecting SCs from UV light. Besides absorbing harmful UV light, the integration of Eu^{3+} into $MAPbI_3$ and $CsPbI_2Br$ stabilized the perovskite phase under ambient conditions and finally dramatically increased the service lifetime of as-prepared SCs.[95] The as-prepared device achieved a PCE of 21.52% (certified 20.52%), and reserved 92% and 89% of the peak PCE under 1-sun consecutive irradiation or heating at 85°C for 1,500 h, respectively.[95] Moreover, DS nanophosphors have also been developed as high-efficiency emitters in luminescent solar concentrators (LSCs). For instance, Mn^{2+}/Yb^{3+} co-doped $CsPbCl_3$ perovskite NPs have been used as efficient emitters embedded in polymer matrices for LSC applications (Figure 13.8a–c), achieving distinct triplet emission in the UV/blue, visible, and NIR regions by DC and DS processes.[96]

Briefly, DC refers to the process of converting one high-energy photon into two low-energy photons, which can be realized by exploiting the rich-level structure of RE elements doped for energy conversion.[2] For PV applications based on DC, RE ions (e.g., Pr^{3+}, Tm^{3+}, Eu^{3+}, and Yb^{3+}) were incorporated into various hosts for realizing from UV to NIR emissions and increasing the photocurrent of SC devices.[97,98] Wang et al. reported that a wideband Ce^{3+}-sensitized quantum-cutting process was integrated into the NPs for enhancing the absorption cross-section. The as-obtained hybrid crystalline silicon (c-Si) SCs based on the quantum-cutting $NaGdF_4:Ce@NaGdF_4:Nd/Yb@NaYF_4$ NPs resulted in 1.4-fold enhancement in PCE.[99] In addition, quantum-cutting emission of $^2F_{5/2}$-$^2F_{7/2}$ for Yb^{3+}-doped perovskite NCs, with over 100% efficiency, have been used as spectral converters for photovoltaic applications.[97,100] For example, Song et al. employed Yb^{3+}(6%)-Pr^{3+}(4%)-Ce^{3+}(3%)-tridoped $CsPbClBr_2$ nanophosphors with an optimum PLQY of 173% as efficient DC layers for $CuIn_{1-x}Ga_xSe_2$ (CIGS) and the silicon SCs, resulting in an ~20% enhancement in PCE. The smartphone was charged with the modified CIGS SC, which resulted in a reduced charging time from 180 to 150 min (Figure 13.8d and e).[97] These works demonstrate that RE-doped perovskite NCs, which contribute to the breakdown of the Shockley–Queisser limitation, are attractive for photovoltaic applications.

Additionally, UCNPs show attractive application prospects in the PV fields due to their ability to convert sub-bandgap photos into above-bandgap photons and reduce the transmission loss of PVs. In the early days, the Er^{3+}-doped $NaYF_4$ UCNPs were employed for the narrow bandgap c-Si SCs, leading to the peak EQE (2.5±0.2)% under 1523 nm laser irradiation (5.1 mW), and the corresponding internal quantum efficiency of 3.8%.[101] Afterwards, UC nanophosphors co-doped with Yb^{3+}/Ln^{3+} (Ln = Er, Tm, Ho) have been used for broad-bandgap DSSCs and PSCs application.[2] Jiang et al. introduced β-$NaYF_4:Yb^{3+}$, Er^{3+} as a UC mesoporous layer into $CH_3NH_3PbI_3$ PSCs to harvest NIR sunlight, resulting in an enhancement PCE (16%) contrasted with traditional TiO_2 NP-based PSCs (14.1%).[102] A conspicuous restriction of UC-based PVs is that UC luminescence typically needs concentrated excitation, such as high-power laser irradiation. To address the limitation, strategies to improve the UC luminescence efficiency have been implemented. For example, Ågren et al. used microlens arrays (MLAs) as dimensional light modulators to improve UC luminescence. The performances of DSSCs incorporated with both multicolor UCNPs and MLA were significantly improved.[103] Moreover, surface plasmon resonance of noble-metal NRs was also used to enhance the QY of UC emissions (Figure 13.8f–h).[104]

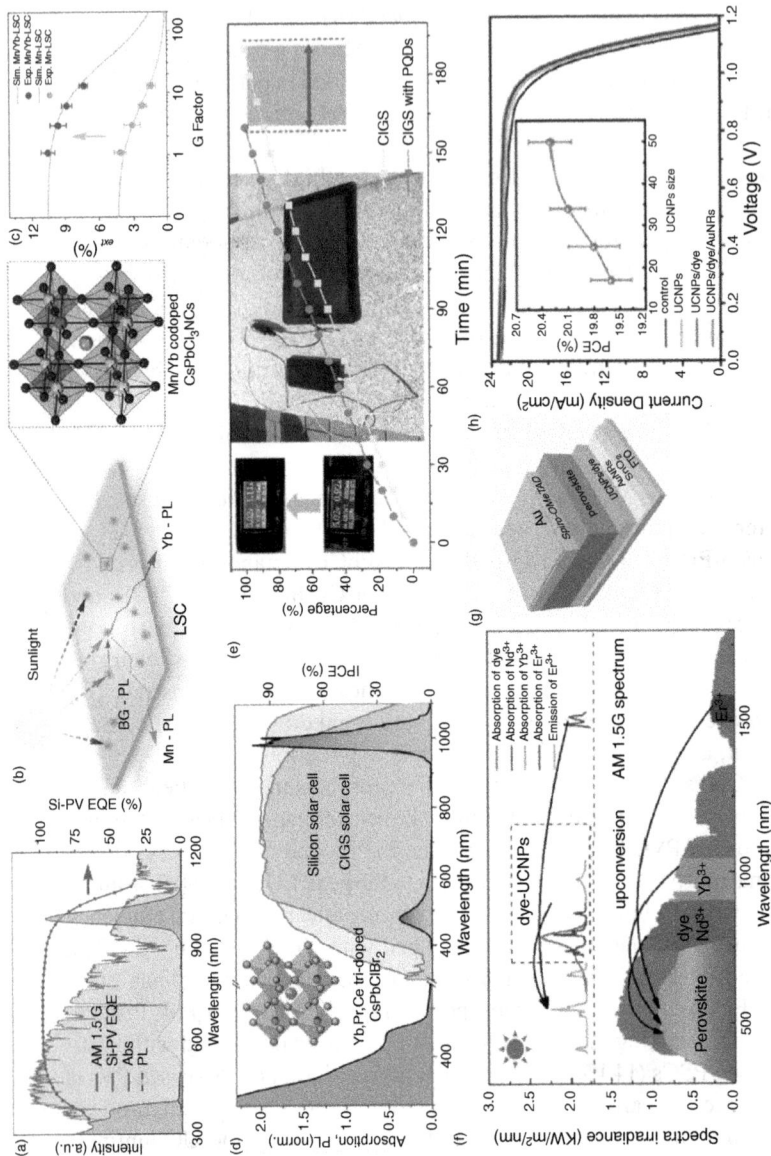

FIGURE 13.8 (a and b) Mn^{2+}/Yb^{3+} co-doped $CsPbCl_3$ perovskite NPs for LSCs application, absorption and photoluminescence spectra of LSCs (a), schematic diagram of LSCs model (b), (c) external optical efficiency (η_{ext}) of LSCs under sunlight illumination. (Reproduced with permission from ref. 96. Copyright 2020 Wiley.) (d and e) $Yb^{3+}(6\%)$-$Pr^{3+}(4\%)$-$Ce^{3+}(3\%)$-tridoped $CsPbClBr_2$ nanophosphors for $CuIn_{1-x}Ga_xSe_2$ (CIGS) and the silicon SC applications, absorption and photoluminescence spectra of nanophosphors, incident PCE of CIGS and silicon SCs (d), comparison of the time for fully charging a mobile phone with a CIGS SC with nanophosphors and a CIGS SC alone under sunlight irradiation (e). (Reproduced with permission from ref. 97. Copyright 2019 ACS.) (f–h) AM 1.5G spectrum of the section absorbed by PSCs and the spectral regions can be exploited through the UC process (f), structure of PSCs (g), reverse J-V curves of PSCs containing various composites (h). (Reproduced with permission from ref. 104. Copyright 2020 ACS.)

13.5 CARBON DOTS (CDS) AND THEIR OPTOELECTRONIC APPLICATIONS

CDs have attracted much attention and have rapidly become a research hotspot since their first discovery in 2004. A variety of research has been reported, focusing on the synthesis of CDs, the manipulation of optical properties, diverse luminescence mechanisms, and various applications. In turn, three types of CDs, including carbon quantum dots (CQDs), graphene quantum dots (GQDs), and carbonized polymer dots (CPDs), are obtained through the two synthesis methods of "top-down" and "bottom-up."[105] Multiple luminous categories including fluorescence, room-temperature phosphorescence (RTP), thermally activated decay fluorescence (TADF), and up-conversion luminescence are demonstrated. Based on these advantages of multicolor emissions of multiple luminous categories and other fascinating merits including excellent electron conductivity, broadband absorption, low toxicity, high photostability, and cost-effective sources, endowing CDs have great potential in diverse applications such as optoelectronic, bioimaging, and anti-counterfeiting. In the aspect of optoelectronic applications, mainly involving the W-LEDs, SCs, PDs, and so on, they arise great interest due to close relation to the energy sources and cutting-edge technology.

13.5.1 W-LED DISPLAYS BASED ON CDS

CDs, a significant member of the luminescence family, exhibit remarkable advantages for W-LED display and illumination. Current research has demonstrated that CDs exhibit tunable absorption covering UV and visible regions due to the existence of $\pi-\pi^*$ and $n-\pi^*$ transitions, implying that both blue and UV chips can be used to combine CDs with various emission colors.[106] Relying on tunable absorption and emission in a broadband region, some progress has been reported. The use of broadband yellow emission of CDs can replace the role of Ce^{3+}-doped YAG to fabricate W-LED.[107,108] In a W-LED composed of a blue chip and a yellow phosphor, it possesses a relatively low CRI owing to the lack of red emission. To improve this shortcoming, developing CDs with the longer-wavelength emissions is necessary. With the participation of orange CDs centered at 570 nm, a warm phosphor-based W-LEDs device was obtained under a blue chip excitation, in which the CRI is larger than 81.[109] By adding red CDs with an emission peaking at 620 nm, the CRI and CCT can be further improved. As shown in Figure 13.9a, a warm W-LED device with an enhanced CRI of 92.7 and a CCT of 3827 K was obtained in the combination of a blue chip, Ce^{3+}-doped $Y_3Al_5O_{12}$, and red CDs.[110] Alternatively, a combination of a blue chip, green CDs, and red CDs is also suitable for preparing a W-LED device. The high QYs of green CDs and red CDs are beneficial to the fabrication of high-efficiency W-LED devices. A high CRI of 92 was reported as adding green and red emission CDs with QYs of 36% and 25%, respectively.[111] A warm W-LED with a luminous efficiency of 68.58 lm W^{-1} and a CRI of 90.2 was manufactured by using red and green CDs to achieve the QYs of 80% and 49%, respectively.[112] The as-fabricated W-LED device exhibits a CRI of 92.9 and a luminous efficiency of 71.75 lm W^{-1}, as shown in Figure 13.9b, based on green and red CDs with high QYs up to 80%.[113] In addition, for the UV chip-activated W-LED devices, high-efficiency blue, green,

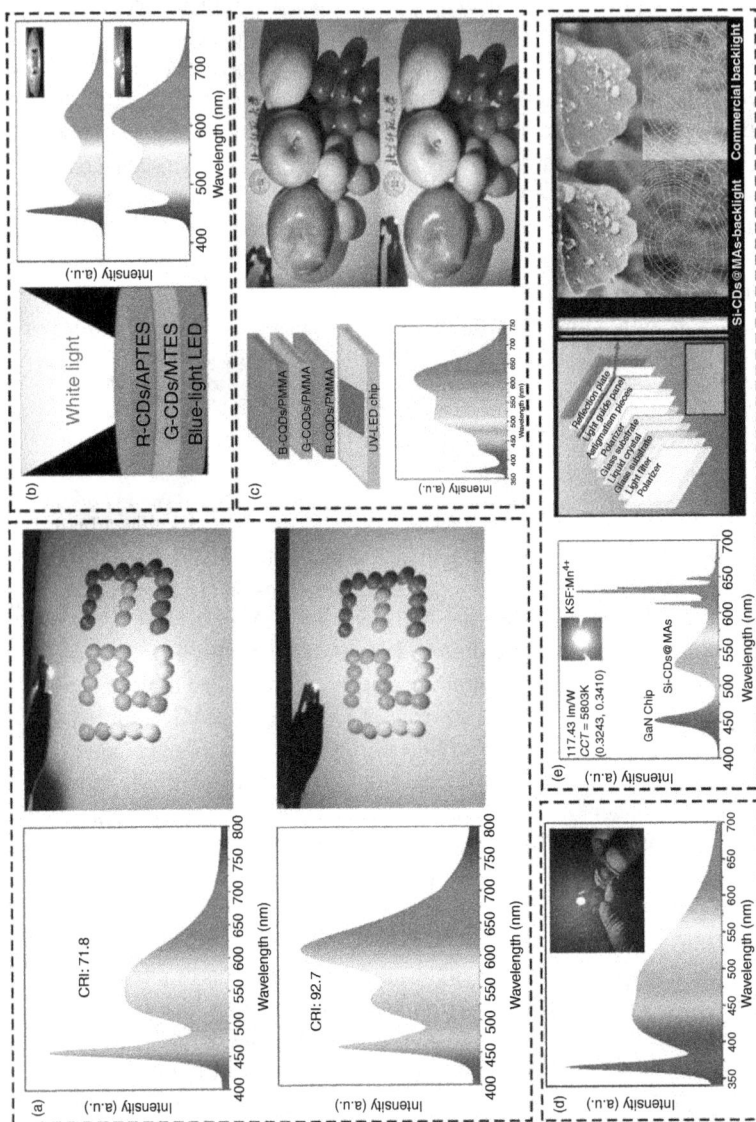

FIGURE 13.9 (a) Contrastive emission spectra and images of W-LED devices with and without red CDs. (Reproduced with permission from ref. 110. Copyright 2020 Wiley.) (b) The W-LED structure, contrastive emission spectra, and actual pictures of cold and warm W-LED devices. (Reproduced with permission from ref. 113. Copyright 2018 ACS.) (c) The W-LED structure, emission spectra, and contrastive images of fruit color are based on the commercial and prepared W-LED devices. (Reproduced with permission from ref. 115. Copyright 2017 Wiley.) (d) Emission spectrum and photograph (inset) of W-LED. (Reproduced with permission from ref. 237. Copyright 2020 Wiley.) (e) Emission spectrum and image of w-LED, LCD structure, as well as working images of the Si-CDs@MAs LCD and commercial LCD screens. (Reproduced with permission from ref. 118. Copyright 2022 Wiley.)

and red emissions of CDs are of great importance. Note that the QYs of blue and green emissions of CDs are generally higher than that of red emissions of CDs and short-wavelength emissions of CDs are more easily obtained than long-wavelength emissions of CDs, demonstrating a key role of high-efficiency red emission of CDs in achieving excellent W-LED devices.[114] It is typical that a warm W-LED device with a high CRI of 97 was reported as using red CDs peaking at 628 nm and a QY of 53%, as shown in Figure 13.9c.[115] In addition to these familiar strategies, another method concerning the achievement of single-component white emission of CDs was also proposed based on a combination of fluorescence and RTP. For instance, a single-component white output was obtained in a type of CQDs composed of a blue fluorescence and yellow RTP. The corresponding single-component W-LEDs derived by a UV chip exhibit a photoluminescence quantum efficiency of 25% and a CRI of 85.[116] Efficient single-component white CPDs were also reported, which exhibit a QY of ~41% of white emission and a QY of ~23% RTP. The as-fabricated cool W-LED consisting of a UV chip and white CPDs possesses a high luminous efficacy of 18.7 lm W^{-1}, as shown in Figure 13.9d.[117] However, such a cool W-LED possesses a high CCT, which is a common shortcoming of single-component W-LED devices due to the absence of the red composition of CDs. Comparatively speaking, the performance of W-LEDs based on single-component white emission is more unsatisfactory than the other types of W-LEDs.

On the other hand, the W-LED device is also applied as the backlight sources for liquid crystal display (LCD) application. As shown in Figure 13.9e, the Si-CDs@MAs (MAs refers to mesoporous alumina) composite exhibits a narrow FWHM of 51 nm, high thermal stability (104.1%@423 K), and an internal QY of 64.46%. The W-LED was made up of the GaN chip with a blue emission, the green-emitting part of Si-CDs@MAs composite, and the sharp red emission of KSF:Mn^{4+} (K_2SiF_6:Mn^{4+}). Compared with the commercial LCD screen, the as-fabricated LCD screen has remarkable color rendition and higher saturation.[118] It is noteworthy that the width of FWHM is crucial for LCD and the FWHM of CDs is larger than the current inorganic narrow-band phosphor. Further development of narrow-band multicolor CDs is necessary for LCD applications.

13.5.2 SCs Based on CDs

CDs have many characteristics, which make them with the adjusting functions of the incident light, electron, and hole, as well as the modifying function of active materials and electrodes, endowing CDs with attractive potential in producing SC devices. They act in various roles in different layers, including but not limited to active material, interfacial modifier, and light conversion agent. In the active layer, CDs yield photogenerated carriers after independently absorbing incident light and play the two roles of electron acceptor and electron donor. For instance, using GCDs as the poly-3-hexylthiophene (P3HT) and poly(3,4-ethylenedioxythiophene):poly(styrenesulfonate) (PEDOT:PSS) electron acceptors leads to an electronic transfer process from P3HT and PEDOT:PSS to GCDs to metals and PCBM.[68,119,120] The N-doped CQDs (N-CQDs) are employed to act as an electron donor, achieving electronic migration from the N-CQDs to TiO_2.[121] Unfortunately, the devices exhibit low PCE as CDs are selected as active materials in the active layer. In order to obtain fascinating SCs, the research community strives to boost the performance of the device through different strategies.[122,123] CDs are selected as interfacial modifiers to improve the performances of devices and their functions are constructed based on the merits of CDs. In the electrode interlayer, the work function of electrodes is an important parameter for optoelectronic devices. Previous studies have demonstrated that the addition of surface modifiers such as the polymers that possess aliphatic amine groups can reduce the work function of electrodes.[124] Because CDs possess abundant function groups and sp^2-conjugated structure, the introduction of CDs leads to the lower WF of electrodes. As shown in Figure 13.10a, the WF of ZnO and Al-doped ZnO/C-dots (AZO) decrease from −3.89 eV (ZnO) to −3.50 eV (ZnO/C-dots) and −3.80 eV (AZO) to −3.39 eV (AZO/C-dots). Similarly, the WF of ITO is altered from −4.74 eV (ITO) to −3.95 eV (ITO/C-dots)). Owing to the formation of the interfacial dipole in the interface, the higher PCEs of SC devices are obtained.[125,126] In the HTL, CDs are attempted to as HTL independently. For instance, a PCE of 3.00% is realized in a device that has an independent CQD HTL.[127] Compared with widely used materials such as spiro-OMeTAD, the performance of devices composed of independent CQDs HTL is unsatisfactory. Thereby, CDs are applied to act as modifiers alike polymers and small molecules. As shown in Figure 13.10b, the oxygen vacancy-abundant CQDs are employed to modify the efficiency of hole extraction and transfer of the $BiVO_4$. The CQDs can provide a superfast hole-transport channel when they are presented in the HTL, resulting in obvious enhancement of the performances of SCs.[128] In addition, a typical hole extraction material of PEDOT:PSS is widely used as the HTL, but its exorbitant conglutination results in a decrease in the electrical and structural homogeneity. A high PCE of 18.03% is achieved due to the improvement of conductivity and structural homogeneity caused by the cooperation of CDs and PEDOT: PSS.[129] In the ETL, an appropriate energy level is necessary for electron injection and hole blocking. The tunable energy gap of CDs exhibits an obvious advantage in the manipulation of the energy level of the ETL. As shown in Figure 13.10c, the ETL composed of the GCD-modified SnO_2 shows a more suitable energy level and higher electron mobility, leading to a champion PCE of 19.6%.[130] Using the SnO_2 modified by red CQDs with rich carboxylic acid and hydroxyl as the ETL shows a large electron mobility of $1.73 \times 10^{-2} cm^2/V$ s, thus achieving a high PCE of 22.77%.[131]

The remarkable stability of SCs is a basic condition for practical applications and always troubles researchers because the devices often suffer from low stability caused by the decomposition of active materials, high-energy UV illumination, high working temperature, moisture, and so on. Various strategies are employed to overcome and reduce the effects of these factors on device performance.[132,133] Among these strategies, CDs are added to the devices because they possess the capacity of hydrophobicity, defect passivation, bonding with decomposed elements, and light conversion from high-energy to low-energy lights based on abundant function groups. As shown in Figure 13.10d, an efficiency of 16.76% and good stability that retains 95.33% of initial PCE against air for 1000 h of all-inorganic $CsPbI_2Br$ perovskite SCs are obtained due to the function of defect passivation, new pathways for hole transfer and electron blocking, and the better hydrophobicity, as introducing functionalized p-type blue CDs.[134] Introducing CDs to decrease the defects in the surface or grain

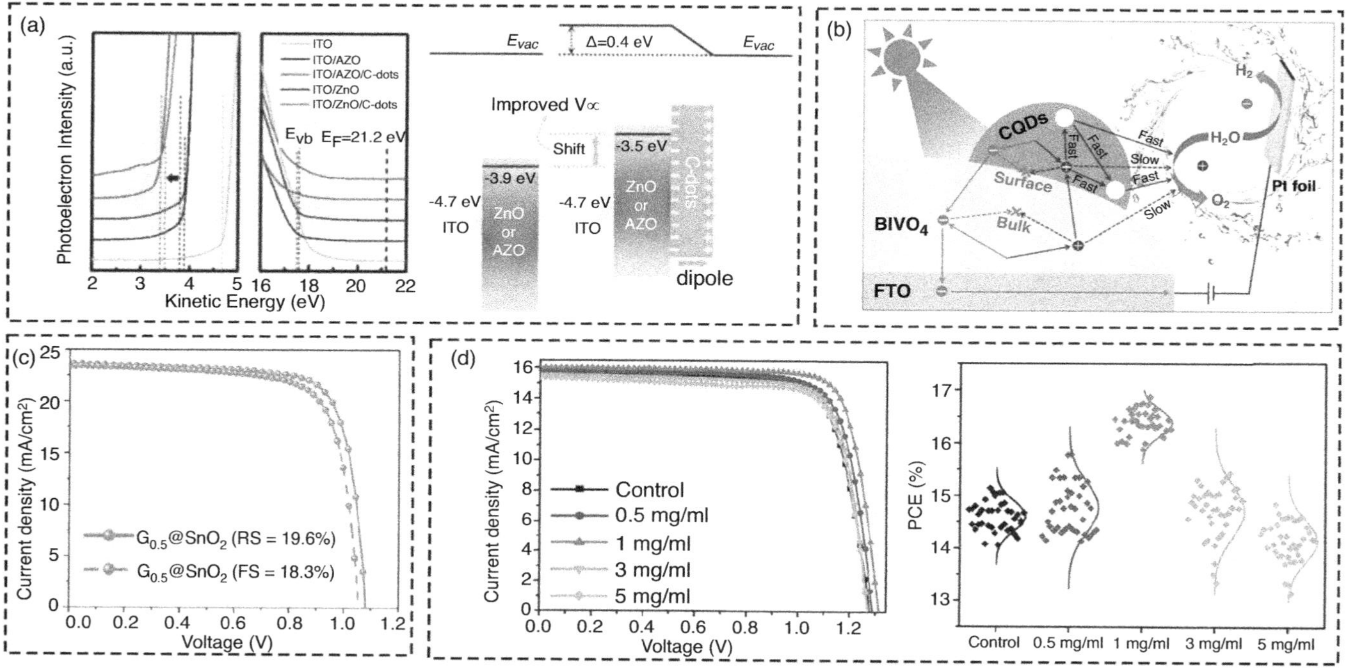

FIGURE 13.10 (a) The ultraviolet photoelectron spectroscopy spectra, vacuum level shift, and reduced WF of the film samples (Glass/ITO/interlayer). (Reproduced with permission from ref. 126. Copyright 2016 Elsevier.) (b) The schematic mechanism of the oxygen vacancy-abundant CQDs/BiVO₄ photoanode. (Reproduced with permission from ref. 128. Copyright 2022 Elsevier.) (c) The J–V curves and schematic illustration of energy levels (the inset) of SCs device with the ETL of GQD/ SnO₂ composite. (Reproduced with permission from ref. 130. Copyright 2019 RSC.) (d) The J–V curves, device structure (the inset), and the reproducibility of the CsPbI₂Br perovskite SCs. (Reproduced with permission from ref. 134. Copyright 2021 Wiley.)

boundaries is also beneficial for high-performance devices. The PCE of the device is enhanced and the optimal PCE reaches 19.38% because the carboxylic groups, hydroxyl groups, and amino groups of CQDs bond with the uncoordinated Pb in MAPbI$_3$.[135] Relying on the light conversion ability from high-energy UV light to the lower energy visible emissions, the introduction of CDs that can convert UV to blue light results in enhanced PCE from 14.6% to 16.4% and a higher efficiency retention from 20% to 70%.[136]

CDs have been applied in different SCs such as DSSCs, PSCs, and organic SCs, and excellent device properties have been reported. However, the device's performances, stability, and cost are still unsatisfactory for meeting the needs of practical applications. Much effort with respect to the CD-modified SCs is still needed in the future to design and develop SCs with comprehensive properties.

13.6 SUMMARY AND OUTLOOK

The study of optoelectronic applications based on fluorescent nanomaterials is a multidisciplinary research area involving physics, chemistry, optics, and other aspects of knowledge. In this chapter, the main focus is on the common optoelectronic properties of fluorescence nanomaterials (semiconductor QDs, perovskite nanomaterials, and RE-doped NPs and CDs) and their use in optoelectronic devices including LEDs, PVs, and PDs. A lot of progress has been made in the last decade toward better efficient, cost-effective, sustainable, and clean energy technologies, but more attempt needs to be done in this area. Due to the effect of composition, structure, optical properties, and electronic transfer ability on optoelectronic device performances, further explorations should focus on structural engineering design for developing more excellent fluorescence nanomaterials, especially possessing high PLQY, narrow FWHM, high-efficiency long-wavelength emissions (especially red and NIR emissions), and broad absorption ranging from UV to visible and NIR. Chemical and thermal stability should also be energetically optimized and enhanced to meet better practical requirements. These developments will drive the discovery of novel and improved fluorescent nanomaterials for commercial application in optoelectronics.

REFERENCES

1. F. P. García de Arquer, D. V. Talapin, V. I. Klimov, Y. Arakawa, M. Bayer, and E. H. Sargent. 2021. Semiconductor quantum dots: Technological progress and future challenges. *Science* 373 (6555): eaaz8541.
2. B. Zheng, J. Fan, B. Chen, X. Qin, J. Wang, F. Wang, R. Deng, and X. Liu. 2022. Rare-earth doping in nanostructured inorganic materials. *Chem. Rev.* 122 (6): 5519–5603.
3. C. Wei, W. Su, J. Li, B. Xu, Q. Shan, Y. Wu, F. Zhang, M. Luo, H. Xiang, Z. Cui, and H. Zeng. 2022. A universal ternary-solvent-ink strategy toward efficient inkjet-printed perovskite quantum dot light-emitting diodes. *Adv. Mater.* 34 (10): 2107798.
4. Y. Zhai, B. Zhang, R. Shi, S. Zhang, Y. Liu, B. Wang, K. Zhang, G. I. N. Waterhouse, T. Zhang, and S. Lu. 2022. Carbon dots as new building blocks for electrochemical energy storage and electrocatalysis. *Adv. Energy Mater.* 12 (6): 2103426.
5. C. R. Kagan, E. Lifshitz, E. H. Sargent, and D. V. Talapin. 2016. Building devices from colloidal quantum dots. *Science* 353 (6302): aac5523.

6. V. I. Klimov. 2014. Multicarrier interactions in semiconductor nanocrystals in relation to the phenomena of auger recombination and carrier multiplication. *Annu. Rev. Condens. Matter Phys.* 5 (1): 285–316.

7. A. I. Ekimov, A. L. Efros, and A. A. Onushchenko. 1985. Quantum size effect in semiconductor microcrystals. *Solid State Commun.* 56 (11): 921–924.

8. R. Osovsky, D. Cheskis, V. Kloper, A. Sashchiuk, M. Kroner, and E. Lifshitz. 2009. Continuous-wave pumping of multiexciton bands in the photoluminescence spectrum of a single CdTe-CdSe core-shell colloidal quantum dot. *Phys. Rev. Lett.* 102 (19): 197401.

9. M. C. Beard, K. P. Knutsen, P. Yu, J. M. Luther, Q. Song, W. K. Metzger, R. J. Ellingson, and A. J. Nozik. 2007. Multiple exciton generation in colloidal silicon nanocrystals. *Nano Lett.* 7 (8): 2506–2512.

10. L. Brus. 1986. Electronic wave functions in semiconductor clusters: Experiment and theory. *J. Phys. Chem.* 90 (12): 2555–2560.

11. J. J. Jasieniak, M. Califano, and S. E. Watkins. 2011. Size-dependent valence and conduction band-edge energies of semiconductor nanocrystals. *ACS Nano* 5 7: 5888–902.

12. A. H. Ip, S. M. Thon, S. Hoogland, O. Voznyy, D. Zhitomirsky, R. Debnath, L. Levina, L. R. Rollny, G. H. Carey, A. Fischer, K. W. Kemp, I. J. Kramer, Z. Ning, A. J. Labelle, K. W. Chou, A. Amassian, and E. H. Sargent. 2012. Hybrid passivated colloidal quantum dot solids. *Nat. Nanotechnol.* 7 (9): 577–582.

13. Y.-S. Park, W. K. Bae, T. Baker, J. Lim, and V. I. Klimov. 2015. Effect of auger recombination on lasing in heterostructured quantum dots with engineered core/shell interfaces. *Nano Lett.* 15 (11): 7319–7328.

14. M. G. Debije, and P. P. C. Verbunt. 2012. Thirty years of luminescent solar concentrator research: Solar energy for the built environment. *Adv. Energy Mater.* 2 (1): 12–35.

15. J. S. Steckel, J. T. Ho, C. Hamilton, J. Xi, C. Breen, W. Liu, P. M. Allen, and S. Coe-Sullivan. 2015. Quantum dots: The ultimate down-conversion material for LCD displays. *J. Soc. Inf. Disp.* 23: 294–305.

16. W. K. Bae, S. Brovelli, and V. I. Klimov. 2013. Spectroscopic insights into the performance of quantum dot light-emitting diodes. *MRS Bull.* 38 (9): 721–730.

17. O. Yarema, D. Bozyigit, I. Rousseau, L. Nowack, M. Yarema, W. Heiss, and V. Wood. 2013. Highly luminescent, size- and shape-tunable copper indium selenide based colloidal nanocrystals. *Chem. Mater.* 25 (18): 3753–3757.

18. A. M. Smith, and S. Nie. 2004. Chemical analysis and cellular imaging with quantum dots. *Analyst* 129 (8): 672–677.

19. R. Zhou, J. Xu, P. Luo, L. Hu, X. Pan, J. Xu, Y. Jiang, and L. Wang. 2021. Near-infrared photoactive semiconductor quantum dots for solar cells. *Adv. Energy Mater.* 11 (40): 2101923.

20. W. M. Girma, M. Z. Fahmi, A. Permadi, M. A. Abate, and J.-Y. Chang. 2017. Synthetic strategies and biomedical applications of I-III-VI ternary quantum dots. *J. Mater. Chem. B* 5 (31): 6193–6216.

21. X. Dai, Z. Zhang, Y. Jin, Y. Niu, H. Cao, X. Liang, L. Chen, J. Wang, and X. Peng. 2014. Solution-processed, high-performance light-emitting diodes based on quantum dots. *Nature* 515 (7525): 96–99.

22. T. Lee, B. J. Kim, H. Lee, D. Hahm, W. K. Bae, J. Lim, and J. Kwak. 2022. Bright and stable quantum dot light-emitting diodes. *Adv. Mater.* 34 (4): 2106276.

23. Y. Yang, Y. Zheng, W. Cao, A. Titov, J. Hyvonen, J. R. Manders, J. Xue, P. H. Holloway, and L. Qian. 2015. High-efficiency light-emitting devices based on quantum dots with tailored nanostructures. *Nat. Photonics* 9 (4): 259–266.

24. Y. Deng, F. Peng, Y. Lu, X. Zhu, W. Jin, J. Qiu, J. Dong, Y. Hao, D. Di, Y. Gao, T. Sun, M. Zhang, F. Liu, L. Wang, L. Ying, F. Huang, and Y. Jin. 2022. Solution-processed green and blue quantum-dot light-emitting diodes with eliminated charge leakage. *Nat. Photonics* 16 (7): 505–511.

25. Z. Yang, M. Gao, W. Wu, X. Yang, X. W. Sun, J. Zhang, H.-C. Wang, R.-S. Liu, C.-Y. Han, H. Yang, and W. Li. 2019. Recent advances in quantum dot-based light-emitting devices: Challenges and possible solutions. *Mater. Today* 24: 69–93.

26. J.-H. Kim, D.-Y. Jo, K.-H. Lee, E.-P. Jang, C.-Y. Han, J.-H. Jo, and H. Yang. 2016. White electroluminescent lighting device based on a single quantum dot emitter. *Adv. Mater.* 28 (25): 5093–5098.

27. W. Shockley, and H. J. Queisser. 1961. Detailed balance limit of efficiency of p-n junction solar cells. *J. Appl. Phys.* 32 (3): 510–519.

28. K. W. Johnston, A. G. Pattantyus-Abraham, J. P. Clifford, S. H. Myrskog, D. D. MacNeil, L. Levina, and E. H. Sargent. 2008. Schottky-quantum dot photovoltaics for efficient infrared power conversion. *Appl. Phys. Lett.* 92 (15): 151115.

29. C. Piliego, L. Protesescu, S. Z. Bisri, M. V. Kovalenko, and M. A. Loi. 2013. 5.2% efficient PbS nanocrystal Schottky solar cells. *Energy Environ. Sci.* 6 (10): 3054–3059.

30. G. H. Carey, L. Levina, R. Comin, O. Voznyy, and E. H. Sargent. 2015. Record charge carrier diffusion length in colloidal quantum dot solids via mutual dot-to-dot surface passivation. *Adv. Mater.* 27 (21): 3325–3330.

31. M. J. Choi, F. P. Garcia de Arquer, A. H. Proppe, A. Seifitokaldani, J. Choi, J. Kim, S. W. Baek, M. Liu, B. Sun, M. Biondi, B. Scheffel, G. Walters, D. H. Nam, J. W. Jo, O. Ouellette, O. Voznyy, S. Hoogland, S. O. Kelley, Y. S. Jung, and E. H. Sargent. 2020. Cascade surface modification of colloidal quantum dot inks enables efficient bulk homojunction photovoltaics. *Nat. Commun.* 11 (1): 103.

32. Y. Liu, H. Wu, G. Shi, Y. Li, Y. Gao, S. Fang, H. Tang, W. Chen, T. Ma, I. Khan, K. Wang, C. Wang, X. Li, Q. Shen, Z. Liu, and W. Ma. Merging passivation in synthesis enabling the lowest open-circuit voltage loss for pbs quantum dot solar cells. *Adv. Mater.* 373 (6555): 2207293.

33. H. Song, Y. Lin, Z. Zhang, H. Rao, W. Wang, Y. Fang, Z. Pan, and X. Zhong. 2021. Improving the efficiency of quantum dot sensitized solar cells beyond 15% via secondary deposition. *J. Amer. Chem. Soc.* 143 (12): 4790–4800.

34. O. Almora, D. Baran, G. C. Bazan, C. Berger, C. I. Cabrera, K. R. Catchpole, S. Erten-Ela, F. Guo, J. Hauch, A. W. Y. Ho-Baillie, T. J. Jacobsson, R. A. J. Janssen, T. Kirchartz, N. Kopidakis, Y. Li, M. A. Loi, R. R. Lunt, X. Mathew, M. D. McGehee, J. Min, D. B. Mitzi, M. K. Nazeeruddin, J. Nelson, A. F. Nogueira, U. W. Paetzold, N.-G. Park, B. P. Rand, U. Rau, H. J. Snaith, E. Unger, L. Vaillant-Roca, H.-L. Yip, and C. J. Brabec. 2021. Device performance of emerging photovoltaic materials (Version 1). *Adv. Energy Mater.* 11 (11): 2002774.

35. J. Du, R. Singh, I. Fedin, A. S. Fuhr, and V. I. Klimov. 2020. Spectroscopic insights into high defect tolerance of Zn: CuInSe2 quantum-dot-sensitized solar cells. *Nat. Energy* 5 (5): 409–417.

36. S. Yue, L. Li, S. C. McGuire, N. Hurley, and S. S. Wong. 2019. Metal chalcogenide quantum dot-sensitized 1D-based semiconducting heterostructures for optical-related applications. *Energy Environ. Sci.* 12 (5): 1454–1494.

37. L. Sun, Z. Y. Koh, and Q. Wang. 2013. PbS quantum dots embedded in a ZnS dielectric matrix for bulk heterojunction solar cell applications. *Adv. Mater.* 25 (33): 4598–4604.

38. Y. Gao, J. Zheng, W. Chen, L. Yuan, Z. L. Teh, J. Yang, X. Cui, G. Conibeer, R. Patterson, and S. Huang. 2019. Enhancing PbS colloidal quantum dot tandem solar cell performance by graded band alignment. *J. Phys. Chem. C* 10 (19): 5729–5734.

39. H. I. Kim, S.-W. Baek, M.-J. Choi, B. Chen, O. Ouellette, K. Choi, B. Scheffel, H. Choi, M. Biondi, S. Hoogland, F. P. García de Arquer, T. Park, and E. H. Sargent. 2020. Monolithic organic/colloidal quantum dot hybrid tandem solar cells via buffer engineering. *Adv. Mater.* 32 (42): 2004657.

40. Y. Mu, Z. He, K. Wang, X. Pi, and S. Zhou. 2022. Recent progress and future prospects on halide perovskite nanocrystals for optoelectronics and beyond. *iScience* 25 (11): 105371.

41. L. Cao, X. Liu, Y. Li, X. Li, L. Du, S. Chen, S. Zhao, and C. Wang. 2020. Recent progress in all-inorganic metal halide nanostructured perovskites: Materials design, optical properties, and application. *Front. Phys.* 16 (3): 33201.

42. J. Jeong, M. Kim, J. Seo, H. Lu, P. Ahlawat, A. Mishra, Y. Yang, M. A. Hope, F. T. Eickemeyer, M. Kim, Y. J. Yoon, I. W. Choi, B. P. Darwich, S. J. Choi, Y. Jo, J. H. Lee, B. Walker, S. M. Zakeeruddin, L. Emsley, U. Rothlisberger, A. Hagfeldt, D. S. Kim, M. Grätzel, and J. Y. Kim. 2021. Pseudo-halide anion engineering for α-FAPbI3 perovskite solar cells. *Nature* 592 (7854): 381–385.

43. L. Lu, X. Pan, J. Luo, and Z. Sun. 2020. Recent advances and optoelectronic applications of lead-free halide double perovskites. *Chem. Eur. J.* 26 (71): 16975–16984.

44. J.-P. Correa-Baena, M. Saliba, T. Buonassisi, M. Grätzel, A. Abate, W. Tress, and A. Hagfeldt. 2017. Promises and challenges of perovskite solar cells. *Science* 358 (6364): 739–744.

45. P. Zhu, and J. Zhu. 2020. Low-dimensional metal halide perovskites and related optoelectronic applications. *InfoMat* 2 (2): 341–378.

46. V. Malgras, S. Tominaka, J. W. Ryan, J. Henzie, T. Takei, K. Ohara, and Y. Yamauchi. 2016. Observation of quantum confinement in monodisperse methylammonium lead halide perovskite nanocrystals embedded in mesoporous silica. *J. Amer. Chem. Soc.* 138 (42): 13874–13881.

47. M. V. Kovalenko, L. Protesescu, and M. I. Bodnarchuk. 2017. Properties and potential optoelectronic applications of lead halide perovskite nanocrystals. *Science* 358 (6364): 745–750.

48. R. E. Brandt, J. R. Poindexter, P. Gorai, R. C. Kurchin, R. L. Z. Hoye, L. Nienhaus, M. W. B. Wilson, J. A. Polizzotti, R. Sereika, R. Žaltauskas, L. C. Lee, J. L. MacManus-Driscoll, M. Bawendi, V. Stevanović, and T. Buonassisi. 2017. Searching for "Defect-Tolerant" photovoltaic materials: Combined theoretical and experimental screening. *Chem. Mater.* 29 (11): 4667–4674.

49. D. N. Dirin, L. Protesescu, D. Trummer, I. V. Kochetygov, S. Yakunin, F. Krumeich, N. P. Stadie, and M. V. Kovalenko. 2016. Harnessing defect-tolerance at the nanoscale: Highly luminescent lead halide perovskite nanocrystals in mesoporous silica matrixes. *Nano Lett.* 16 (9): 5866–5874.

50. H. Huang, M. I. Bodnarchuk, S. V. Kershaw, M. V. Kovalenko, and A. L. Rogach. 2017. Lead halide perovskite nanocrystals in the research spotlight: Stability and defect tolerance. *ACS Energy Lett.* 2 (9): 2071–2083.

51. H. Li, H. Lin, D. Ouyang, C. Yao, C. Li, J. Sun, Y. Song, Y. Wang, Y. Yan, Y. Wang, Q. Dong, and W. C. H. Choy. 2021. Efficient and stable red perovskite light-emitting diodes with operational stability >300 h. *Adv. Mater.* 33 (15): 2008820.

52. Y. Liu, Y. Dong, T. Zhu, D. Ma, A. Proppe, B. Chen, C. Zheng, Y. Hou, S. Lee, B. Sun, E. H. Jung, F. Yuan, Y.-k. Wang, L. K. Sagar, S. Hoogland, F. P. García de Arquer, M.-J. Choi, K. Singh, S. O. Kelley, O. Voznyy, Z.-H. Lu, and E. H. Sargent. 2021. Bright and stable light-emitting diodes based on perovskite quantum dots in perovskite matrix. *J. Amer. Chem. Soc.* 143 (38): 15606–15615.

53. J. Song, J. Li, L. Xu, J. Li, F. Zhang, B. Han, Q. Shan, and H. Zeng. 2018. Room-temperature triple-ligand surface engineering synergistically boosts ink stability, recombination dynamics, and charge injection toward EQE-11.6% perovskite QLEDs. *Adv. Mater.* 30 (30): 1800764.

54. L. Xu, J. Li, B. Cai, J. Song, F. Zhang, T. Fang, and H. Zeng. 2020. A bilateral interfacial passivation strategy promoting efficiency and stability of perovskite quantum dot light-emitting diodes. *Nat. Commun.* 11 (1): 3902.

55. Y. Liu, J. Cui, K. Du, H. Tian, Z. He, Q. Zhou, Z. Yang, Y. Deng, D. Chen, X. Zuo, Y. Ren, L. Wang, H. Zhu, B. Zhao, D. Di, J. Wang, R. H. Friend, and Y. Jin. 2019. Efficient blue light-emitting diodes based on quantum-confined bromide perovskite nanostructures. *Nat. Photonics* 13 (11): 760–764.

56. Y. Dong, Y.-K. Wang, F. Yuan, A. Johnston, Y. Liu, D. Ma, M.-J. Choi, B. Chen, M. Chekini, S.-W. Baek, L. K. Sagar, J. Fan, Y. Hou, M. Wu, S. Lee, B. Sun, S. Hoogland, R. Quintero-Bermudez, H. Ebe, P. Todorovic, F. Dinic, P. Li, H. T. Kung, M. I. Saidaminov, E. Kumacheva, E. Spiecker, L.-S. Liao, O. Voznyy, Z.-H. Lu, and E. H. Sargent. 2020. Bipolar-shell resurfacing for blue LEDs based on strongly confined perovskite quantum dots. *Nat. Nanotechnol.* 15 (8): 668–674.

57. Y. Liu, Z. Li, J. Xu, Y. Dong, B. Chen, S. M. Park, D. Ma, S. Lee, J. E. Huang, S. Teale, O. Voznyy, and E. H. Sargent. 2022. Wide-bandgap perovskite quantum dots in perovskite matrix for sky-blue light-emitting diodes. *J. Amer. Chem. Soc.* 144 (9): 4009–4016.

58. G. Pan, X. Bai, W. Xu, X. Chen, D. Zhou, J. Zhu, H. Shao, Y. Zhai, B. Dong, L. Xu, and H. Song. 2018. Impurity ions codoped cesium lead halide perovskite nanocrystals with bright white light emission toward ultraviolet-white light-emitting diode. *ACS Appl. Mat. Interfaces* 10 (45): 39040–39048.

59. H. Wu, S. Wang, F. Cao, J. Zhou, Q. Wu, H. Wang, X. Li, L. Yin, and X. Yang. 2019. Ultrastable inorganic perovskite nanocrystals coated with a thick long-chain polymer for efficient white light-emitting diodes. *Chem. Mater.* 31 (6): 1936–1940.

60. Y. Zhang, Y. Ma, Y. Wang, X. Zhang, C. Zuo, L. Shen, and L. Ding. 2021. Lead-free perovskite photodetectors: Progress, challenges, and opportunities. *Adv. Mater.* 33 (26): 2006691.

61. L. Zhang, T. Yang, L. Shen, Y. Fang, L. Dang, N. Zhou, X. Guo, Z. Hong, Y. Yang, H. Wu, J. Huang, and Y. Liang. 2015. Toward highly sensitive polymer photodetectors by molecular engineering. *Adv. Mater.* 27 (41): 6496–6503.

62. X. Li, D. Yu, F. Cao, Y. Gu, Y. Wei, Y. Wu, J. Song, and H. Zeng. 2016. Healing all-inorganic perovskite films via recyclable dissolution-recyrstallization for compact and smooth carrier channels of optoelectronic devices with high stability. *Adv. Funct. Mater.* 26 (32): 5903–5912.

63. Y. H. Lin, W. Huang, P. Pattanasattayavong, J. Lim, R. Li, N. Sakai, J. Panidi, M. J. Hong, C. Ma, N. Wei, N. Wehbe, Z. Fei, M. Heeney, J. G. Labram, T. D. Anthopoulos, and H. J. Snaith. 2019. Deciphering photocarrier dynamics for tuneable high-performance perovskite-organic semiconductor heterojunction phototransistors. *Nat. Commun.* 10 (1): 4475.

64. C. Xie, C.-K. Liu, H.-L. Loi, and F. Yan. 2020. Perovskite-based phototransistors and hybrid photodetectors. *Adv. Funct. Mater.* 30 (20): 1903907.

65. T. Bai, T. Yang, J. Chen, D. Zheng, Z. Tang, X. Wang, Y. Zhao, R. Lu, and K. Han. 2021. Efficient luminescent halide quadruple-perovskite nanocrystals via trap-engineering for highly sensitive photodetectors. *Adv. Mater.* 33 (8): 2007215.

66. D. Wu, Y. Xu, H. Zhou, X. Feng, J. Zhang, X. Pan, Z. Gao, R. Wang, G. Ma, L. Tao, H. Wang, J. Duan, H. Wan, J. Zhang, L. Shen, H. Wang, and T. Zhai. 2022. Ultrasensitive, flexible perovskite nanowire photodetectors with long-term stability exceeding 5000 h. *InfoMat* 4 (9): e12320.

67. J. Feng, C. Gong, H. Gao, W. Wen, Y. Gong, X. Jiang, B. Zhang, Y. Wu, Y. Wu, H. Fu, L. Jiang, and X. Zhang. 2018. Single-crystalline layered metal-halide perovskite nanowires for ultrasensitive photodetectors. *Nat. Electron.* 1 (7): 404–410.

68. S.-X. Li, Y.-S. Xu, C.-L. Li, Q. Guo, G. Wang, H. Xia, H. H. Fang, L. Shen, and H. B. Sun. 2020. Perovskite single-crystal microwire-array photodetectors with performance stability beyond 1 year. *Adv. Mater.* 32 (28): 2001998.

69. F. Li, S. Zhou, J. Yuan, C. Qin, Y. Yang, J. Shi, X. Ling, Y. Li, and W. Ma. 2019. Perovskite quantum dot solar cells with 15.6% efficiency and improved stability enabled by an α-CsPbI3/FAPbI3 bilayer structure. *ACS Energy Lett.* 4 (11): 2571–2578.

70. M. Hao, Y. Bai, S. Zeiske, L. Ren, J. Liu, Y. Yuan, N. Zarrabi, N. Cheng, M. Ghasemi, P. Chen, M. Lyu, D. He, J. H. Yun, Y. Du, Y. Wang, S. Ding, A. Armin, P. Meredith, G. Liu, H. M. Cheng, and L. Wang. 2020. Ligand-assisted cation-exchange engineering for high-efficiency colloidal Cs1–xFAxPbI3 quantum dot solar cells with reduced phase segregation. *Nat. Energy* 5 (1): 79–88.

71. J. Kim, S. Cho, F. Dinic, J. Choi, C. Choi, S. M. Jeong, J.-S. Lee, O. Voznyy, M. J. Ko, and Y. Kim. 2020. Hydrophobic stabilizer-anchored fully inorganic perovskite quantum dots enhance moisture resistance and photovoltaic performance. *Nano Energy* 75: 104985.

72. X. Zhang, H. Huang, Y. M. Maung, J. Yuan, and W. Ma. 2021. Aromatic amine-assisted pseudo-solution-phase ligand exchange in CsPbI3 perovskite quantum dot solar cells. *Chem. Commun.* 57 (64): 7906–7909.

73. X. Ling, S. Zhou, J. Yuan, J. Shi, Y. Qian, B. W. Larson, Q. Zhao, C. Qin, F. Li, G. Shi, C. Stewart, J. Hu, X. Zhang, J. M. Luther, S. Duhm, and W. Ma. 2019. 14.1% CsPbI3 perovskite quantum dot solar cells via cesium cation passivation. *Adv. Energy Mater.* 9 (28): 1900721.

74. K. Chen, Q. Zhong, W. Chen, B. Sang, Y. Wang, T. Yang, Y. Liu, Y. Zhang, and H. Zhang. 2019. Short-chain ligand-passivated stable α-CsPbI3 quantum dot for all-inorganic perovskite solar cells. *Adv. Funct. Mater.* 29 (24): 1900991.

75. J. B. Hoffman, G. Zaiats, I. Wappes, and P. V. Kamat. 2017. CsPbBr3 solar cells: Controlled film growth through layer-by-layer quantum dot deposition. *Chem. Mater.* 29 (22): 9767–9774.

76. Q. Tian, W. Yao, W. Wu, and C. Jiang. 2019. NIR light-activated upconversion semiconductor photocatalysts. *Nanoscale Horiz.* 4 (1): 10–25.

77. Y. Liu, D. Tu, H. Zhu, and X. Chen. 2013. Lanthanide-doped luminescent nanoprobes: Controlled synthesis, optical spectroscopy, and bioapplications. *Chem. Soc. Rev.* 42 (16): 6924–6958.

78. X. Li, F. Zhang, and D. Zhao. 2015. Lab on upconversion nanoparticles: Optical properties and applications engineering via designed nanostructure. *Chem. Soc. Rev.* 44 (6): 1346–78.

79. D. Chen, Y. Wang, and M. Hong. 2012. Lanthanide nanomaterials with photon management characteristics for photovoltaic application. *Nano Energy* 1 (1): 73–90.

80. K. Lingeshwar Reddy, R. Balaji, A. Kumar, and V. Krishnan. 2018. Lanthanide doped near infrared active upconversion nanophosphors: Fundamental concepts, synthesis strategies, and technological applications. *Small* 14 (37): e1801304.

81. W. Yao, Q. Tian, and W. Wu. 2019. Tunable emissions of upconversion fluorescence for security applications. *Adv. Opt. Mater.* 7 (6): 1801171.

82. J. McKittrick, and L. E. Shea-Rohwer. 2014. Review: Down conversion materials for solid-state lighting. *J. Am. Ceram. Soc.* 97 (5): 1327–1352.

83. Z. Xia, Z. Xu, M. Chen, and Q. Liu. 2016. Recent developments in the new inorganic solid-state LED phosphors. *Dalton Trans.* 45 (28): 11214–11232.

84. K. A. Denault, J. Brgoch, M. W. Gaultois, A. Mikhailovsky, R. Petry, H. Winkler, S. P. DenBaars, and R. Seshadri. 2014. Consequences of optimal bond valence on structural rigidity and improved luminescence properties in SrxBa$_2$-xSiO^4:Eu^{2+} orthosilicate phosphors. *Chem. Mater.* 26 (7): 2275–2282.

85. Z. Xia, and A. Meijerink. 2017. Ce^{3+}-Doped garnet phosphors: Composition modification, luminescence properties and applications. *Chem. Soc. Rev.* 46 (1): 275–299.

86. M. Zhao, H. Liao, M. S. Molokeev, Y. Zhou, Q. Zhang, Q. Liu, and Z. Xia. 2019. Emerging ultra-narrow-band cyan-emitting phosphor for white LEDs with enhanced color rendition. *Light: Sci. Appl.* 8 (1): 38.

87. S. Bao, H. Yu, G. Gao, H. Zhu, D. Wang, P. Zhu, and G. Wang. 2022. Rare-earth single atom based luminescent composite nanomaterials: Tunable full-color single phosphor and applications in WLEDs. *Nano Res.* 15 (4): 3594–3605.

88. J. Qiao, G. Zhou, Y. Zhou, Q. Zhang, and Z. Xia. 2019. Divalent europium-doped near-infrared-emitting phosphor for light-emitting diodes. *Nat. Commun.* 10 (1): 5267.

89. L. Zhang, D. Wang, Z. Hao, X. Zhang, G.-h. Pan, H. Wu, and J. Zhang. 2019. Cr3+-doped broadband NIR garnet phosphor with enhanced luminescence and its application in NIR spectroscopy. *Adv. Opt. Mater.* 7 (12): 1900185.

90. M. Zhao, Q. Zhang, and Z. Xia. 2020. Structural engineering of Eu^{2+}-doped silicates phosphors for LED applications. *Acc. Mater. Res.* 1 (2): 137–145.
91. Y. Zhang, S. Miao, Y. Liang, C. Liang, D. Chen, X. Shan, K. Sun, and X.-J. Wang. 2022. Blue LED-pumped intense short-wave infrared luminescence based on Cr3+-Yb3+-co-doped phosphors. *Light: Sci. Appl.* 11 (1): 136.
92. F. Auzel. 2004. Upconversion and anti-stokes processes with f and d ions in solids. *Chem. Rev.* 104 (1): 139–174.
93. N. Chander, A. F. Khan, and V. K. Komarala. 2015. Improved stability and enhanced efficiency of dye sensitized solar cells by using europium doped yttrium vanadate down-shifting nanophosphor. *RSC Adv.* 5 (81): 66057–66066.
94. J. Jin, H. Li, C. Chen, B. Zhang, W. Bi, Z. Song, L. Xu, B. Dong, H. Song, and Q. Dai. 2018. Improving efficiency and light stability of perovskite solar cells by incorporating YVO4:Eu3+, Bi3+ nanophosphor into the mesoporous TiO2 layer. *ACS Appl. Energy Mater.* 1 (5): 2096–2102.
95. L. Wang, H. Zhou, J. Hu, B. Huang, M. Sun, B. Dong, G. Zheng, Y. Huang, Y. Chen, L. Li, Z. Xu, N. Li, Z. Liu, Q. Chen, L.-D. Sun, and C.-H. Yan. 2019. A Eu^{3+}-Eu^{2+} ion redox shuttle imparts operational durability to Pb-I perovskite solar cells. *Science* 363 (6424): 265–270.
96. T. Cai, J. Wang, W. Li, K. Hills-Kimball, H. Yang, Y. Nagaoka, Y. Yuan, R. Zia, and O. Chen. 2020. Mn2+/Yb3+ codoped CsPbCl3 perovskite nanocrystals with triple-wavelength emission for luminescent solar concentrators. *Adv. Sci.* 7 (18): 2001317.
97. D. Zhou, R. Sun, W. Xu, N. Ding, D. Li, X. Chen, G. Pan, X. Bai, and H. Song. 2019. Impact of host composition, codoping, or tridoping on quantum-cutting emission of ytterbium in halide perovskite quantum dots and solar cell applications. *Nano Lett.* 19 (10): 6904–6913.
98. D. C. Yu, R. Martín Rodríguez, Q. Y. Zhang, A. Meijerink, and F. T. Rabouw. 2015. Multi-photon quantum cutting in Gd2O2S:Tm3+ to enhance the photo-response of solar cells. *Light: Sci. Appl.* 4 (10): e344–e344.
99. T. Sun, X. Chen, L. Jin, H.-W. Li, B. Chen, B. Fan, B. Moine, X. Qiao, X. Fan, S.-W. Tsang, S. F. Yu, and F. Wang. 2017. Broadband Ce(III)-sensitized quantum cutting in core-shell nanoparticles: Mechanistic investigation and photovoltaic application. *J. Phys. Chem. C* 8 (20): 5099–5104.
100. X. Luo, T. Ding, X. Liu, Y. Liu, and K. Wu. 2019. Quantum-cutting luminescent solar concentrators using ytterbium-doped perovskite nanocrystals. *Nano Lett.* 19 (1): 338–341.
101. A. Shalav, B. S. Richards, T. Trupke, K. W. Krämer, and H. U. Güdel. 2004. Application of NaYF4:Er3+ up-converting phosphors for enhanced near-infrared silicon solar cell response. *Appl. Phys. Lett.* 86 (1): 013505.
102. J. Roh, H. Yu, and J. Jang. 2016. Hexagonal β-NaYF4:Yb3+, Er3+ nanoprism-incorporated upconverting layer in perovskite solar cells for near-infrared sunlight harvesting. *ACS Appl. Mat. Interfaces* 8 (31): 19847–19852.
103. Q. Liu, H. Liu, D. Li, W. Qiao, G. Chen, and H. Ågren. 2019. Microlens array enhanced upconversion luminescence at low excitation irradiance. *Nanoscale* 11 (29): 14070–14078.
104. W. Bi, Y. Wu, C. Chen, D. Zhou, Z. Song, D. Li, G. Chen, Q. Dai, Y. Zhu, and H. Song. 2020. Dye sensitization and local surface plasmon resonance-enhanced upconversion luminescence for efficient perovskite solar cells. *ACS Appl. Mat. Interfaces* 12 (22): 24737–24746.
105. Z. Wu, Z. Liu, and Y. Yuan. 2017. Carbon dots: Materials, synthesis, properties and approaches to long-wavelength and multicolor emission. *J. Mater. Chem. B* 5 (21): 3794–3809.
106. D. Trapani, R. Macaluso, I. Crupi, and M. Mosca. 2022. Color conversion light-emitting diodes based on carbon dots: A review. *Materials* 15 (15): 5450.

107. Q. Lou, Q. Ni, C. Niu, J. Wei, Z. Zhang, W. Shen, C. Shen, C. Qin, G. Zheng, K. Liu, J. Zang, L. Dong, and C. X. Shan. 2022. Carbon nanodots with nearly unity fluorescent efficiency realized via localized excitons. *Adv. Sci.* 9 (30): 2203622.

108. J. Guo, Y. Lu, A. Q. Xie, G. Li, Z. B. Liang, C. F. Wang, X. Yang, and S. Chen. 2022. Yellow-emissive carbon dots with high solid-state photoluminescence. *Adv. Funct. Mater.* 32 (20): 2110393.

109. H. Guo, Z. Liu, X. Shen, and L. Wang. 2022. One-pot synthesis of orange emissive carbon quantum dots for all-type high color rendering index white light-emitting diodes. *ACS Sustain. Chem. Eng.* 10 (26): 8289–8296.

110. X. Zhang, H. Yang, Z. Wan, T. Su, X. Zhang, J. Zhuang, B. Lei, Y. Liu, and C. Hu. 2020. Self-quenching-resistant red emissive carbon dots with high stability for warm white light-emitting diodes with a high color rendering index. *Adv. Opt. Mater.* 8 (15): 2000251.

111. Y. Zhai, Y. Wang, D. Li, D. Zhou, P. Jing, D. Shen, and S. Qu. 2018. Red carbon dots-based phosphors for white light-emitting diodes with color rendering index of 92. *J. Colloid Interface Sci.* 528: 281–288.

112. B. Yuan, Z. Xie, P. Chen, and S. Zhou. 2018. Highly efficient carbon dots and their nanohybrids for trichromatic white LEDs. *J. Mater. Chem. C* 6 (22): 5957–5963.

113. B. Yuan, S. Guan, X. Sun, X. Li, H. Zeng, Z. Xie, P. Chen, and S. Zhou. 2018. Highly efficient carbon dots with reversibly switchable green-red emissions for trichromatic white light-emitting diodes. *ACS Appl. Mat. Interfaces* 10 (18): 16005–16014.

114. J. Wang, J. Zheng, Y. Yang, X. Liu, J. Qiu, and Y. Tian. 2022. Tunable full-color solid-state fluorescent carbon dots for light emitting diodes. *Carbon* 190: 22–31.

115. Z. Wang, F. Yuan, X. Li, Y. Li, H. Zhong, L. Fan, and S. Yang. 2017. 53% efficient red emissive carbon quantum dots for high color rendering and stable warm white-light-emitting diodes. *Adv. Mater.* 29 (37): 1702910.

116. T. Yuan, T. Meng, P. He, Y. Shi, Y. Li, X. Li, L. Fan, and S. Yang. 2019. Carbon quantum dots: An emerging material for optoelectronic applications. *J. Mater. Chem. C* 7 (23): 6820–6835.

117. Z. Wang, Y. Liu, S. Zhen, X. Li, W. Zhang, X. Sun, B. Xu, X. Wang, Z. Gao, and X. Meng. 2020. Gram-scale synthesis of 41% efficient single-component white-light-emissive carbonized polymer dots with hybrid fluorescence/phosphorescence for white light-emitting diodes. *Adv. Sci.* 7 (4): 1902688.

118. J. Chen, X. Zou, W. Li, H. Zhang, X. Zhang, M. S. Molokeev, Y. Liu, and B. Lei. 2022. Strategy to construct high thermal-stability narrow-band green-emitting Si-CDs@MAs phosphor for wide-color-gamut backlight displays. *Adv. Opt. Mater.* 7 (6): 2200851.

119. Y. Li, Y. Hu, Y. Zhao, G. Shi, L. Deng, Y. Hou, and L. Qu. 2011. An electrochemical avenue to green-luminescent graphene quantum dots as potential electron-acceptors for photovoltaics. *Adv. Mater.* 23 (6): 776–780.

120. W. Kwon, G. Lee, S. Do, T. Joo, and S. W. Rhee. 2014. Size-controlled soft-template synthesis of carbon nanodots toward versatile photoactive materials. *Small* 10 (3): 506–513.

121. S. Chakrabarti, D. Carolan, B. Alessi, P. Maguire, V. Svrcek, and D. Mariotti. 2019. Microplasma-synthesized ultra-small NiO nanocrystals, a ubiquitous hole transport material. *Nanoscale Adv.* 1 (12): 4915–4925.

122. H. Min, D. Y. Lee, J. Kim, G. Kim, K. S. Lee, J. Kim, M. J. Paik, Y. K. Kim, K. S. Kim, M. G. Kim, T. J. Shin, and S. Il Seok. 2021. Perovskite solar cells with atomically coherent interlayers on SnO2 electrodes. *Nature* 598 (7881): 444–450.

123. S. Tan, B. Yu, Y. Cui, F. Meng, C. Huang, Y. Li, Z. Chen, H. Wu, J. Shi, Y. Luo, D. Li, and Q. Meng. 2022. Temperature-reliable low-dimensional perovskites passivated black-phase CsPbI3 toward stable and efficient photovoltaics. *Angew. Chem. Int. Ed.* 61 (23): e202201300.

124. Yinhua Zhou, Canek Fuentes Hernandez, Jaewon Shim, Jens Meyer, Anthony J. Giordano, Hong Li, Paul Winget, Theodoros Papadopoulos, Hyeunseok Cheun, Jungbae Kim, Mathieu Fenoll, Amir Dindar, Wojciech Haske, Ehsan Najafabadi, Talha M. Khan, Hossein Sojoudi, Stephen Barlow, Samuel Graham, Jean Luc Brédas, Seth R. Marder, Antoine Kahn, and B. Kippelen. 2012. A universal method to produce low-work function electrodes for organic electronics. *Science* 336 (6079): 327–332.

125. H. Xu, L. Zhang, Z. Ding, J. Hu, J. Liu, and Y. Liu. 2018. Edge-functionalized graphene quantum dots as a thickness-insensitive cathode interlayer for polymer solar cells. *Nano Res.* 11 (8): 4293–4301.

126. X. Lin, Y. Yang, L. Nian, H. Su, J. Ou, Z. Yuan, F. Xie, W. Hong, D. Yu, M. Zhang, Y. Ma, and X. Chen. 2016. Interfacial modification layers based on carbon dots for efficient inverted polymer solar cells exceeding 10% power conversion efficiency. *Nano Energy* 26: 216–223.

127. S. Paulo, G. Stoica, W. Cambarau, E. Martinez-Ferrero, and E. Palomares. 2016. Carbon quantum dots as new hole transport material for perovskite solar cells. *Synth. Met.* 222: 17–22.

128. T. Zhou, J. Wang, Y. Zhang, C. Zhou, J. Bai, J. Li, and B. Zhou. 2022. Oxygen vacancy-abundant carbon quantum dots as superfast hole transport channel for vastly improving surface charge transfer efficiency of BiVO4 photoanode. *Chem. Eng. J.* 431: 133414.

129. Z. Li, C. Liu, X. Zhang, J. Guo, H. Cui, L. Shen, Y. Bi, and W. Guo. 2019. Using easily prepared carbon nanodots to improve hole transport capacity of perovskite solar cells. *Mater. Today Energy* 12: 161–167.

130. Y. Zhou, S. Yang, X. Yin, J. Han, M. Tai, X. Zhao, H. Chen, Y. Gu, N. Wang, and H. Lin. 2019. Enhancing electron transport via graphene quantum dot/SnO2 composites for efficient and durable flexible perovskite photovoltaics. *J. Mater. Chem. A* 7 (4): 1878–1888.

131. W. Hui, Y. Yang, Q. Xu, H. Gu, S. Feng, Z. Su, M. Zhang, J. Wang, X. Li, J. Fang, F. Xia, Y. Xia, Y. Chen, X. Gao, and W. Huang. 2020. Red-carbon-quantum-dot-doped SnO2 composite with enhanced electron mobility for efficient and stable perovskite solar cells. *Adv. Mater.* 32 (4): 1906374.

132. S. Liu, Y. Sun, L. Chen, Q. Zhang, X. Li, and J. Shuai. 2022. A review on plasmonic nanostructures for efficiency enhancement of organic solar cells. *Mater. Today Phys.* 24: 100680.

133. K. Fan, J. Yu, and W. Ho. 2017. Improving photoanodes to obtain highly efficient dye-sensitized solar cells: A brief review. *Mater. Horiz.* 4 (3): 319–344.

134. X. Guo, B. Zhao, K. Xu, S. Yang, Z. Liu, Y. Han, J. Xu, D. Xu, Z. Tan, and S. F. Liu. 2021. p-type carbon dots for effective surface optimization for near-record-efficiency CsPbI2 Br solar cells. *Small* 17 (37): 2102272.

135. Q. Guo, F. Yuan, B. Zhang, S. Zhou, J. Zhang, Y. Bai, L. Fan, T. Hayat, A. Alsaedi, and Z. Tan. 2018. Passivation of the grain boundaries of CH3NH3PbI3 using carbon quantum dots for highly efficient perovskite solar cells with excellent environmental stability. *Nanoscale* 11 (1): 115–124.

136. J. Jin, C. Chen, H. Li, Y. Cheng, L. Xu, B. Dong, H. Song, and Q. Dai. 2017. Enhanced performance and photostability of perovskite solar cells by introduction of fluorescent carbon dots. *ACS Appl. Mat. Interfaces* 9 (16): 14518–14524.

14 The Applications of Molecular Biosensors in Biotechnology

Meenakshi Gupta and Maryam Sarwat

14.1 BIOSENSORS: AN OVERVIEW

A biosensor is a tool that senses chemical or biological reactions by generating signals related to the amount of an analyte. These sensors are generally employed in the food industry, illness monitoring, and drug development process (Nikhil et al., 2016). Figure 14.1 shows a conventional biosensor work, which includes the following parts.

- **Analyte:** A target substance that is to be detected. For instance, a biosensor made to sense glucose uses glucose as an "analyte."
- **Bioreceptor:** It is a molecule that uniquely recognizes the analyte. Bioreceptors include aptamers, enzymes, cells, antibodies, and deoxyribonucleic acid (DNA). As the bioreceptor comes in contact with the analyte,

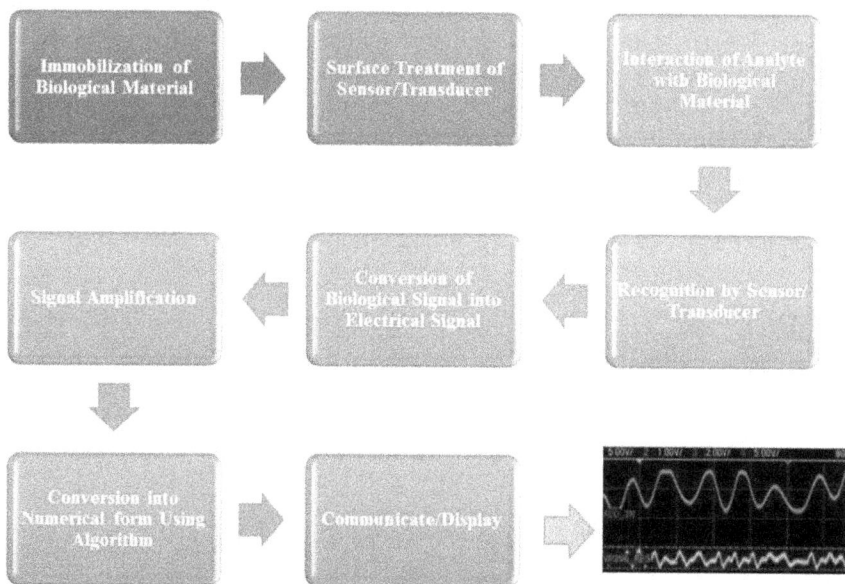

FIGURE 14.1 Working of a biosensor.

DOI: 10.1201/9781003352372-14

a signal is produced in the form of pH, mass shift, light, charge, or heat. This signal production is defined as bio-recognition.

- **Transducer:** A transducer transforms one type of energy into another and this energy conversion is defined as signalization. The transducer transforms the signal produced by the bioreceptor into a quantifiable form (either optical or electrical signals). These quantified signals are proportional to the number of reactions between the bioreceptor and the analyte.
- **Electronics:** Electronics processes the transduced signal and get it ready for display. It is made up of sophisticated electrical circuitry that carries out signal conditioning tasks. It amplifies the signal and converts it from analog to digital form (Figure 14.2).
- **Display:** The display has a user interpretation system that generates numbers or curves that the user can understand. It quantifies the signals generated by electronics. It is frequently a combination of software and hardware that generates user-friendly results. The results can be tabular, graphic, numeric, or image depending on the user (Nikhil et al., 2016).

14.2   HISTORICAL BACKGROUND

The concept of biosensors dates back around 1906 when M. Cremer introduced the theory that the concentration of an acid in a fluid is directly related to the electric potential that exists between parts of the fluid on opposite sides of a glass membrane (1906). Further, in 1909, Sorensen introduced the idea of pH, and then the pH measurement electrode was invented by Hughes (1922). Griffin and Nelson pioneered demonstrating the immobilization of the enzyme invertase on aluminum hydroxide and charcoal between 1909 and 1922 (1916). Leland C. Clark, Jr. (Father of Biosensors) made the first "true" biosensor for the detection of oxygen in 1956 and the oxygen electrode was renamed as "Clark electrode." Later, Clark demonstrated an amperometric enzyme electrode in 1962 for sensing glucose (Figure 14.3). However, the first potentiometric biosensor was developed by Guilbault and Montalvo, Jr. in 1967 for the detection of urea. The biosensors were first commercialized in 1975 by Yellow Spring Instruments (YSI) (Guilbault and Montalvo, 1969).

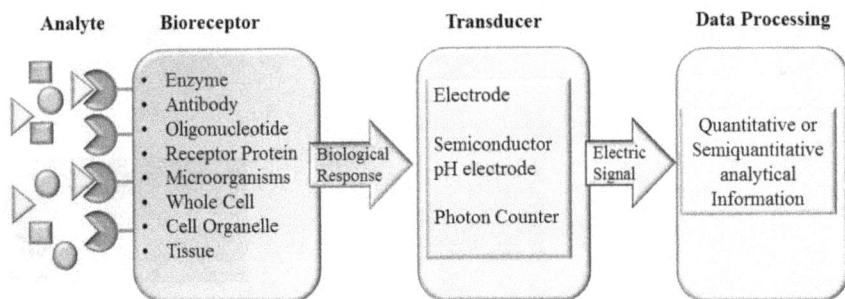

FIGURE 14.2   Biosensor detection process.

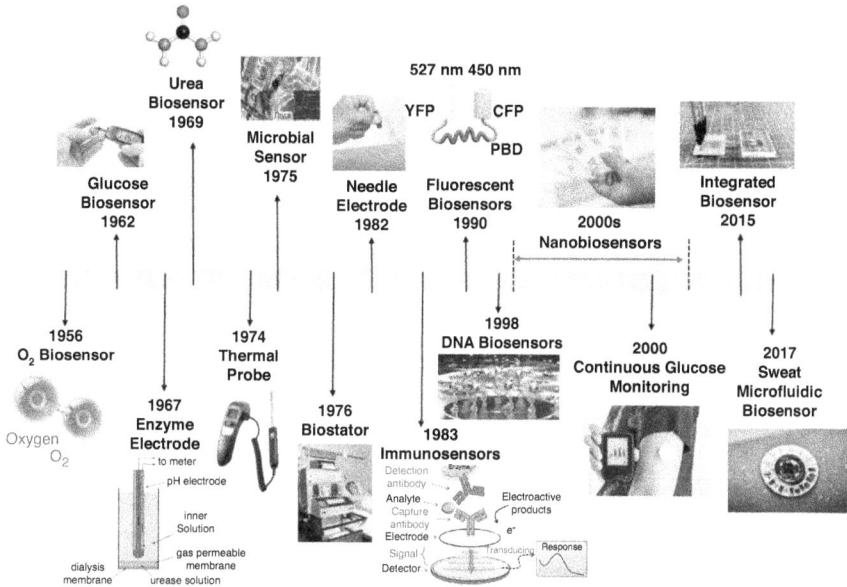

FIGURE 14.3 Biosensor development timeline.

Soon after the discovery of the i-STAT sensor, notable growth in the arena of biosensors has been made. This field has become a multidisciplinary research area, connecting basic science principles (physics, chemistry, and biology) and fundamentals of micro/nanotechnology, electronics, and applied medicine (Nikhil et al., 2016). From 2005 to 2015, "Web of Science" database indexed over 84,000 reports on the topic of "biosensors."

14.3 CHARACTERISTICS OF A BIOSENSOR

Every biosensor has certain static and dynamic characteristics. The working efficiency of the biosensor is affected by the optimization of these properties.

14.3.1 SELECTIVITY

The most important feature is the selectivity of any biosensor. The capability of any bioreceptor to detect the analyte of interest in the presence of other contaminants and admixtures is referred to as selectivity. The best example of selectivity is the interaction of an antibody and an antigen. Antibodies immobilized on the surface of the transducer act as bioreceptors and react with the antigen present in the solution (Nikhil et al., 2016).

14.3.2 REPRODUCIBILITY

The capacity of a biosensor to provide identical responses for a replicated experiment is referred to as reproducibility. It is characterized by the precision and accuracy of both the transducer and electronics. Precision refers to the sensor's ability to generate consistent results and accuracy is the ability of the sensor to deliver a mean value close to the actual value every time a sample is quantified. Reproducible signals give excellent reliability and validity to biosensor response inferences (Nikhil et al., 2016).

14.3.3 Stability

The degree of susceptibility to environmental instabilities in/around the biosensing system is defined as stability. Such disruptions can increase the chance of error and affect the results of the biosensors. Stability is considered the most important feature in areas where continuous monitoring and long incubation steps are used for experiments. The temperature-sensitive responses of transducers and electronics can also affect the stability of a biosensor. Thus, routine and proper calibration of the biosensor is required to confirm a stable sensor response. The affinity of the bioreceptor is another factor that can influence stability. Bioreceptors with high affinities promote either covalent linkage or strong electrostatic bonding of the analyte, improving the biosensor stability. The deprivation of the bioreceptor with time is another factor that affects measurement stability (Nikhil et al., 2016).

14.3.4 Sensitivity

The smallest amount of analyte that a biosensor can sense is called sensitivity or the limit of detection. A biosensor is sometimes required to detect analyte concentrations as low as ng/ml or fg/ml in a sample. For example, a concentration as low as 4 ng/ml of prostate-specific antigen is allied with prostate cancer, for which doctors recommend biopsy tests. As a result, sensitivity is observed as an important property of a biosensor (Nikhil et al., 2016).

14.3.5 Linearity

Linearity can be defined as the accuracy of the measured response and is mathematically represented as

$$y = mc \tag{14.1}$$

where c is the concentration of the analyte, y is the output signal, and m is the biosensor's sensitivity.

The linearity is related to its resolution and the tested range of the analyte. The biosensor's resolution is the smallest change in analyte concentration necessary to bring a change in response. A good resolution is required depending on its application. The linear range is defined as the range of analyte concentrations where the response changes linearly with concentration (Nikhil et al., 2016).

14.4 TYPES OF BIOSENSORS

Biosensors of various types are used, including tissue-based, enzyme-based, DNA biosensors, immunosensors, thermal biosensors, and piezoelectric biosensors (Figure 14.4).

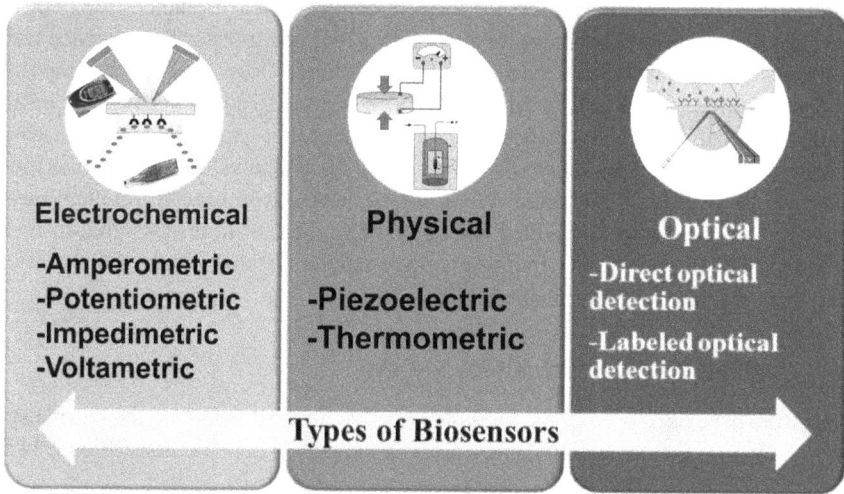

FIGURE 14.4 Types of biosensors.

It was in 1967 when Updike and Hicks published the first enzyme-based sensor (Mehrotra, 2016). Enzyme biosensors are developed based on immobilization methods, like enzyme adsorption via ionic bonding, van der Waals forces, or covalent bonding. Peroxidases, polyphenol oxidases, oxidoreductases, and amino oxidases are common enzymes used for this purpose. Later, the first microbe or cell-based biosensor was developed by Divies (Mohankumar et al., 2021). Tissues for tissue-based sensors are derived from plants and animals. The analyte of interest might be a process substrate or inhibitor. Rechnitz developed the first tissue-based sensor for detecting arginine, an amino acid. Membranes, mitochondria, microsomes, and chloroplast were used to create organelle-based sensors. The stability of these biosensors was high but the detection time was longer, hence, reducing the specificity (Mehrotra, 2016).

The DNA biosensors are based on the fact that a single-strand nucleic acid molecule of the sample can identify and bind to its complementary strand by stable hydrogen bonding between the two strands (Kavita, 2017). Magnetic biosensors are miniaturized biosensors that utilize the magnetoresistance effect to detect magnetic micro- and nanoparticles in microfluidic channels (Nabaei et al., 2018). Thermal/calorimetric biosensors are created by incorporating the biosensor materials into a physical transducer. There are two types of piezoelectric biosensors: quartz crystal microbalances and surface acoustic wave devices. They detect variations in a piezoelectric crystal's resonance frequency occurring due to mass changes in the crystal structure (Ramanathan and Danielsson, 2001).

Optical biosensors have a light source and optical components that collectively produce a light beam with particular characteristics and direct it to a modulating agent, a modified sensing head, and a photodetector (Chen and Wang, 2020). The development of genetically encoded biosensors has been aided by green fluorescent protein and subsequent auto-fluorescent protein (AFP) variants and genetic fusion reporters. This type of biosensor is simple to engineer and integrate into cells

(Mehrotra, 2016). Another example is a single-chain Förster resonance energy transfer (FRET) biosensor. They have a pair of AFPs that can transfer fluorescence resonance energy between themselves when brought close. Depending on the intensity, ratio, or lifetime of AFPs, different methods can be used to control changes in FRET signals. Synthetic chemistry is used to create protein and peptide biosensors, which are further enzymatically labeled with synthetic fluorophores. Because of the independence of genetically encoded AFPs, they can be easily used to regulate target activity. They also have the added benefit of being able to improve signal-to-noise ratio and response sensitivity by introducing photoactivatable groups and chemical quenchers (Mehrotra, 2016; Shaner, 2005).

14.5   APPLICATIONS OF BIOSENSORS

Biosensors have been used in the food industry, the medical field, and the marine sector, as they offer greater stability and sensitivity than traditional methods (Figure 14 5).

14.5.1   In Food Processing, Monitoring, Food Authenticity, Quality, and Safety

Quality and safety, food product maintenance, and processing are all difficult issues in the food processing industry. Traditional techniques for performing wet lab experiments have many limitations including human fatigue, more time, and extra money. Food authentication and monitoring alternatives that provide unbiased and reliable

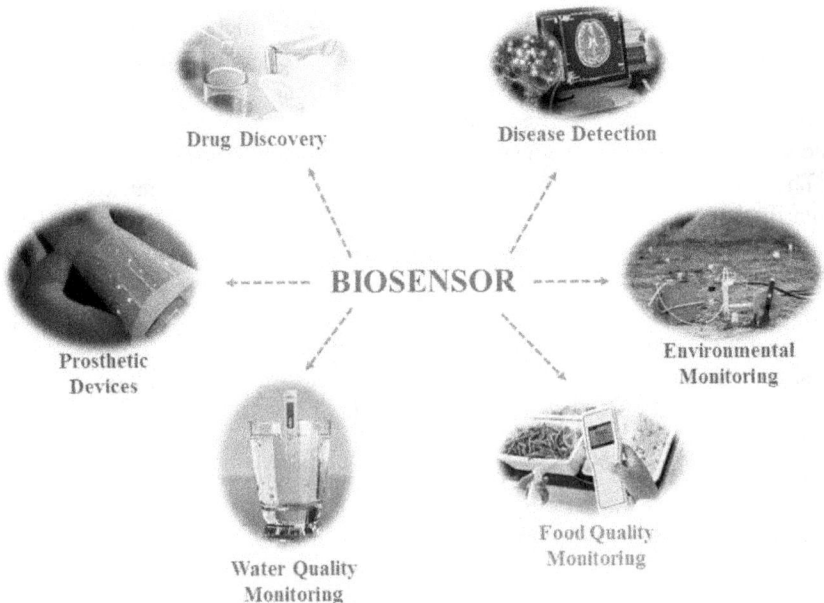

FIGURE 14.5   Major areas of application of biosensors.

measurement of food products economically are therefore looked for the food industry. Thus, the advancement of biosensors according to the demand for real-time, simple, selective, and low-cost techniques appears to be advantageous (Mishra et al., 2018; Neethirajan et al., 2018).

Ghasemi-Varnamkhasti et al. (2012) investigated beer aging using enzymatic biosensors based on cobalt phthalocyanine. These biosensors demonstrated an excellent ability to observe the aging of beer throughout storage. Further, for detecting pathogens in food products, biosensors can be used. The presence of *E. coli* in vegetables is due to fecal contamination in food. *E. coli* can be detected using potentiometric alternating biosensing systems to detect alterations in pH caused by ammonia. The liquid phase, obtained after washing vegetables with peptone water, is separated by amalgamating it in a sonicator to detach bacterial cells from food (Ercole et al., 2003).

In the dairy industry, enzymatic biosensors are also used. A flow cell is fixed with a screen-printed carbon electrode. On a photocrosslinkable polymer, enzymes were immobilized on electrodes. The organophosphate pesticides in milk could be quantified using an automated flow-based biosensor (Mishra et al., 2012).

Sweeteners are commonly used food additives today, and they are linked to a variety of undesirable diseases such as dental caries, cardiovascular disease, obesity, and type 2 diabetes. Artificial sweeteners are considered addictive, coaxing us to eat high-energy foods unconsciously and causing unintentional weight gain. As a result, sensing and measuring them is critical. Ion chromatographic methods have been used in the past to distinguish between the two types of sweeteners. Multichannel biosensors have been investigated as a more effective method of combining lipid films with electrochemical techniques for the rapid screening of sweeteners. The signals are analyzed in MATLAB using spatiotemporal techniques. Because all sweeteners are controlled by heterodimeric GPCRs in Type-II cells in the bud, they have a variety of binding sites to recognize sweet stimuli of various structures. According to research, there are two categories of sweet stimuli for signal transduction: the cyclic adenosine monophosphate pathway (for natural sugars) and the inositol triphosphate and diacylglycerol pathway (for artificial sweeteners). The response to artificial sweeteners is heavily reliant on ligand-binding sites in taste receptor amino-terminal domains. Taste receptor cell signal responses to natural and artificial sweeteners are distinct. When glucose was applied, the taste epithelium biosensor showed sparse signals with positive waveforms, whereas sucrose sustained signals with negative spikes. The taste epithelium produced more intense signals in response to artificial sweeteners, demonstrating that the responses to artificial sweeteners differed significantly from those to natural sugars in both time and frequency (Mehrotra, 2016).

14.5.2 In Fermentation Processes

Quality control and assurance are critical in the fermentation industry. Thus, effective fermentation process monitoring is required to create, optimize, and operate biological reactors at peak efficiency. Biosensors can detect process conditions indirectly by checking the number of products, enzymes, biomass, antibodies, or by-products of the process. Because of their simple instrumentation, strong selectivity, low pricing, and ease of automation, biosensors perfectly control the fermentation sector and produce reproducible results (Shi et al., 2017). Several types of commercial biosensors are now available, which can detect biochemical parameters (glucose, lactate, lysine, ethanol, etc.) and are widely utilized in China, accounting for around 90% of the market. Saccharification was observed during the fermentation phase using the traditional Fehling's method. Because this procedure includes lowering sugar titration, the results were erroneous. However, since the commercialization of the glucose biosensor in 1975, the fermentation sectors have profited. Glucose biosensors are being used successfully in the factories to produce output in the saccharification and fermentation workshop. The bio-enzymatic approach is used to create glucose. Biosensors are also used in ion exchange retrieval, which detects changes in biological composition. For example, a glutamate biosensor was utilized to research the ion exchange retrieval of an isoelectric glutamate liquor supernatant. Monitoring key metabolites in real time is vital for the timely optimization and control of biological processes (Yan et al., 2014; Mehrotra, 2016).

14.5.3 Biosensing Technology for Sustainable Food Safety

Food quality is defined by its appearance, aroma, nutritive quality, purity, taste, texture, and chemicals (Scognamiglio, 2014). To maintain safety and quality, smart nutrient monitoring and rapid screening for chemical and biological pollutants are critical. Material science, nanotechnology, electromechanical, and microfluidic technologies are making sensing technology more marketable. Efforts are being undertaken to build control systems that ensure food quality and safety, and hence human health. Because food composition might change during storage, glucose monitoring becomes essential. German investigated the electrochemical properties of glucose oxidase mounted on a graphite rod that had been modified with gold nanoparticles (AuNPs) to boost its sensitivity (German et al., 2010).

Glutamine is essential for critical processes including transport and signaling and acts as a precursor for amino acids, protein, and sugar. Patients who are glutamine deficient experience various pathologies such as malabsorptive diseases and must be supplemented to promote immunological responses, maintain intestinal functionality, and reduce bacterial translocation. For detection in the fermentation process, a glutaminase-based microfluidic biosensor chip featuring flow-injection analysis for electrochemical sensing was adopted (Chen et al., 2014; Mehrotra, 2016).

Because of their capacity to interact with only the dangerous fractions of metal ions, biosensors can detect overall toxicity and particular toxic metals. Pesticides are extremely dangerous to the environment. Organophosphates and carbamic insecticides are the most commonly used pesticides (Tudi et al., 2021). Immunosensors have proven their worth as agrifood and environmental monitoring devices. Acetylcholinesterase and butyrylcholinesterase biosensors for carbaryl, aldicarb, chlorpyrifos-methyl, paraoxon, and other chemicals have been developed (Tothill, 2001). Arduini and colleagues invented oxy with screen-printed electrodes. Pesticides in wines and juices are detected using a similar sort of biosensor. Arsenic can be tested using bacteria-based bioassays (Arduini et al., 2006).

14.5.4 In Medical Field

Biosensor applications are escalating quickly in medical sciences. Glucose biosensors are extensively used in medical applications for diabetes mellitus diagnosis, which demands accurate management of blood-glucose levels. At-home usage of blood-glucose biosensors represents 85% of the substantial global market (Martinkova and Pohanka, 2015). Biosensors are widely utilized in the healthcare profession to detect infectious diseases. A potential biosensor technology for the diagnosis of urinary tract infections, identification of the pathogen, and antimicrobial susceptibility is being researched (Mehrotra, 2016). It is critical to identify end-stage patients with heart failure who are susceptible to unfavorable outcomes during the early stages of left ventricular-aided device placement. A new biosensor based on hafnium oxide (HfO_2) was used for human interleukin (IL)-10 detection at an early stage (Lee et al., 2012). The interaction of recombinant IL-10 with matching monoclonal antibodies is being investigated for rapid cytokine detection following device installation. Electromechanical impedance spectroscopy and fluorescence patterns are used to assess the interaction between the antibody and the antigen, and fluorescence patterns are used to achieve bio-recognition of the protein (Kulkarni et al., 2022). HfO_2 is a very sensitive bio-field-effect transistor designed and synthesized for antibody deposition with electrochemical impedance spectroscopy detection of a human antigen (Chen et al., 2010). The most pressing issue now is heart failure, which affects around one million people. Immunoaffinity column assay, fluorometric, and enzyme-linked immunosorbent test are all techniques for detecting cardiovascular disorders. These are time-consuming, labor-intensive, and demand qualified workers. Biosensors based on electric measurement use biochemical molecule recognition to obtain intended selectivity with a specific biomarker of interest (Mehrotra, 2016).

Other biosensor uses include immunosensor array for clinical immunophenotyping of acute leukemias, biochip for rapid and accurate detection of various markers of cancer, histone deacylase inhibitor assay from resonance energy transfer, microfluidic impedance assay for endothelin-induced cardiac hypertrophy, quantitative measurement of cardiac markers in undiluted serum, neurochemical detection by diamond microneedle electrodes, and effect of oxazaborolidines on immobilized fructosyl transferase in dental diseases (Mehrotra, 2016).

14.5.5 Fluorescent Biosensors

Fluorescent biosensors are scanning agents used in cancer research and therapeutic development. They have provided new insights into the regulation and function of enzymes at the cellular level. Fluorescent biosensors are tiny scaffolds on which one or more fluorescent probes are placed via a receptor. The receptor recognizes a certain analyte or target, resulting in the transmission of a fluorescent signal that can be easily detected and analyzed (Wang et al., 2009). Fluorescent biosensors may detect ions, metabolites, and protein biomarkers in complicated solutions and indicate the presence, activity, or condition of the target. These sensors are used to detect arthritis, inflammatory illnesses, cardiovascular and neurological diseases, viral infection, cancer, and metastases (Mehrotra, 2016). Fluorescent biosensors are used for drug discovery programs and lead optimization. These are thought to be effective strategies for the preclinical and clinical evaluation of candidate molecules. Fluorescent biosensors are used efficiently in clinical and molecular diagnostics for early diagnosis of biomarkers, monitoring disease development and responsiveness to treatment, intravital imaging, and image-guided surgery. A genetically encoded FRET biosensor was utilized on cancer patient cells to measure Bcr-Abl kinase activity and establish a link with the disease state in chronic myeloid leukemia. This probe was also used to regulate therapy response and to watch the emergence of drug-resistant cells, allowing for the prediction of alternative medicines (Morris, 2013).

14.5.6 BIODEFENSE BIOSENSING APPLICATIONS

Biosensors can be used for military purposes. The primary goal of such biosensors is to detect and select organisms that pose a danger in real time, known as biowarfare agents. Several attempts have been made to design these biosensors using molecular approaches based on the chemical markers of BWAs. Nucleic acid-based sensing systems are more effective than antibody-based detection approaches as they provide gene-based specificity without extra amplification stages to attain the needed levels of detection sensitivity. The human papillomavirus (dsDNA virus) has been classified into two types: HPV 16 and 18. Both are linked to aggressive cervical cancer. A unique acoustic wave peptide nucleic acid (PNA) biosensor detects HPV genomic DNA directly without PCR amplification and can attach to target DNA sequences with high efficacy and precision (Mehrotra, 2016).

14.5.7 IN METABOLIC ENGINEERING

Environmental concerns and the lack of sustainable practices in petroleum-derived commodities are steadily emphasizing the importance of developing microbial cell factories for chemical synthesis. Researchers see metabolic engineering as a key enabler of a sustainable bioeconomy (Woolston and Edgar, 2013). They also anticipate that a significant portion of pharmaceuticals, commodity chemicals, and fuels will be created from renewable feedstocks by utilizing microbes rather than depending on petroleum refining or plant extraction. The enormous potential for diversity production necessitates efficient screening procedures for identifying individuals with the required phenotype. The previous approaches, however, had a low throughput. To overcome this barrier, genetically encoded biosensors can be used as they offer the potential for high-throughput screening and fluorescence-activated cell sorting. FRET sensors were made out of a pair of donor and acceptor fluorophores sandwiched together with a ligand-binding peptide. When it was attached to a ligand, the peptide suffered a conformational shift, resulting in a FRET change. Despite its great orthogonality, temporal precision, and ease of assembly, FRET sensors could only report the abundance of metabolites involved and could not regulate the signal downstream (Lindenburg and Merkx, 2014).

Transcription factors are naturally occurring sensory proteins that regulate gene expression in response to environmental conditions. It is performed by hacking into the host transcription system and using a synthetic condition-specific promoter to induce the production of a reporter gene. These have weak orthogonality and background noise (Costa et al., 2022).

Another type of biosensor is riboswitches, which are regulatory domains of mRNA and can selectively bind to a ligand and change its structure, thereby regulating the transcription of its encoded protein. In comparison to TF-based biosensors, they are faster since the RNA has been already transcribed, and they do not depend on protein–metabolite or protein–protein interactions. Ribosomes in bacterial systems have been extensively altered in recent decades (Findeiß et al., 2017).

14.5.8 BIOSENSORS IN PLANT BIOLOGY

Traditional mass spectroscopy approaches for determining cellular and subcellular localization, as well as measuring ion and metabolite levels, offered exceptional precision but lacked critical information about the position and dynamics of receptors, enzyme substrates, and transporters. This information, however, can be easily tapped by utilizing biosensors. To quantify a dynamic process under a physiological environment, we must devise techniques to see the real process, such as the conversion of one metabolite to another or the activation of signaling events. Sensors that respond dynamically can help with this visualization (Mehrotra, 2016).

Roger Tsien's lab pioneered to create protein prototype sensors for measuring caspase activity and controlling calcium levels in living cells (Groher and Suess, 2014). FRET between two spectrum variations of GFP was used to create these sensors. *In vivo* biosensor applications include high temporal resolution imaging of calcium oscillations employing chameleon sensors. Biosensors can be used to identify missing components related to analyte transport, regulation, or metabolism. A transportation step in phloem loading-sucrose efflux from the mesophyll is performed by the FRET sensor for sucrose, which is responsible for protein identification (Okumoto, 2012).

14.6 CONCLUSION AND FUTURE PROSPECTS

Molecular biosensors are currently widely used in biomedical diagnostics as well as a variety of other applications. Biosensor devices necessitate the association of various disciplines and rely on distinct aspects. The fast development of biosensors over the last decades has been mainly due to: (1) advancement in miniaturizing and microfabrication technologies; (2) application of innovative bio-recognition molecules; (3) novel nanostructures and nanomaterial devices; and (4) improved communication among life scientists and engineering/physical researchers.

A variety of target compounds and affinity reagents could be utilized in biosensors. The gold standard in biosensors is antibody-based systems. Novel affinity reagents such as synthetic receptors, in particular, aptamers (peptide aptamers and oligonucleotide aptamers), are now being developed to replace antibodies on biosensors. As probe molecules for DNA and microRNA sensing, DNA and nucleotide analogs including PNAs and locked nucleic acids are frequently utilized. Determining the amounts of protein glycosylation using lectins is presently of considerable interest in medical diagnosis, as is toxin detection in monitoring systems. Appropriate bioconjugation techniques and biomolecule stabilization on electrodes are required for the creation of economically viable biosensors. In biosensing devices, a variety of transduction mechanisms can be used. Beyond traditional pregnancy tests, lateral flow systems are extremely promising for the development of simple and inexpensive sensors, whereas lab-on-chip devices combine various microfabrication techniques, allowing biosensors to be used in a broad range of applications with minute volume and minimal sample preparation. Cell and tissue-based biosensors have genetically modified proteins that are introduced into cells. They enable scientists to continually and non-invasively detect hormones, toxin, or medication levels utilizing biophotonics or even other physical principles. Biosensors, such as nitrite and nitrate sensors, are utilized in marine applications to detect eutrophication.

Sensing devices based on nucleic acid hybridization detection have evolved for organism detection. The goal of the Monterey Bay Aquarium Research Institute's "Environmental Sample Processor" to automatically identify the toxic algae *in situ* from moorings utilizing ribosomal RNA probes is a promising development in this field.

One of the primary goals is to detect pollutants, pesticides, and heavy metals using biosensors. The use of nanostructures in biosensors opens possibilities for a new type of biosensor technology. Nanomaterials enhance the electrochemical, mechanical, magnetic, and optical features of biosensors, and they are progressing toward single-molecule biosensors having high-throughput biosensor arrays. Biological molecules have unique structures and activities and figuring out how to properly utilize the structure and function of nanomaterials and biomolecules to produce single-molecule multifunctional nanofilms, nanocomposites, and nanoelectrodes remains a significant issue. Processing, characterization, interface issues, the availability of high-quality nanomaterials, nanomaterial tailoring, and the principles dictating the performance of these nanoscale compounds on the electrode surface are all significant obstacles to the currently available approaches. Ways to improve the signal-to-noise ratio, as well as signal transduction and amplification, are important hurdles. Future research should focus on elucidating the method of interaction among nanomaterials and biomolecules on the electrode surface or nanofilms, as well as on leveraging unique features to create a new generation of biosensors. However, nanomaterial-based biosensors have very promising future applications, including food analysis, clinical diagnosis, process control, and environmental monitoring.

REFERENCES

Arduini, F., Ricci, F., Tuta, C.S., Moscone, D., Amine, A. and Palleschi, G. 2006. Detection of carbamic and organophosphorous pesticides in water samples using a cholinesterase biosensor based on Prussian Blue-modified screen-printed electrode. *Anal Chim Acta.* 580: 155–162.
Chen, C. and Wang, J. 2020. Optical biosensors: An exhaustive and comprehensive review. *Analyst.* 145: 1605–1628.
Chen, Q.H., Yang, Y., He, H.L., Xie, J.F., Cai, S.X., Liu, A.R., Wang, H.L. and Qiu, H.B. 2014. The effect of glutamine therapy on outcomes in critically ill patients: A meta-analysis of randomized controlled trials. *Crit Care.* 18: 1–13.

Chen, Y.W., Liu, M., Kaneko, T. and McIntyre, P.C. 2010. Atomic layer deposited hafnium oxide gate dielectrics for charge-based biosensors. *Electrochem. Solid-State Lett.* 13: G29.

Costa, V.G., Costa, S.M., Saramago, M., Cunha, M.V., Arraiano, C.M., Viegas, S.C. and Matos, R.G. 2022. Developing new tools to fight human pathogens: A journey through the advances in RNA technologies. *Microorganisms.* 10: 2303.

Cremer, M. 1906. Über die Ursache der elektromotorischen Eigenschaften der Gewebe, zugleich ein Beitrag zur Lehre von den polyphasischen Elektrolytketten. *Z Biol.* 47: 562–608.

Ercole, C., Del Gallo, M., Mosiello, L., Baccella, S. and Lepidi, A. 2003. Escherichia coli detection in vegetable food by a potentiometric biosensor. *Sens Actuators B Chem.* 91: 163–168.

Findeiß, S., Etzel, M., Will, S., Mörl, M. and Stadler, P.F. 2017. Design of artificial ribo-switches as biosensors. *Sensors.* 17: 1990.

German, N., Ramanaviciene, A., Voronovic, J. and Ramanavicius, A. 2010. Glucose biosensor based on graphite electrodes modified with glucose oxidase and colloidal gold nanoparticles. *Microchim Acta.* 168: 221–229.

Ghasemi-Varnamkhasti, M., Rodríguez-Méndez, M.L., Mohtasebi, S.S., Apetrei, C., Lozano, J., Ahmadi, H., Razavi, S.H. and de Saja, J.A. 2012. Monitoring the aging of beers using a bioelectronic tongue. *Food Control.* 25: 216–224.

Griffin, E.G. and Nelson, J.M. 1916. The influence of certain substances on the activity of invertase. *J Am Chem Soc.* 38: 722–730.

Groher, F. and Suess, B. 2014. Synthetic riboswitches-a tool comes of age. *Biochim Biophys Acta (BBA)-Gene Regul Mech.* 1839: 964–973.

Guilbault, G.G. and Montalvo, Jr., J.G. 1969. Urea-specific enzyme electrode. *J Am Chem Soc.* 91: 2164–2165.

Hughes, W.S. 1922. The potential difference between glass and electrolytes in contact with the glass. *J Am Chem Soc.* 44: 2860–2867.

Kavita, V. 2017. DNA biosensors-a review. *J Bioeng Biomed Sci.* 7: 1–5.

Kulkarni, M.B., Ayachit, N.H. and Aminabhavi, T.M. 2022. Biosensors and microfluidic bio-sensors: From fabrication to application. *Biosensors.* 12: 543.

Lee, M., Zine, N., Baraket, A., Zabala, M., Campabadal, F., Caruso, R., Trivella, M.G., Jaffrezic-Renault, N. and Errachid, A. 2012. A novel biosensor based on hafnium oxide: Application for early stage detection of human interleukin-10. *Sens Actuators B Chem.* 175: 201–207.

Lindenburg, L. and Merkx, M. 2014. Engineering genetically encoded FRET sensors. *Sensors.* 14: 11691–11713.

Martinkova, P. and Pohanka, M., 2015. Biosensors for blood glucose and diabetes diagnosis: Evolution, construction, and current status. *Anal Lett.* 48: 2509–2532.

Mehrotra, P. 2016. Biosensors and their applications-A review. *J Oral Biol Craniofac Res.* 6: 153–159.

Mishra, G.K., Barfidokht, A., Tehrani, F. and Mishra, R.K. 2018. Food safety analysis using electrochemical biosensors. *Foods.* 7: 141.

Mishra, R.K., Dominguez, R.B., Bhand, S., Muñoz, R. and Marty, J.L. 2012. A novel auto-mated flow-based biosensor for the determination of organophosphate pesticides in milk. *Biosens Bioelectron.* 32: 56–61.

Mohankumar, P., Ajayan, J., Mohanraj, T. and Yasodharan, R. 2021. Recent developments in biosensors for healthcare and biomedical applications: A review. *Measurement.* 167: 108293.

Morris, M.C. 2013. Fluorescent biosensors-probing protein kinase function in cancer and drug discovery. *Biochim Biophys Acta Proteins Proteom.* 1834: 1387–1395.

Nabaei, V., Chandrawati, R. and Heidari, H. 2018. Magnetic biosensors: Modelling and simulation. *Biosens Bioelectron.* 103: 69–86.

Neethirajan, S., Ragavan, V., Weng, X. and Chand, R. 2018. Biosensors for sustainable food engineering: Challenges and perspectives. *Biosensors.* 8: 23.

Nikhil, B., Pawan, J., Nello, F. and Pedro, E. 2016. Introduction to biosensors. *Essays Biochem,* 60: 1–8.

Okumoto, S. 2012. Quantitative imaging using genetically encoded sensors for small molecules in plants. *Plant J.* 70: 108–117.

Ramanathan, K. and Danielsson, B. 2001. Principles and applications of thermal biosensors. *Biosens. Bioelectron.* 16: 417–423.

Scognamiglio, V., Arduini, F., Palleschi, G. and Rea, G. 2014. Biosensing technology for sustainable food safety. *Trends Anal Chem.* 62: 1–10.

Shaner, N.C., Steinbach, P.A. and Tsien, R.Y. 2005. A guide to choosing fluorescent proteins. *Nat Methods.* 2: 905–909.

Shi, J., Feng, D. and Li, Y. 2017. Biosensors in fermentation applications. *Ferment Process.* 145. doi: 10.5772/65077.

Tothill, I.E. 2001. Biosensors developments and potential applications in the agricultural diagnosis sector. *Comput Electron Agric.* 30: 205–218.

Tudi, M., Daniel Ruan, H., Wang, L., Lyu, J., Sadler, R., Connell, D., Chu, C. and Phung, D.T. 2021. Agriculture development, pesticide application and its impact on the environment. *Int J Environ Res Public Health.* 18: 1112.

Wang, H., Nakata, E. and Hamachi, I. 2009. Recent progress in strategies for the creation of protein-based fluorescent biosensors. *Chem BioChem.* 10: 2560–2577.

Woolston, B.M., Edgar, S. 2013. Stephanopoulos G. Metabolic engineering: Past and future. *Annu Rev Chem Biomol Eng.* 4: 259–288.

Yan, C., Dong, F., Chun-yuan, B., Si-rong, Z. and Jian-guo, S. 2014. Recent progress of commercially available biosensors in china and their applications in fermentation processes. *J Northeast Agric Univ.* 21: 73–85.

15 Polymer Nanocomposites Applications in Molecular Recognition

*Maninder Singh, Aniruddha Nag,
and Mohammad Asif Ali*

15.1 INTRODUCTION

Molecular recognition with organic capsule receptors through supramolecular chemistry and nanotechnology based on molecular recognition chemistry has been paid much attention in science and technology (Ariga et al., 2012). In advanced molecular recognition systems, polymer and nanostructured supports with properly engineered interface structures play crucial roles (Zhou et al., 2021). The host–guest encapsulation gives rise through the chemical and physical processes. Now, supramolecular concepts are being realized through micro/nanofabrication (Lu et al., 2023). Therefore, researchers are seeking alternative dynamic combinatorial libraries that are an appealing component of molecular recognition chemistry since no laborious and time-consuming techniques are required (Dumartin et al., 2020). Molecular imprinting polymers (MIPs) have long been utilized to synthesize tailor-made recognition materials by forming a polymer network around a template molecule. Molecularly imprinted polymers are advantageous over their natural counterparts because of their physical/chemical stability, the high selectivity of recognition, and reusability, and they are useful in biosensors, chromatographic separations, drug delivery techniques, etc. (Guo et al., 2015). However, the recognitions are limited to a small molecular level, inhibiting its application for macromolecules and biological molecules. Further, the selectivity toward the aqueous phase and the usage of functional monomers is restricted in the case of biomacromolecules. However, surface imprinting, epitope imprinting, etc. can resolve such issues. For the MIPs, it has been affected by their chemical and biological recognition system, which affects their robustness and storage endurance (Mahony et al., 2005). Although the bulk MIPs prepared by conventional method exhibit high selectivity, inevitable drawbacks are still suffered, such as a time-consuming and complicated preparation process, low binding capacity, poor site accessibility, and slow binding kinetics (Gao et al., 2010; Wu et al., 2016). In another case, water pollution is a severe concern of the modern world, which can be solved by combining membrane separation techniques along with MIPs to create molecularly imprinted membranes to recognize pollution resistance properties selectively.

Furthermore, this chapter presents the promising development of molecular recognition frontiers, organic capsule receptors, metallo-capsule receptors, helical

DOI: 10.1201/9781003352372-15

receptors, dendrimer receptors, and the design of future receptors, the type of polymer, their mechanistic approach and their effects on the environment, etc. Therefore, due to their behavior, it is widely used in molecular recognition selectively. The first section includes a description of nanotechnology with specific molecular recognition, and the second section discusses polymer nanocomposites.

15.2   SPECIFIC APPLICATION OF MIPs

15.2.1   Metallo-Capsule Receptors

Inorganic metals combined with aromatic ligands to prepare metallo-capsule receptors and they are traditionally cage-shaped with polycyclic ligands. Pd^{2+}-pyridine coordination compounds are enclosed metallo-capsules consisting of square-planar geometry. Negatively charged such receptors are suitable for organic cations, whereas, in other cases, metallo-capsules are useful to bind guests and control their reactivity and mechanism. Mateus et al. (2011) reported Cu^{2+}-coordination molecules for organic guests functioning based on H-bonding and electrostatic interactions. In another example, Ca^{2+}–cyclen complex with helical structure was reported by Misaki et al. (2009) where chiral guest anions form a diastereomeric complex with enantiomeric metallo-capsule receptors. Large-size metallo-receptors were reported, which can facilitate biomolecular reactions of the guests, for example, Clever et al. (2009) have combined two Pd^{2+} centers with banana-shaped ligands to accommodate dianion guests through electrostatic interactions. M2L4-type metallo receptor was reported by Yoshizawa et al. using two Pd^{2+} centers with bisanthracene ligands which creates a 1 nm space enough large to encapsulate fullerene while protecting the guest from external environments (Kishi et al., 2011).

15.2.1.1   Chemiresistor Sensor

Chemiresistors are useful as a gas sensor to detect trace concentrations of explosive material vapors; they come along with high sensitivity, wide linear response range, minimum installation space, and are cost-effective as well. Dinitrotoluene (DNT) is one of the most lethal materials to aquatic environment during explosion among all other nitroaromatics. Due to high vapor pressure, it is difficult to detect it using normal sensors. Nitro-aromatic structures are feasible to be measured by electrochemical method in a faster manner with high sensitivity, and many reported studies could be found toward developing electrode elements for DNT recognition. Recently, glassy carbon electrode (GCE), GCE-coated carbon nanotubes, and organic polymers have been reported for such applications; however, such materials are either expensive or non-sustainable for the environmental aspects. Inorganic clays are naturally abundant materials (montmorillonite, etc.) and can be reduced using sodium dithionate to prepare Fe^{2+}-enriched clay. Furthermore, they were drop cast over the in-situ N-doped phenol/formaldehyde polymer dispersed with rGO to prepare Fe^{2+}-clay-rGO/P. Fe^{2+}-rich environment plays a key role in the recognition of DNT in the electrochemical measurement method even when present in a trace amount to justify its high sensitivity. Such materials are applicable to a wide range of materials, including salts, aromatics, and a range of concentrations (Koudehi &

Pourmortazavi, 2018). As with other options, conductive polymers (such as polyaniline, polypyrrole (PPy), and their derivatives) already gained attention. Among them, PPy exhibits high conductivity, biocompatibility, gas measurement ability, and environmental stability. However, PPy suffers limitations from structural brittleness, which restricts their applications and can be overcome when combined with flexible polymers such as polyvinyl alcohol (PVA). The MIP technology works based on the target molecule shape and size, and is used to develop sensors with predetermined selective gas molecules. A chemiresistor sensor with a PVA/PPy/MIP nanocomposite coating could successfully detect high vapor pressure substances such as DNT. The key factor behind the success of this material is the H-bonding formation between PPy and nitroaromatic moieties such as DNT. High selectivity sensing could also be observed for this material while checked, among other organic matters. Even with a trace amount (concentration range between 0.1 and 70 ppm), the detection efficiency comes out with a linear calibration curve ($R^2 = 0.9974$) (Bairagi et al., 2018).

15.2.1.2   Drug Recognition

Biocompatible polymers play a significant role in biomedical applications. Polylactide is one of the famous examples, which have already been adopted for drug delivery and various therapeutic applications. Alongside, metallic nanoparticles (for example, Au and Ag) are useful for specific applications in electronics, optics, and analytical areas. High surface area, porous morphology, desired size controllability, etc. are advantageous for their applications in the biomedical field. Titanium dioxide (TiO_2) is used in biomedical applications (photodynamic therapies in cancer, redox chemical reactions with biomolecules using separate photogenerated charges) due to its good biocompatibility and high reactivity. Similarly, magnetic nanoparticles (MNPs) are being studied for biomedical and biotechnological applications, including drug delivery, biological separations, biosensor, and MRI contrast enhancers. Surface modification of such MNPs is important for specific applications, such as MIPs coated with MNPs help separate or concentrate chemicals using external magnetic fields. Poly(nisopropylacrylamide) (PNIPAM)-based polymer colloids, while in aqueous solution, exhibit low critical solution temperature and show interfacial property alteration through the coil-globule transition. PNIPAMs are useful in drug delivery, biomedical, tissue engineering, and protective containers for biologically active molecules. Furthermore, such polymers are functional while attached to a flat surface to control cell adhesion and protein adsorption using thermo-controlled wettability and surface strength (Wang et al., 2009).

Considering the above discussions, biomedical applications of PLA-TiO_2 nanocomposites were investigated specifically for drug recognition. NanoTiO$_2$-PLA composite effectively influences the binding capacity of the anticancer drug daunorubicin to DNA and increases the detectability of the same. Such blending nanocomposites have a large surface area and unique properties to stand out in the bioanalysis method and show increased sensitivity and binding affinity of the specific anticancer drug (Song et al., 2006). Furthermore, this nano TiO_2-PLA system is advantageous as drug molecules can self-assemble on the nanosystem's surface and enhance the drug molecules' detectability through DNA binding for relative biorecognition (Song, Pan, Chen, et al., 2008). In another aspect of a similar approach, copolymer of poly

(N-isopropylacrylamide)-co-polystyrene (PNIPAM-co-PS) was fabricated into nano-fibers using electrospinning followed by blending with nanoTiO$_2$. Such a functional nanocomposite system readily binds with the anticancer drug daunorubicin using self-assembly on its surface. Thus, detection sensitivity is enhanced for biomolecular recognition and such a system is useful for future applications in bioanalysis and as a targeted drug carrier (Song, Pan, Li, et al., 2008).

15.2.2 ORGANIC CAPSULE RECEPTOR

Multidimensional capsule receptors consist of distinctive cavities for molecular and mesoscale guests. The organic capsules come with a broad range of specific interactions to accommodate specific guests. Some of the reported organic capsule receptor approaches are as follows: Cram (1988) synthesized carceplex type receptors, which give rotational mobility to the guests, whereas Rebek (2005) reported partially exposed guests encapsulated in bowl-shaped receptors with the characteristic fast guest exchange. Such capsule receptors were designed based on covalent and noncovalent (H-bonding, coordination, van der Waals, electrostatic) bond interactions. Kumar Dey and Das et al. (2011) developed selective F⁻ ion detectable receptors based on multiple H-bonding interactions of triamide receptor. Ballaster et al. reported about calix[4]pyrrole receptors by attaching aryl substituents on meso-carbons (Chas & Ballester , 2012), and such receptors can effectively bind with pyridine N-oxides using H-bonding and hydrophobic interaction. Chas and Ballester (2012) developed an interlocked capsule based on the H-bonding and π-interactions between calix-pyrrole and calixarene moieties to encapsulate specific size ion-pairs into their cavities. Bao et al. (2010) presented a tripodal receptor that can accommodate cyclohexane guests into their basket-type cavity prepared using intermolecular H-bonding. Cresswell et al. (2010) reported anion selective receptor which is water soluble as well; in this case, two zwitterionic receptors in their crystal form were prepared by placing carboxylate anions of one receptor in the middle of two pyridinium cation of another receptor. Also, other research groups reported metal–organic framework-based receptors that are selective for a specific gas, and such molecules are also classified as organic capsule-type receptors (Ariga et al., 2012).

15.2.2.1 Biosensors

Recently, numerous attempts have been made to replace biological receptors with synthetic counterparts as the recognition element in chemo/biosensors. A molecular imprinting strategy is advantageous in using polymeric materials with specific recognition and binding capacity (Mao et al., 2011). Further, compared to the biochemical/biological recognition systems, MIPs are robust, economical, and have higher storage capacity. However, MIPs come with a few negative points, such as poor binding ability and site accessibility makes them less feasible on purpose after structural optimizations with suitable forms and templates to sit on the surface site with the closest proximity. Surface binding sites are the materials in demand for their excellent accessibility, template detachment, and reduced mass transfer resistance. Nanostructure imprinted materials are very useful owing to their high surface area and advantages such as higher binding capacity, easier installation, and faster binding kinetics. Another well-known approach is polymer-based recognition element preparation using polymerization with a cross-linker around a template and finally extracting the template to leave the recognition molecules at the target sites. Targeted applications of such MIPs range from separation media, antibody mimicking, and biosensors. Considering the broad spectrum of advantages of MNPs, they were incorporated into MIPs to prepare composites by Ansell and Mosbach (1998; Wang et al., 2009). Magnetic MIP nanowires were reported by Li et al. (2006), whereas bovine serum-albumin surface-imprinted submicron particles were prepared with magnetic susceptibility by Tan et al. (2008). Fe_3O_4 MNPs are coated with estrone imprinted polymer, providing better recognition specificity and magnetization. Such magnetically modified polymer composites have high affinity and selectivity toward their target molecules, making them ideal for chemical and biochemical separation, biosensor recognition and drug delivery, selective cell sorting, etc.

In another approach, silicon nanoparticles can act as fluorescent probes in the visible region in biological systems. Silicon nanoparticles are usually chemically modified to prevent aggregation and enhance hydrophilicity in a biological medium, for example, allylamine modification using surface modification through catalysis (Warner et al., 2005). However, a more straightforward approach can be more preferable in this case, such as electrostatic adsorption of polyelectrolytes on Si nanoparticle surfaces. It is a dual-fluorescent composite containing inorganic moiety with high fluorescence yield and high surface density of organic ionic moieties, making it ideal for trace analysis of biological molecules and particularly effective for cytochrome (Gu et al., 2010). Graphene is a material of special interest for its unique characteristics such as large surface area, electronic properties, and mechanical properties. They are being used as a reinforcement element in high-performance polymers. Graphene sheet/Congo red and MIP composites were prepared using dopamine as the template to be used for molecular recognition (Mao et al., 2011). Such composite exhibits rapid adsorption-desorption dynamics, high selectivity, trace sensitivity, and molecular separations. For further advancement, graphene and Fe_3O_4 MNPs were combined with being applied in MIPs as they provide superior magnetism and higher adsorption ability (Ning et al., 2014). Magnetic graphene MIPs were prepared and applied for the recognition of bovine hemoglobin (Guo et al., 2015), which shows excellent selectivity and advantages over other MIP composites in terms of separation ability due to the Fe_3O_4 magnetism along with the specific affinity toward recognition site for the target molecule, bovine hemoglobin in this case.

15.3   MOLECULAR RECOGNITION AND RECEPTOR ARCHITECTURE

Molecular recognition events are some of the most important aspects of biological and chemical systems, providing unique and diversified structures and functions of numerous molecules, assemblies, and materials (Chakrabarty et al., 2011; Fathalla et al., 2009; Kubik, 2010). Several effective methodologies for designing receptors have been devised, including computer-based approach, template formation, molecular imprinting protocols, and microwave-assisted methods (Clever et al., 2009; Collin et al., 2010; Miyake et al., 2012; Pernites et al., 2010). To classify based on the size receptors, including crown ethers, cryptands are usually molecular scale receptors that are typically used for guests as metal cations and amino acids (Mutihac et al., 2011; Ravikumar & Ghosh, 2012). Currently, meso-scale receptors (dendrimers, helical capsules) are getting research attention as they are suitable for nanometer-sized molecules such as fullerenes, proteins, and polymeric drugs. Finally, nanoscale-based molecular recognition chemistry is helpful for precisely patterning nanomaterials (<100 nm).

Dynamic combinatorial libraries have been used successfully for receptor syntheses over the last decade (Li et al., 2013; Ludlow & Otto, 2008; Peyralans & Otto, 2009). They are realized through the reaction of complementary components that selectively interact with one another in a predetermined manner (Corbett et al., 2006; Cougnon & Sanders, 2012; Otto, 2012). In contrast to traditional combinatorial synthesis, which involves the covalent preparation of many different compounds, dynamic combinatorial synthesis allows for the selective formation of the thermodynamically most stable combination from an equilibrated mixture of various components (Herrmann, 2014; Zhang et al., 2005). Dynamic combinatorial libraries are appealing to molecular recognition chemistry since no laborious and time-consuming techniques are required. A guest template in the dynamic library interacts with specific components and influences the equilibrium like standard template reactions (Stefankiewicz et al., 2012). Because the templates range from simple anions and tiny molecules to proteins and other macromolecules, dynamic libraries have many applications for synthesizing receptors at the molecular, meso-, and nanoscale. Three reversible interactions between library components are available in dynamic combinatorial receptor syntheses: hydrogen bonds, metal-coordination bonds, and covalent bonds. Hydrogen and coordination bonds, for example, are relatively weak and thus labile enough to allow dynamic equilibration ideal for library creation. Hydrogen bonds prefer a linear arrangement of the three atoms involved, whereas metal-coordination bonds exhibit ligand substitution rates that vary depending on the metal center, ligand, and environment. Covalent bond formation/breaking, on the other hand, has slow kinetics and requires a suitable catalyst to accelerate.

Thus, we can use the appropriate combination of these weak interactions to construct a dynamic combinatorial library to synthesize the desired receptor (Corbett et al., 2006; Li et al., 2013; Otto, 2012; Peyralans & Otto, 2009). A typical application of a dynamic combinatorial library has been presented for catenane synthesis. Sanders et al. used acetylcholine as a guest template in a prolinyl-phenylalanine hydrazine solution (Corbett et al., 2006; Cougnon & Sanders, 2012; Stefankiewicz et al., 2012). When trifluoroacetic acid was introduced to a dipeptide hydrazine solution without a template, the ensuing linear intermediates were transformed over several hours into a sequence of cyclic oligomers. The addition of an acetylcholine template affected the response profiles dramatically. Six dipeptides were constructed around acetylcholine to form [2] catenane diastereomers composed of two interlocked macrocyclic trimers. The [2] catenanes continued to accumulate for several days before being separated in 67% yield. According to Nuclear Magnetic Resonance (NMR) spectroscopy results, the trimethylammonium moiety of acetylcholine served as a leading recognition site in the catenane synthesis. Anion-templated receptor syntheses were made available by dynamic combinatorial libraries. Kubik et al. combined bis(cyclopeptide) and dithiols in their study (Kubik, 2010; Rodriguez-Docampo et al., 2011).

In an aqueous acetonitrile solution, these receptors demonstrated excellent selectivity for SO_4^{2-} anion over SeO_4^{2-}, I^-, and other anions, which depended on the receptor's rigidity and flexibility. The dynamic combinatorial library was also used to create a water-soluble cyclophane. Otto et al. transformed a naphthalenedithiol into a water-soluble combination of octameric [2] catenanes composed of two interlocking molecular squares. The catenanes decomposed into the cyclophane that bound the template with nanomolar affinity when the mixture was re-equilibrated in the presence of an adamantyl ammonium cation template. Cryo-transmission electron micrographs revealed that the free tetramer and its guest complex aggregated into sheet-like aggregates and remained stable in an aqueous solution for a week.

Using disulfide equilibrium, a new series of donor–acceptor [2] catenanes were synthesized from electron-donors (D) and electron-acceptors (A). Conventional catenane syntheses of this type typically yielded an alternating DADA stack, whereas Pantos et al. got unexpected DAAD, DADD, and ADAA stacks. All of the library components were created with two cysteine-terminated hydrophilic side chains of varying lengths connected by a big hydrophobic 1,5-dialkoxy-naphthalene (D) or 1,4,5,8-naphthalenenetetracarboxylic diimide (A). The disulfide library was made by dissolving the components at pH 8 and stirring them for 5 days in air to complete their thiol-oxidation. There are 22 different types of [2] catenanes with different DAAD, DDAD, DADA, and AADA arrangements.

To prepare unique catenanes, using a highly polar media helped overcome donor–acceptor interactions. To polymerize, Fujii and Lehn used reversible imine bond formation between ao-diamines and dialdehydes. For alkali metal complexation, the polymers contained imine and pyridine nitrogen atoms, as well as ether oxygen atoms, and 1,5-dialkoxynaphthalene and 1,4,5,8-naphthalene-tetracarboxylic diimide for donor–acceptor interactions. Because specific alkali metal complexation enhanced chain folding and donor–acceptor stacking interactions, the resulting polymer allowed for excellent optical detection of alkali metal cations. A dynamic combinatorial library was also used in a self-recognition reaction involving covalent and complementary hydrogen bonds. As a crucial chemical, Xu and Giuseppone created imine Al1Am1 by condensation of aldehyde Al1 with adenosine amine Am1.63. When five different types of building blocks, Al1, Al2, Al3, Am1, and Am2, were combined in $CDCl_3$, a homodimer [Al1Am1]$_2$ was mostly formed. The complementary hydrogen bonds formed by the library components increased the synthesis of the optimal combination creating the most favorable hydrogen bonds (Herrmann, 2014).

The combinatorial design of luminous lanthanide complex-type sensors was demonstrated to be an effective alternative to rational design. Because of their long-lived, well-resolved, and predictable luminescence patterns, some lanthanide complexes are used in the labeling and sensing of biological targets. Lanthanide centers are highly labile and have large ionic radii; therefore, they frequently produce dynamic mixes of multi-component complexes with diverse ligands in solution. In the library, Shinoda et al. mixed seven types of N-aromatics, four lanthanide cations, and seven amino acid guests. Among the 196 combinations tested, a Yb^{3+} complex with picolinic acid functioned as a near-infrared luminophore specific for zwitterionic amino acids, whereas a Tb^{3+} complex with bipyridine strongly responded to anionic Glu. A similar combinatorial screening resulted in the discovery of a novel class of chirality probes for specific amino acids. Several examples of combinatorial syntheses of medicinal ligands were recently developed, and their methodologies are relevant in dynamic combinatorial syntheses of effective receptors (Ludlow & Otto, 2008; Zhang et al., 2005). A number of synthetic receptors have been studied in organic solvents, but high efficiency in water remains elusive. Hall et al. created a receptor library to manufacture receptors targeting Thomsen–Friedenreich (TF) disaccharide, a tumor-associated carbohydrate antigen. Because boronic acids have been shown to identify these carbohydrates, a library of 400 peptidyl bis(boroxole) receptors was generated using split-pool synthesis and evaluated for TF disaccharide binding using the ELISA method. The chosen receptor bound the TF disaccharide via cooperative boronate synthesis, hydrogen bonding, hydrophobic packing, and CH-p interactions. The combinatorial libraries are efficient ways to find receptors for biomacromolecules created high-affinity ligands for myelin-associated glycoprotein using a three-step fragment-based combinatorial method (MAG). This protein is an immunoglobulin-like lectin that binds to sialic acid and hinders axonal regeneration following damage. Although its crystal structure is unknown, the tested receptor's binding behavior was well defined by NMR monitoring of the transverse magnetization decay. Following the selection of a sialic acid derivative as a first-site ligand, the library for second-site ligand selection was made up of 80 low-molecular-weight ($M_w < 300$), moderately lipophilic ($\log P < 3$) and highly soluble organic compounds. The spin-labeled first-site ligand improved paramagnetic relaxation when 5-nitro indole was coupled to the second site next to the first site. Because paramagnetic relaxation is distance dependent, the two bound ligands' binding locations and spatial orientations were determined. Finally, the target protein was incubated with a library of acetylene and azide reactive first- and second-site ligands, respectively. The triazole synthesis between the two compatible fragments bound to the MAG protein surface produced the high-affinity ligand. Using interfaces developed within materials is essential for more practical applications such as detection, separation, removal, and delivery based on molecular recognition, even though model interfaces such as the air–water interface provide fundamental results under relatively simple conditions. In advanced molecular recognition systems, polymer and nanostructured supports with properly engineered interface structures play crucial roles.

Yashima and colleagues examined dynamic helical polymers, such as poly(ethylene oxide), in one compelling example of chiral sensing with polymeric materials (phenylacetylene). A strong induced circular dichroism was brought on by binding chiral guests to side chains, which enhanced chiral effects by twisting the polymer leading chains. Due to these effects, chiral compounds might be detected compassionately, and this method could be used to calculate absolute configuration and measure enantiomeric excess. Often, polymers serve as suitable media for molecular recognition. For molecular recognition, Kitano et al. created polymer brush structures. The highly hydrophilic nature of the glass plate that had been treated with a brush of sulfobetaine telomer was helpful for resistance against the nonspecific adsorption of proteins like lysozyme and albumin. N-methacryloyloxysuccinimide residues were added to this surface, and Concanavalin A was then added to the modified glass chip. The Concanavalin

A-fixed zwitterion telomer brush selectively identified mannose residues collected on the surface of Au colloidal particles. The subsequent increase in absorbance at 550 nm was brought on by localized surface plasmon resonance. Even in the presence of hydrophobic and hydrophilic substances, the fluorous-core star polymers efficiently and selectively detected fluorous compounds. Huber and colleagues created polymeric colloids that contained 2,4-diamino triazine residues for functionalization. The later recognition site has a donor–acceptor–donor pattern with a triple hydrogen bond motif. Due to its ability to easily recognize uracil with a triple hydrogen bond motif of the type's acceptor–donor–acceptor, colloidal materials are useful in drug discovery. Molecular imprinting into polymeric matrices is another effective approach for identifying specially shaped visitor molecules. Li and colleagues created macro-porous hydrogels using silica colloid as a template and L-DOPA as a catalyst (Rodriguez-Docampo et al., 2011). These hydrogels showed hypsochromic changes to a blue color when L-DOPA was added. D-DOPA, in contrast, exhibited a modest binding to the imprinted gel and did not result in a color change. Qin et al. recently developed a molecularly imprinted polymer-based potentiometric sensing technique. The technique could be utilized to identify neutral species by using a charged chemical with the analyte's structural similarity as an indicator ion for signal transduction.

Advanced applications can be produced by combining numerous recognition sites at the polymer interface. Bunz, Rotello, and others used conjugated fluorescent polymers containing pendant-charged residues to create array-based sensing of different cells. The latter functional groups offered contact with cell membranes that were multivalent. Cells have unique surface characteristics; hence, surface molecular recognition allows for detecting minute variations between various cell kinds. The created system can distinguish between various cell types as well as between cancerous and non-cancerous mammalian cells. Their approach could be used to create polymer-based imaging agents and delivery systems.

Non-covalently connected polymer structures can be created based on molecular recognition between relatively remote units.

Many people refer to them as supramolecular polymers. In forming a Fe^{2+}-terpyridine 1:2 metal–ligand complex and the dimerization of a self-complementary guanidino carbonyl pyrrole carboxylate zwitterion, Grohn, Schmuck, and colleagues reported the preparation of supramolecular polymers through self-assembly of small molecular units based on orthogonal binding interactions. Small cyclic oligomers were created due to the utilized molecule's incredible flexibility.

The traditional ring chain polymerization paradigm was used in this supramolecular polymerization. Nair and colleagues synthesized several complimentary hydrogen-bonded networks based on various hydrogen bonding recognition patterns. As hydrogen bonding motifs, thymine/2,4-diamino triazine and cyanuric acid/Hamilton wedge pairs were chosen in this instance. The type of hydrogen bonding motif used for inter-chain cross-linking and the cross-linking agent concentration significantly impacted the mechanical properties of these materials. As a result, their rheological characteristics can range from viscous to highly elastic, and at room temperature, the dynamic modulus can vary by up to five orders of magnitude (Rodriguez-Docampo et al., 2011; Stefankiewicz et al., 2012).

15.3.1 HELICAL RECEPTORS

The primary mechanism by which biological helical receptors achieve extremely selective chemical recognition is folding lengthy peptide sequences into specific conformations. In addition to peptide-based receptors, numerous synthetic helical compounds can host cationic, anionic, and neutral visitors inside their folded structures. As schematically, single, double, triple, and related helical receptors wrap around spherical and rod-like visitors.

Effective receptors can be created to enclose particular visitors in their interior cavities because the size and form of the helical cavity are regulated by changing the identity of modular construction blocks and other functional components. Recent research suggests that certain helical receptor types can facilitate dynamic binding and release of visitors by helical association, dissociation, or folding.

The helical receptor can be used in transport, signaling, catalysis, and sensing processes by adjusting the conformation to exert the most significant interaction with the guest.

It is known that a variety of synthetic ligands can create helical metal complexes. Several polynuclear complexes showed helical chirality and were the building blocks for molecular catenanes, knots, and related topological structures. These complexes have guest metal cations aligned with the orientation of the helical cavities and are wrapped by one, two, or more multidentate ligands. DNAs are excellent, programmable scaffolding for metal arrays, as Shionoya et al. have shown. They created DNA–metal complexes with double and triple strands substituting metal-coordinated base pairs for biological hydrogen-bonded base pairs. One-dimensional metal arrays were produced when pyridine- or hydroxy pyridone-containing nucleosides were added to the DNA multiplexes (Otto, 2012).

Numerous helical lanthanide compounds have been introduced since the ground-breaking works of Bunzli, Piguet, and others. Flexible thioalkyl chains were joined by Piguet et al. using a bis-2-benzimidazole-8-hydroxyquinoline ligand. A D3-symmetrical triple-stranded complex was quantitatively produced when AgI was added to a mixed solution of lanthanide metal salt (La^{3+}, Eu^{3+}, or Lu^{3+}) and ligand at a millimolar concentration. An enantiomerically pure triple-stranded Eu^{3+} complex was created by Gunnlaugsson et al., in which pyridine-bisamide units worked together to produce a chiral triplex structure. Upon helicate production, this displayed chiral and persistent red emission.

15.4 CONCLUDING REMARKS

The chapter has covered the significant concern about polymer nanocomposites applications in molecular recognition. Some polymers are also used for molecular recognition, which contributes significantly to a sustainable society. Thus, some biopolymers follow degradation via environmental or biological pathways, which manage the natural ways to dispose of and maintain the carbon footprint and participate in molecular recognition. A new tuneable supramolecular nanocomposites polymeric system has been designed through noncovalent interaction, π–π stacking interactions, and host–guest interaction used in medical treatments, so it is well-established as a form of molecular recognition in medicine. However, this subject is still immature, and molecular recognition chemistry targets are continuously being developed. The development of molecular recognition polymeric materials is an exciting area, especially in studying cells, where drug-induced pH and temperature changes can be better understood. Also, to understand how small molecules act in normal or cancer cells at the nanoscale, the material activity is greatly improved due to its high surface areas. Without a doubt, developing new materials to achieve high quantum yield and stability for advanced applications will require the collaborative efforts of chemists, biologists, and physicists. Systems integration between macroscopic length scales and molecular-level phenomena is one of the near-future targets. The future goals of molecular recognition might include the self-organization-based rational design that becomes a more realistic aim to accomplish in the future that may play a significant role.

REFERENCES

Ansell, R. J., & Mosbach, K. (1998). Magnetic molecularly imprinted polymer beads for drug radioligand binding assay. *Analyst*, *123*(7), 1611–1616. https://doi.org/10.1039/A801903G.

Ariga, K., Ito, H., Hill, J. P., & Tsukube, H. (2012). Molecular recognition: From solution science to nano/materials technology. *Chemical Society Reviews*, *41*(17), 5800–5835. https://doi.org/10.1039/C2CS35162E.

Bairagi, P. K., Gupta, G. S., & Verma, N. (2018). Fe-enriched clay-coated and reduced graphene oxide-modified N-doped polymer nanocomposite: A natural recognition element-based sensing electrode for DNT. *Electroanalysis*. https://doi.org/10.1002/elan.201800585.

Bao, X., Rieth, S., Stojanović, S., Hadad, C. M., & Badjić, J. D. (2010). Molecular recognition of a transition state. *Angewandte Chemie*, *122*(28), 4926–4929. https://doi.org/10.1002/ange.201000656.

Chakrabarty, R., Mukherjee, P. S., & Stang, P. J. (2011). Supramolecular coordination: Self-assembly of finite two- and three-dimensional ensembles. *Chemical Reviews*, *111*(11), 6810–6918. https://doi.org/10.1021/cr200077m.

Chas, M., & Ballester, P. (2012). A dissymmetric molecular capsule with polar interior and two mechanically locked hemispheres. *Chemical Science*, *3*(1), 186–191. https://doi.org/10.1039/C1SC00668A.

Clever, G. H., Tashiro, S., & Shionoya, M. (2009). Inclusion of anionic guests inside a molecular cage with palladium(II) centers as electrostatic anchors. *Angewandte Chemie International Edition*, *48*(38), 7010–7012. https://doi.org/10.1002/anie.200902717.

Collin, J.-P., Durola, F., Frey, J., Heitz, V., Reviriego, F., Sauvage, J.-P., Trolez, Y., & Rissanen, K. (2010). Templated synthesis of cyclic [4] rotaxanes consisting of two stiff rods threaded through two bis-macrocycles with a large and rigid central plate as spacer. *Journal of the American Chemical Society*, *132*(19), 6840–6850. https://doi.org/10.1021/ja101759w.

Corbett, P. T., Leclaire, J., Vial, L., West, K. R., Wietor, J.-L., Sanders, J. K. M., & Otto, S. (2006). Dynamic combinatorial chemistry. *Chemical Reviews*, *106*(9), 3652–3711. https://doi.org/10.1021/cr020452p.

Cougnon, F. B. L., & Sanders, J. K. M. (2012). Evolution of dynamic combinatorial chemistry. *Accounts of Chemical Research*, *45*(12), 2211–2221. https://doi.org/10.1021/ar200240m.

Cram, D. J. (1988). The design of molecular hosts, guests, and their complexes (Nobel Lecture). *Angewandte Chemie International Edition in English*, *27*(8), 1009–1020. https://doi.org/10.1002/anie.198810093.

Cresswell, A. L., Piepenbrock, M.-O. M., & Steed, J. W. (2010). A water soluble, anion-binding zwitterionic capsule based on electrostatic interactions between self-complementary hemispheres. *Chemical Communications*, *46*(16), 2787–2789. https://doi.org/10.1039/B926149D.

Dumartin, M., Septavaux, J., Maréchal, M. D., Jeamet, E., Dumont, E., Perret, F., Vial, L., & Leclaire, J. (2020). The dark side of disulfide-based dynamic combinatorial chemistry. *Chemical Science*, *11*, 8151–8156. https://doi.org/10.1039/D0SC02399J.

Fathalla, M., Lawrence, C. M., Zhang, N., Sessler, J. L., & Jayawickramarajah, J. (2009). Base-pairing mediated non-covalent polymers. *Chemical Society Reviews*, *38*(6), 1608–1620. https://doi.org/10.1039/B806484A.

Gao, R., Kong, X., Su, F., He, X., Chen, L., & Zhang, Y. (2010). Synthesis and evaluation of molecularly imprinted core-shell carbon nanotubes for the determination of triclosan in environmental water samples. *Journal of Chromatography A*, *1217*(52), 8095–8102. https://doi.org/10.1016/j.chroma.2010.10.121.

Gu, Z., Chen, X.-Y., Shen, Q.-D., Ge, H.-X., & Xu, H.-H. (2010). Hybrid nanocomposites of semiconductor nanoparticles and conjugated polyelectrolytes and their application as fluorescence biosensors. *Polymer*, *51*(4), 902–907. https://doi.org/10.1016/j.polymer.2009.12.035.

Guo, J., Wang, Y., Liu, Y., Zhang, C., & Zhou, Y. (2015). Magnetic-graphene based molecularly imprinted polymer nanocomposite for the recognition of bovine hemoglobin. *Talanta*, *144*, 411–419. https://doi.org/10.1016/j.talanta.2015.06.057.

Herrmann, A. (2014). Dynamic combinatorial/covalent chemistry: A tool to read, generate and modulate the bioactivity of compounds and compound mixtures. *Chemical Society Reviews*, *43*(6), 1899–1933. https://doi.org/10.1039/C3CS60336A.

Kishi, N., Li, Z., Yoza, K., Akita, M., & Yoshizawa, M. (2011). An M2L4 molecular capsule with an anthracene shell: Encapsulation of large guests up to 1 nm. *Journal of the American Chemical Society*, *133*(30), 11438–11441. https://doi.org/10.1021/ja2037029.

Koudehi, M. F., & Pourmortazavi, S. M. (2018). Polyvinyl alcohol/polypyrrole/molecularly imprinted polymer nanocomposite as highly selective chemiresistor sensor for 2,4-DNT vapor recognition. *Electroanalysis*, *30*(10), 2302–2310. https://doi.org/10.1002/elan.201700751.

Kubik, S. (2010). Anion recognition in water. *Chemical Society Reviews*, *39*(10), 3648–3663. https://doi.org/10.1039/B926166B.

Kumar Dey, S., & Das, G. (2011). A selective fluoride encapsulated neutral tripodal receptor capsule: Solvatochromism and solvatomorphism. *Chemical Communications*, *47*(17), 4983. https://doi.org/10.1039/c0cc05430e.

Li, Y., Yin, X.-F., Chen, F.-R., Yang, H.-H., Zhuang, Z.-X., & Wang, X.-R. (2006). Synthesis of magnetic molecularly imprinted polymer nanowires using a nanoporous alumina template. *Macromolecules*, *39*(13), 4497–4499. https://doi.org/10.1021/ma0526185.

Li, J., Nowak, P., & Otto, S. (2013). Dynamic combinatorial libraries: From exploring molecular recognition to systems chemistry. *Journal of the American Chemical Society*, *135*(25), 9222–9239. https://doi.org/10.1021/ja402586c.

Lu, S., Morrow, D.J., Li, Z., Guo, C., Yu, X., Wang, H., Schultz, J. D., O'Connor, J.P., Jin, N., Fang, F., Wang, W., Cui, R., Chen, O., Su, C., Wasielewski, M.R., Ma, X., & Li, X. (2023). Encapsulating semiconductor quantum dots in supramolecular cages enables ultrafast guest-host electron and vibrational energy transfer. *Journal of the American Chemical Society*, *145*(9), 5191–5202. https://doi.org/10.1021/jacs.2c11981.

Ludlow, R. F., & Otto, S. (2008). Systems chemistry. *Chemical Society Reviews*, *37*(1), 101–108. https://doi.org/10.1039/B611921M.

Mahony, J. O., Nolan, K., Smyth, M. R., & Mizaikoff, B. (2005). Molecularly imprinted polymers-potential and challenges in analytical chemistry. *Analytica Chimica Acta*, *534*(1), 31–39. https://doi.org/10.1016/j.aca.2004.07.043.

Mao, Y., Bao, Y., Gan, S., Li, F., & Niu, L. (2011). Electrochemical sensor for dopamine based on a novel graphene-molecular imprinted polymers composite recognition element. *Biosensors and Bioelectronics, 28*(1), 291–297. https://doi.org/10.1016/j.bios.2011.07.034.

Mateus, P., Delgado, R., Brandão, P., & Félix, V. (2011). Recognition of oxalate by a copper(II) polyaza macrobicyclic complex. *Chemistry - A European Journal, 17*(25), 7020–7031. https://doi.org/10.1002/chem.201100428.

Misaki, H., Miyake, H., Shinoda, S., & Tsukube, H. (2009). Asymmetric twisting and chirality probing properties of quadruple-stranded helicates: Coordination versatility and chirality response of Na^+, Ca^{2+}, and La^{3+} complexes with octadentate cyclen ligand. *Inorganic Chemistry, 48*(24), 11921–11928. https://doi.org/10.1021/ic901496s.

Miyake, H., Ueda, M., Murota, S., Sugimoto, H., & Tsukube, H. (2012). Helicity inversion from left- to right-handed square planar Pd(ii) complexes: Synthesis of a diastereomer pair from a single chiral ligand and their structure dynamism. *Chemical Communications, 48*(31), 3721. https://doi.org/10.1039/c2cc18154a.

Mutihac, L., Lee, J. H., Kim, J. S., & Vicens, J. (2011). Recognition of amino acids by functionalized calixarenes. *Chemical Society Reviews, 40*(5), 2777–2796. https://doi.org/10.1039/C0CS00005A.

Ning, F., Peng, H., Li, J., Chen, L., & Xiong, H. (2014). Molecularly imprinted polymer on magnetic graphene oxide for fast and selective extraction of 17β-Estradiol. *Journal of Agricultural and Food Chemistry, 62*(30), 7436–7443. https://doi.org/10.1021/jf501845w.

Otto, S. (2012). Dynamic molecular networks: From synthetic receptors to self-replicators. *Accounts of Chemical Research, 45*(12), 2200–2210. https://doi.org/10.1021/ar200246j.

Pernites, R. B., Ponnapati, R. R., & Advincula, R. C. (2010). Surface plasmon resonance (SPR) detection of theophylline via electropolymerized molecularly imprinted polythiophenes. *Macromolecules, 43*(23), 9724–9735. https://doi.org/10.1021/ma101868y.

Peyralans, J. J. P., & Otto, S. (2009). Recent highlights in systems chemistry. *Current Opinion in Chemical Biology, 13*(5), 705–713. https://doi.org/10.1016/j.cbpa.2009.08.006.

Ravikumar, I., & Ghosh, P. (2012). Recognition and separation of sulfate anions. *Chemical Society Reviews, 41*(8), 3077–3098. https://doi.org/10.1039/C2CS15293B.

Rebek Jr., J. (2005). Simultaneous encapsulation: Molecules held at close range. *Angewandte Chemie International Edition, 44*(14), 2068–2078. https://doi.org/10.1002/anie.200462839.

Rodriguez-Docampo, Z., Eugenieva-Ilieva, E., Reyheller, C., Belenguer, A. M., Kubik, S., & Otto, S. (2011). Dynamic combinatorial development of a neutral synthetic receptor that binds sulfate with nanomolar affinity in aqueous solution. *Chemical Communications, 47*(35), 9798–9800. https://doi.org/10.1039/C1CC13451E.

Song, M., Pan, C., Li, J., Wang, X., & Gu, Z. (2006). Electrochemical study on synergistic effect of the blending of nano TiO2 and PLA polymer on the interaction of antitumor drug with DNA. *Electroanalysis, 18*(19–20), 1995–2000. https://doi.org/10.1002/elan.200603613.

Song, M., Pan, C., Chen, C., Li, J., Wang, X., & Gu, Z. (2008). The application of new nanocomposites: Enhancement effect of polylactide nanofibers/nano-TiO2 blends on biorecognition of anticancer drug daunorubicin. *Applied Surface Science, 255*(2), 610–612. https://doi.org/10.1016/j.apsusc.2008.06.131.

Song, M., Pan, C., Li, J., Zhang, R., Wang, X., & Gu, Z. (2008). Blends of TiO2 nanoparticles and poly (N-isopropylacrylamide)-co-polystyrene nanofibers as a means to promote the biorecognition of an anticancer drug. *Talanta, 75*(4), 1035–1040. https://doi.org/10.1016/j.talanta.2008.01.005.

Stefankiewicz, A. R., Sambrook, M. R., & Sanders, J. K. M. (2012). Template-directed synthesis of multi-component organic cages in water. *Chemical Science, 3*(7), 2326–2329. https://doi.org/10.1039/C2SC20347B.

Tan, C. J., Chua, H. G., Ker, K. H., & Tong, Y. W. (2008). Preparation of bovine serum albumin surface-imprinted submicrometer particles with magnetic susceptibility through core–shell miniemulsion polymerization. *Analytical Chemistry*, *80*(3), 683–692. https://doi.org/10.1021/ac701824u.

Wang, X., Wang, L., He, X., Zhang, Y., & Chen, L. (2009). A molecularly imprinted polymer-coated nanocomposite of magnetic nanoparticles for estrone recognition. *Talanta*, *78*(2), 327–332. https://doi.org/10.1016/j.talanta.2008.11.024.

Warner, J. H., Hoshino, A., Yamamoto, K., & Tilley, R.D. (2005). Water-soluble photoluminescent silicon quantum dots. *Angewandte Chemie International Edition*, *44*(29), 4550–4554. https://doi.org/10.1002/anie.200501256.

Wu, Y., Zhao, J., Wang, C., Lu, J., Meng, M., Dai, X., Yan, Y., & Li, C. (2016). A novel approach toward fabrication of porous molecularly imprinted nanocomposites with bio-inspired multilevel internal domains: Application to selective adsorption and separation membrane. *Chemical Engineering Journal*, *306*, 492–503. https://doi.org/10.1016/j.cej.2016.07.089.

Zhang, J., Jing, B., Tokutake, N., & Regen, S. L. (2005). Transbilayer complementarity of phospholipids in cholesterol-rich membranes. *Biochemistry*, *44*(9), 3598–3603. https://doi.org/10.1021/bi048258f.

Zhou, Y., Liu, J., Li, H., Zhang, H., Guan, Z., & Jiang, Y. (2021). Molecular recognition of the self-assembly mechanism of glycosyl amino acetate-based hydrogels. *ACS Omega*, *6*, 21801–21808. https://doi.org/10.1039/C2CS35162E.

16 Biological Images and Application of Multiple Fluorescent Materials

Rishi Singh, Syed Sibtay Razi, and Vivek Mishra

16.1 INTRODUCTION

Fluorescent materials have evolved into the analytical instrument of choice in a variety of study fields due to their ease of use in preparation, sensitivity, and selectivity and industrial domains, making it possible to research crucial physiological and pathological processes at the cellular level as well as quickly identify chemical molecules of interest.[1] These materials are widely used in various fields, including biology, chemistry,[2–4] materials science, and imaging technologies.[2] Basically designing of fluorescent materials consists of a chromophore, which is the part of the molecule responsible for absorbing and emitting light, and often includes additional chemical groups that enhance their solubility, stability, and targeting capabilities.[2] These materials can be organic or inorganic in nature. Organic fluorescent dyes are typically small molecules with a conjugated system of double bonds that allows for efficient absorption and emission of light.[5,6] Examples of organic fluorescent dyes include fluorescein, rhodamine, cyanine dyes, Bodipy, Indocyanine green, and Alexa Fluor dyes.[6] These dyes are commonly used in biological imaging applications due to their high brightness, photostability, and compatibility with various labelling techniques.[6] Inorganic fluorescent materials are typically composed of semiconductor nanoparticles, commonly known as quantum dots (QDs).[7] QDs are nanoscale crystals made of semiconductor materials such as cadmium selenide or indium phosphide. They exhibit unique optical properties due to quantum confinement effects, allowing precise control over their emission colour by tuning their size.[8] QDs have high brightness, excellent photostability, and narrow emission spectra, making them valuable tools in biological imaging, sensing, and optoelectronic applications.[9] Besides organic dyes and QDs, other fluorescent materials include fluorescent proteins (e.g., green fluorescent protein or GFP), fluorescent nanoparticles, and fluorescent polymers.[10–12] Each of these materials has its own set of properties and advantages, and their selection depends on the specific application requirements.[12] Fluorescent materials are used extensively in techniques such as fluorescence microscopy, flow cytometry, immunofluorescence, fluorescence in situ hybridization, and fluorescent biosensors.[13–15] Their ability to emit light upon excitation provides researchers with a powerful tool for visualizing and studying biological processes at the cellular and molecular level. Fluorescent materials are used in a wide variety of biological imaging applications.[15]

DOI: 10.1201/9781003352372-16

They can be used to label specific molecules or structures, which can then be visualized under a fluorescent microscope. This allows researchers to study the distribution and behaviour of these molecules in cells and tissues. Biological imaging techniques play a crucial role in studying various biological processes at the cellular and molecular level. One common approach in biological imaging involves the use of multiple fluorescent materials to label different cellular components or molecules of interest.[15] This technique, known as multicolour or multi-channel fluorescence imaging, allows researchers to visualize and study multiple targets simultaneously within a biological sample.[15] One of the most common applications of fluorescent materials in biological imaging is the labelling of DNA. This can be done by using fluorescent probes that bind to specific sequences of DNA. When these probes are excited with light, they emit light of a different wavelength. This allows researchers to visualize the location of specific genes or DNA sequences in cells.[16] Fluorescent materials can also be used to label proteins. This can be done by using fluorescent antibodies that bind to specific proteins. When these antibodies are excited with light, they emit light of a different wavelength, which may visualize the location of specific proteins in cells and tissues. In addition to DNA and proteins, fluorescent materials can also be used to label other biological molecules, such as lipids, carbohydrates, and RNA, which is a specific study for the structure and function of these molecules in cells and tissues.[17–19] The use of multiple fluorescent materials in biological imaging can provide even more detailed information about the distribution and behaviour of biological molecules.[19] This is because each fluorescent material emits light of a different wavelength (visible to NIR). By using multiple fluorescent materials, researchers can create images that show the location of multiple molecules in the same cell or tissue. This technique is known as multicolour fluorescence microscopy.[20] It is a powerful tool that is used to study a wide variety of biological processes, including cell division, protein trafficking, and gene expression.[20–22]

16.2 TARGETED STRATEGY FOR BIOLOGICAL IMAGES

Targeted optical imaging methods are very useful because they are non-invasive, quick, extremely sensitive, and reasonably priced, and play a significant role in basic life science research as well as therapeutic applications.[17] There are numerous cutting-edge imaging technologies that have emerged with the goal of enhancing optical imaging's resolution in order to produce the most accurate representation of biological objects and/or processes. To image biological molecules more sensitively or beyond the diffraction limit of light, techniques such as stimulated emission depletion (STED) microscopy, stochastic optical reconstruction microscopy, photoactivated localization microscopy, and total internal reflection fluorescence microscopy have been developed. These high-resolution or super-resolution imaging techniques have been successfully used to visualize cellular structures in an in vitro environment, but the temporal resolution still has to be increased in order to capture dynamic processes that take place in a living cell.[23–25] Because the spatial resolution is frequently inversely correlated with the amount of photons gathered from the probes, fluorescent probes with high brightness and photostability are crucial for these new imaging modalities.[26] The development of various imaging agents has also received a lot of interest, with fluorescent materials

Biological Images and Application 307

receiving particular focus. Here, we basically focused on some targeted materials for biological imaging agents and materials (Figure 16.1).

16.2.1 Mitochondrial Targeted Materials for Imaging

In addition to supplying the cell with energy, the double-membrane organelles known as mitochondria serve critical functions in controlling the life cycle of cancer cells. The creation of reactive oxygen species (ROS), the transition from proliferation to differentiation, the biosensing of stimuli to the outer or inner membranes, and the control of autophagy and apoptosis are all functions of the mitochondria.[27] Changes in mitochondrial dynamics, number, shape, mitochondrial DNA, and protein mutation are all associated with mitochondrial alterations in cancer. Because of this, it is crucial to create new fluorescence techniques for the detection and comprehension of mitochondrial processes, particularly for the early detection and treatment of disorders. From Tang's work's probe 1 (466 nm emission), Yoon et al. created a new TPP containing AIE Luminogen probe 2 (530 nm emission), noting that probe 2 had a longer emission wavelength that could overcome the cell's autofluorescence and that it was photostable in the presence of both biothiols and various ROS (Figure 16.2).[28,29] In the Dwm study, Probe 2 was discovered to be more sensitive than the commercially available indicator Rhodamine 123. AIEgens can also be used to identify mitochondrial components including ROS,[30] enzymes,[31] and viscosity,[32] as well as their characteristics.

16.2.2 Lysosome-Targeted Materials for Imaging

Many biomolecules are broken down by the numerous hydrolytic enzymes found in lysosomes, which also provide an interorganelle communication for preserving cellular homeostasis. Various disorders, including cancer, can be brought on by lysosomal

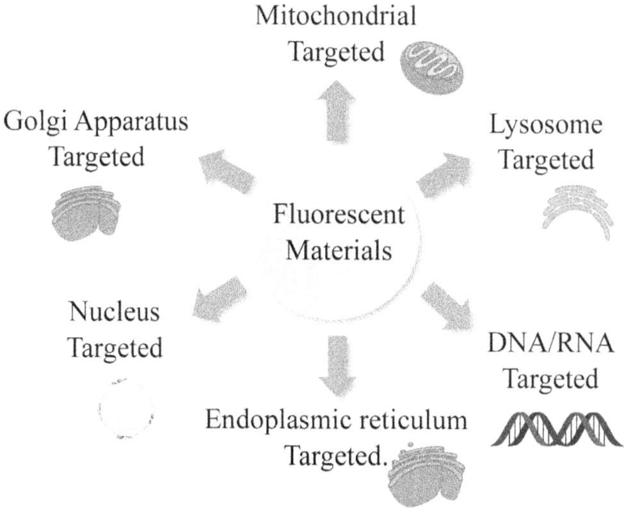

FIGURE 16.1 Schematic diagram of fluorescent material applications.

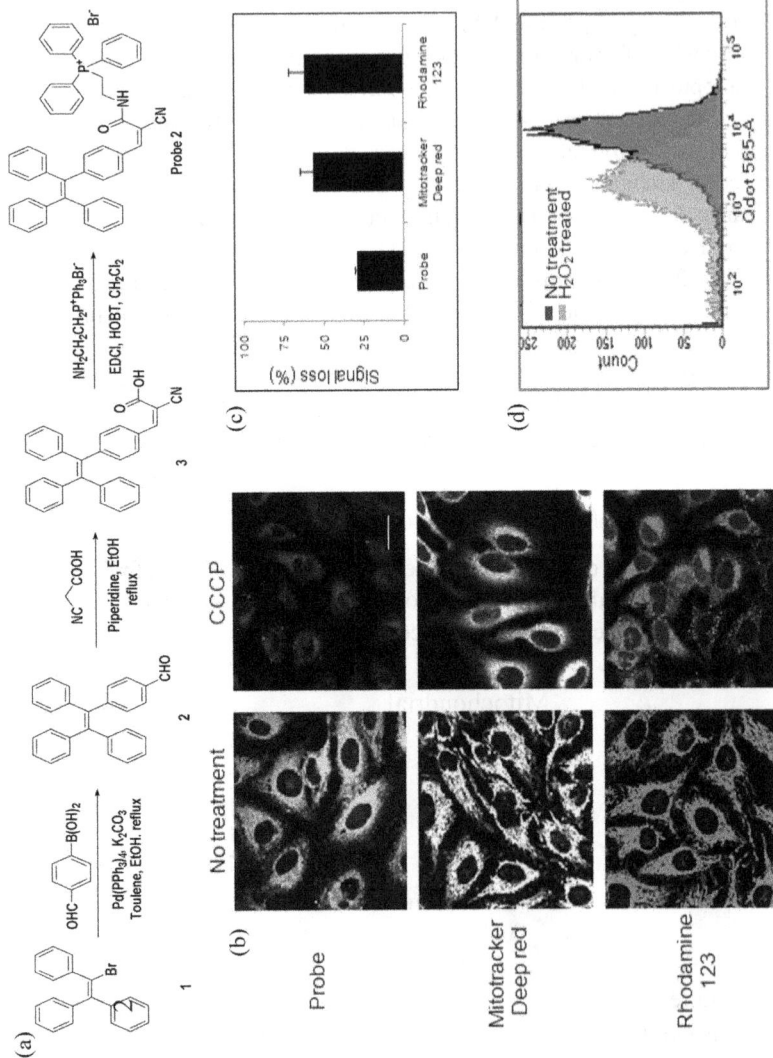

FIGURE 16.2 (a) The creation of fluorescent probe 2 and its use in tracking alterations in mitochondrial potential. (b) Confocal fluorescence images show HeLa cells that have been treated with 20 mM CCCP for 20 min and then exposed to 5 mM probe, 50 nM Mitotracker Deep Red, or 500 nM Rhodamine 123 for 30 min. The excitation wavelengths used were 405, 635, and 470 nm for the probe (emission: 490–590 nm), Mitotracker Deep Red (emission: 655–755 nm), and Rhodamine 123 (emission: 490–590 nm), respectively. A scale bar of 20 m is shown for reference. (c) A graph comparing the amount of fluorescent signal loss to the change in mitochondrial membrane potential is displayed. (d) Cells treated with H_2O_2 and normal cells were analysed using FACS. (Reprinted with permission from ref. 28. Copyright@ 2018 ACS.)

dysfunction, which includes aberrant contents and damaged lysosomal membranes.[33] As a result, lysosomal activity monitoring is important for cancer identification.

AIEgens have been successfully used for lysosomal monitoring because of their excellent photostability and capacity to counteract ACQ.[34,35] Tetraphenylethene (TPE)-based probes were created by Zhao et al. for the detection of lysosomal contents. These probes include polyelectrolyte-conjugated AIEgens for heparin detection[36] and neutral conjugated polymers and polyelectrolytes containing TPE for the assessment of 2,4,6-trinitrophenol.[37] Detection limit for these AIEgen probes can be as low as 5 nM, and they also have very low cytotoxicity. A new photoswitchable AIE nanoprobe (DNBS-DCM-SP) was created by Chen et al. on the basis of H_2S's capacity to identify the photochromic spiropyran moiety in AIEgens. When exposed to H_2S, the DNBS moiety withdraws electrons and separates from the complex by cleaving an O-S bond, which causes fluorescence to be released at a wavelength of 592 nm.[38] As a result of alternating UV and visible light irradiation, the H_2S-activated DCM-SP transitions reversibly by selective fluorescence resonance energy transfer (FRET) from DCM to the open-ring structure of spiropyran which displays reversible dual-colour fluorescence for H_2S detection with good sensitivity and selectivity (Figure 16.3).[38]

16.2.3 ENDOPLASMIC RETICULUM–TARGETED MATERIALS FOR IMAGING

In eukaryotic cells, the ER is a crucial single-layer organelle located next to the nucleus. Depending on whether a ribosome coating is present, the ER can be divided into smooth and rough types and makes up more than half of the cell's total membrane material. The main site of protein translation, modification, and maturation, as well as the production of ROS, is the rough ER, which is covered in numerous ribosomes. The ER is also involved in controlling the homeostasis of calcium ions and oxidative stress, and it has been linked to the emergence of numerous malignancies.[39]

For the detection of ER and ER components, various AIEgens have been developed thus far, and the results exhibit great specificity and sensitivity. A unique two-photon probe (ER-NPA) was created by Gao et al. for the fluorescence-based detection of HOCl in the ER. In this case, the ER targeting was accomplished by methyl sulphonamide, and the fluorophore was 4-hydroxy-1,8-naphthalimide, which was identified by paminophenyl ether. The detection limit of HOCl was reported to be as low as 6.2 nM when detecting HOCl fluctuations in zebrafish and HeLa cells subjected to ER stress. HOCl exposure to ER-NAP resulted in enhanced fluorescence at 550 nm after 60 s.[40] When reacting with unfolded ER proteins in vitro (DL can be as low as 50 lM) and in living cells, TPE-NMI, a TPE derivative made from maleimide, displayed red fluorescence (emission at 561 nm), allowing measurement and observation by flow cytometry and confocal microscopy.[41] A better understanding of cancer necessitates the detection of changes in the ER because it is known that unfolded proteins are linked to ER modifications in cancer (Figure 16.4).[42]

16.2.4 NUCLEUS-TARGETED MATERIALS FOR IMAGING

Given that it houses most of the cell's genetic material, the nucleus is the centre of the cell. When healthy cells develop into cancerous cells, the nucleus exhibits a few distinctive physicochemical traits, including modifications to DNA, RNA, and proteins as

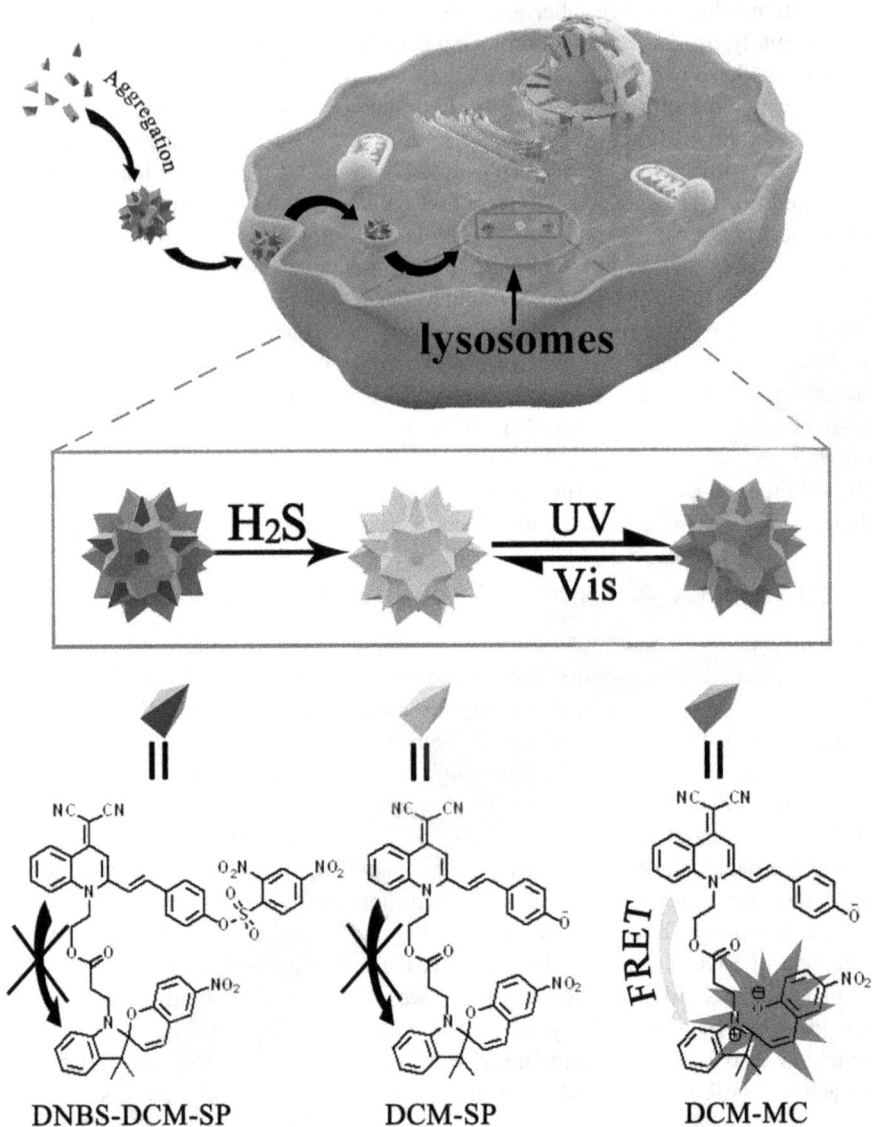

FIGURE 16.3 HeLa cells were used to create the fluorescence pictures, which were first treated with L-Cys at concentrations of 0 M (a–c) and 200 M (d–f), respectively, before being exposed to DNBS-DCM-SP (10 M). (II) Images of HeLa cells that had been exposed to Cys (200 M) and then DNBS-DCM-SP (10 M) under visible light (g–i) or UV irradiation (j–l) are photoswitchable, reversible fluorescence. The signals come from nanoparticles in the cell, not interference, according to the overlay images. Ex = 468 nm, em1 = 550–600 nm (Channel 1), and em2 = 650–700 nm (Channel 2). 20 m for the scale bar. (Reprinted with permission from ref. 38. Copyright@ 2018 Elsevier.)

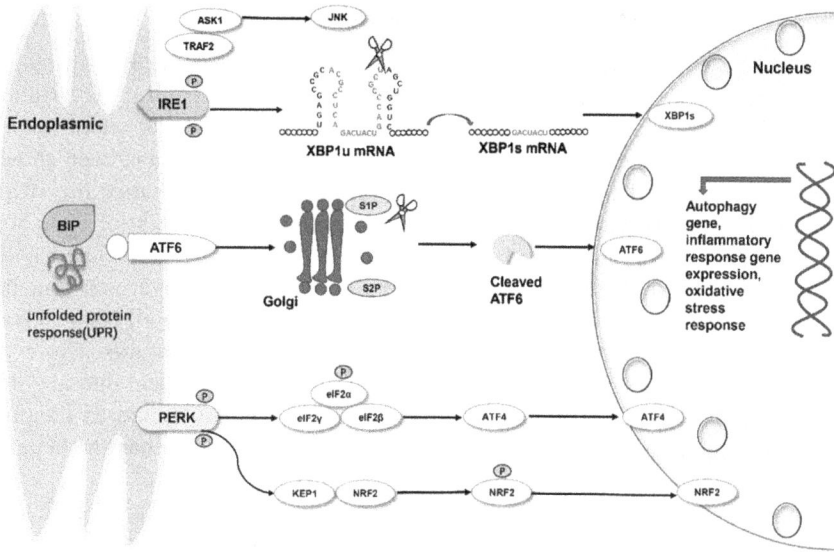

FIGURE 16.4 Membranes and secreted proteins are synthesized by ER and translocated into ER lumen. Accumulation of unfolded proteins in the ER lumen results in activation of the UPR response in ER stress, dissociation of UPR sensors PERK, ATF6, and IRE1. PERK activates the cytosolic domain by dimerization and autophosphorylation. PERK phosphorylation of eIF2a inhibits general protein synthesis and promotes translation of ATF4 mRNA. The active PERK also phosphorylates NRF2, and the phosphorylated NRF2 dissociates from KEAP1 and translocates to the nucleus. IRE1 contains an endoribonuclease domain that is activated by dimerization and autophosphorylation. After IRE1 is activated, the unspliced XBP1 u mRNA is processed, and the spliced XBP1s mRNA is translated into an active transcription factor. IRE1 also activates the kinase domain of TRAF2 and ASK1, resulting in activation of JNK. ATF6 is activated and translocated to the Golgi apparatus and is cleaved by site 1 and site 2 proteases (S1P and S2P) in the Golgi apparatus, and the cleaved ATF6 translocate to the nucleus. (Reprinted with permission from ref. 42. Copyright@ 2019 Elsevier.)

well as morphological abnormalities that can be used as markers for cancer and abnormal biological processes.[43] Electrostatic interactions with nucleic acids are a key component of several AIEgens created for nuclear imaging. To encourage cellular uptake, Tang et al. created two AIEgens that emit in the orange-red spectrum: TPE-TPA-BTD (TTB) and TPE-NPA-BTD (TNB). They did this by fusing the AIEgen dots with the Tat peptide (YGRKKRRQRRRC, Tat-AIE). Both Tat-AIEs demonstrated significant Stokes shifts and high emission efficiencies in aqueous solutions (55% and 57%, respectively); they were also discovered to be photostable, biocompatible, and capable of long-term nuclear imaging (>5 days).[44] The cNGR-CPPNLS-RGD-PyTPE, TCNTP AIEgen probe was developed by the Xia group to improve membrane penetration and nuclear targeting. It contains the peptides cNGR and RGD for effective cell targeting, CPP for cell penetration, NLS for nuclear localization, and a TPE derivative (PyTPE, AIEgens, yellow emission) as the fluorescent reagent.

16.2.5 GOLGI APPARATUS–TARGETED MATERIALS FOR IMAGING

A single-layer membrane surrounds the highly polar organelle known as the Golgi apparatus. The organelle participates in the secretion and transportation of cellular components and is made up of proteins and lipids that are dispersed between the endoplasmic reticulum and the cell membrane. It has been shown that, among other disorders, cancer[45] and vascular disease[46] change the shape and function of the Golgi apparatus. Therefore, imaging is essential for monitoring the Golgi apparatus and managing disorders that it is associated with. Currently, sphingophosphates (NBDCeramide), quinoline derivatives, and the boron-dipyrromethane derivative Golgi-Tracker Red/Green have all been created as commercially accessible probes for imaging the Golgi apparatus. However, the imaging findings obtained with these probes are far from ideal due to quick bleaching and limited specificity. AIEgen probes for Golgi apparatus imaging provide better advantages and promising application prospects when the benefits of AIEgens are taken into account.

Two AIEgen probes have recently been created. By combining TPE and 20,30-O-isopropylideneadenosine (Ad), Zhao et al. created a novel AIEgen called TPE-Ad; the Ad moiety served as an active site for several molecules in the Golgi apparatus. The quantum yield of TPE-Ade was 2.3% in a solution of 5% water and 95% ethanol but increased to 24.2% in a solution of 95% water and 5% ethanol due to the suppression of intramolecular rotation. TPE-Ad was also very stable when exposed to laser radiation (405 nm). The TPE-Ad effectively targeted the Golgi apparatus when treated with HeLa cells (Pearson's coefficient was 0.86), demonstrating that TPE-Ad was a superior probe for Golgi apparatus imaging (Figure 16.5).[47]

16.3 APPLICATIONS OF MULTIPLE FLUORESCENT MATERIALS IN BIOLOGICAL IMAGING

16.3.1 CO-LOCALIZATION STUDIES FOR SPECIFIC BIOLOGICAL IMAGING

Several fluorescent materials (biological markers) are used in co-localization investigations in biological imaging to label various molecules or cellular components of interest (targeted therapy).[48] Using this method, scientists can examine the spatial arrangement and potential interactions of these molecules within a biological sample. In co-localization investigations, various fluorescent materials are used as follows.

16.3.1.1 Choosing the Right-Targeted Fluorophores

The fluorescent materials used by researchers all have unique emission spectra (visible to NIR range) that make them easy to identify from one another. They might decide on fluorophores like green, red, and blue, for instance.[48] The target molecule or biological component (morpholine for lysosomes, piperazine and triphenylphosphine for mitochondria, etc.) that each fluorophore is typically attached to is a specific antibody or probe.[48,49]

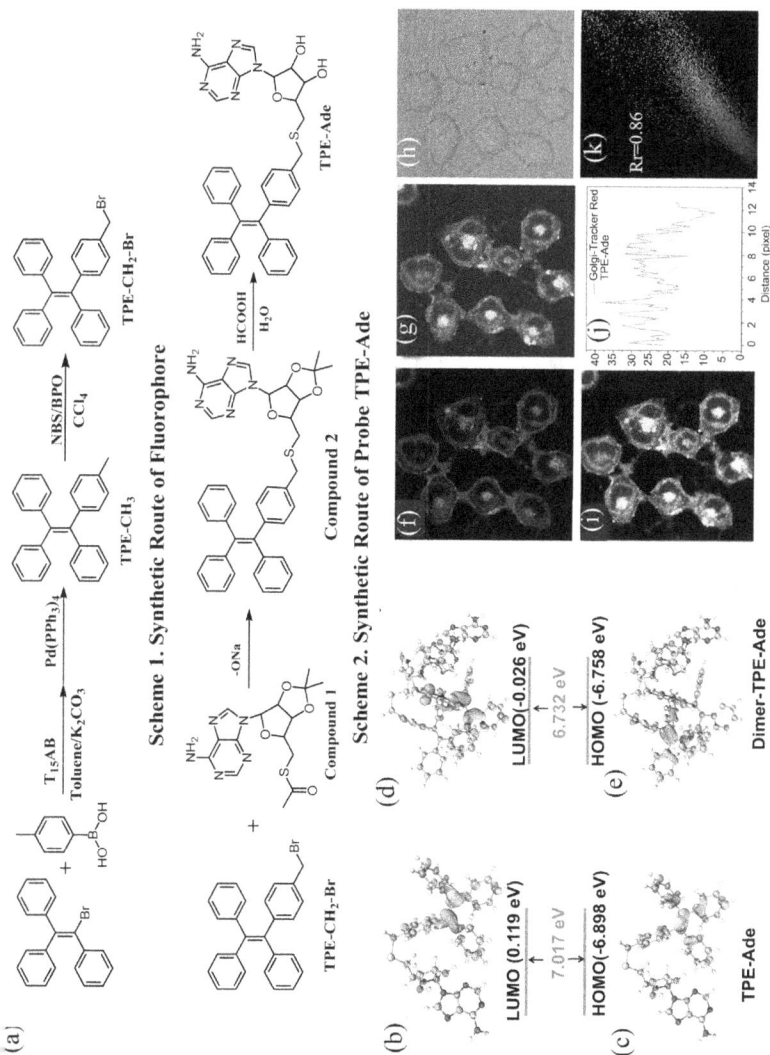

FIGURE 16.5 (a) Synthetic route of TPE and TPE-Ade. (b)–(e) Transition energy and frontier molecular orbitals of TPE-Ade and Dimer-TPE-Ade. The calculations are determined using CAM-B3LYP/6–31 g(d) exchange correlation functionals and basis sets. (f) Co-localization fluorescence images of TPE-Ade (10 μM) with commercial Golgi-Tracker Red (50 nM) in HL-7402 cells. (g) Fluorescence image of HeLa cells stained with Golgi-Tracker Red. The excitation wavelength was 543 nm, and the emission was collected at 562–662 nm. (h) Fluorescence image of cells stained with TPE-Ade (10 μM) for 15 min. The excitation wavelength was 405 nm, and the emission was collected at 425–485 nm. (i) Bright-field images, (j) merge image of probe TPE-Ade and Golgi-Tracker Red-treated cells, (k) intensity profile of cross Costain image, and (i) Pearson's coefficient graph of overlay. (Reprinted with permission from ref. 47. Copyright@ 2021 Elsevier.)

16.3.1.2 Synchronized Labelling Methods for Targeted Part

A combination of the labelled probes, each of which is directed towards a distinct chemical or cellular component, is incubated with the biological material, such as cells or tissue slices. The fluorophores are localized specifically within the sample as a result of the probes' binding to their particular targets.[17,49]

16.3.1.3 Sample Imaging

The sample is then photographed using a fluorescence microscope outfitted with the proper filter sets to record each fluorophore's emission. Each channel used by the microscope to record images corresponds to a certain emission wavelength range.[17]

16.3.1.4 Image Analysis for Different Range

To measure the co-localization of the various fluorophores, image analysis techniques are applied to the obtained images at different wavelength ranges (from visible to NIR). The degree of overlap between the various fluorescence signals can be determined using a variety of methods and software tools.[50–52]

16.3.1.5 Co-localization Analysis for Different Methods

Researchers can evaluate the level of co-localization between the molecules or cellular components they labelled by comparing the signals from the various fluorophores. By using statistical metrics like Pearson's correlation coefficient, Manders' coefficients, or spatial overlap coefficients, co-localization can be analysed. The degree and spatial relationship of co-localization between the various targets are quantified by these metrics.[53,54]

16.3.1.6 Interpretation and Conclusions

Researchers can make inferences about potential interactions, spatial organization, or functional linkages between the labelled molecules or cellular components based on the co-localization analyses. Insights into molecular interactions in biological systems can be gained via co-localization studies, which can also help discover co-localized proteins, ascertain subcellular localization patterns, illuminate signalling pathways, and find co-localized proteins.[17,54] Researchers have a great tool to visualize and quantify the spatial interactions between various molecules or cellular components when using numerous fluorescent materials in co-localization experiments. It adds to our understanding of cellular processes and mechanisms by providing insightful information on the structure and functional relationships seen in biological systems.[17,50,52,54]

16.3.2 LIVE-CELL IMAGING

Multiple fluorescent materials are used in live-cell imaging to explore dynamic cellular processes in real time. Researchers can concurrently monitor and visualize various chemicals or cellular components within living cells by utilizing different fluorophores. How several fluorescent pigments are used in live-cell imaging is as follows:

16.3.2.1 Selection of Specific Fluorophores for Binding

To make sure that the signals from various targets (Vis to NIR) can be differentiated, researchers chose fluorescent materials (organic scaffolds, nanomaterials, QDs, protein labelling fluorophores, imaging dyes, etc.) with various emission spectra.

Typically, a probe or antibody that has been precisely designed to bind to the target chemical or biological component is attached to each fluorophore.[17,25,47–48]

16.3.2.2   Cell Labelling Experiments

Combining the labelled probes in living cells allows the fluorophores to bind to certain targets in a targeted manner. This causes various chemicals or cellular structures within the cells to be specifically labelled.

16.3.2.3   Live-Cell Imaging Setup

The tagged cells are put in a plate or imaging chamber that has been designed to maintain the ideal temperature, humidity, and CO_2 levels for cell survival. A fluorescence microscope with a sensitive camera, suitable excitation light sources, and filter sets for collecting the emission from each fluorophore may be a part of the imaging equipment.

16.3.2.4   Image Acquisition

The imaging of the living cells takes place throughout time as successive images are recorded in several fluorescence channels that correspond to the emission wavelengths of the various fluorophores. Depending on the exact study question, the imaging can be done in 2D or 3D.

16.3.2.5   Data Analysis

Using specialized software, the captured image sequences are analysed to track and examine the behaviour of the labelled molecules or cellular components over time. This may entail quantifying dynamic events like protein trafficking or organelle dynamics, tracking the motion of particular structures, monitoring variations in fluorescence intensity, or all three.[17,25]

16.3.2.6   Interpretation and Visualization

To acquire insights into the dynamic processes taking place inside the living cells, the data from live-cell imaging investigations are displayed and analysed. Multiple targets can be seen simultaneously to get a full picture of their interactions, spatial linkages, and temporal dynamics. Researchers can examine intricate cellular processes like cell signalling, protein dynamics, organelle connections, and cell behaviour in response to stimuli or environmental changes by employing a variety of fluorescent materials in live-cell imaging. Real-time monitoring of multiple targets simultaneously within living cells enables a more thorough comprehension of the dynamics and coordination of cellular activities.

16.3.3   Molecular Profiling

For molecular profiling in biological imaging, other fluorescent materials are also used. The goal of molecular profiling is to examine the location and expression patterns of different biomarkers or genes within a tissue or cell sample. The use of several fluorescent materials in molecular profiling is demonstrated as follows:

16.3.3.1 Selection of Fluorophores from Visible to NIR Range for Specifically Binding

Different biomarkers or genes of interest are labelled using different fluorophores with unique emission spectra. Each fluorophore is typically coupled to an antibody or probe that specifically targets the target molecule.

16.3.3.2 Multiplexed Labelling for Analysis

A variety of labelled probes, each of which is directed towards a distinct biomarker or gene, are incubated with the sample. This makes it possible to simultaneously label several targets inside of a single sample.

16.3.3.3 Imaging Pattern and Analysis

Using a fluorescence microscope fitted with the proper filter sets, the sample is photographed to record the emission from each fluorophore. Each channel used by the microscope to record images corresponds to a particular range of emission wavelengths. The signals from each tagged biomarker or gene are recognized and quantified using image analysis algorithms on the obtained images. The signals from various fluorophores are separated and distinguished using software tools and algorithms.

16.3.3.4 Expression Profiling

Researchers can ascertain the expression levels and spatial distribution of several biomarkers or genes inside the sample by analysing the signals from many fluorophores. These details shed light on the sample's cellular heterogeneity, marker co-expression, and molecular profiles.

16.3.3.5 Data Interpretation

Understanding the links between various biomarkers or genes, identifying cell types or subpopulations, or detecting abnormal expression patterns linked to diseases or disorders can all be done by interpreting the molecular profiling data. Characterizing biological samples, clarifying illness mechanisms, and locating prospective diagnostic or therapeutic targets are all made easier with the help of this knowledge. The simultaneous detection and analysis of numerous biomarkers or genes inside a single sample is made possible using various fluorescent materials in molecular profiling. By enabling the detection of intricate patterns and connections between various molecules, this multiplexing capacity offers a complete and in-depth perspective of the molecular landscape and advances our knowledge of cellular and molecular biology.

16.3.4 Super-Resolution Microscopy for Imaging

A potent method that enables scientists to view the structure and function of biological molecules with unheard-of detail is the employment of several fluorescent materials in biological imaging through super-resolution microscopy. A type of microscopy called super-resolution microscopy is capable of capturing images of biological structures at resolutions that are less than the diffraction limit of light. Numerous methods, including single-molecule localization microscopy and STED microscopy, are used to accomplish this. Fluorescent materials are used in super-resolution microscopy to label specific biological molecules. When these molecules are excited with light, they emit light of a different wavelength. This allows researchers to visualize the location of these molecules in cells and tissues. Researchers can produce images that indicate the locations of numerous molecules in the same cell or tissue by using multiple fluorescent materials in super-resolution microscopy. Numerous biological processes, including cell division, protein trafficking, and gene expression, can be studied using this information.

16.3.5 CELL-CYCLE STUDY

Different cell-cycle stages can be marked with fluorescence probes. This enables scientists to see how the cell cycle develops and spot any anomalies.

16.3.6 PROTEIN TRAFFICKING STUDY

Different proteins involved in protein transport can be marked with fluorescent antibodies. This enables scientists to see how proteins flow throughout tissues and cells.

16.3.7 GENE EXPRESSION ANALYSIS

Specific genes can be marked with fluorescence probes. This makes it possible for scientists to see how these genes are expressed in tissues and cells.

A potent method for studying a wide range of biological processes is super-resolution microscopy, which makes use of several fluorescent materials. It is a useful tool for scientists who want to comprehend the composition and operation of cells and tissues. Here are some of the benefits of using multiple fluorescent materials in biological imaging through super-resolution microscopy:

16.3.8 ENHANCED RESOLUTION

Biological structures can be imaged with super-resolution microscopy at resolutions that are less than the light diffraction limit. As a result, scientists can now see biological molecules' structure and operation in unprecedented detail.

16.3.9 IMPROVED CONTRAST IMAGES

Images can have better contrast when several luminous materials are used. This is because fluorescent materials all generate light with unique wavelengths. Researchers can produce images with great contrast that show the locations of numerous molecules in the same cell or tissue by combining several fluorescent materials.

16.3.10 REDUCED PHOTOBLEACHING

Fluorescent molecules lose their capacity to emit light through the process of photobleaching. Using several fluorescent materials at once helps lessen photobleaching. This is because light of a different wavelength excites each fluorescent substance differently. Researchers can use numerous fluorescent materials to shorten the time each fluorescent material is activated by light, which can lessen photobleaching.

16.4 CONCLUSION AND FUTURE POTENTIAL

In conclusion, the current chapter provides a thorough overview of recent developments in the areas of biological imaging detection, methods, and applications. This overview will be useful for directing future research efforts in these areas. Numerous innovations have been made in a variety of disciplines, supporting fluorescent materials bright future. The existing biological constraints must be overcome in the future through cross-disciplinary research collaborations to improve medical theranostics uses and create new applications.

REFERENCES

1. H. Lusic, M. W. Grinstaff, X-ray-computed tomography contrast agents, *Chem. Rev.* 2013, 113, 1641.
2. N. Yadav, R. P. Gaikwad, V. Mishra, M.B. Gawande, Synthesis and photocatalytic Applications of functionalized carbon quantum dots, *Bull. Chem. Soc. Jpn.* 2022, 95(11):1638–79.
3. N. Yadav, D. Mudgal, V. Mishra, In-situ synthesis of ionic liquid-based-carbon quantum dots as fluorescence probe for hemoglobin detection, *Anal. Chim. Acta* 2023, 7:341502.
4. S. Jindal, R. Anand, N. Sharma, N. Yadav, D. Mudgal, R. Mishra, V. Mishra, Sustainable Approach for Developing Graphene-Based Materials from Natural Resources and Biowastes for Electronic Applications, *ACS Appl. Electron. Mater.* 2022, 16(5):2146–74.
5. H. Kobayashi, M. Ogawa, R. Alford, P. L. Choyke, Y. Urano, New strategies for fluorescent probe design in medical diagnostic imaging, *Chem. Rev.* 2010, 110, 2620.
6. S.K. Panda, I. Aggarwal, H. Kumar, L. Prasad, A. Kumar, A. Sharma, D. V. N. Vo, D. V. Thuan, V. Mishra, Magnetite nanoparticles as sorbents for dye removal: a review, *Environ. Chem. Lett.* 2021, 19:2487–525.
7. J. Mei, N. L. C. Leung, R. T. K. Kwok, J. W. Y. Lam, B. Z. Tang, Aggregation-induced emission: together we shine, united we soar!, *Chem. Rev.* 2015, 115, 11718.
8. *Soft Nanomaterials*, ed. H. S. Nalwa, American Scientific Publishers, Lewis Way, 2009.
9. M. K. So, C. J. Xu, A. M. Loening, S. S. Gambhir, J. H. Rao, Self-illuminating quantum dot conjugates for in vivo imaging, *Nat. Biotechnol.* 2006, 24, 339–343.
10. M. Swierczewska, S. Lee, X. Y. Chen, Inorganic nanoparticles for multimodal molecular imaging, *Mol. Imaging* 2011, 10, 3–16.
11. L. Tang, J. Cheng, Nonporous silica nanoparticles for nanomedicine application, *Nano Today* 2013, 8, 290–312.
12. C. R. Maldonado, L. Salassa, N. Gomez-Blanco, J. C. Mareque-Rivas, Nano-functionalization of metal complexes for molecular imaging and anticancer therapy, *Coord. Chem. Rev.* 2013, 257, 2668–88.
13. R. Benya, J. Quintana, B. Brundage, Adverse reactions to indocyanine green: a case report and a review of the literature, *Cathet. Cardiovasc. Diagn.* 1989, 17, 231–3.
14. L. H. Feng, C. L. Zhu, H. X. Yuan, L. B. Liu, F. T. Lv, S. Wang, Conjugated polymer nanoparticles: preparation, properties, functionalization and biological applications, *Chem. Soc. Rev.* 2013, 42, 6620–33.

15. K. Li, B. Liu, Polymer-encapsulated organic nanoparticles for fluorescence and photo-acoustic imaging, *Chem. Soc. Rev.* 2014, 43, 6570–97.

16. Q.Y. Xiao, X. Zhao, H. Xiong, Biocompatible and noncytotoxic nucleoside-based AIEgens sensor for lighting-up nucleic acids, *Chin. Chem. Lett.* 2021, 32, 1687–90.

17. S. Erbas-Cakmak, S. Kolemen, A. C. Sedgwick, T. Gunnlaugsson, T. D. James, J. Yoon, E U. Akkaya, Molecular logic gates: The past, present and future. *Chem. Soc. Rev.* 2018, 47(7): 2228–48.

18. A. C. Sedgwick, J. E. Gardiner, G. Kim, M. Yevglevskis, M. D. Lloyd, A. T. A. Jenkins, S. D. Bull, J. Yoon, T. D. James, Long-wavelength TCF-based fluorescence probes for the detection and intracellular imaging of biological thiols. *Chem. Commun.* 2018, 54(38): 4786–89.

19. J. L. Kolanowski, F. Liu, E. J. New, Fluorescent probes for the simultaneous detection of multiple analytes in biology. *Chem. Soc. Rev.* 2018, 47(1): 195–208.

20. O. Kolokythas, T. Gauthier, A. T. Fernandez, H. Xie, B. A. Timm, C. Cuevas, M. K. Dighe, L. M. Mitsumori, M. F. Bruce, D. A. Herzka, G. K. Goswami, R. T. Andrews, K. M. Oas, T. J. Dubinsky, B. H. Warren, Ultrasound-based elastography: A novel approach to assess radio frequency ablation of liver masses performed with expandable ablation probes: A feasibility study, *J. Ultrasound Med.* 2008, 27, 935.

21. J. Mei, N. L. C. Leung, R. T. K. Kwok, J. W. Y. Lam, B. Z. Tang, Aggregation-induced emission: together we shine, united we soar!, *Chem. Rev.* 2015, 115, 11718.

22. A. Gnach, T. Lipinski, A. Bednarkiewicz, J. Rybka, J. A. Capobianco, Upconverting nanoparticles: assessing the toxicity, *Chem. Soc. Rev.* 2015, 44, 1561.

23. H.T. Bai, W. He, J.H.C. Chau, Z. Zheng, R.T.K. Kwok, J.W.Y. Lam, B.Z. Tang, AIEgens for microbial detection and antimicrobial therapy, *Biomaterials* 2021, 268, 120598.

24. R. Zhang, G. Niu, Q. Lu, X. Huang, J.H.C. Chau, R.T.K. Kwok, X. Yu, M.H. Li, J.W.Y. Lam, B.Z. Tang, Cancer cell discrimination and dynamic viability monitoring through wash-free bioimaging using AIEgens, *Chem. Sci.* 2020, 11, 7676–84.

25. D. Wang, B.Z. Tang, Aggregation-Induced Emission Luminogens for Activity-Based Sensing, *Acc. Chem. Res.* 2019, 52, 2559–70.

26. J. Liang, B. Tang, B. Liu, Specific light-up bioprobes based on AIEgen conjugates, *Chem. Soc. Rev.* 2015, 44, 2798–11.

27. D.C. Wallace, Mitochondria and cancer, *Nat. Rev. Cancer* 2012, 12, 685–98.

28. J. Li, N. Kwon, Y. Jeong, S. Lee, G. Kim, J. Yoon, Aggregation-Induced Fluorescence Probe for Monitoring Membrane Potential Changes in Mitochondria, *ACS Appl. Mater. Interfaces* 2018, 10, 12150–4.

29. C.W.T. Leung, Y.N. Hong, S.J. Chen, E.G. Zhao, J.W.Y. Lam, B.Z. Tang, A Photostable AIE Luminogen for Specific Mitochondrial Imaging and Tracking, *J. Am. Chem. Soc.* 2013, 135, 62–5.

30. Q. Zhang, P. Zhang, Y. Gong, C.F. Ding, Two-photon AIE based fluorescent probe with large stokes shift for selective and sensitive detection and visualization of hypochlorite, *Sens. Actuators B Chem.* 2019, 278, 73–81.

31. H. Li, Q. Yao, F. Xu, Y. Li, D. Kim, J. Chung, G. Baek, X. Wu, P.F. Hillman, E.Y. Lee, H. Ge, J. Fan, J. Wang, S.J. Nam, X. Peng, J. Yoon, An Activatable AIEgen Probe for High-Fidelity Monitoring of Overexpressed Tumor Enzyme Activity and Its Application to Surgical Tumor Excision, *Angew. Chem. Int. Ed. Engl.* 2020, 59, 10186–95.

32. X.D. Wang, L. Fan, S.H. Wang, Y.W. Zhang, F. Li, Q. Zan, W.J. Lu, S.M. Shuang, C. Dong, Real-time monitoring mitochondrial viscosity during mitophagy using a mitochondria-immobilized near-infrared aggregation-induced emission probe, *Anal. Chem.* 2021, 93, 3241–9.

33. D. Dersh, Y. Iwamoto, Y. Argon, Tay–Sachs disease mutations in HEXA target the α chain of hexosaminidase A to endoplasmic reticulum–associated degradation, *Mol. Biol. Cell* 2016, 27, 3813–7.

34. C.W.T. Leung, Z.M. Wang, E.G. Zhao, Y.N. Hong, S.J. Chen, R.T.K. Kwok, A.C.S. Leung, R.S. Wen, B.S. Li, J.W.Y. Lam, B.Z. Tang, A Lysosome-Targeting AIEgen for Autophagy Visualization, *Adv. Healthcare Mater.* 2016, 5, 427–31.

35. Y.L. Mu, L. Pan, Q. Lu, S. Xing, K.Y. Liu, X. Zhang, A bifunctional sensitive fluorescence probe based on pyrene for the detection of pH and viscosity in lysosome, *Spectrochim. Acta Part A Mol. Biomol. Spectrosc.* 2022, 264, 120228.

36. Y.A. Wang, H.M. Yao, Z.Y. Zhuang, J.Y. Yao, J.Y. Zhou, Z.J. Zhao, High photoluminescence quantum yield of 18.7% by using nitrogen-doped Ti3C2 MXene quantum dots, *J. Mater. Chem. B* 2018, 6, 6360–4.

37. J. Yao, Z. Zhuang, H. Yao, R. Shi, C. Chang, J. Zhou, Z. Zhao, Tetraphenylethene-based polymeric fluorescent probes for 2, 4, 6-trinitrophenol detection and specific lysosome labelling, *Dyes Pigm.* 2020, 182,108588.

38. Y.X. Hong, P.S. Zhang, H. Wang, M.L. Yu, Y. Gao, J. Chen, Photoswitchable AIE nanoprobe for lysosomal hydrogen sulfide detection and reversible dual-color imaging, *Sens. Actuat. B Chem.* 2018, 272, 340–7.

39. J.R. Cubillos-Ruiz, S.E. Bettigole, L.H. Glimcher, Tumorigenic and immunosuppressive effects of endoplasmic reticulum stress in cancer, *Cell* 2017, 168, 692–706.

40. T.T. Yang, J.Y. Sun, W. Yao, F. Gao, A two-photon fluorescent probe for turn-on monitoring HOCl level in endoplasmic reticulum, *Dyes Pigm.* 2020, 180, 108435.

41. S.X. Zhang, M.J. Liu, L.Y.F. Tan, Q.T. Hong, Z.L. Pow, T.C. Owyong, S.Y. Ding, W. W.H. Wong, Y.N. Hong, A Maleimide-functionalized Tetraphenylethene for Measuring and Imaging Unfolded Proteins in Cells, *Chem. Asian J.* 2019, 14, 904–9.

42. Y.N. Lin, M. Jiang, W.J. Chen, T.J. Zhao, Y.F. Wei, Cancer and ER stress: Mutual crosstalk between autophagy, oxidative stress and inflammatory response, *Biomed. Pharmacother.* 2019, 118, 109249.

43. H. Liu, Z.H. Zhang, Y.J. Zhao, Y.X. Zhou, B. Xue, Y.C. Han, Y.L. Wang, X.L.E. Mu, S.L. Zang, X.F. Zhou, Z.B. Li, Asymmetric selenophene-based non-fullerene acceptors for high-performance organic solar cells, *J. Mater. Chem. B* 2019, 7, 1435–41.

44. W. Qin, K. Li, G.X. Feng, M. Li, Z.Y. Yang, B. Liu, B.Z. Tang, Bright and photostable organic fluorescent dots with aggregation-induced emission characteristics for noninvasive long-term cell imaging, *Adv. Funct. Mater.* 2014, 24, 635–43.

45. S.M. Shen, C. Zhang, M.K. Ge, S.S. Dong, L. Xia, P. He, N. Zhang, Y. Ji, S. Yang, Y. Yu, J.K. Zheng, J.X. Yu, Q. Xia, G.Q. Chen, PTENα and PTENβ promote carcinogenesis through WDR5 and H3K4 trimethylation, *Nat. Cell Biol.* 2019, 21, 1436–48.

46. J.E. Lee, K. Patel, S. Almodovar, R.M. Tuder, S.C. Flores, P.B. Sehgal, Dependence of Golgi apparatus integrity on nitric oxide in vascular cells: implications in pulmonary arterial hypertension, *Am. J. Physiol. Heart Circ. Physiol.* 2011, 300, H1141–58.

47. X.Y. Xing, Y. Jia, J.R. Zhang, Z.B. Wu, M.M. Qin, P. Li, X. Feng, Y. Sun, G.J. Zhao, A novel aggregation induced emission (AIE) fluorescence probe by combining tetraphenylethylene and 2′, 3′-O-isopropylideneadenosine for localizing Golgi apparatus, *Sens. Actuat. B Chem.* 2021, 329, 129245.

48. Z. Wang, Y. Zhou, R.H. Xu, Y.Z. Xu, D.F. Dang, Q.F. Shen, L.J. Meng, B.Z. Tang, Seeing the unseen: AIE luminogens for super-resolution imaging, *Coord. Chem. Rev.* 2022, 451, 214279.

49. H.B. Cheng, Y.Y. Li, B.Z. Tang, J. Yoon, Assembly strategies of organic-based imaging agents for fluorescence and photoacoustic bioimaging applications, *Coord. Chem. Rev.* 2020, 49, 21–31.

50. M. Li, Y. Gao, Y. Y. Yuan, Y. Z. Wu, Z. F. Song, B. Z. Tang, B. Liu, Q. C. Zheng, One-step formulation of targeted aggregation-induced emission dots for image-guided photodynamic therapy of cholangiocarcinoma, *ACS Nano* 2017, 11, 3922.

51. B. Suarez-Pena, L. Negral, L. Castrillon, L. Megido, E. Maranon, Y. Fernandez-Nava, Imaging techniques and scanning electron microscopy as tools for characterizing a Si-based material used in air monitoring applications, *Materials* 2016, 9, 109.

52. P. K. Garg, S. K. Singh, G. Prakash, A. Jakhetiya, D. Pandey, Role of positron emission tomography-computed tomography in non-small cell lung cancer, *World J. Methodol.* 2016, 6, 105.

53. V. Biju, Chemical modifications and bioconjugate reactions of nanomaterials for sensing, imaging, drug delivery and therapy, *Chem. Soc. Rev.* 2014, 43, 744.

54. Y. Liu, C. M. Deng, L. Tang, A. J. Qin, R. R. Hu, J. Z. Sun, B. Z. Tang, Specific Detection of d-Glucose by a Tetraphenylethene-Based Fluorescent Sensor, *J. Am. Chem. Soc.* 2011, 133, 660.

Index

Note: **Bold** page numbers refer to tables and *italic* page numbers refer to figures.

For Product Safety Concerns and Information please contact our EU
representative GPSR@taylorandfrancis.com
Taylor & Francis Verlag GmbH, Kaufingerstraße 24, 80331 München, Germany

www.ingramcontent.com/pod-product-compliance
Lightning Source LLC
Chambersburg PA
CBHW060807220326
41598CB00022B/2562